Cell-Surface Peptidases

in health and disease

Cell-Surface Peptidases

in health and disease

Edited by A.J. Kenny and C.M. Boustead

βIOS SCIENTIFIC PUBLISHERS

First published in 1997

A CIP catalogue record for this book is available from the British Library.

ISBN 1 859962 75 0

BIOS Scientific Publishers Ltd
9 Newtec Place, Magdalen Road, Oxford OX4 1RE, UK.
Tel. +44 (0) 1865 726286. Fax. +44 (0) 1865 246823
World-Wide Web home page: http://www.Bookshop.co.uk/BIOS/

DISTRIBUTORS

Australia and New Zealand
Blackwell Science Asia
54 University Street
Carlton, South Victoria 3053

India
Viva Books Private Limited
4325/3 Ansari Road, Daryaganj
New Delhi 110002

Singapore and South East Asia
Toppan Company (S) PTE Ltd
38 Liu Fang Road, Jurong
Singapore 2262

USA and Canada
BIOS Scientific Publishers
PO Box 605, Herndon,
VA 20172-0605

Production Editor: Rachel Robinson.
Typeset by Saxon Graphics Ltd, Derby, UK.
Printed by Biddles Ltd, Guildford, UK.

Contents

Contributors

Ait-Slimane, T. CJM-INSERM 96.07, CHU Saint-Antoine, 27 rue de Chaligny, 75012 Paris, France

Albino, A.P. Genitourinary Oncology Research Laboratory, Division of Solid Tumor Oncology, Department of Medicine, Memorial Sloan-Kettering Cancer Center and Cornell University Medical Center, New York, NY 10021, USA

Alhenc-Gelas, F. INSERM, Unité 367, 17 rue du Fer à Moulin, 75005 Paris, France

Baricault, L. CJM-INSERM 96.07, CHU Saint-Antoine, 27 rue de Chaligny, 75012 Paris, France

Bauer, K. Max-Planck-Institut für Experimentelle Endokrinologie, Feodor-Lynen Strasse 7, Hannover D-30625, Germany

Bauvois, B. INSERM Unité 365, Institut Curie, Section de Biologie, 26 rue d'Ulm, Paris 75231, Cedex 05, France

Beaumont, A. Unité de Pharmacochimie Moléculaire et Structurale, INSERM U266, CNRS URA D 1500, UFR des Sciences Pharmaceutiques et Biologiques, 4 Avenue de l'Observatoire, 75270 Paris, France

Berbar, T. CJM-INSERM 96.07, CHU Saint-Antoine, 27 rue de Chaligny, 75012 Paris, France

Bernard, A.-M. Centre d'Immunologie de Marseille-Luminy, Case Postale 906, 13288 Marseille, Cedex 9, France

Boileau, G. Département de Biochimie, Faculté de Médecine, Université de Montréal, Case 6128, Succursale 'Centre-Ville', Montréal H3C 3J7, Canada

Cooper, M.D. Division of Developmental and Clinical Immunology, Departments of Medicine, Pediatrics and Microbiology, University of Alabama at Birmingham and the Howard Hughes Medical Institute, Birmingham, AL 35294, USA

Corvol, P. INSERM Unité 36, Laboratoire de Médecine Expérimentale, Collège de France, 3 rue d'Ulm, 75005 Paris, France

Costerousse, O. INSERM Unité 367, 17 rue du Fer à Moulin, 75005 Paris, France

Crine, P. Département de Biochimie, Faculté de Médecine, Université de Montréal, Case Postale 6128, Succursale 'Centre-Ville', Montréal H3C 3J7, Canada

Danielsen, E.M. Department of Medical Biochemistry and Genetics, Biochemistry Laboratory C, The Panum Institute, University of Copenhagen, Blegdamsvej 3, DK 2200 Copenhagen N, Denmark

Danilov, S. INSERM Unité 367, 17 rue du Fer à Moulin, 75005, Paris, France

David, F. CJM-INSERM 96.07, CHU Saint-Antoine, 27 rue de Chaligny, 75012 Paris, France

Delmas, B. Unité de Virologie et Immunologie Moléculaires, Domaine de Vilvert, Institut National de la Recherche Agronomique, 78350 Jouy-en-Josas, France

Dion, N. Département de Biochimie, Faculté de Médecine, Université de Montréal, Case Postale 6128, Succursale 'Centre-Ville', Montréal H3C 3J7, Canada

Fournié-Zaluski, M.-C. Unité de Pharmacochimie Moléculaire et Structurale, INSERM U266, CNRS URA D 1500, UFR des Sciences Pharmaceutiques et Biologiques, 4 Avenue de l'Observatoire, 75270 Paris, France

Greaves, A. Division of Immunology and Cancer Research, Hospital for Sick Children, and University of Toronto, 555 University Avenue, Toronto M5G 1X8, Canada

Hahn, D. Institute of Biochemistry and Molecular Biology, University of Bern, Buehlstrasse 28, CH-3012 Bern, Switzerland

Hall, L. Department of Biochemistry, University of Bristol School of Medical Sciences, University Walk, Bristol BS8 1TD, UK

Heuer, H. Max-Planck-Institut für Experimentelle Endokrinologie, Feodor-Lynen Strasse 7, D-30625 Hannover, Germany

Hooper, N.M. Department of Biochemistry and Molecular Biology, University of Leeds, Leeds LS2 9JT, UK

Iffländer, F. Max-Planck-Institut für Experimentelle Endokrinologie, Feodor-Lynen Strasse 7, Hannover D-30625, Germany

Ishii, E. Division of Pediatrics, Saga Prefectural Hospital, Koseikan 1-12-9, Mizugae, Saga 840, Japan

Kenny, A.J. Le Presbytère, 34390 St Etienne d'Albagnan, France

Kurachi, K. Department of Human Genetics, 3712 Medical Science II Building, University of Michigan Medical School, Ann Arbor, MI 48109-0618, USA

Lenoir, C. CJM-INSERM 96.07, CHU Saint-Antoine, 27 rue de Chaligny, 75012 Paris, France

Letarte, M. Division of Immunology and Cancer Research, Hospital for Sick Children, 555 University Avenue, Toronto, Ontario M5G 1X8, Canada

Lottaz, D. Institute of Biochemistry and Molecular Biology, University of Bern, Buehlstrasse 28, CH-3012 Bern, Switzerland

Maldonado, R. Unité de Pharmacochimie Moléculaire et Structurale, INSERM U266, CNRS URA D 1500, UFR des Sciences Pharmaceutiques et Biologiques, 4 Avenue de l'Observatoire, 75270 Paris, France

Marguet, D. Centre d'Immunologie de Marseille-Luminy, Case 906, 13288 Marseille, Cedex 9, France

Nanus, D.M. Genitourinary Oncology Research Laboratory, Division of Solid Tumor Oncology, Department of Medicine, Memorial Sloan-Kettering Cancer Center, 1275 York Avenue, New York, NY 10021, USA

Noble, F. Unité de Pharmacochimie Moléculaire et Structurale, INSERM U266, CNRS URA D 1500, UFR des Sciences Pharmaceutiques et Biologiques, 4 Avenue de l'Observatoire, 75270 Paris, France

Norén, O. Department of Medical Biochemistry and Genetics, Biochemistry Laboratory C, The Panum Institute, University of Copenhagen, Blegdamsvej 3, DK-2200 Copenhagen N, Denmark

Olsen, J. Department of Medical Biochemistry and Genetics, Biochemistry Laboratory C, The Panum Institute, University of Copenhagen, Blegdamsvej 3, DK-2200 Copenhagen N, Denmark

Papandreou, C.N. Genitourinary Oncology Research Laboratory, Division of Solid Tumor Oncology, Department of Medicine, Memorial Sloan-Kettering Cancer Center, 1275 York Avenue, New York, NY 10021, USA

Peters, A. Max-Planck-Institut für Experimentelle Endokrinologie, Feodor-Lynen Strasse 7, Hannover D-30625, Germany

Pierres, M. Centre d'Immunologie de Marseille-Luminy, Case 906, 13288 Marseille, Cedex 9, France

Roques, B.P. Unité de Pharmacochimie Moléculaire et Structurale, INSERM U266, CNRS URA D 1500, UFR des Sciences Pharmaceutiques et Biologiques, 4 Avenue de l'Observatoire, 75270 Paris, France

Sapin, C. CJM-INSERM 96.07, CHU Saint-Antoine, 27 rue de Chaligny, 75012 Paris, France

Schmitmeier, S. Max-Planck-Institut für Experimentelle Endokrinologie, Feodor-Lynen Strasse 7, Hannover D-30625, Germany

Schomburg, L. Max-Planck-Institut für Experimentelle Endokrinologie, Feodor-Lynen Strasse 7, Hannover D-30625, Germany

Sjöström, H. Department of Medical Biochemistry and Genetics, Biochemistry Laboratory C, The Panum Institute, University of Copenhagen, Blegdamsvej 3, DK-2200 Copenhagen N, Denmark

Sterchi, E.E. Institute of Biochemistry and Molecular Biology, University of Bern, Buehlstrasse 28, CH-3012 Bern, Switzerland

Stöcker, W. Institut für Zoophysiologie, Westfälische Wilhelms-Universität Münster, Hindenburgplatz 55, D-48143 Münster, Germany

Trugnan, G. CJM-INSERM 96.07, CHU Saint-Antoine, 27 rue de Chaligny, 75012 Paris, France

Turner, A.J. Department of Biochemistry and Molecular Biology, University of Leeds, Leeds LS2 9JT, UK

Turwitt, S. Max-Planck-Institut für Experimentelle Endokrinologie, Feodor-Lynen Strasse 7, Hannover D-30625, Germany

Vogel, L.K. Department of Medical Biochemistry and Genetics, Biochemistry Laboratory C, The Panum Institute, University of Copenhagen, Blegdamsvej 3, DK-2200 Copenhagen N, Denmark

Wang, J. Medical Institute of Bioregulation, Kyushu University, 3-1-1 Maidashi, Higashi-ku, Fukuoka 812, Japan

Wilkens, M. Max-Planck-Institut für Experimentelle Endokrinologie, Feodor-Lynen Strasse 7, Hannover D-30625, Germany

Wilkins, M.R. Department of Clinical Pharmacology, Royal Postgraduate Medical School, Hammersmith Hospital, Du Cane Road, London W12 0NN, UK

Williams, T.A. INSERM Unité 36, Laboratoire de Médecine Expérimentale, Collège de France, 3 rue d'Ulm, 75005 Paris, France

Abbreviations

α-hANP	α-human atrial natriuretic peptide
ACDP	*Acinetobacter calcoaceticus* dipeptidase
ACE	angiotensin-converting enzyme
AcSDKP	*N*-acetyl-seryl-aspartyl-lysyl-proline
ADAM	a disintegrin and metalloprotease
AI	angiotensin I
AII	angiotensin II
AIDS	acquired immune deficiency syndrome
ALL	acute lymphoblastic leukaemia
AML	acute myeloblastic (myeloid) leukaemia
ANP	atrial natriuretic peptide
AP-A	aminopeptidase A
AP-N	aminopeptidase N
AP-P	aminopeptidase P
AUC	area-under-curve
β-APP	β-amyloid precursor protein
BK	bradykinin
BLP	bombesin-like peptides
BMP-1	bone morphogenetic protein 1
BNP	brain natriuretic peptide
BRK	baby rat kidney
CALLA	common acute lymphoblastic leukaemia antigen
CCK	cholecystokinin
CD	cluster differentiation
CHAPS	3-[(3-cholamidopropyl)-dimethylammonia]-1-propanesulphonate
CHO	Chinese hamster ovary
CMC	critical micellar concentration
CNBr	cyanogen bromide
CNP	C-type natriuretic peptide
CNS	central nervous system
CP-M	carboxypeptidase M
CRD	cross-reacting determinant
CRE	cAMP-responsive element
CVS	cardiovascular system
DAP-IV	dipeptidyl aminopeptidase IV
DEAE	diethylaminoethyl
DFP	di-isopropylfluorophosphate
DMSO	dimethyl sulphoxide
DNase	deoxyribonuclease
Dnp-F	1-fluoro-2,4-dinitrobenzene
DOCA	deoxycorticosterone acetate
DPP-IV	dipeptidyl peptidase IV
DPP-X	dipeptidyl peptidase X
E-24.11	endopeptidase-24.11
E-24.18	endopeptidase-24.18
EAP	epididymal apical protein
ECE	endothelin-converting enzyme
EDTA	ethylenediamine tetraacetic acid
EGF	epidermal growth factor
EGTA	ethylene glycol bis-(β-aminoethylether) *N,N,N′,N′*-tetraacetic acid
ER	endoplasmic reticulum
ET	endothelin
FAP-α	fibroblast-activating protein α
γ-GT	γ-glutamyl transpeptidase
GEMSA	guanidinoethylmercaptosuccinic acid
GIP	gastric inhibitory peptide
GLP	glucagon-like peptide
GM-CSF	granulocyte-macrophage colony-stimulating factor
GP-2	glycoprotein 2
GPI	glycosyl phosphatidylinositol
GRF	growth-releasing factor
HCMV	human cytomegalovirus
HCV	human coronavirus
HIV	human immunodeficiency virus
HPLC	high performance liquid chromatography
i.p.	intraperitoneal
i.v.	intravenous
IFN-α	interferon α
IFN-γ	interferon γ
IGF-1	insulin-like growth factor 1
IL-2	interleukin-2
IL-7	interleukin-7
LHRH	luteinizing hormone-releasing hormone
LPS	lipopolysaccharide
LTD_4	leukotriene D_4
mAb	monoclonal antibody
MDC	metalloproteinase-like, disintegrin-like, cysteine-rich (proteins)
MDCK	Madin–Darby canine kidney

MDP	membrane dipeptidase
MGTA	D, L-2-mercaptomethyl-3-guanidinoethylthiopropanoic acid
MHC	major histocompatibility complex
MMP	matrix metalloproteinase
NAAG	*N*-acetyl-L-aspartyl-L-glutamate
NAALA	*N*-acetylated α-linked acidic (dipeptidase)
NAD	nicotinamide adenine dinucleotide
NMDA	*N*-methyl-D-aspartate
NMR	nuclear magnetic resonance
NSF	*N*-ethylmaleimide-sensitive factor
p.o.	*per os* (orally)
paba-peptide	*N*-benzoyl-L-tyrosyl-*p*-aminobenzoic acid
PCR	polymerase chain reaction
PGA	pyroglutamyl aminopeptidase
PI-PLC	phosphatidylinositol-specific phospholipase C
PMA	phorbol 12-myristate 13-acetate
PMSF	phenylmethylsulphonyl fluoride
pNA	*p*-nitroanilide
PNS	peripheral nervous system
PPH	paba-peptide hydrolase
PTH	parathyroid hormone
PTHrP	parathyroid hormone-related peptide
RACE	rapid amplification of 5′ cDNA ends
RCC	renal cell carcinoma
RCE	retinoblastoma control element
RER	rough endoplasmic reticulum
RGP	radial growth phase
RNase	ribonuclease
RT-PCR	reverse transcriptase-polymerase chain reaction
SCLC	small cell lung cancer
SDS-PAGE	sodium dodecyl sulphate polyacrylamide gel electrophoresis
SH3	Src homology 3
SHR	spontaneous hypertensive rat
SNAP	synaptosomal-associated protein
TGEV	transmissible gastroenteritis virus
TGF	transforming growth factor
TGN	trans-Golgi network
THAM	thymocyte activation molecule
TLN	thermolysin
TNF	tumour necrosis factor
TPP	tripeptidyl peptidase
TRH	thyrotrophin-releasing hormone
TSH	thyrotrophin-stimulating hormone
VGP	vertical growth phase
VIP	vasoactive intestinal polypeptide
WT	Wilms' tumour

Preface

Ten years have passed since the publication of *Mammalian Ectoenzymes* (edited by A.J. Kenny and A.J. Turner, Elsevier, 1987), a work that merited only two chapters specifically devoted to the cell-surface peptidases. At that time only one of the peptidases, endopeptidase-24.11, had been cloned; we were just beginning to understand their topology, including the presence of a glycolipid anchor on one of them; none had been identified as a CD antigen on leukocytes and only a few tentative steps had been taken towards the illumination of their function.

The last decade has seen many major advances. More enzymes have been added to the list of ectopeptidases, in particular the endothelin-converting enzymes, almost all of the group have now been cloned and knowledge of their amino acid sequences has revealed their relationships within the superfamilies of proteinases. Several are now known to be CD antigens, some double as virus receptors and real advances have been achieved in understanding their roles, for example in terminating peptide signals in the nervous system and at peripheral sites. Many highly selective inhibitors have been synthesised, some of which are already well established in clinical practice. Although we still await the completion of crystallographic studies, our knowledge of their catalytic sites has been greatly extended by the application of site-directed mutagenesis and molecular modelling using the crystal structures of some related soluble peptidases.

These advances have brought this group of ectoenzymes to the attention of a wide range of scientists in many different disciplines. The topic has been the focus of several international meetings, including two Jacques Monod Conferences in 1991 and 1994, but no comprehensive review has emerged to provide an up-to-date account of the cell-surface peptidases. For this reason we have invited a number of those actively involved in the field to contribute chapters to this book. We are very grateful to all our authors for their scholarly contributions and it is a pleasure to acknowledge the help we have had from the publisher, especially from our production editor, Rachel Robinson.

A.J. Kenny (*St Etienne d'Albagnan, France*)
C.M. Boustead (*Dijon, France*)

Chapter 1

Ectoenzymes, ectopeptidases and CD antigens

A. John Kenny

1.1 Cell-surface peptidases are ectoenzymes

1.1.1 Background

Enzymology developed in three stages. The first enzymologists focused upon activities observed in secreted fluids, such as saliva, gastric juice or the intestinal contents, where the function was apparent from the outset and the activities could be examined without complex extraction methods. Next, biochemists began tackling the bewildering complexity of the enzymes packaged within cells, which posed more difficult technical problems in defining distinct activities and in isolating individual enzymes. Interest in biochemical events at cell surfaces developed relatively late. After all, so much was going on inside the cell that there seemed to be no obvious reason to look for enzymes at the surface. Transport systems were acceptable membrane activities, but metabolic processes at the cell surface were another matter, poorly understood and often greeted with disbelief. A pioneer in this field was Rothstein, who showed that the hydrolysis of glucose 1-phosphate and other phosphate esters took place at the intestinal surface, not in the intestinal lumen, nor within the enterocyte (Rothstein *et al.*, 1953). In spite of this and other papers pointing to the existence of enzymes at the surface of cells, the concept of ectoenzymes metabolizing substrates at the surface was painfully slow to take hold. Perhaps the most puzzling example of this collective mental block can be seen in the juxtaposition of two papers from the same laboratory and submitted on the same date. The first (Miller and Crane, 1961a) is a study of the intestinal hydrolysis of disaccharides and sugar phosphate esters which states "the only conclusion that it seems possible to draw is that the hydrolases are intracellular enzymes". The second (Miller and Crane, 1961b) is a paper, rightly regarded as a classic, reporting the first preparation of a brush border fraction from intestinal mucosa in which invertase and maltase were enriched 10-fold, but again concluding that "hydrolysis takes place within the brush border of the mucosal epithelium". The reluctance to accept the existence of ectoenzymes was in part sustained, for those working on intestinal physiology, by an obsession with philosophical arguments as to whether hydrolysis of dipeptides and disaccharides preceded or followed absorption. The experimental systems then available were inadequate to provide clear answers and the battles were largely concerned with the promotion of one or other model (see, for example, Ciba Foundation Symposium, 1972; Ugolev, 1965). Electron microscopy was at this

Cell-Surface Peptidases in Health and Disease, edited by A.J. Kenny and C.M. Boustead.

time pointing the right way by demonstrating, by negative staining, the presence of knobs covering the surface of microvilli from intestine and kidney (Oda and Seki, 1965; Wong-Leung and Kenny, 1968). One study in particular correlated the removal of these knobs (by treating a brush border preparation with papain) with the release of invertase and maltase from the membrane (Johnson, 1967).

Although the existence of hydrolases active at the luminal surface of the intestine was at the outset grudgingly accepted, the term *ectoenzyme* (or more strictly *ecto-enzyme*) had been coined in 1957 by Engelhardt to explain the rapid hydrolysis of ATP by intact avian erythrocytes (Engelhardt, 1957). He excluded the possibility that the hydrolysis was due to 'exo-enzymes' (i.e. secreted enzymes) or to 'endo-enzymes' acting within the cell, and he rightly emphasized the strict asymmetry of ecto-ATPase, facing the surrounding medium.

1.1.2 Experimental criteria needed for a peptidase to be an ectoenzyme (an ectopeptidase)

Ectoenzymes are, by definition, integral membrane proteins, anchored by hydrophobic interactions with the lipids of the plasma membrane, in such a way that the active site is exposed only to the external, non-cytosolic, surface of the membrane. Thus a putative ectoenzyme must be shown to be an integral membrane protein, exhibiting amphipathic properties (for a fuller discussion, see Kenny and Maroux, 1982). The behaviour with detergents is crucial. The detergent forms a protective micelle around the hydrophobic tail of the protein, leading to a number of changes in physical properties. The apparent M_r (Stoke's radius) determined by gel filtration is increased by an amount corresponding to the size of the micelle. The mobility in charge-shift electrophoresis may be altered when a non-ionic detergent is partially replaced by a cationic or anionic detergent. The detergent-solubilized peptidase may be adsorbed by hydrophobic chromatography (e.g. phenyl- or octyl-Sepharose) and eluted by high concentrations of the detergent. Removal of the detergent micelle leads to the formation of aggregates (protein micelles) of high M_r, in which the hydrophobic tails themselves associate in the interior of a micelle. Finally, the amphipathic protein has the ability to be reconstituted into liposomes.

The second essential criterion concerns the orientation in the membrane: the peptidase must be asymmetric in orientation, facing outwards from the cell surface. This may be demonstrated by showing the accessibility to non-permeable reagents (e.g. a substrate, an antibody or a proteinase capable of clipping the enzyme from its anchor). These approaches would be undermined if the cells were permeable to the reagent or if the membrane preparation did not form sealed right-side-out vesicles. The simple demonstration of peptidase activity when intact cells are incubated with a substrate is inadequate proof that the peptidase is an ectoenzyme. For example, the aminopeptidase activity of Jurkat cells (a human T-cell lymphoma line) had been assumed to be that of aminopeptidase N (CD13), until a more detailed study of this point showed that an intracellular aminopeptidase was responsible for the phenomenon, the confusion arising from the ability of the substrates and inhibitors to cross the plasma membrane (Murray *et al.*, 1994).

These criteria are important in assessing possible physiological roles of peptidases. Potential substrates, such as neuropeptides, are confined within membrane-bound systems inside the cell before their release by exocytosis at the surface. They are never accessible to cytoplasmic peptidases but may be hydrolysed by an appropriate ectoenzyme after release. Too often the demonstration of *in vitro* hydrolysis of a neuropeptide by a peptidase is extrapolated to indicate a physiological role. Before credence can be given to this interpretation one should look for evidence that the peptidase is an ectoenzyme. Such evidence is, of course, a necessary but not a sufficient condition to support a putative physiological role. Endopeptidase-24.15 and endopeptidase-24.16 are two enzymes that illustrate this concern; both are efficient in hydrolysing a range of neuropeptides (see, for example, Checler, 1993), but both are predominantly cytosolic in location, with only a minor component in each case bound to the particulate material of brain homogenate and this is not indicative of an integral membrane association, still less of an ectoenzyme (Millican *et al.*, 1991). The hydropathy plot of the cDNA-derived sequence of endopeptidase-24.15 (Pierotti *et al.*, 1990) reveals no putative anchor sequence. A loose association of a minor proportion with membranes is not diagnostic of an ectoenzyme (see Chapter 19 for a fuller discussion).

1.2 Cell-surface peptidases (ectopeptidases)

1.2.1 Nomenclature and terminology

The terminology of proteolytic enzymes, which have been studied for more than a century, is inevitably complicated and demands the imposition of some order and consistency. All these enzymes are correctly designated as *peptidases*, indicating that they hydrolyse peptide bonds (which, in all but a few cases, are α-peptide bonds). Those enzymes that require the presence of an unsubstituted N or C terminus in the substrate are *exopeptidases*, those that do not are *endopeptidases* (see *Figure 1.1*). The terms proteinase and protease (both implying the ability to hydrolyse macromolecular substrates) are roughly equivalent to the term endopeptidase, the secondary and tertiary structure of protein substrates usually preventing attack by exopeptidases. The action of an endopeptidase is generally not favoured by the presence of a free N or C terminus close to the scissile bond. Exopeptidases remove a single amino acid, a dipeptide or a tripeptide from one or other terminus, which actions form the basis of the classification of the exopeptidases. Similar considerations of specificity cannot be applied to the endopeptidases, which are best distinguished by their active sites yielding the four main classes: *serine, cysteine, aspartic* and *metallo*-endopeptidases. The same catalytic mechanisms operate among the exopeptidases, so that some (e.g. carboxypeptidases) can be further divided on this basis.

1.2.2 The Enzyme Commission (E.C.) classification

All the hydrolases are designated by the International Union of Biochemistry and Molecular Biology (1992) as E.C. 3.- and the peptidases as E.C. 3.4.-. The main classes of peptidases are defined by a third numeral (11–24) as noted in *Figure 1.1*. The exopeptidases are classified mainly on the basis of their actions, aminopeptidases, carboxypeptidases, dipeptidases, and so on. In most cases only peptides with an unsubstituted terminus are attacked, with the exception of a very small number, grouped as *omega peptidases* (3.4.19.-) which can release certain modified terminal residues (e.g. acylaminoacyl peptidase, pyroglutamyl peptidase, peptidyl glycinamidase). The nomenclature, in most cases,

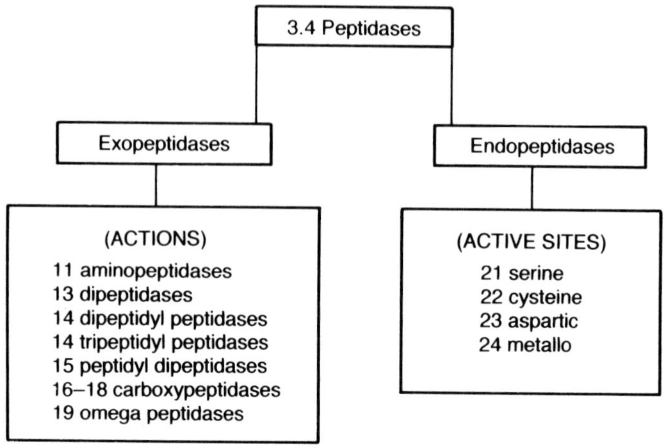

Figure 1.1. The Enzyme Commission classification of the peptidases, showing the number allocated to each of the main classes.

derives directly from this classification, but the E.C. gives major emphasis to the identity of the P_1 or P_1' (following the convention for identifying residues in the peptide substrate introduced by Schechter and Berger (1967) (see *Figure 1.2*), yielding names such as membrane alanyl aminopeptidase (3.4.11.2), glutamyl aminopeptidase (3.4.11.7) or X-Pro aminopeptidase (3.4.11.9). In the case of the last named, also known as aminopeptidase P, the name accurately describes the restricted specificity of the peptidase, but for the other two the term is misleading. The striking feature of alanyl aminopeptidase is its very broad specificity which is in no way limited to alanyl peptides, hence its other name, aminopeptidase N, from its preference for neutral (uncharged) side chains. Similarly, glutamyl aminopeptidase also releases aspartyl residues, hence its other name, aminopeptidase A (acidic side chains). In this book we have favoured unambiguous but less restrictive names, such as aminopeptidase N, carboxypeptidase P.

Enzymes that attack 'interior' rather than terminal peptide bonds defy classification on the principles that are appropriate for the exopeptidases, where the properties of the substrate are paramount. It is rare to find an endopeptidase that is specific for a bond

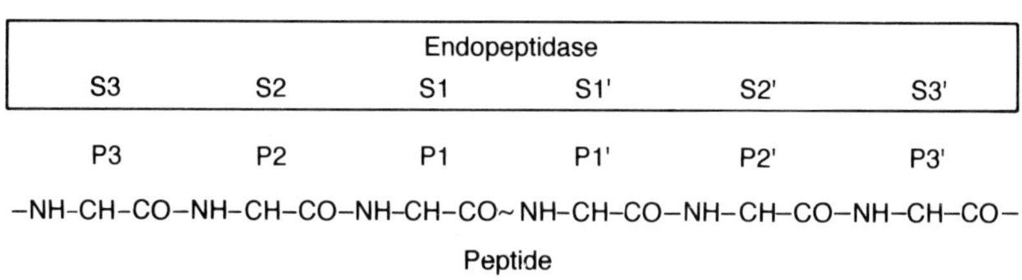

Figure 1.2. The numbering of the amino acid residues around the scissile bond (shown ~) of a peptide substrate and of the residues at the active site of the peptidase which may interact with the side chains of the peptide substrate (Schechter and Berger, 1967).

involving a single type of side chain and so there is no logic in attempting to devise a nomenclature or classification relating to specificity. Hence the current division into the four groups indicating the catalytic mechanism: serine endopeptidases (3.4.21.), cysteine endopeptidases (3.4.22.), aspartate endopeptidases (3.4.23.) and metallo-endopeptidases (3.4.24.) while a final group (3.4.99.) contains a handful of enzymes for which the mechanism remains undefined. In referring to the two main cell-surface endopeptidases discussed in this book we have preferred the names endopeptidase-24.11 (E-24.11) and endopeptidase-24.18 (E-24.18) (which names immediately evoke the concept of *metallo*-peptidases) rather than the terms neprilysin and meprin, but the alternative names for the peptidases discussed in this book are given in *Table 1.1*.

1.2.3 Peptidase families and clans

The relative ease by which cDNA-derived sequences can now be obtained has added another level of sophistication to the classification of peptidases, one based on evolutionary considerations. At an early stage, the pentapeptide motif, HEXXH, was observed to be a characteristic of zinc binding in a number of metallopeptidases, the motif providing two of the three ligands for the zinc atom. The process has advanced rapidly, enabling Rawlings and Barrett (1993) to examine over 600 sequences of peptidases, separating out families and groups of families, designated as clans. They recognized 84 families, some of which were related to each other, thus creating family clans. The largest class was that of the metallopeptidases, 25 families, of which 13 were in clan A ('zincins'), possessing the common zinc-binding motif, HEXXH. Other families had HXXE and HXXEH binding motifs and some others bound two zinc atoms. A further classification of the families of the zinc metallopeptidases has been developed recently (Hooper, 1994). The zincins (those containing the HEXXH consensus sequence) can be subdivided into the gluzincins including the E-24.11 family (which family includes endothelin-converting enzyme), angiotensin-converting enzyme (ACE), thermolysins, aminopeptidases N and A. E-24.18, a member of the astacin family, is classed with the metzincins.

1.3 Cluster differentiation (CD) antigens

One of the unexpected, but highly rewarding developments in the last few years has been the demonstration that certain surface molecules identified by immunologists and haematologists as CD antigens are identical to some well-researched cell-surface peptidases. Thus, CD10 (CALLA, the common acute lymphoblastic leukaemia antigen) is identical to E-24.11 (Jongeneel *et al.*, 1989; Letarte *et al.*, 1988; Shipp *et al.*, 1989). CD13 is identical to aminopeptidase N (Look *et al.*, 1989) and CD26 to dipeptidyl peptidase IV (Ulmer *et al.*, 1990) and the murine BP-1/6C3 antigen is aminopeptidase A (Wu *et al.*, 1990, 1991).

1.4 A growth area

My interest began nearly 30 years ago with the observation that there was intense peptidase activity in particulate fractions of kidney homogenates. These early observations led

Table 1.1. Nomenclature of the cell-surface peptidases described in this book

Name	Enzyme Commission (E.C.) no.	[Abbreviation] and other names	Chapter
Endopeptidase-24.11	3.4.24.11	[E-24.11] Neutral endopeptidase Neprilysin Enkephalinase CALLA CD10	6
Angiotensin-converting enzyme	3.4.15.1	[ACE] Peptidyl dipeptidase A Kininase II	7, 8
Endothelin-converting enzyme	3.4.24.-	[ECE]	9
Endopeptidase-24.18	3.4.24.18	[E-24.18] Meprin A PABA-peptide hydrolase	10
Aminopeptidase N	3.4.11.2	[AP-N] Membrane alanyl aminopeptidase Microsomal aminopeptidase Aminopeptidase M CD13	11
Aminopeptidase A	3.4.11.7	[AP-A] Glutamyl aminopeptidase Aspartate aminopeptidase Angiotensinase A	12
Dipeptidyl-peptidase IV	3.4.14.5	[DPP-IV] X-Pro-dipeptidyl-aminopeptidase CD26	13
Membrane dipeptidase	3.4.13.19	[MDP] Renal dipeptidase Dehydropeptidase 1	14
Aminopeptidase P	3.4.11.9	[AP-P] X-Pro-aminopeptidase Proline aminopeptidase	14
Carboxypeptidase M	3.4.17.12	[CP-M]	14
Thyrotropin-releasing hormone-degrading enzyme	3.4.19.6	[TRH-degrading enzyme] Pyroglutamyl peptidase II Thyroliberinase Pyroglutamyl aminopeptidase II	15
Aminopeptidase W	3.4.11.16	[AP-W] X-Trp aminopeptidase	16
Hepsin	3.4.21.-		17
Fertilin	(3.4.24.-)		18

to the isolation and characterization of E-24.11 (Kerr and Kenny, 1974a,b) and to the development of methods for the preparation of renal brush borders and microvilli (Booth and Kenny, 1974; George and Kenny, 1973). This interest in the ectopeptidases then developed into an obsession, which took several directions, even to synaptic membranes in the brain (Barnes *et al.*, 1993; Matsas *et al.*, 1983). Many others have shared the obsession, with the result that a minority interest began to attract a very large following with a consequent explosion in publications (see *Figure 1.3.*). ACE leads the field in publications, stimulated by the introduction of oral inhibitors of this enzyme for the treatment of hypertension. Pharmaceutical companies have since invested heavily in the development

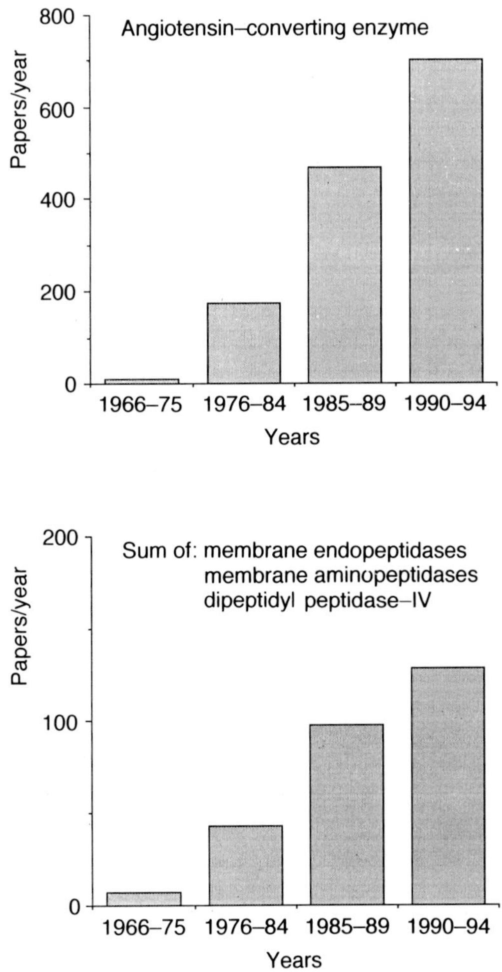

Figure 1.3. Growth in the numbers of published papers on cell-surface peptidases, 1966–1994. The numbers are derived from a Medline search. Each of the three ectopeptidases in the lower panel had comparable numbers of papers/year in this survey; for simplicity they are given as the sum of the three. Publications on ACE continue to out-strip those on other ectopeptidases.

of inhibitors of E-24.11, but publications on this enzyme are at present an order of magnitude fewer than those for ACE. It is too early to plot the rise of endothelin-converting enzyme, but it may one day rival the position of ACE. Apart from the interest in certain enzymes as therapeutic targets, the identification of several ectopeptidases with CD antigens has now brought immunologists and haematologists into the field.

1.5 The organization of this book

The topics are grouped in three sections. First there are chapters (1–5) on general aspects of ectopeptidases covering topics common to all or many. None of the group have yet been successfully analysed by X-ray crystallography, but much can be learned by studying the structures of certain related soluble peptidases. Important cell-biological questions and the different forms of membrane anchorage are also tackled, as is the pharmacology of inhibitors to some of the peptidases. The second section (Chapters 6–18) is devoted to detailed accounts of the individual enzymes, including some identified only by molecular cloning and whose present status is that of putative peptidases. The third section (Chapters 19–25) describes some of the roles that this group of ectoenzymes may play in health and disease.

References

Barnes K, Turner AJ, Kenny AJ. (1993) An immuno-electron microscopic study of pig substantia nigra shows colocalization of endopeptidase-24.11 with substance P. *Neuroscience* **53**: 1073–1082.

Booth AG, Kenny AJ. (1974) A rapid method for the preparation of microvilli from rabbit kidney. *Biochem. J.* **142**: 575–581.

Checler F. (1993) Neuropeptide-degrading peptidases. In: *Methods in Neurotransmitter and Neuropeptide Research* (eds S.H. Parvez, M. Naoi, T. Nagatsu and S. Parvez), pp. 375–418. Elsevier Science Publishers, Amsterdam.

Ciba Foundation Symposium. (1972) Peptide transport and hydrolysis in bacteria and mammalian gut. *Excerpta Med.* **50**: 302.

Engelhardt WA. (1957) Enzymes as structural elements of physiological mechanisms. *Proc. Int. Symp. Enzyme Chem. (Tokyo) IUB Symposium* **2**: 163–166.

George SG, Kenny AJ. (1973) Studies on the enzymology of purified preparations of brush border from rabbit kidney. *Biochem. J.* **134**: 43–57.

Hooper NJ. (1994) Families of zinc metalloproteases. *FEBS Lett.* **354**: 1–6.

Johnson CF. (1967) Disaccharidase localization in hamster intestine brush borders. *Science* **155**: 1670–1672.

Jongeneel CV, Quackenbush EJ, Ronco P, Verroust P, Carrel S, Letarte M. (1989) Common acute lymphoblastic leukaemia antigen expressed on leukaemia and melanoma cell lines has neutral endopeptidase activity. *J. Clin. Invest.* **83**: 713–717.

Kenny AJ, Maroux S. (1982) Topology of the microvillar membrane hydrolases of kidney and intestine. *Physiol. Rev.* **62**: 91–128.

Kerr MA, Kenny AJ. (1974a) The purification and specificity of a neutral endopeptidase from rabbit kidney brush border. *Biochem. J.* **137**: 477–488.

Kerr MA, Kenny AJ. (1974b) The molecular weight and properties of a neutral metallo-endopeptidase from rabbit kidney brush border. *Biochem. J.* **137**: 489–495.

Letarte M, Vera S, Tran R, Addis JBL, Onizuka RJ, Quackenbush EJ, Jongeneel CV, McInnes RR. (1988) Common acute lymphocytic leukemia antigen is identical to neutral endopeptidase. *J. Exp. Med.* **168**: 1247–1253.

Look AT, Ashmun RA, Shapiro LH, Peiper SC. (1989) Human myeloid plasma membrane glycoprotein CD13 (gp 150) is identical to aminopeptidase N. *J. Clin. Invest.* **83**: 1299–1307.

Matsas R, Fulcher IS, Kenny AJ, Turner AJ. (1983) Substance P and [Leu]enkephalin are hydrolysed by an enzyme in pig caudate synaptic membranes that is identical with the endopeptidase of kidney microvilli. *Proc. Natl Acad. Sci. USA* **80**: 3111–3115.

Miller D, Crane RK. (1961a) The digestive system of the epithelium of the small intestine. I. The intracellular locus of disaccharide and sugar phosphate ester hydrolysis. *Biochim. Biophys. Acta* **52**: 281–293.

Miller D, Crane RK. (1961b) The digestive system of the epithelium of the small intestine. II. Localization of disaccharide hydrolysis in the isolated brush border portion of intestinal epithelial cells. *Biochim. Biophys. Acta* **52**: 294–298.

Millican PE, Kenny AJ, Turner AJ. (1991) Purification and properties of neurotensin-degrading endopeptidases from pig brain. *Biochem. J.* **276**: 583–591.

Murray H, Turner AJ, Kenny AJ. (1994) The aminopeptidase activity in the human T-cell lymphoma line (Jurkat) is not at the cell-surface and is not aminopeptidase-N (CD-13). *Biochem. J.* **298**: 353–360.

Oda T, Seki S. (1965) Molecular structure and biochemical function of the microvillus membrane of intestinal cells with special emphasis on the elementary particles. *J. Electron Microsc. (Japan)* **14**: 210–217.

Pierotti A, Dong KW, Glucksman MJ, Orlowski M, Roberts JL. (1990) Molecular cloning and primary structure of rat testis metalloendopeptidase EC 3.4.24.15. *Biochemistry* **29**: 10323–10329.

Rawlings ND, Barrett AJ. (1993) Evolutionary families of peptidases. *Biochem. J.* **290**: 205–218.

Rothstein A, Meier R, Scharff TG. (1953) Relation of cell surface to metabolism. IX. Digestion of phosphorylated compounds by enzymes located on the surface of intestinal cells. *Am. J. Physiol.* **137**: 41–46.

Schechter I, Berger A. (1967) On the size of the active site in proteases. I. Papain. *Biochem. Biophys. Res. Commun.* **27:** 157–162.

Shipp MA, Vijayaraghavan J, Schmidt EV, Masteller EL, D'Adamio L, Hersh LB, Reinherz EL. (1989) Common acute lymphoblastic leukemia antigen (CALLA) is active neutral endopeptidase 24.11 ('enkephalinase'): direct evidence by cDNA transfection analysis. *Proc. Natl Acad. Sci. USA* **86**: 297–301.

Ugolev AM. (1965) Membrane (contact) digestion. *Physiol. Rev.* **45**: 555–595.

Ulmer AJ, Mattern T, Feller AC, Heyman E, Flad HD. (1990) CD26 antigen is a surface dipeptidyl peptidase IV (DPPIV) as characterised by monoclonal antibodies clone TII-19-4-7 and 4EL1C7. *Scand. J. Immunol.* **31**: 429–435.

Wong-Leung YL, Kenny AJ. (1968) The intracellular location of some particulate peptidases in the kidney of the rat and the rabbit. *Biochem. J.* **110**: 5P-6P.

Wu Q, Lahti JM, Air GM, Burrows PD, Cooper MD. (1990) Molecular cloning of the murine BP-1/6C3 antigen: a member of the zinc-dependent metallopeptidase family. *Proc. Natl Acad. Sci. USA* **87**: 993–997.

Wu Q, Cooper MD, Pierres M, Gorvel JP. (1991) Aminopeptidase A activity of the murine B differentiation antigen BP-1/6C3. *Proc. Natl Acad. Sci. USA* **88**: 676–680.

Chapter 2

Structure and function of some soluble zinc peptidases related to those at the cell surface

Walter Stöcker

2.1 Introduction

Cell-surface peptidases are involved in a variety of physiological processes ranging from morphogenesis, tissue differentiation and brain function to the general regulation of haemostasis. These enzymes are believed to activate or inactivate hormones, transmitters, growth factors and other proteins. However, at the molecular level the most versatile functions of many ectopeptidases are still poorly understood, since knowledge of the corresponding three-dimensional structures is rare. This is probably due to the fact that cell-surface peptidases in general exhibit more complex structures than their soluble counterparts. They are often highly glycosylated and many of them are composed of multiple modules, implying that a proteolytic domain is joined by additional functional units which serve as membrane anchors or which provide sites for substrate recognition and oligomerization. Many surface peptidases occur as complex homo- or hetero-oligomeric structures. As membrane-bound enzymes they are not easily crystallizable for X-ray structure analysis and their limited solubility and large size precludes the use of nuclear magnetic resonance (NMR) techniques for structural analysis.

An exception to the rule is the recently solved X-ray crystal structure of leishmanolysin, a zinc endopeptidase identical with the major surface antigen of *Leishmania major*, the elicitor of leishmaniasis (Bouvier *et al.,* 1995; Schlagenhauf *et al.*, 1995; P. Metcalf and R. Etges, personal communication). This enzyme is linked to the outer membrane of the promastigote form of this protozoan parasite by a glycosyl phosphatidylinositol (GPI) anchor. Leishmanolysin is one example of the large number of zinc metalloproteinases that carry a HEXXH-zinc binding motif in their active site (Jongeneel *et al.,* 1989; McKerrow, 1987; Rawlings and Barrett, 1993; Stöcker *et al.,* 1988; Vallee and Auld, 1993). This signature was first observed in the structure of the bacterial zinc peptidase thermolysin, where it is part of an α-helix in the active site (Matthews, 1988; Matthews *et al.,* 1972). The HEXXH-motif is also present in a considerable number of cell-surface peptidases known to be zinc-dependent (*Table 2.1*). Prominent examples are membrane aminopeptidase N (AP-N) and aminopeptidase A (AP-A), peptidyl dipeptidase A (angiotensin I-converting enzyme, ACE), endopeptidase-24.11 (E-24.11, neprilysin,

Cell-Surface Peptidases in Health and Disease, edited by A.J. Kenny and C.M. Boustead.

Table 2.1. HEXXH-peptidases of known structure and their cell-surface counterparts

Soluble peptidase	Cell-surface peptidase
Zincins, short zinc-binding motif HEXXH	
Thermolysin (*B. thermoproteolyticus*)[a]	Endopeptidase-24.11[b]
Bacillolysin (*B. cereus*)[c]	Peptidyl dipeptidase A (ACE, human, rabbit)[d]
Pseudolysin (*P. aeruginosa*)[e]	Endothelin-converting enzyme (ECE, human)[f]
	Aminopeptidase N (AP-N; human)[g]
	Aminopeptidase A (AP-A; human)[h]
	KELL blood group protein (human)[i]
Metzincin superfamily, long zinc-binding motif HEXXHXXGXXH	
Astacin (crayfish)[j]	Endopeptidase 24.18, meprin A/B (human, mouse, rat)[k]
Neutrophil collagenase (human)[l]	Membrane type matrix metalloproteinase[m]
Fibroblast collagenase (human)[n]	(MT1–4 MMP, human, chicken)
Stromelysin I (human)[o]	
Matrilysin (human)[p]	
Adamalysin II, atrolysin C[q]	PH30 (guinea pig sperm)[r]
(rattlesnake venom)	Cyritestin (mouse sperm)[s]
	EAP I (rat sperm)[t]
	Myelin basic protein-cleaving enzyme (bovine)[u]
Leishmanolysin (*L. major*)[v]	Leishmanolysin (*L. major*)

[a] Matthews (1988); [b] Li and Hersh (1995); Roques *et al.* (1995); [c] Pauptit *et al.* (1988), [d] Soubrier *et al.* (1988), Ehlers *et al.* (1989); [e] Thayer *et al.* (1991), Morihara (1995); [f] Xu *et al.* (1994), Schmidt *et al.* (1994), Ikura *et al.* (1994); [g] Olsen *et al.* (1988); [h] Nanus *et al.* (1993); [i] Lee *et al.* (1991); [j] Bode *et al.* (1992), Gomis-Rüth *et al.* (1993a), Stöcker *et al.* (1993); [k] Bond and Beynon (1995); [l] Bode *et al.* (1994), Reinemer *et al.* (1994), Stams *et al.* (1994); [m] Sato *et al.* (1994), Okada *et al.* (1995), Takino *et al.* (1995), Nagase (1996); [n] Gooley *et al.* (1994), Becker *et al.* (1995); [o] Browner *et al.* (1995); [p] Gomis-Rüth *et al.* (1993b, 1994); [q] Zhang *et al.* (1994); [r] Blobel *et al.* (1992); [s] Heinlein *et al.* (1994); [t] Perry *et al.* (1992); [u] Howard and Glynn (1995); [v] Button and McMaster (1988), Bouvier *et al.* (1995), Schlagenhauf *et al.* (1995), Schlagenhauf (1995).

enkephalinase, neutral endopeptidase, CD10) or endothelin-converting enzyme (ECE) (for overview, see Rawlings and Barrett, 1995). Attempts have been made to correlate these enzymes to zinc peptidases of known structure, in order to get an idea of the spatial arrangement of their active sites as a basis for rational drug design. For about two decades the only available lead structures were thermolysin (Matthews, 1988) and carboxypeptidase (Christianson and Lipscomb, 1989). The latter does not contain a HEXXH-motif, but its active site structure proved to be similar geometrically to that of thermolysin. However, despite some slight sequence similarities between, for example, E-24.11, AP-N, AP-A and ECE, on the one hand and thermolysin on the other, it proved to be difficult or almost impossible to model with sufficient reliability the active sites of the enzymes of interest on to the known structures (Benchetrit *et al.,* 1988).

From the beginning of this decade, the database of known template structures of HEXXH-proteinases has grown tremendously. In 1992, the structure of astacin, a prototypical metalloproteinase from crayfish revealed a mode of zinc binding different from that of thermolysin (Bode *et al.,* 1992; Gomis-Rüth *et al.,* 1993a; Stöcker *et al.,* 1993). In astacin, the zinc-binding motif is elongated to HEXXHXXGXXH, with all three histidine residues being zinc ligands. It turned out that this elongated motif is the signature for a superfamily of proteins comprising at least four families of zinc peptidases, the astacins, the matrix metalloproteinases (MMPs), the bacterial serralysins and the adamalysins/reprolysins (Hooper, 1994; Jiang and Bond, 1992; Murphy *et al.,* 1991; Rawlings and Barrett, 1993; Stöcker *et al.,* 1990; Vallee and Auld, 1993). The three-dimensional structures of these peptidases are strikingly similar; they all contain the zinc-binding motif **HEXXHXXGXXH**, and a conserved methionine-containing β turn, the 'Met-turn', beneath their active sites; these features together led to the designation 'metzincins' for this superfamily of zinc peptidases (Bode *et al.,* 1993; Stöcker and Bode, 1995; Stöcker *et al.,* 1995).

This chapter provides a short account of structural and functional features of soluble zinc peptidases that may be applied as well to cell-surface enzymes, for which three-dimensional structures are not yet available. Special emphasis is put on astacin and members of the metzincin superfamily.

2.2 Thermolysin: an archetypal zinc endopeptidase

2.2.1 Overall structure

Thermolysin from *Bacillus thermoproteolyticus* was the first HEXXH-protease whose three-dimensional structure was solved by X-ray crystallography (Matthews *et al.,* 1972; structures can be withdrawn from, e.g. http: //expasy.hcuge.ch). Since then, two closely related enzymes, bacillolysin (Pauptit *et al.,* 1988) and pseudolysin (Thayer *et al.,* 1991) have been analysed at the molecular level. Generally, thermolysins are made up of two lobes which are separated by the central substrate binding cleft. The upper lobe (*Figure 2.1*) is made up of a mixed α/β structure consisting of a highly twisted β-pleated sheet and two prominent helices (A and B), whereas the lower lobe is predominantly composed of a series of α-helices. Both lobes are connected by helix B, the central active site helix containing the typical HEXXH motif. Its two histidine residues (His142 and His146) are separated by a single helix turn and thus point out at the same side of the helix, perfectly positioned for the binding of the catalytic zinc ion. Such an arrangement of helix-borne histidinyl residues is frequently encountered in metallo-proteins even of completely different evolutionary origin (for reviews, see Holm *et al.,* 1996; Lipscomb and Sträther, 1996). The third zinc ligand of thermolysin is glutamic acid residue Glu166, which is located 20 residues downstream of the second histidine ligand, embedded in the conserved motif NEXXSD. In the absence of a substrate or inhibitor, the fourth coordination site of the active site metal is occupied by a water molecule, that is clamped between the zinc and Glu143 of the HEXXH motif. In the extensively studied catalytic mechanism of thermolysin (Auld, 1987; Matthews, 1988), this glutamic acid residue is believed to act as the catalytic base, which polarizes the zinc-bound water for nucleophilic attack of the scissile peptide bond. Besides the catalytically active zinc ion, thermolysins contain calcium ions, which are important for the stability and the structural integrity of the protein (Fontana *et al.,* 1986).

Despite their low overall sequence similarities, there are some significant common

Figure 2.1. α-Carbon backbone model of the structure of thermolysin (stippled; Matthews *et al.*, 1972) overlaid by the corresponding structure of astacin (solid lines; Gomis-Rüth *et al.* 1993a).

features between the thermolysins and cell-surface zinc peptidases like E-24.11, ECE, ACE and AP-N/AP-A, since in addition to the HEXXH zinc-binding motif, glutamate residues have been identified by site-directed mutagenesis as potential third zinc ligands, which are located 58 (E-24.11; Le Mouhal *et al.*, 1991), 54 (ECE; Shimada *et al.*, 1994), 23 (C-terminal domain of ACE; Williams *et al.*, 1994) and 19 (AP-N/AP-A; Wang and Cooper, 1996) residues downstream of the second His in the HEXXH sequence.

2.2.2 Phosphoramidon binding: a feature of some thermolysin-like enzymes

The cleavage specificities of thermolysins and a series of thermolysin-like enzymes are similar. This is reflected by their inhibition by phosphoramidon (i.e. *N*-[(α-rhamnopyranosyloxy)-hydroxy-phosphinyl]-L-Leu-L-Trp; Matthews, 1988; Morihara, 1995). This naturally occuring phosphonamidate binds to the active site of thermolysin as a transition state analogue inhibitor with its phosphinyl group mimicking the water-attacked, tetragonal peptide carbonyl group of a substrate during cleavage (Matthews, 1988). The apolar leucine and tryptophan residues of the inhibitor occupy the S_1' and S_2'-subsites of the substrate-binding region (according to the nomenclature of Schechter and Berger, 1967; see *Figure 1.2*).

The inhibition of zinc peptidases by phosphoramidon has been useful in categorizing cell-surface peptidases. As an inhibitor of endopeptidases it binds to E-24.11 (Kenny, 1977) and ECE (Ikura *et al.*, 1994; Schmidt *et al.*, 1994; Xu *et al.*, 1994) with inhibition constants in the nanomolar range, but is not potent against exopeptidases such as

ACE, AP-N and AP-A. Phosphoramidon is also inactive towards zinc endopeptidases of the astacin family [e.g. endopeptidase-24.18 (E-24.18, meprin, PABA-peptide hydrolase); see Chapter 10], indicating that the substrate-binding region of these enzymes is fundamentally different from the thermolysin-like peptidases.

2.3 The metzincins: a new superfamily of zinc peptidases

The metzincins were recognized after the three-dimensional structures of a series of peptidases other than thermolysin became available. The members of this protein superfamily are characterized by the elongated zinc-binding motif, **HEXXHXXGXXH** and, based on their sequences, they belong to four different protein families: (i) the astacins, named after the crayfish peptidase astacin, (ii) the serralysins, comprising large bacterial enzymes, (iii) the adamalysins/reprolysins (or ADAMs) from snake venom and mammalian tissues, and (iv) the matrixins (i.e. mammalian MMPs, collagenases).

2.3.1 The astacins

The astacins have been named after the prototypical endopeptidase astacin, a major collagenolytic enzyme in the digestive tract of the freshwater crayfish *Astacus astacus* L. (Dumermuth *et al.,* 1991; Geier *et al.,* 1997; Stöcker *et al.,* 1993). Apart from astacin, this protein family comprises E-24.18, a membrane-bound zinc enzyme from mammalian tissues, presumably involved in the processing of biologically active peptides (see also Chapter 10; Bond and Beynon, 1995; Dumermuth *et al.*, 1991), and also molecules which are active during morphogenesis and tissue differentiation, such as bone morphogenetic protein 1 (BMP1; Kessler *et al.,* 1996; Li *et al.,* 1996; Suzuki *et al.,* 1996; Wozney *et al.,* 1988), the tolloid gene product from the fly *Drosophila*, hatching enzymes from fish and birds, the UVS.2 gene product from frog, SPAN/BP10 from sea urchin, HMP from the freshwater polyp *Hydra* and flavastacin from bacteria. Generally, these are mosaic proteins, which in addition to their proteolytic module contain further modules serving as membrane anchors or as regulatory sites (cf. Bond and Beynon, 1995; Stöcker *et al.,* 1993, 1995).

2.3.2 The serralysins

The serralysins (Häse and Finkelstein, 1994; Maeda and Morihara, 1995) are secreted by various pathogenic bacteria. These proteinases consist of two modules, an N-terminal proteolytic module and a C-terminal, calcium-binding regulatory module (Baumann, 1994; Baumann *et al.,* 1993), which is thought to be responsible for export and activation.

2.3.3 The adamalysins and reprolysins / ADAMs

A series of very aggressive peptidases has been identified in the venom glands of snakes of the family Crotalidae (adders), which cause haemorrhage and tissue necrosis (Bjarnasson and Fox, 1995; Fox and Bjarnasson, 1995). The toxins have been named adamalysins after adamalysin II, an enzyme from the rattle snake *Crotalus adamanteus*

(Gomis-Rüth *et al.,* 1993b, 1994a). The adamalysins are secreted as proenzymes. An N-terminal propeptide is removed during activation (Grams *et al.,* 1994). Some venom metalloproteinases contain additional regulatory modules, which are presumably responsible for interactions with the extracellular matrix. Interestingly, the adamalysins are homologous to cell-surface proteins which have been recently discovered in the mammalian reproductive tract and found to be involved in cellular adhesion during sperm/egg fusion (Wolfsberg *et al.,* 1993). This finding gave rise to the alternative designation 'reprolysins' (for reviews, see Bjarnasson and Fox, 1995; Fox and Bjarnasson, 1995). These proteins contain disintegrin-like and cysteine-rich modules attached to a proteinase module and were therefore named the ADAMs-family (a disintegrin-like and metalloproteinase), which is believed to have a general function in cell–cell fusion (Huovila *et al.,* 1996). Some ADAMs also possess membrane spanning and cytoplasmic domains, and hence are true cell-surface peptidases. Interestingly, some ADAMs like cyritestin (Heinlein *et al.,* 1994) and the epididymal apical protein (EAP I; Perry *et al.,* 1992) have lost their zinc-binding ligands and are most likely proteolytically inactive (see *Table 2.2*).

2.3.4 The matrixins

The matrixins of vertebrates are believed to be involved in the regulated degradation of the extracellular matrix during development and differentiation. They are also expressed in certain diseases like arthritis or tumour invasion and metastasis (Birkedal-Hansen, 1995; Birkedal-Hansen *et al.,* 1993; Nagase, 1996; Stetler-Stevenson *et al.,* 1993). In the simplest case, the mature matrixins consist of a single, zinc-binding catalytic module. Others comprise additional haemopexin- and collagen-like modules believed to be important for enzyme–substrate interactions and/or regulatory purposes (Bode, 1995).

Recently, membrane-bound matrixins (membrane type MMPs, or MT-MMPs) have been identified on the surface of invasive tumour cells (Okada *et al.,* 1995; Sato *et al.,* 1994). MT-MMPs are also expressed in placenta, lung and kidney (Takino *et al.,* 1995). Like other matrixins the mature MT-MMPs are composed of an N-terminal catalytic domain joined by a haemopexin-like domain. In addition, they possess a C-terminal transmembrane region. The MT-MMPs are believed to be responsible for the activation of pro-gelatinase A (MMP-2) in the interstitium.

2.4 The structure of the metzincins

2.4.1 Topology and topology-based sequence alignment

The structures of astacin, adamalysin, the alkaline proteinase and the neutrophil collagenase were superimposed and analysed quantitatively. This allowed for the detection of topologically equivalent sequence positions and for the rearrangement of the sequence alignment *(Table 2.2).* The overall sequence identities between the four prototypes are low. The closest relationship is observed between alkaline proteinase/collagenase (27% identity) and between alkaline proteinase/astacin (26% identity), whereas a more distant relationship is seen between astacin/collagenase (16% identity) (Bode *et*

Table 2.2. Sequences around the zinc-binding motif and the Met-turn of the metzincins

		Zinc-binding motif	
alk	149	**QSWYLINS**SYSANVNPANGN**YGRQTLTHEIGHTLGLSHPG**DYNAGEGDPT	199
Erw	b	**SAWF--NY**NQSSIRNPDTDE**YGRHSFTHEIGHALGLSHPA**EYNAGEGDISY	
serr		**QTWY--NV**NQSNVKHPATED**YGRQTFTHEIGHALGLSHPG**DYNAGEGNPTY	
ada	123	**VGIVKDHS**-------PINL-**LVAVTMAHELGHNLGMEHDG**KD---------	156
EAP	I	**CSVVKDLL**-------PDVN-**IIGNRMAHQLGHSLGMRHDD**FP---------	
cyr		**AGIALHPK-------**TLAVE**GFAIVL**SQLLGINLGLAYDDVY---------	
col	173	**DAHFDAE**ETWT--NT-SANYN**LFLVAAHEFGHSLGLAHSS**D----------	210
str		**DAHFDDD**EQWT--KD-TTGTN**LFLVAAHEIGHSLGLFHSA**N----------	
mat		**DAHFDED**ERWT--DGSSLGIN**FLYAATHELGHSLGMGHSS**D----------	
ast	76	**QVSLQAN**G-----------CV**YHGTIIHELMHAIGFYHEH**TRMDRDNYVTI	
mep		**NISIGE**-G-----------CD**FKATIEHEILHALGFFHEQ**SRTDRDDYVNI	115
BMP1		**AISIGKN**------------CD**KFGIVVHELGHVVGFWHEH**TRPDRDRHVSI	
lei		VINIPAANIASRY-----DQLVTRVVT**HEMAHALG** ▲ **H**IKMRNA-------	

AGYYTALTMA*FSGPFFEDARIVANVPNVRGKNFDVP*
VINSSTAVAKAREQYGCDTLEYLEVEDQGGAGSAGS

		Met-turn	
alk	200	A-------------------DATYAEDTRA**YSVMSY**WEEQNTG--------	223
Erw	b	K------------------NSAAYAEDSRQ**FSIMSY**WEVENTG--------	
serr		R-------------------DVTYAEDTRQ**FSLMSY**WSETNTG--------	
ada	157	--------------------CLRGAS----**LCIMRPGL**-------------	170
EAP	I	--------------------CTCPLG----**KCVMGAGS**-------------	
cyr		-------------------NCFCPGS----**TCIMNPSA**-------------	
col	211	-----------------------------**PGALMYPNY**-------------	219
str		-----------------------------**TEALMYPLY**-------------	
mat		-----------------------------**PNAVMYPTY**-------------	
ast	116	NYQNVDPSMTSNFD-IDT-YSRYVGEDYQ**YYSIMHY**GKYSFSIQWGVLETI	164
mep		WWDQIITDYEHNFNTYDDNTITDLNTPYD**YESLMHY**GPFSFN-KNESIPTI	
BMP1		VRENIQPGQEYNFLKMEPQEVESLGETYD**FDSIMHY**ARNTFS-RGIFLDTI	
lei		-----------------------------Q**DELMAPA**AAIFQDLGFYQADF	

Residues which are topologically equivalent are in bold face; the zinc binding **HEXXHXXGXXH**-motif and the Met-turn are indicated. **Serralysins**: alk = alkaline proteinase from *Pseudomonas aeruginosa* (Okuda *et al.*, 1990); Erw b = *Erwinia chrysanthemi* proteinase b (Delepelaire and Wandersman, 1989); serr = *Serratia marcenscens* proteinase (Nakahama *et al.*, 1986). **Adamalysins/reprolysins**: ada = adamalysin II from *Crotalus adamanteus* (Gomis-Rüth *et al.*, 1994a); EAP I = epididymal apical protein I from rat epididymis (Perry *et al.*, 1992); cyr = cyritestin from mouse testis (Heinlein *et al.*, 1994). **Matrixins**: col = collagenase from human neutrophil granulocytes (Hasty *et al.*, 1990); mat = matrilysin, human punctuated (or putative) matrix metalloproteinase, PUMP1 (Quantin *et al.*, 1989); str = human stromelysin 1 (Whitham *et al.*, 1986). **Astacins**: ast = astacin from crayfish (Titani *et al.*, 1987); mep = meprin α from mouse kidney (Jiang *et al.*, 1992); BMP1 = human bone morphogenetic protein 1 (Wozney *et al.*, 1988); lei = leishmanolysin; the 40-residue insert disrupting the HEXXHXXGXXH-motif between the second and third histidine residue is boxed (Button and McMaster, 1988).

al., 1993; Stöcker and Bode, 1995; Stöcker *et al.,* 1995). However, in spite of large differences on the sequence level, the four metzincin prototypes astacin, adamalysin II, and the catalytic modules of neutrophil collagenase and the *Pseudomonas* alkaline proteinase display a surprisingly high topological similarity. They are all globular entities, subdivided into two subdomains by the substrate binding cleft, with the zinc at its bottom *(Figure 2.2).* The upper subdomains are constructed of a twisted β-sheet and two long α-helices (hA, hB). The sheet comprises four parallel (sI, sII, sIII, sV) and one antiparallel (sIV) strands *(Figure 2.2)*. The only antiparallel β-strand sIV ('edge strand') forms the upper borderline of the substrate binding cleft. It is complementary to an incoming substrate which upon binding would enlarge the β-sheet by just another strand. In the alkaline proteinase (Baumann *et al.,* 1993) and adamalysin (Gomis-Rüth *et al.,* 1993b, 1994a) additional structural elements are placed on top of the β-sheet.

In collagenase, the loop between strands sIII and sIV *(Figure 2.2)* bulges over a second zinc ion and a calcium ion, which probably play structural roles (Bode *et al.,* 1994). Smaller bulge structures are seen in adamalysin (Gomis-Rüth *et al.,* 1993b,

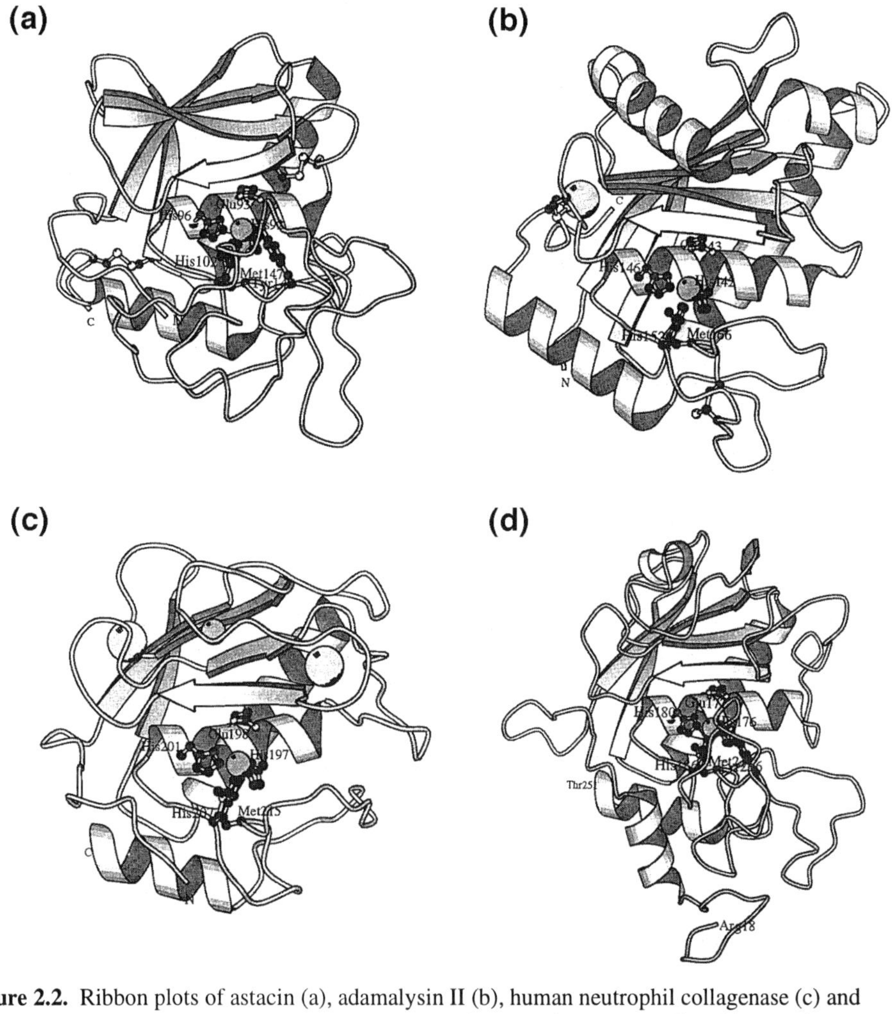

Figure 2.2. Ribbon plots of astacin (a), adamalysin II (b), human neutrophil collagenase (c) and *Pseudomonas* alkaline protease (d). Zinc and calcium ions are shown as small and large spheres, respectively. The zinc ligands, the methionine of the Met-turn, and the catalytic glutamic acid are labelled. In the standard orientation the cleft lies horizontally in the paper plane (Stöcker and Bode, 1995).

1994a) and in the alkaline proteinase (Baumann *et al.*, 1993), but are absent in astacin (Bode *et al.,* 1992; Gomis-Rüth *et al.,* 1993a).

In a similar fashion to thermolysin, the 'active site helix' (helix hB), contains the two histidine zinc ligands of the **HEXXH** motif. However, in contrast to thermolysin, the helix is terminated at a conserved glycine residue, three residues downstream of the second histidine zinc ligand *(Figure 2.3).* The peptide chain leaves the upper subdomain after the glycine for the positioning of the third histidine zinc ligand of the **HEXXHXXGXXH** motif.

Subsequently the polypeptide enters the more irregularly folded C-terminal subdomain, whose only common regular secondary structure element is a long α-helix (hC) close to the C terminus *(Figure 2.2).* However, the lower subdomain contains a unique structural feature in the form of a strictly conserved 1,4-tight turn located just beneath the zinc binding active site. This turn has been named the 'Met-turn', since it contains an invariant methionine residue, whose ϵ-methyl group is about 4 Å distant from the faces of the first and second zinc-binding imidazoles *(Figure 2.4)* (Bode *et al.,* 1993; Stöcker and Bode, 1995; Stöcker *et al.,* 1995). Before and after the Met-turn there are surface loops, which shape the lower rim of the substrate binding cleft.

In some metzincins the polypeptide chain leaves the catalytic module to enter the β-sandwich module (serralysins), the haemopexin-like module (most matrixins), or disintegrin-like structures (some adamalysins/reprolysins) or into other intervening domains (some astacins), respectively. In E-24.18, the chain first runs into the so-called MAM domain to end up finally in the EGF-like module and the membrane-spanning module (see Chapter 10, and Bond and Beynon, 1995).

There are several remarkable conserved sequence motifs, which may serve as signatures of the metzincin superfamily. In **HEBXHXBGBXHZ**, including the three histidine zinc ligands and the catalytic Glu, the B residue is bulky and anchored in the hydrophobic core of the protein, and Z labels a conserved residue, which is typical for

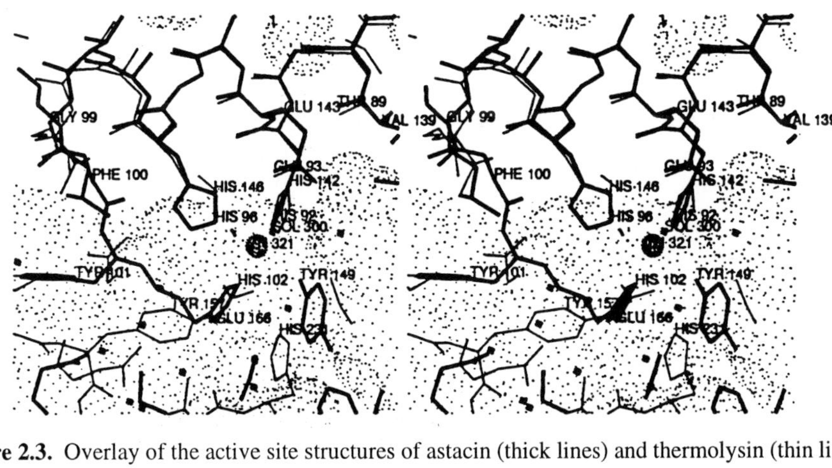

Figure 2.3. Overlay of the active site structures of astacin (thick lines) and thermolysin (thin lines), displayed together with the Connolly dot surface corresponding to astacin, the zinc ion (sphere of radius 0.5 Å) and a few localized solvent molecules (crosses) (Bode *et al.*, 1992).

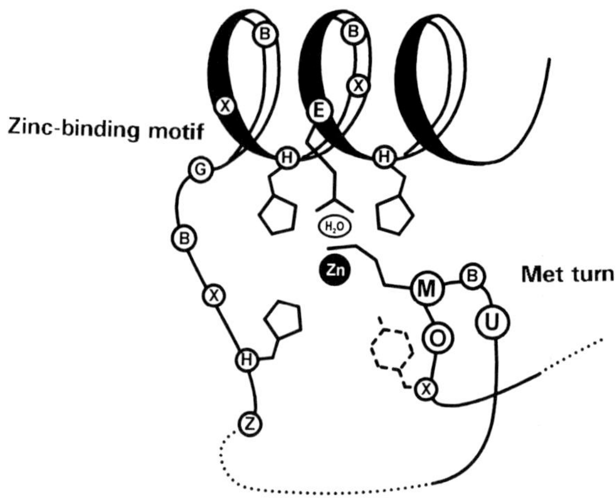

Figure 2.4. Schematic depiction of the active site of the metzincins. The critical methionine in the Met-turn, the histidine zinc ligands and the conserved other residues around the zinc-centre are shown. Variable segments are represented as dashed lines (Stöcker and Bode, 1995).

each of the families. In the Met-turn consensus sequence (**UBMO**X), B is again a hydrophobic side chain, while U and O are conserved family-specific residues, and X is the tyrosine zinc-ligand of the astacins and serralysins *(Table 2.2; Figure 2.4).*

2.4.2 The catalytic metal-binding centre of the metzincins

The zinc-binding regions of the four metzincin prototypes are strikingly similar. The catalytic zinc is ligated by the three His residues of the **HEXXHXXGXXH** zinc-binding motif with zinc-(Nϵ)-imidazole distances between 2.0 and 2.2 Å. In the absence of a substrate or inhibitor, a water molecule acting as a fourth zinc ligand is inserted between the catalytic Glu of the HEXXHXXGXXH-motif and the metal, forming a trigonal pyramidal coordination sphere around the zinc, with the Nϵ-nitrogen of the second histidine at the pyramid vertex, and the zinc in the centre of the triangular base formed by the other ligands. Astacin and the alkaline proteinase are a special case, where the hydroxyl oxygen of a conserved tyrosine residue following the Met-turn acts as a fifth zinc ligand, resulting in a trigonal-bipyramidal coordination sphere in the free enzymes (Baumann, 1994; Gomis-Rüth *et al.,* 1993a, 1994b).

2.4.3 Leishmanolysin: a special case

Leishmanolysin has been characterized as the major surface antigen of the flagellate protozoan *Leishmania* and recognized to be a zinc-dependent peptidase with a relatively broad specificity (Bouvier *et al.,* 1995; Button and McMaster, 1988; Etges *et al.,* 1986). It was also identified as a HEXXH-protein (Jongeneel *et al.,* 1989). Its three-dimensional structure was solved recently by X-ray crystallography (Schlagenhauf, 1995; Schlagenhauf *et al.,* 1995). The protein was crystallized in a truncated version lacking the C-terminally attached GPI-anchor, which was removed by phospholipase C-treatment. This is so far the only cell-surface peptidase whose structure has been solved. Leishmanolysin is a three-domain protein, with the catalytic zinc located between the first and the second subdomain. Most interestingly, the topology of the N-terminal subdomain identifies this peptidase as a member of the metzincin superfamily. The catalytic zinc ion is typically complexed by three histidines and a glutamic acid-bound water molecule. Moreover there is a methionine residue located beneath the metal binding site as part of a 1,4-β-tight turn, the Met-turn. The active site of leishmanolysin is perfectly superimposable on to the structure of astacin. However, leishmanolysin is not retrieved from the database by looking for a **HEXXHXXGXXH** sequence, because a long intervening spacer separates the second from the third zinc ligand (see *Table 2.2*; Schlagenhauf, 1995; P. Metcalf and R. Etges, personal communication). The two C-terminal subdomains of leishmanolysin are unrelated to corresponding parts of other metzincins.

2.4.4 Structural relationship of the metzincins and thermolysins

Although the similarity scores between the metzincins and the thermolysin-like zinc peptidases are extremely low, there are striking similarities in the topologies, which are confined to the zinc-binding active site helix, the β-strands sI, sIII, sV and sIV (edge strand) and to the helix hA (Bode *et al.,* 1993; Gomis-Rüth *et al.,* 1993a, 1994a, Stöcker and Bode, 1995; Stöcker *et al.,* 1995). Hence, the metzincins and the thermolysin-like enzymes most likely have arisen by divergent evolution from a common ancestor. The term 'zincins' was suggested for proteins which contain a HEXXH-zinc-binding motif and which are related to each other due to the similar fold of their polypeptide chains (Bode *et al*., 1993).

2.5 Activation of pro-metzincins

The N-terminal section of adamalysin is rather short and solvent accessible, whereas astacin, collagenase and the alkaline proteinase have N-terminal segments running parallel to helix hC before they enter the molecular body at a conserved tryptophan residue *(Figure 2.2).* In the alkaline proteinase there are additional structural elements preceding this region, which are in contact with the C-terminal β-sandwich module (Baumann *et al.,* 1993).

Two versions of collagenase are known, a Met80 form, in which the first six residues are unordered, and a one residue longer Phe79 form with a perfectly ordered N-terminal region (Bode *et al.,* 1994; Knäuper *et al.,* 1990). The N-terminal ammonium group of the Phe79 form and the side chain of the conserved Asp232 of helix hC form a salt bridge (Reinemer *et al.,* 1994), and the next residue, another invariant aspartate, Asp233, is indirectly linked via a hydrogen bonding network to the Met-turn and to the first histidine zinc ligand. Hence, the N terminus of collagenase and of other matrixins is in touch with the active centre, which might explain the increased activity of the Phe form compared with the truncated Met form (Bode *et al.,* 1994; Reinemer *et al.*, 1994).

In human pro-stromelysin-1, Pro86 is the first residue which has a topologically equivalent counterpart in the mature enzyme. A cysteine residue as part of a conserved motif PRCGXPD, the 'cysteine switch', coordinates the catalytic zinc (K. Appelt, personal communication). The cysteine-containing propeptide can be removed proteolytically or chemically to yield the active enzymes (Bläser *et al.*, 1991; Nagase *et al.*, 1990; Springman *et al.*, 1990).

Helix hC of the alkaline proteinase contains a topologically equivalent conserved tandem of aspartates, similar to the situation in collagenase *(Table 2.2).* The first (Asp237) is salt bridged to the side chain of Arg18 linking the N-terminal stretch to the catalytic module *(Figure 2.2).* The second (Asp238) is likewise engaged in a hydrogen bonding system with the active site. A special role in the activation of the proserralysins is attributed to the calcium-binding of the calcium β-sandwich module (Baumann *et al.*, 1993).

In the astacins, the single conserved aspartate residue (Asp186) in the helix hC is structurally and functionally equivalent to the second Asp in collagenase and the alkaline proteinase *(Table 2.2).* However, activation of proastacins is most likely achieved in a completely different fashion. In the mature form of astacin the two N-terminal residues Ala1 and Ala2 are completely buried in a water-filled cavity of the lower subdomain *(Figure 2.5).* At the bottom of this cavity, the ammonium group of Ala1 is connected by a water-mediated salt bridge with the carboxylate of Glu103, the direct neighbour of the zinc-ligand His102. Thus, the N terminus is in physical contact with the active site. This is reminiscent of the trypsin-like serine proteinases (Huber and Bode, 1978), which become activated after proteolytic removal of a pro-peptide from the inactive zymogen, through the interaction of the new N terminus with the carboxylate of Asp194, neighbouring the catalytic Ser195. In astacin, the arrangement of the N terminus indicates an analogous activation mechanism, since N-terminally extended proforms could not exhibit such a conformation (Bode *et al.*, 1992; Stöcker *et al.*, 1993). The necessity of pro-protein processing in the astacins has been demonstrated for E-24.18, from mouse, rat and humans (Chapter 10; Bond and Beynon, 1995; Geier *et al.*, 1997).

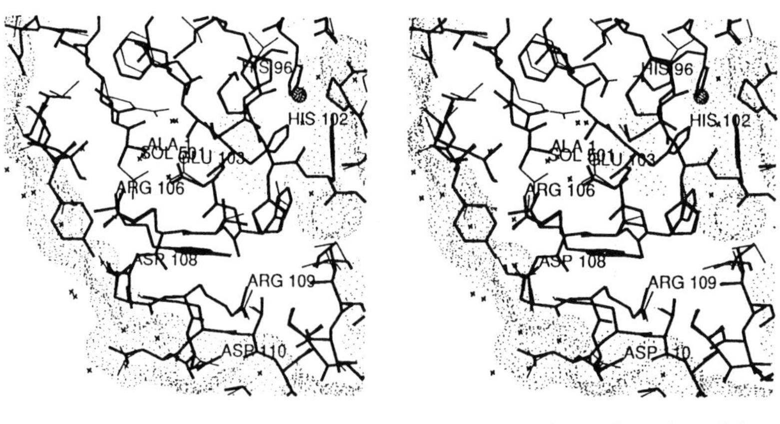

Figure 2.5. Overlay of the location of the N terminus in the mature form of astacin and the computer model of mouse meprin α subunit. A water-mediated salt bridge is formed in astacin between Ala1 and Glu103, the neighbour of zinc ligand His102 (Stöcker *et al.*, 1993).

An aspartate tandem present in the helix hC of adamalysin is shifted by one residue *(Table 2.2)* and is not conserved in the other members of the adamalysin family (Gomis-Rüth *et al.,* 1994a). However, the activation of proadamalysins may occur in a similar fashion as in the collagenases, since they also contain a cysteine-switch-like propeptide (PKMCGV) in their N-terminal regions (Bjarnasson and Fox, 1995; Fox and Bjarnasson, 1995; Grams *et al.,* 1994).

2.6 Catalytic properties of the metzincins

2.6.1 General features

Structures of metzincin inhibitor complexes revealed that substrates are presumably bound in an extended conformation, which is different from thermolysin (Becker *et al.,* 1995; Bode *et al.,* 1994; Borkakoti *et al.,* 1994; Dhanaraj *et al.,* 1995; Gooley *et al.,* 1994; Grams *et al.,* 1995, 1996; Lovejoy *et al.,* 1994; Matthews, 1988; Stams *et al.,* 1994; Stöcker *et al.,* 1995). It appears that the same kind of enzyme–substrate interactions are principally involved N-terminally of the cleavage site in the different structures, because the substrate aligns with the edge strand under formation of two (or three) hydrogen bonds in the known structures (Bode *et al.,* 1994). However, the C-terminal half of the substrate is bound in different ways. In the matrixins (Grams *et al.,* 1995) and probably in the adamalysins, it is fenced by the antiparallel bulge segment from above, and by the parallel wall-forming segment from below by two pairs of hydrogen bonds. Comparable main chain interactions are not observed in astacin and are expressed less in the alkaline proteinase (Baumann *et al.,* 1993).

Several metzincins, like the collagenases, gelatinases (Birkedal-Hansen, 1995; Birkedal-Hansen *et al.,* 1993; Stetler-Stevenson *et al.,* 1993), snake venom proteinases (Bjarnasson and Fox, 1995; Fox and Bjarnasson, 1995) and astacins (Kessler *et al.,* 1996; Li *et al.,* 1996; Stöcker and Zwilling, 1995) are able to cleave proteins of the extracellular matrix. Correspondingly, there is a preference for Pro in P_3 and, with the exception of astacin (Stöcker *et al.,* 1990), large, mainly hydrophobic residues in P_1' (Grams *et al.,* 1995; Netzel-Arnett *et al.,* 1993a, b). These similar specificities are in good accordance with structural constraints. For example, the S_1' pockets in general are rather large (Grams *et al.,* 1995), but blocked in astacin (Stöcker *et al.,* 1993). A groove in the northern wall of the active site cleft of each of the four prototypes corresponds to a specific S_3 subsite very suitable for proline, and P_2 residues are harboured in shallow depressions at the bottom of the cleft, whereas the P_1 side chains are projecting from the cleft and in contact with the southern wall of the substrate binding region.

2.6.2 The tyrosine switch in the catalytic mechanism of the astacins

The transition state analogue phosphinic peptide Z-Pro-Lys-Pheψ(PO_2CH_2)Ala-Pro-OMe binds into the active site of astacin by antiparallel alignment along the edge β-strand of the upper subdomain *(Figure 2.6)* (Grams *et al.,* 1996). The methyl side chain of the P_1' Ala points toward the edges of Pro176 and Thr89, indicating that the primary specificity of astacin for small aliphatic residues in the P_1' position (Stöcker

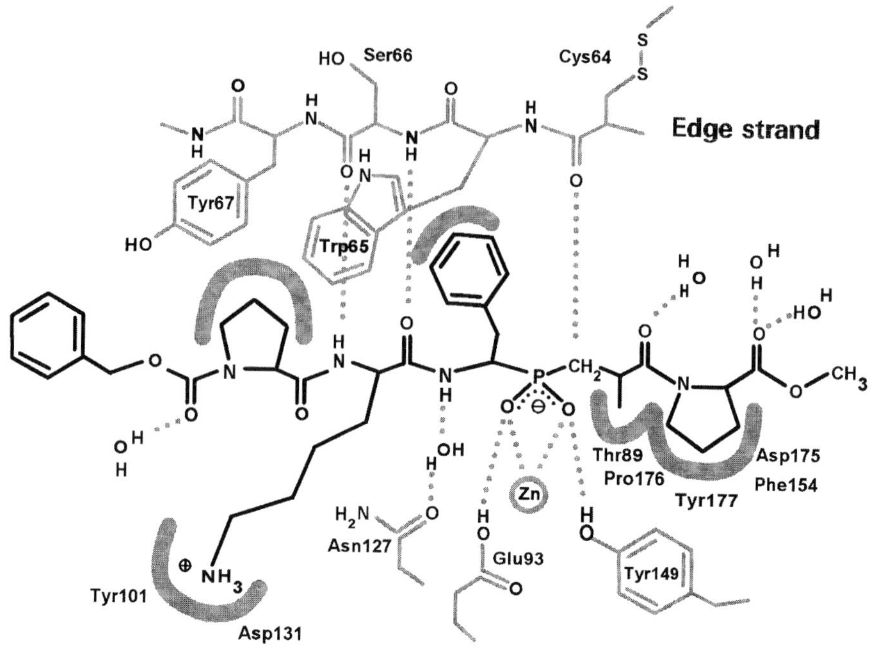

Figure 2.6. Scheme of the binding interactions between astacin and the transition state analogue inhibitor Z-Pro-Lys-PheΨ(PO_2CH_2)Ala-Pro-OMe (Grams *et al.*, 1996).

et al., 1990) is based primarily on sterical hindrance rather than productive binding interactions. The phosphinyl group of the inhibitor, mimicking a water-attacked carbonyl group, is bound to the zinc ion, Glu93 and Tyr149 *(Figure 2.6)*. In the astacin–inhibitor complex, the phosphinyl group inserts between the zinc and the hydroxyl group of the zinc ligand Tyr149, which forms a hydrogen bond to the lower phosphinyl oxygen. Thereby, the distance between the zinc and the phenolic oxygen increases from 2.5 Å to 5.0 Å *(Figure 2.6)*. The shift of Tyr149 is due to a rotation around the backbone and a concomitant counter movement of the side chain of Phe154. The induced conformational change propagates along the surface loops following and preceding the Met-turn resulting in a hinge movement, which decreases by about 1 Å the (backbone) distance between the lower and the upper subdomain, defining the active site cleft.

Tyr149 seems to be analogous in both position and function (i.e. transition state stabilization) to Arg127 of carboxypeptidase (Christianson and Lipscomb, 1989; Kim and Lipscomb, 1990; Phillips *et al.,* 1990) and His231 of thermolysin (Beaumont *et al.,* 1995; Matthews, 1988).

Glu93 of astacin is thought to act as a proton shuttle between the zinc-bound activated water and the nitrogen of the scissile peptide bond (Grams *et al.,* 1996), as suggested for Glu143 of thermolysin (Matthews, 1988) or Glu270 of carboxypeptidase (Auld, 1987; Christianson and Lipscomb, 1989) *(Figure 2.7)*.

Taken together, the hinge movement of the lower subdomain strengthens the binding

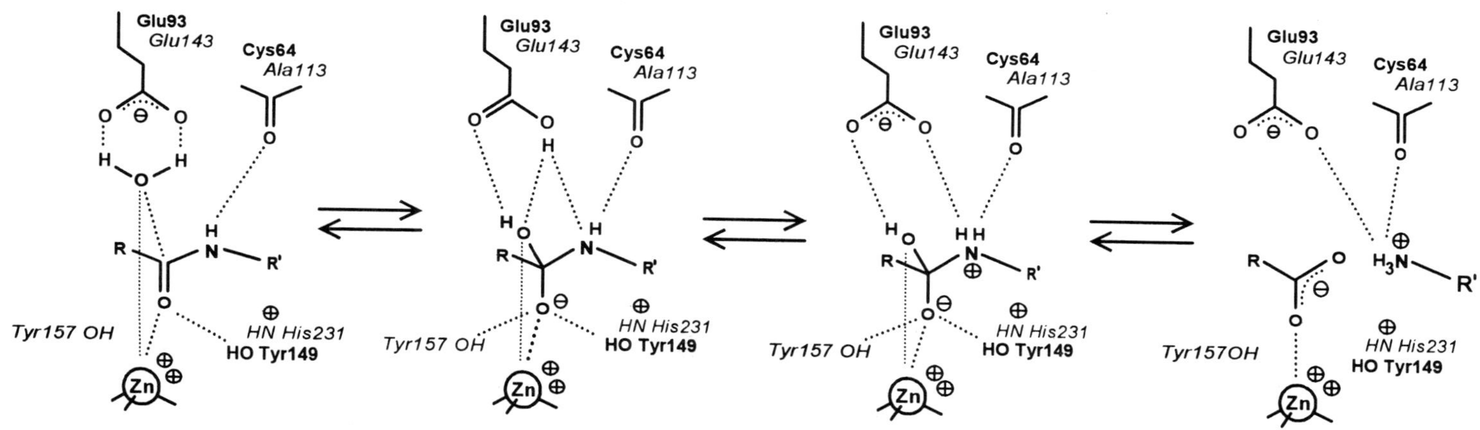

Figure 2.7. Catalytic mechanism of astacin compared to thermolysin; astacin residues are in bold face and the corresponding thermolysin residues are in italics (Grams *et al.*, 1996).

of the transition state mimetic and is in agreement with induced fit and transition state theory. This conformational change is triggered by the unique 'tyrosine switch' mechanism of astacin, which most likely is applicable to the astacin family as a whole (Grams *et al.,* 1996) and also to the serralysin-like enzymes (Baumann *et al.,* 1995; Miyatake *et al.,* 1995).

2.7 Comparison of astacin and E-24.18 by molecular modelling

Astacin and the catalytic domain of E-24.18 (meprin) share about 35% identical sequence positions. Thus, the knowledge about the structure and function of astacin can also be used for a better understanding of the catalytic apparatus of the surface peptidase. In contrast with other members of the astacin family, there is a sufficient amount of kinetic data on astacin and mouse meprin A that can be used for comparisons (Stöcker and Zwilling, 1995; Wolz and Bond, 1995). Both astacin and mouse meprin A prefer substrates comprising five or more residues. Both efficiently catalyse the hydrolysis of peptides carrying bulky residues in P_1 and P_2 and proline residues in P_3 and P_2', respectively. However, astacin and meprin A differ markedly in their preference for residues in the P_1' position. In this subsite, astacin tolerates only small uncharged residues (see above), but meprin A also hydrolyses substrates with Arg, Phe, or Lys residues in this position (Stöcker *et al.,* 1990, 1991; Wolz, 1994; Wolz *et al.,* 1991).

This striking difference has been investigated by molecular modelling of the α-chain of mouse meprin A based on the X-ray crystal structure of astacin (Stöcker *et al.,* 1993). The model reveals that the peptide backbones of the two enzymes can be superimposed. Due to the similar chain length of astacin and the meprin α catalytic subdomain, only five single residue deletions, one double deletion and three single residue insertions are introduced. However, the marked difference in the P_1'-specificity can be explained, since in the presumed S_1'-subsite, there is a proline ring (Pro176) in astacin that seals the bottom of the substrate-binding cleft. The counterpart of this proline residue is a glycine in meprin α. In addition, the immediately following residue, Tyr177 of astacin, is deleted in meprin α. These two changes lead to a more open configuration of the S_1'-subsite in meprin α, allowing the binding of extended substrate side chains in this region. The upper edge of the S_1'/S_2'-subsites is lined by Cys64 and Thr89 in both astacin and meprin α. However, the lower edge of the cleft is different due to the replacement of Asp175 of astacin by isoleucine or leucine in meprin and other astacins, respectively. Furthermore, the binding of aromatic substrate side chains in the S_3' subsite might be promoted by Tyr177 and/or Trp158 in astacin, which are deleted and replaced by Asn158 in meprin, respectively. In summary, the subtle differences of the specificities of astacin and meprin α can be explained at least in part by differences in their active sites as deduced from modelling the mammalian enzyme on the prototypical crayfish peptidase.

Acknowledgements

The author thanks Wolfram Bode (Martinsried) and Frank Grams (Mannheim) for discussions and for providing computer graphics artwork. The support of this research by grants from the Deutsche Forschungsgemeinschaft (Sto 185/3–1 and Sto 185/3–2) is gratefully acknowledged.

References

Auld DS. (1987) Acyl group transfer – metalloproteinases. In: *Enzyme Mechanisms* (eds M.I. Page and A. Williams), pp. 240–258. Royal Society of Chemistry, Letchworth.

Baumann U. (1994) Crystal structure of the 50 kDa metallo protease from *Serratia marcescens*. *J. Mol. Biol.* **242:** 244–251.

Baumann U, Wu S, Flaherty KM, McKay DB. (1993) Three-dimensional X-ray crystallographic structure of the alkaline protease of *Pseudomonas aeruginosa*. *EMBO J.* **12:** 3357–3364.

Baumann U, Bauer M, Létoffé S, Delepelaire P, Wandersman C. (1995) Crystal structure of a complex between *Serratia marcescens* metallo-protease and an inhibitor from *Erwinia chrysanthemi*. *J. Mol. Biol.* **248:** 653–661.

Beaumont A, O'Donohue MJ, Paredes N, Rousselet N, Assicot M, Bohuon C, Fournié-Zaluski MC, Roques BP. (1995) The role of histidine 231 in thermolysin-like enzymes. A site directed mutagenesis study. *J. Biol. Chem.* **270:** 16803–16808.

Becker JW, Marcy AI, Rokosz LL *et al.* (1995) Stromelysin-1: three-dimensional structure of the inhibited catalytic domain and of the C-truncated proenzyme. *Protein Sci.* **4:** 1966–1976.

Benchetrit T, Bissery V, Mornon JP, Devault A, Crine P, Roques BP. (1988) Primary structure homologies between two zinc metallopeptidases, the neutral endopeptidase 24.11 ('enkephalinase') and thermolysin through clustering analysis. *Biochemistry* **27:** 592–596.

Birkedal-Hansen H. (1995) Proteolytic remodeling of extracellular matrix. *Curr. Opin. Cell Biol.* **7:** 728–735.

Birkedal-Hansen H, Moore WGI, Bodden MK, Windsor LJ, Birkedal-Hansen B, DeCarlo A, Engler JA. (1993) Matrix metalloproteinases: a review. *Crit. Rev. Oral Biol. Med.* **4:** 197–250.

Bjarnasson JB, Fox JW. (1995) Snake venom metalloproteinases: reprolysins. *Methods Enzymol.* **248:** 345–368.

Bläser J, Knäuper V, Osthues A, Reinke H, Tschesche H. (1991) Mercurial activation of human polymorphonuclear leukocyte procollagenase. *Eur. J. Biochem.* **202:** 1223–1230.

Blobel CP, Wolfsberg TG, Turck CW, Myles DG, Primakoff P, White JM. (1992) A potential fusion protein and an integrin-ligand domain in a protein active in sperm–egg fusion. *Nature* **356:** 248–252.

Bode W. (1995) A helping hand for collagenases. *Structure* **3:** 527–530.

Bode W, Gomis-Rüth F-X, Huber R, Zwilling R, Stöcker W. (1992) Structure of astacin and implications for activation of astacins and zinc ligation of collagenases. *Nature* **358:** 164–167.

Bode W, Gomis-Rüth F-X, Stöcker W. (1993) Astacins, serralysins, snake venom and matrix metalloproteinases exhibit identical zinc-binding environments (HEXXHXXGXXH and Met-turn) and topologies and should be grouped into a common family, the 'metzincins'. *FEBS Lett.* **331:** 134–140.

Bode W, Reinemer P, Huber R, Kleine T, Schnierer S, Tschesche H. (1994) The crystal structure of human neutrophil collagenase inhibited by a substrate analog reveals the essentials for catalysis and specificity. *EMBO J.* **13:** 1263–1269.

Bond JS and Beynon RB. (1995) The astacin family of metalloendopeptidases. *Protein Sci.* **4:** 1247–1261.

Borkakoti N, Winkler FK, Williams DH, D'Arcy A, Broadhurst MJ, Brown PA, Johnson WH, Murray EJ. (1994) Structure of the catalytic domain of human fibroblast collagenase complexed with an inhibitor. *Nature Struct. Biol.* **1:** 106–110.

Bouvier J, Schneider P, Etges R. (1995) Leishmanolysin: surface metalloproteinase of Leishmania. *Methods Enzymol.* **248:** 614–633.

Browner MF, Smith WW, Castelhano AL. (1995) Matrilysin–inhibitor complexes; common themes among metalloproteases. *Biochemistry* **34:** 6602–6610.

Button LL, McMaster WR. (1988) Molecular cloning of the major surface antigen of *Leishmania*. *J. Exp. Med.* **167:** 724–729.

Christianson DW, Lipscomb WN. (1989) Carboxypeptidase A. *Acc. Chem. Res.* **22:** 62–69.

Delepelaire P, Wandersman C. (1989) Protease secretion by *Erwinia chrysanthemi*. *J. Biol. Chem.* **264:** 9083–9089.

Dhanaraj V, Ye QZ, Johnson LL *et al.* (1996) X-ray structure of a hydroxamate inhibitor complex of stromelysin catalytic domain and its comparison with members of the zinc metalloproteinase superfamily. *Structure* **4:** 375–386.

Dumermuth E, Sterchi EE, Jiang W, Wolz RL, Bond JS, Flannery AV, Beynon RJ. (1991) The astacin family of metalloendopeptidases. *J. Biol. Chem.* **266:** 21381–21385.

Ehlers MRW, Fox EA, Strydom DJ, Riordan JF. (1989) Molecular cloning of human testicular angiotensin converting enzyme: the testis isoenzyme is identical to the C-terminal half of endothelial angiotensin converting enzyme. *Proc. Natl Acad. Sci. USA* **86:** 7741–7745.

Etges R, Bouvier J, Bordier C. (1986) The major surface protein of *Leishmania* promastigotes is a protease. *J. Biol. Chem.* **261:** 9098–9101.

Fontana A, Fassina G, Vita C, Dalzoppo D, Zamai M, Zambonin M. (1986) Structural and stability features of thermolysin revealed by its autolytic process. In: *Zinc Enzymes* (eds I. Bertini, C. Luchinat, W. Maret and M. Zeppezauer), pp. 225–237. Birkhäuser Verlag, Basel.

Fox JW, Bjarnasson JB. (1995) Atrolysins: metalloproteinases from *Crotalus atrox* venom. *Methods Enzymol.* **248:** 368–387.

Geier G, Jacob E, Stöcker W, Zwilling R. (1997) Genomic organization of the zinc-endopeptidase astacin. *Arch. Biochem. Biophys.* **337:** 300–307.

Gomis-Rüth F-X, Stöcker W, Huber R, Zwilling R, Bode W. (1993a) The refined 1.8 Å X-ray crystal structure of astacin, a zinc-endopetidase from the crayfish *Astacus astacus* L. Structure determination, refinement, molecular structure, and comparison to thermolysin. *J. Mol. Biol.* **229:** 945–968.

Gomis-Rüth F-X, Kress LF, Bode W. (1993b) First structure of a snake venom metalloproteinase: a prototype for matrix metalloproteinases/collagenases. *EMBO J.* **12:** 4151–4157.

Gomis-Rüth F-X, Kress LF, Kellermann J, Mayr I, Lee X, Huber R, Bode W. (1994a) Refined 2.0 Å X-ray crystal structure of the snake venom zinc endopeptidase adamalysin II. Primary and tertiary structure determination, refinement, molecular structure and comparison with astacin, collagenase and thermolysin. *J. Mol. Biol.* **239:** 513–544.

Gomis-Rüth F-X, Grams F, Yiallouros I, Nar H, Küsthardt U, Zwilling R, Bode W, Stöcker W. (1994b) Crystal structures, spectroscopic features and catalytic properties of cobalt (II)-, copper (II)-, nickel (II)- and mercury (II)-derivatives of the zinc-endopeptidase astacin. A correlation of structure and proteolytic activity. *J. Biol. Chem.* **269:** 17111–17117.

Gooley PR, O'Connell JF, Marcy AI *et al.* (1994) The NMR structure of the inhibited catalytic domain of human stromelysin-1. *Nature Struct. Biol.* **1:** 111–118.

Grams F, Huber R, Kress LF, Moroder L, Bode W. (1994) Activation of snake venom metalloproteinases by a cysteine-switch-like mechanism. *FEBS Lett.* **335:** 76–80.

Grams F, Reinemer P, Powers JC, Kleine T, Pieper M, Tschesche H, Huber R, Bode W. (1995) X-ray structures of human neutrophil collagenase complexed with peptide hydroxamate and peptide thiol inhibitors – implications for substrate-binding and rational drug design. *Eur. J. Biochem.* **228:** 830–841.

Grams F, Dive V, Yiotakis A, Yiallouros I, Vassilou S, Zwilling R, Bode W, Stöcker W. (1996) Structure of astacin with a transition state analogue inhibitor. *Nature Struct. Biol.* **3:** 671–675.

Häse CC, Finkelstein RA. (1994) Bacterial extracellular zinc-containing metalloproteases. *Microbiol. Rev.* **57:** 823–837.

Hasty KA, Pourmotabbed TF, Goldberg GI, Thompson JP, Spinella DG, Stephens RM, Mainardi CL. (1990) Human neutrophil collagenase. A distinct gene product with homology to other matrix metalloproteinases. *J. Biol. Chem.* **265:** 11421–11424.

Heinlein UAO, Wallat S, Senftleben A, Lemaire L. (1994) Male germ cell-expressed mouse gene TAZ83 encodes a putative cysteine-rich transmembrane protein (cyritestin) sharing homologies with snake toxins and sperm–egg fusion proteins. *Dev. Growth Diff.* **36:** 49–58.

Holm RH, Kennepohl P, Solomon EI. (1996) Structural aspects of metal sites in biology. *Chem. Rev.* **96:** 2239–2314.

Hooper NM (1994) Families of zinc metalloproteases. *FEBS Lett.* **354:** 1–6.

Howard L, Glynn P. (1995) Membrane-associated metalloproteinase recognized by characteristic cleavage of myelin basic protein: assay and isolation. *Methods Enzymol.* **248:** 388–395.

Huber R, Bode W. (1978) Structural basis of the activation and action of trypsin. *Acc. Chem. Res.* **11:** 114–122.

Huovila APJ, Almeida EAC, White JM. (1996) ADAMs and cell fusion. *Curr. Opin. Cell Biol.* **8:** 692–699.

Ikura T, Sawamura T, Shiraki T *et al.* (1994) cDNA cloning and expression of bovine endothelin converting enzyme. *Biochem. Biophys. Res. Comm.* **203:** 1417–1422.

Jiang W, Bond JS. (1992) Families of metallopeptidases and their relationships. *FEBS Lett.* **312:** 110–114.

Jiang W, Gorbea CM, Flannery AV, Beynon RJ, Grant GA, Bond JS. (1992) The α-subunit of meprin A. Molecular cloning and sequencing, differential expression in inbred mouse strains, and evidence for divergent evolution of the α and β subunits. *J. Biol. Chem.* **267:** 9185–9193.

Jongeneel CV, Bouvier J, Bairoch A. (1989) A unique signature identifies a family of zinc-dependent metallopeptidases. *FEBS Lett.* **242:** 211–214.

Kenny AJ. (1977) Endopeptidases in the brush border of the kidney proximal tubule. *CIBA Found. Symp.* **50:** 209–215.

Kessler E, Takahara K, Biniaminov L, Brusel M, Greenspan DS. (1996) Bone morphogenetic protein-1: the type I procollagen C-proteinase. *Science* **271:** 360–362.

Kim H, Lipscomb WN. (1990) Crystal structure of the complex of carboxypeptidase A with a strongly bonded phosphonate in a new crystalline form: comparison with structures of other complexes. *Biochemistry* **29:** 5546–5555.

Knäuper V, Krämer S, Reinke H, Tschesche H. (1990) Characterization and activation of procollagenase from human polymorphonuclear leucocytes – N-terminal sequence determination of the proenzyme and various proteolytically activated forms. *Eur. J. Biochem.* **189:** 295–300.

Le Moual H, Devault A, Roques BP, Crine P, Boileau G. (1991) Identification of glutamic acid 646 as a zinc-coordinating residue in endopeptidase 24.11. *J. Biol. Chem.* **266:** 15670–15675.

Lee S, Zambas ED, Marsh WL, Redman CM. (1991) Molecular cloning and primary structure of Kell bloodgroup protein. *Proc. Natl Acad. Sci. USA* **88:** 6353–6357.

Li C, Hersh LB. (1995) Neprilysin: assay methods, purification and characterization. *Methods Enzymol.* **248:** 253–263.

Li S-W, Sieron, AL, Fertala A, Hojima Y, Arnold WV, Prockop DJ. (1996) The C-proteinase that processes procollagens to fibrillar collagen is identical to the protein previously identified as bone morphogenetic protein-1. *Proc. Natl Acad. Sci. USA* **93:** 5127–5130.

Lipscomb WN, Sträther N. (1996) Recent advances in zinc enzymology. *Chem. Rev.* **96:** 2375–2433.

Lovejoy B, Cleasby A, Hassell AM *et al.* (1994) Structure of the catalytic domain of fibroblast collagenase complexed with an inhibitor. *Science* **263:** 375–377.

Maeda H, Morihara K. (1995) Serralysin and related bacterial proteinases. *Methods Enzymol.* **248:** 395–413.

Matthews BW. (1988) Structural basis of the action of thermolysin and related zinc peptidases. *Acc. Chem. Res.* **21**: 333–340.

Matthews BW, Jansonius JN, Colman PM, Schoenborn BP, Dupourque D. (1972) Three-dimensional structure of thermolysin. *Nature (London) New Biol.* **238:** 37–41.

McKerrow JH. (1987) Human fibroblast collagenase contains an amino acid sequence homologous to the zinc-binding site of *Serratia* protease. *J. Biol. Chem.* **262:** 5943.

Miyatake H, Hata Y, Fuji T, Hamada K, Morihara K, Katsube Y. (1995) Crystal structure of the unliganded alkaline protease from *Pseudomonas aeruginosa* IFO3080 and its conformational changes on ligand binding. *J. Biochem.* **118:** 474–479.

Morihara K. (1995) Pseudolysin and other pathogen endopeptidases of the thermolysin family. *Methods Enzymol.* **248:** 242–253.

Murphy GJP, Murphy G, Reynolds JJ. (1991) The origin of matrix metalloproteinases and their familial relationships. *FEBS Lett.* **289:** 4–7.

Nagase H. (1996) Matrix metalloproteinases. In: *Zinc Metalloproteinases in Health and Disease* (ed. N. M. Hooper), pp. 153–204. Taylor and Francis, London.

Nagase H, Enghild JJ, Suzuki K, Salvesen G. (1990) Stepwise activation mechanisms of the precursor of matrix metalloproteinase 3 (stromelysin) by proteinases and (4-aminophenyl) mercuric acetate. *Biochemistry* **29:** 5783–5789.

Nakahama K, Yoshimura K, Marumoto R, Kikuchi M, Lee IS, Hase T, Matsubara H. (1986) Cloning and sequencing of *Serratia* protease gene. *Nucleic Acids Res.* **14:** 5843- 5855.

Nanus DM, Engelstein D, Gastl GA, Gluck L, Vidal MJ, Morrison M, Finstad CL, Bander NH, Albino AP. (1993) Molecular cloning of the human kidney differentiation antigen gp160: human aminopeptidase A. *Proc. Natl Acad. Sci. USA* **90:** 7089–7073.

Netzel-Arnett S, Fields G, Birkedal-Hansen H, Van Wart H. (1993a) Sequence specificities of human fibroblast and neutrophil collagenases. *J. Biol. Chem.* **266:** 6747–6755.

Netzel-Arnett S, Sang Q-X, Moore WGI, Navre M, Birkedal-Hansen H, Van Wart H. (1993b) Comparative specificity of human 72- and 92-kDa gelatinases (type IV collagenases) and PUMP (matrilysin). *Biochemistry* **32:** 6427–6432.

Okada A, Bellocq JP, Rouyer N, Chenard MP, Rio MC, Chambon P, Basset P. (1995) Membrane type matrix metalloproteinase (MT-MMP) gene is expressed in stromal cells of human colon, breast, head and neck carcinomas. *Proc. Natl Acad. Sci. USA* **92:** 2730–2734.

Okuda K, Morihara K, Atsumi Y, Takeuchi H, Kawamoto S, Kawasaki H, Suzuki K, Fukushima J. (1990) Complete nucleotide sequence of the structural gene for alkaline proteinase from *Pseudomonas aeruginosa. Infect. Immun.* **58:** 4083–4088.

Olsen J, Cowell GM, Konigshofer E *et al.* (1988) Complete amino acid sequence of human intestinal aminopeptidase N as deduced from cloned cDNA. *FEBS Lett.* **238:** 307–317.

Pauptit RA, Karlsson R, Picot D, Jenkins JA, Niklaus-Reimer A-S, Jansonius JN. (1988) Crystal structure of neutral protease from *Bacillus cereus* refined at 3.0 Å resolution and comparison with the homologous, but more thermostable enzyme thermolysin. *J. Mol. Biol.* **199:** 525–537.

Perry ACF, Jones R, Barker PJ, Hall L. (1992) A mammalian epididymal protein with remarkable sequence similarity to snake venom hemorrhagic peptides. *Biochem. J.* **286**: 671–674.

Phillips MA, Fletterick R, Rutter WJ. (1990) Arginine 127 stabilizes the transition state in carboxypeptidase. *J. Biol. Chem.* **265:** 20692–20698.

Quantin B, Murphy G, Breathnach R. (1989) Pump-1 cDNA codes for a protein with characteristics similar to those of classical collagenase family members. *Biochemistry* **28:** 5327–5334.

Rawlings ND, Barrett AJ. (1993) Evolutionary families of peptidases. *Biochem. J.* **290:** 205–218.

Rawlings ND, Barrett AJ. (1995) Evolutionary families of metallopeptidases. *Methods Enzymol.* **248:** 183–228.

Reinemer P, Grams F, Huber R, Kleine T, Schnierer S, Pieper M, Tschesche H, Bode W. (1994) Structural implications for the role of the N terminus in the 'superactivation' of collagenases. *FEBS Lett.* **338:** 227–233.

Roques BP, Noble F, Crine P, Fournié-Zaluski M-C. (1995) Inhibitors of neprilysin: design, pharmacological and clinical applications. *Methods Enzymol.* **248:** 263–283.

Sato H, Takino T, Okada Y, Cao J, Shinagawa A, Yamamoto E, Seiki M. (1994) A matrix metalloproteinase expressed on the surface of invasive tumour cells. *Nature* **370:** 61–65.

Schechter I, Berger A. (1967) On the size of the active site in proteases. I. Papain. *Biochem. Biophys. Res. Commun.* **27:** 157–162.

Schlagenhauf E. (1995) X-ray crystallographic studies of leishmanolysin, the major surface metalloproteinase from *Leishmania major*. Thesis, University of Heidelberg.

Schlagenhauf E, Etges R, Metcalf P. (1995) Crystallization and preliminary X-ray diffraction studies of leishmanolysin, the major surface metalloproteinase from *Leishmania major. Proteins: Struct. Funct. Genet.* **22:** 58.

Schmidt M, Kröger B, Jacob E, Seulberger H, Subkowski T, Otter R, Meyer T, Schmalzling G, Hillen H. (1994) Molecular characterization of human and bovine enothelin converting enzyme (ECE-1). *FEBS Lett.* **356:** 238–243.

Shimada K, Takahashi M, Tanzawa K. (1994) Cloning and functional expression of endothelin converting enzyme from rat endothelial cells. *J. Biol. Chem.* **269:** 18275–18278.

Soubrier F, Alhenc-Gelas F, Hubert C, Allegrini MJ, Tregar G, Corvol P. (1988) Two putative active centers in human angiotensin I-converting enzyme revealed by molecular cloning. *Proc. Natl Acad. Sci. USA* **85:** 9386–9390.

Springman EB, Angleton EL, Birkedal-Hansen H, Van Wart HE. (1990) Multiple modes of activation of latent human fibroblast collagenase: evidence for the role of a Cys73 active site zinc complex in latency and a 'cysteine switch' mechanism for activation. *Proc. Natl Acad. Sci. USA* **87:** 364–368.

Stams T, Spurlino JC, Smith DL, Wahl RC, Ho TF, Qorronfleh MW, Banks TM, Rubin B. (1994) Structure of human neutrophil collagenase reveals large S_1' specificity pocket. *Nature Struct. Biol.* **1:** 119–123.

Stetler-Stevenson WG, Liotta LA, Kleiner DE Jr. (1993) Extracellular matrix 6: role of matrix metalloproteinases in tumor invasion and metastasis. *FASEB J.* **7:** 1434–1441.

Stöcker W, Bode W. (1995) Structural features of a superfamily of zinc-endopeptidases: the metzincins. *Curr. Opin. Struct. Biol.* **5:** 383–390.

Stöcker W, Zwilling R. (1995) Astacin. *Methods Enzymol.* **248:** 305–325.

Stöcker W, Wolz RL, Zwilling R, Strydom DJ, Auld DS. (1988) Astacus protease – a zinc metalloenzyme. *Biochemistry* **27**, 5026–5032.

Stöcker W, Ng M, Auld DS. (1990) Fluorescent oligopeptide substrates for kinetic characterization of the specificity of *Astacus* protease. *Biochemistry* **29:** 10418–10425.

Stöcker W, Sauer B, Zwilling R. (1991) Kinetics of nitroanilide cleavage by astacin. *Biol. Chem. Hoppe-Seyler* **372:** 385–392.

Stöcker W, Gomis-Rüth F-X, Bode W, Zwilling R. (1993) Implications of the three-dimensional structure of astacin for the structure and function of the astacin family of zinc-endopeptidases. *Eur. J. Biochem.* **214:** 215–231.

Stöcker W, Grams F, Baumann U, Reinemer P, Gomis-Rüth F-X, McKay DB, Bode W. (1995) The metzincins – topological and sequential relations between the astacins, adamalysins, serralysins, and matrixins (collagenases) define a superfamily of zinc-peptidases. *Protein Sci.* **4:** 823–840.

Suzuki N, Labosky PA, Furuta Y *et al.* (1996) Failure of ventral body wall closure in mouse embryos lacking a procollagen C-proteinase encoded by BMP1, a mammalian gene related to *Drosophila* tolloid. *Development* **122:** 3587–3595.

Takino T, Sato H, Yamamoto E, Seiki M. (1995) Cloning of a human gene potentially encoding a novel matrix metalloproteinase having a C-terminal transmembrane domain. *Gene* **155:** 239–298.

Thayer MM, Flaherty KM, McKay DB. (1991) Three-dimensional structure of the elastase of *Pseudomonas aeruginosa* at 1.5 Å resolution. *J. Biol. Chem.* **266:** 1864–2871.

Titani K, Torff H-J, Hormel S, Kumar S, Walsh KA, Rödl J, Neurath H, Zwilling R. (1987) Amino acid sequence of a unique protease from the crayfish *Astacus fluviatilis. Biochemistry* **26:** 222–226.

Vallee BL, Auld DS. (1993) Zinc: biological functions and coordination motifs. *Acc. Chem. Res.* **26:** 543–551.

Wang J, Cooper MD. (1996) Aminopeptidases: structure and biological function. In: *Zinc Metalloproteinases in Health and Disease* (ed. N.M. Hooper), pp. 131–151. Taylor and Francis, London.

Whitham SE, Murphy G, Angel P *et al.* (1986) Comparison of human stromelysin and collagenase by cloning and sequence analysis. *Biochem. J.* **240:** 913–916.

Williams TA, Corvol P, Soubrier F. (1994) Identification of two active site residues in human angiotensin I-converting enzyme. *J. Biol. Chem.* **269:** 29430–29434.

Wolfsberg TG, Bazan JF, Blobel CP, Myles DG, Primakoff P, White JM (1993) The precursor region of a protein active in sperm–egg fusion contains a metalloprotease and a disintegrin domain: structural, functional, evolutionary implications. *Proc. Natl Acad. Sci. USA* **90:** 10783–10787.

Wolz RL. (1994) A kinetic comparison of the homologous proteases astacin and meprin A. *Arch. Biochem. Biophys.* **310:** 144–151.

Wolz RL, Bond JS. (1995) Meprins A and B. *Methods Enzymol.* **248:** 325–345.

Wolz RL, Harris RB, Bond JS. (1991) Mapping the active site of meprin-A with peptide substrates and inhibitors. *Biochemistry* **30:** 8488–8493.

Wozney JM, Rosen V, Celeste AJ, Mitsock LM, Whitters MJ, Kriz RW, Hewick RM, Wang EA. (1988) Novel regulators of bone formation: molecular clones and activities. *Science* **242:** 1528–1534.

Xu D, Emoto N, Giaid A, Slaughter C, Kaw S, DeWit D, Yanagisawa M. (1994) ECE-1: a membrane bound metalloproteinase that catalyzed the proteolytic activation of big endothelin-1. *Cell* **78:** 473–485.

Zhang D, Botos I, Gomis-Rüth F-X, Doll R, Blood C, Njoroge FG, Fox JW, Bode W, Meyer EF. (1994) Structural interaction of natural and synthetic inhibitors with the venom metalloproteinase atrolysin C (form d). *Proc. Natl Acad. Sci. USA* **91:** 8447–8451.

Chapter 3

Membrane anchorage of cell-surface peptidases

Nigel M. Hooper

3.1 Introduction

Cell-surface peptidases are ectoenzymes that are asymmetrically oriented across the lipid bilayer, with the bulk of their mass, including their active site, located on the extracellular side. This distribution places them in the correct position for hydrolysing, for example, peptide hormones in the circulation and the kidney tubules, neuropeptides in the synaptic cleft and dietary peptides in the intestine. Although the enzymes are predominantly present at the cell surface, in some cases intracellular forms can be found in either exocytic or endocytic vesicles where they have the same relative orientation as at the plasma membrane, with the active site on the extra-cytoplasmic face. Three modes of attachment to the plasma membrane are known (*Figure 3.1* and *Table 3.1*). Some cell-surface peptidases are anchored by a hydrophobic sequence of amino acids positioned close to the N terminus which doubles as the signal sequence for directing the

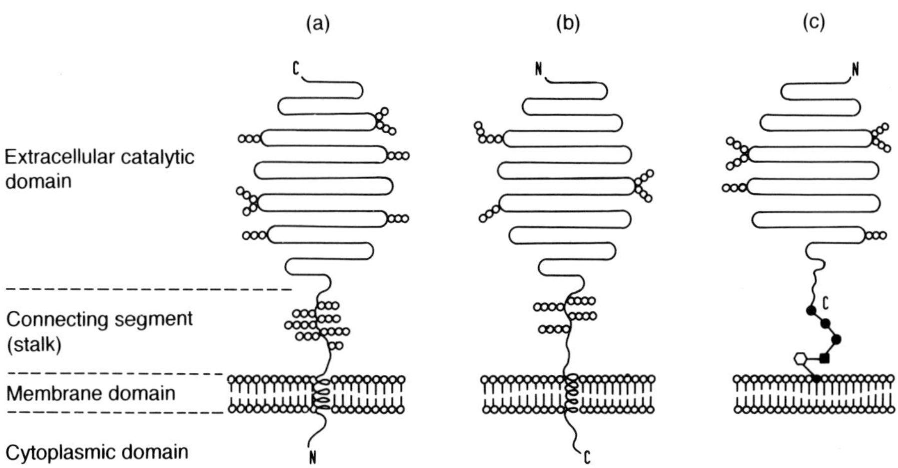

Figure 3.1. The three types of membrane anchorage of cell-surface peptidases. (a) Type II integral membrane protein; (b) type I integral membrane protein; (c) GPI-anchored protein. See text for details.

Cell-Surface Peptidases in Health and Disease, edited by A.J. Kenny and C.M. Boustead.

Table 3.1. Topology of cell-surface peptidases

Topology	Peptidase	EC number	References
Type II	Aminopeptidase A	3.4.11.7	Wu *et al.* (1990), Li *et al.* (1993), Hesp and Hooper (1997)
	Aminopeptidase N	3.4.11.2	Olsen *et al.* (1988), Malfroy *et al.* (1989)
	Dipeptidyl peptidase IV	3.4.14.5	Ogata *et al.* (1989)
	Endopeptidase-24.11	3.4.24.11	Devault *et al.* (1987), Malfroy *et al.* (1987)
	Endothelin-converting enzyme-1	3.4.24.-	Shimada *et al.* (1994), Xu *et al.* (1994)
	Endothelin-converting enzyme-2	3.4.24.-	Emoto and Yanagisawa (1995)
	Glut4 aminopeptidase	3.4.11.-	Keller *et al.* (1995)
	γ-Glutamyl transpeptidase	2.3.2.2	Laperche *et al.* (1986)
	Leucine aminopeptidase (oxytocinase)	3.4.11.-	Rogi *et al.* (1996)
	Thyrotropin-releasing hormone-degrading enzyme	3.4.19.6	Schauder *et al.* (1994)
Type I	Angiotensin-converting enzyme	3.4.15.1	Soubrier *et al.* (1988), Lattion *et al.* (1989)
	Endopeptidase-24.18	3.4.24.18	Johnson and Hersh (1992), Corbeil *et al.* (1992), Gorbea *et al.* (1993)
GPI	Aminopeptidase P	3.4.11.9	Hyde *et al.* (1996)
	Carboxypeptidase M	3.4.17.12	Tan *et al.* (1989)
	Membrane dipeptidase	3.4.13.19	Rached *et al.* (1990), Adachi *et al.* (1990)

nascent polypeptide chain into the lumen of the rough endoplasmic reticulum during translation. These are classified as type II integral membrane proteins, and are typified by endopeptidase-24.11 (E-24.11; EC 3.4.24.11) and aminopeptidase N (AP-N; EC 3.4.11.2). Others have a cleavable N-terminal signal sequence followed by a second hydrophobic stop-transfer sequence towards the C terminus. For type I integral membrane proteins, such as angiotensin-converting enzyme (ACE; EC 3.4.15.1), this second hydrophobic sequence acts as the membrane-anchoring domain, whereas in the glycosyl phosphatidylinositol (GPI)-anchored peptidases, such as membrane dipeptidase (MDP; EC 3.4.13.19), this C-terminal hydrophobic sequence is removed and replaced with a complex GPI anchor structure.

3.2 Type II cell-surface peptidases

3.2.1 Structural features of type II peptidases

The majority of cell-surface peptidases are type II integral membrane proteins, being anchored to the plasma membrane by an uncleaved N-terminal signal sequence (*Table*

3.1). The cDNA-derived amino acid sequence of such proteins clearly reveals the hydrophobic region at the N terminus (see *Figure 3.2a*), and hydropathy plot analysis indicates the absence of any additional hydrophobic regions of sufficient length to span the lipid bilayer. Type II integral membrane proteins usually have a small cytoplasmic domain at the N terminus, which can be as few as four amino acids in the case of γ-glutamyl transpeptidase (Laperche *et al.*, 1986) or as many as 82 residues in the case of endothelin-converting enzyme-2 (Emoto and Yanagisawa, 1995). The hydrophobic region of some 20–25 amino acid residues is long enough to span the interior of the bilayer in an α-helical conformation. Either side of this hydrophobic domain are clusters of positively charged residues which interact with negatively charged groups on the headgroups of the phospholipids in the bilayer.

3.2.2 Comparison of proteinase- and detergent-solubilized forms

Prior to the cloning and sequencing of the corresponding cDNAs, the classification of AP-N, aminopeptidase A (AP-A), γ-glutamyl transpeptidase and dipeptidyl peptidase IV (DPP-IV) as type II integral membrane proteins came about as the result of sequencing the N terminus of the detergent- and proteinase-solubilized forms of the proteins (see *Figure 3.3*) (reviewed in Kenny and Maroux, 1982; Semenza, 1986). Even if the cDNA-derived amino acid sequence for a peptidase reveals the presence of an N-terminal signal sequence, it is essential to show directly that the mature membrane-bound form of the protein contains this region. Solubilization with detergents results in an amphipathic form of the protein that retains the hydrophobic anchoring domain, as shown by reconstitution into artificial lipid vesicles or temperature-induced phase separation in Triton X-114 (see Section 3.3.2). In contrast, solubilization from the membrane with a proteinase, such as trypsin or papain, results in a soluble, hydrophilic form that lacks the hydrophobic membrane-anchoring domain (*Figure 3.3*). The observation that the N-terminal sequences of the two solubilized forms were different, and in some

Figure 3.2. Schematic representation of the nascent polypeptide chains of: (a) type II, (b) type I and (c) GPI-anchored peptidases. ■ represents the hydrophobic N-terminal signal sequence; ▒ represents the C-terminal hydrophobic sequence; SP, site of action of signal peptidase; T, site of action of the putative GPI transamidase; ω, site of cleavage of the polypeptide chain and addition of the preformed GPI anchor by the transamidase.

EP-24.11 MGRSESQMDITDINTP**K**P**KKK**Q**R**WTPLE*ISLSVLVLLLTVIAVTMIALYAT*YDDGIC**K**SSDC-62
↓
AP-A MSTDS**KR**YCI**K**T**K**H*VAIICAAVVAVGLIVGLSVGLT***R**SCDS**K**DGGQGT-48
↓
AP-N MA**K**GFYIS**K***ALGILGILLGVAAVATIIALSVV*YAQE**K**N**K**NAEH-43

Figure 3.3. N-terminal amino acid sequences of three type II cell-surface peptidases. The sequences at the N terminus of rabbit E-24.11 (Devault *et al.*, 1987), porcine AP-N (Delmas *et al.*, 1994) and porcine AP-A (Hesp and Hooper, 1996) are aligned centred on the hydrophobic membrane-spanning region (italics). Positively charged residues either side of the hydrophobic region are in bold, and the sites of cleavage of the stalk regions of AP-N and AP-A by trypsin are indicated by (↓).

cases that the sequence of the amphipathic form entered a predominantly hydrophobic stretch of residues (Fulcher *et al.*, 1986), indicated that the membrane-anchoring domain was near the N terminus. In negatively stained electron microgaphs the amphipathic forms of AP-N and E-24.11 appear as 'lollipop' structures, with a distinct stalk region, that varies in length from 2 to 9 nm depending on the protein, separating the large globular extracellular domain from the membrane surface (see *Figure 3.1*) (Kenny and Turner, 1987). It is within this stalk region that the proteases act to generate the hydrophilic form by cleavage at a susceptible peptide bond (*Figure 3.3*). However, not only must there be a susceptible peptide bond in this region but the stalk length must be long enough to allow the access of the protease. Thus, the short-stalked E-24.11 (stalk length 2 nm; Kenny *et al.*, 1983) cannot be released by papain (the shortest dimension of a papain molecule is 3.6 nm) whereas AP-N (stalk length 5 nm; Louvard *et al.*, 1975) is readily cleaved.

3.3 Type I cell-surface peptidases

3.3.1 Structural features of type I peptidases

Only two cell-surface peptidases have been identified as type I integral membrane proteins; ACE and endopeptidase 24.18 (E-24.18) (*Table 3.1*). Like type II proteins, these enzymes have a hydrophobic N-terminal signal sequence (*Figure 3.2b*) that directs the growing polypeptide chain–ribosome–mRNA complex to the rough endoplasmic reticulum. However, the N-terminal signal sequence of type I proteins is cleaved by signal peptidase generating a new N terminus (*Figure 3.2b*). The rest of the polypeptide chain is then translocated across the membrane until a second stretch of hydrophobic amino acids is encountered towards the C terminus (*Figure 3.2b*). This hydrophobic region acts as the stop-transfer–membrane-anchoring domain.

ACE was identified as a type I protein by comparing the N-terminal sequences of the detergent- and protease-solubilized forms of the porcine kidney enzyme (Hooper *et al.*, 1987a). Unlike AP-N and DPP-IV, the two forms of ACE had identical N-terminal sequences. As the difference in molecular weight between the two forms was only 5–10 kDa, this placed the membrane-anchoring domain near the C terminus. The subsequent cloning and sequencing of the cDNA encoding human ACE confirmed this conclusion

(Soubrier *et al.*, 1988). In addition to the cleaved N-terminal signal sequence there is only one other region of significant hydrophobicity to act as a membrane-spanning domain and that is located near the C terminus. In the case of the heterodimeric E-24.18, although the α-subunit contains a putative C-terminal hydrophobic anchoring domain, this is cleaved off during biosynthesis. The cleaved α-subunit is tethered at the membrane by covalent disulphide linkage to the β-subunit whose C-terminal membrane-anchoring domain is not cleaved (Milhiet *et al,* 1994; Marchand *et al.*, 1994, 1995).

3.3.2 Methods used to distinguish between the amphipathic and hydrophilic forms of membrane proteins

When examining the mode of membrane anchorage of a peptidase it is often necessary to distinguish between the amphipathic form that retains the hydrophobic membrane-anchoring domain and the hydrophilic form that has lost this domain. This is highlighted by studies on ACE where, although the enzyme can be solubilized from the membrane by Triton X-100, a procedure that would be expected to solubilize the intact protein, the resultant solubilized form failed to reincorporate into lipid vesicles, indicating that it had lost the hydrophobic anchoring domain (Hooper *et al.*, 1987a). Only when solubilization with Triton X-100 was carried out in the presence of ethylene diamine tetraacetic acid (EDTA) could the amphipathic form be isolated. This is because ACE is susceptible to proteolytic cleavage by an EDTA-sensitive secretase (see Section 3.4). There are a number of techniques that can be used to distinguish between the amphipathic and hydrophilic forms of membrane proteins, including non-denaturing polyacrylamide gel electrophoresis, hydrophobic chromatography and labelling with photoactivatable hydrophobic reagents such as 3-trifluoromethyl-3-(*m*[^{125}I]-iodophenyl)-diazirine ([^{125}I]TID) (for details see Hooper, 1992). However, probably the two most commonly used techniques for distinguishing between the amphipathic and hydrophilic forms of peptidases are reconstitution into lipid vesicles and temperature-induced phase separation in Triton X-114.

The ability of a protein to reincorporate into an artificial lipid bilayer is probably the most conclusive evidence that it possesses a hydrophobic anchoring domain. Several techniques have been described for reconstituting membrane proteins into lipid vesicles. One that we have used extensively has the advantage of being relatively quick and allows the proteins to reincorporate even in the presence of low concentrations of detergent (Hooper, 1992; Scotto and Zakim, 1985). Essentially, the method uses vesicles of dimyristoyl-phosphatidylcholine containing small amounts of myristate as fusogen. Upon incubation at 18°C these vesicles spontaneously incorporate integral membrane proteins. Centrifugation through a sucrose cushion can then be used to separate the proteoliposomes from unincorporated protein. We have used this method for distinguishing between the amphipathic and hydrophilic forms of the type I peptidase ACE and the GPI-anchored peptidase MDP (Hooper *et al.*, 1987a; Hooper and Turner, 1989).

Probably the most rapid and convenient technique for distinguishing between the amphipathic and hydrophilic forms of membrane proteins is temperature-induced phase separation in Triton X-114 (Bordier, 1981; Hooper, 1992). At 30°C this non-ionic detergent partitions into two distinct phases that can be separated by low speed centrifugation:

a detergent-rich phase and a less dense detergent-poor or aqueous phase. The amphipathic form of a cell-surface peptidase will partition into the detergent-rich phase, whereas the hydrophilic form partitions into the aqueous phase. Again, this technique can be used for both transmembrane polypeptide-anchored peptidases and GPI-anchored ones.

3.4 GPI-anchored cell-surface peptidases

3.4.1 Structural features of GPI-anchored peptidases

Three cell-surface peptidases have been identified as being anchored in the plasma membrane by a covalently attached GPI moiety, aminopeptidase P (AP-P), carboxypeptidase M (CP-M) and MDP (*Table 3.1* and Chapter 14). A GPI anchor can be considered as an alternative to the transmembrane polypeptide anchor of type I proteins, both having essentially the same topology (*Figure 3.1b* and *c*). The GPI anchor on porcine kidney MDP is one of the best characterized mammalian GPI anchors, and recently its complete structure has been determined (*Figure 3.4*) (Brewis *et al.*, 1995). Like all other GPI anchors, it consists of the conserved core structure ethanolamine-PO_4-6Manα1–2Manα1–6Manα1–4GlcN$H_2\alpha$1–6*myo*-inositol-1-PO_4-lipid attached to the C-terminal amino acid. In the case of MDP the lipid is unusually diacyl glycerol (predominantly distearoyl-glycerol) as opposed to the more common alkylacyl glycerol structures found in other mammalian GPI-anchored proteins. Attached to the conserved core are various side chain modifications of ethanolamine phosphate and other sugar residues, including sialic acid (*Figure 3.4*). A parallel determination of the glycan core structure of human kidney MDP revealed that the side-chain modifications were essentially the same, indicating that they are probably tissue- and/or protein-specific (*Figure 3.4*) (Brewis *et al.*, 1995).

GPI-anchored proteins are synthesised with a cleavable N-terminal signal sequence (*Figure 3.2c*). The second stretch of hydrophobic amino acids runs right up to the C terminus (*Figure 3.2c*). Separated from this second hydrophobic region by a hydrophilic spacer region is the site of anchor attachment, termed the ω-site (*Figure 3.2c*). Comparison of naturally occurring GPI-anchored proteins and many site-directed mutants of, in particular, human placental alkaline phosphatase has revealed that the amino acid at the ω-site can be only one of six, all with small side chains (Ala, Asn, Asp, Cys, Gly or Ser). Ala or Gly are necessary at the ω+2 site (i.e. two residues downstream; Udenfriend and Kodukula, 1995). In contrast, the ω+1 residue is relatively unrestricted. In the case of MDP the site of anchor attachment was identified by digesting the protein with trypsin and removing all the tryptic peptides other than the C-terminal hexapeptide by chromatography on anhydrotrypsin–Sepharose (Brewis *et al.*, 1995). Sequencing of the isolated C-terminal peptide (with the GPI anchor still attached) revealed that the ω residue was Ser. Alignment of the obtained sequence of the C-terminal peptide with the cDNA-predicted amino acid sequence (Rached *et al.*, 1990) identified Ala as the ω+2 residue, consistent with the above requirements. More recent work on AP-P has indicated that the ω and ω+2 residues are both Ala (Hyde *et al.*, 1996). Within 1–2 min of the end of translation, a putative transamidase in the lumen of the rough endoplasmic reticulum cleaves the nascent polypeptide chain on the C-terminal side of the ω residue

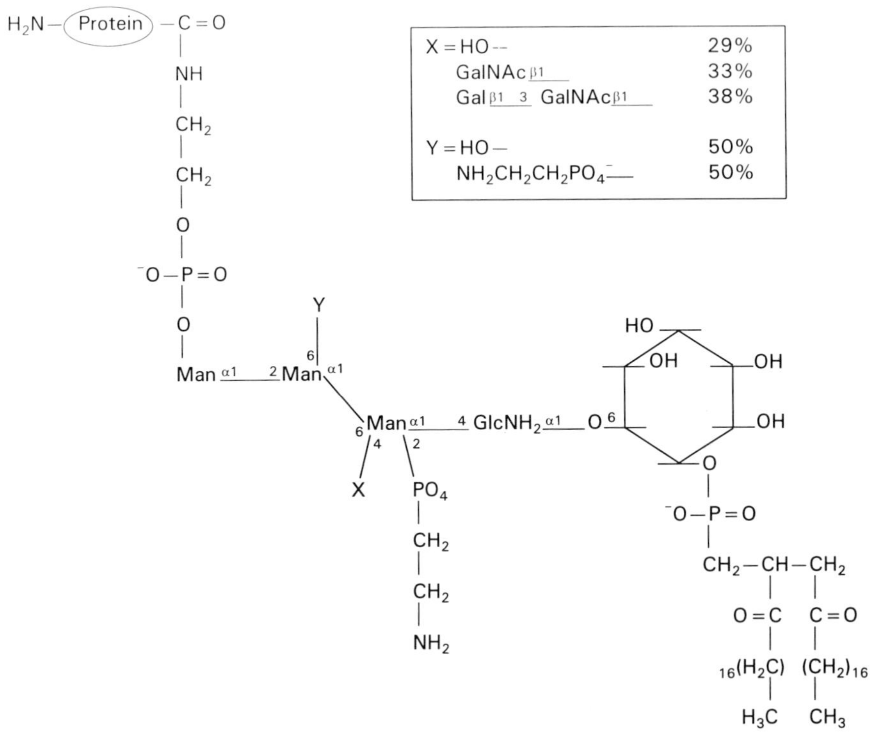

Figure 3.4. Structure of the GPI anchor on porcine kidney membrane dipeptidase. The core structure of the GPI anchor of porcine kidney MDP is shown with the fatty acid chains, the extra ethanolamine phosphate and the relative distribution and structures of the side-chain modifications (X and Y) (Brewis *et al.*, 1995).

and concomitantly adds on a preformed GPI anchor (Udenfriend and Kodukula, 1995). Thus the C-terminal hydrophobic sequence appears to tether the protein transiently in the membrane prior to the addition of the GPI anchor.

3.4.2 Identification of a GPI anchor on a protein

The susceptibility of a protein to release from the membrane by bacterial phosphatidyl-inositol-specific phospholipase C (PI-PLC) was the original method used to identify a GPI anchor on alkaline phosphatase (Ikezawa *et al.*, 1976; Low and Finean, 1977). This is still the most commonly employed and simplest criterion to demonstrate the presence of a GPI anchor on a protein and was the means by which we first demonstrated that MDP and AP-P were GPI-anchored (Hooper *et al.*, 1987b; Hooper and Turner, 1988a). Preparations of recombinant PI-PLC from *Bacillus thuringiensis*, *Bacillus cereus* and *Staphylococcus aureus* are now available from a number of suppliers. (For details of their properties and suppliers see Hooper, 1992.) Essentially, cells or membranes are incubated in the presence of PI-PLC and the dose-dependent appearance of a soluble, hydrophilic form of the membrane protein examined for following high-speed centrifugation or phase separation in Triton X-114 (see for example Brewis *et al.*, 1994; Hooper and Turner, 1988a). Appropriate controls should be included to rule out the possibility of endogenous proteases or phospholipases causing the observed release, and it is prudent to monitor for a known GPI-anchored protein, such as alkaline phosphatase which is widespread in mammalian cells and tissues and which is readily assayed (Hooper, 1992), in order to confirm that the PI-PLC is active under the conditions of the experiment. Although release of a protein by PI-PLC is a good indication of the presence of a GPI anchor, lack of release does not rule out the protein being GPI anchored as several mammalian GPI-anchored proteins, such as human erythrocyte acetylcholinesterase, are resistant to PI-PLC cleavage due to additional acylation of the inositol ring (Toutant *et al.*, 1989). Methods for overcoming this problem have been described in detail (Guther *et al.*, 1994).

Confirmation of the presence of a GPI anchor following either exogenous or endogenous PLC treatment can be obtained by detection of the cross-reacting determinant (CRD) using anti-CRD antibodies. Polyclonal antisera raised against one GPI-anchored protein often cross-react with other unrelated GPI-anchored proteins due to the presence of the CRD (Zamze *et al.*, 1988). The CRD is cryptic in the amphipathic membrane-bound forms of GPI-anchored proteins and is only exposed following cleavage of the anchor by PLC. The CRD epitope is due mainly to the inositol 1,2-cyclic monophosphate that is formed following PLC cleavage of the anchor, although other minor epitopes on the glucosamine and the side-chain sugars may also be involved (Hooper *et al.*, 1991; Zamze *et al.*, 1988). The CRD can be detected by either immunoelectrophoretic blot analysis or enzyme-linked immunoadsorbent assays (for details see Hooper, 1992). The anti-CRD antisera that have been generated against porcine and human MDP have been widely used in studies on GPI-anchored proteins and related structures (see for example Broomfield and Hooper, 1993; Chang *et al.*, 1994; Galasko *et al.*, 1995; Vogel *et al.*, 1992).

Another means of identifying the presence of a GPI anchor on a protein is to biosynthetically label cells expressing the protein of interest with components of the GPI anchor, such as [^{3}H]- or [^{14}C]-ethanolamine, [^{3}H]-inositol, or [^{3}H]- or [^{14}C]-fatty acids (Hooper, 1992). The protein is then immunoprecipitated and the presence of the radiolabel identified following sodium dodecyl sulphate polyacrylamide gel electrophoresis and fluorography. If the protein is labelled with fatty acid, the linkage to the protein can be identified by digestion with PI-PLC or by chemical treatments (nitrous acid and alcoholic hydroxylamine).

Mammalian GPI-anchored proteins are generally resistant to solubilization with non-ionic detergents that have a low critical micellar concentration such as Triton X-100, Triton X-114 and Nonidet P40, but are readily solubilized by detergents that have a high critical micellar concentration such as *n*-octyl-β-D-glucopyranoside (octyl glucoside) and 3-[(3-cholamidopropyl)dimethylammonio]-1-propane-sulphonate (CHAPS) (Hooper and Turner, 1988b). This pattern of detergent insolubility may be due to the proteins clustering in membrane microdomains rich in cholesterol and glycosphingolipids (Brown, 1992; Parkin *et al.*, 1996). In contrast, transmembrane polypeptide-anchored proteins are readily solubilized efficiently by both types of detergent (Hooper and Turner, 1988b). Thus, the pattern of detergent solubility, in particular the ratio of activity solubilized by octyl glucoside to that solubilized by Triton X-100, can be used to predict the presence of a GPI anchor. The observation that GPI-anchored proteins were relatively insoluble in Triton X-114 led us to modify the normal two-phase procedure for phase separation in Triton X-114 (Bordier, 1981) (see Section 3.3.2), to a three-phase procedure in which the GPI-anchored proteins are recovered in a detergent-insoluble pellet following incubation at 0°C (Hooper and Bashir, 1991). The resulting supernatant can then be further fractionated following incubation at 30°C into a detergent-rich phase containing transmembrane polypeptide-anchored proteins and an aqueous phase containing soluble, hydrophilic proteins.

3.5 Release of cell-surface peptidases in soluble forms

A number of the cell-surface peptidases exist in soluble, hydrophilic forms in serum and other body fluids. In some cases, such as E-24.11, the circulating form in plasma is normally present at very low levels, but increases dramatically in certain pathological situations (Almenoff *et al.*, 1984; Johnson *et al.*, 1985). Whether this is due to a gross release of cell-surface proteins or is restricted to E-24.11 is not clear. In other cases, notably ACE, although the level of the circulating form is increased in certain disease states (e.g. sarcoidosis, Gaucher's disease, leprosy and hyperthyroidism; Erdos and Skidgel, 1987; Lieberman, 1975), significant levels of the circulating form are present under normal conditions.

Whilst examining the mode of membrane anchorage of ACE we observed that this peptidase, unlike a number of others, could be released from porcine kidney microvillar membranes in a time- and temperature-dependent manner (Hooper *et al.*, 1987a). The released form of the enzyme lacked the hydrophobic membrane-anchoring domain. Subsequent studies have revealed that ACE is cleaved post-translationally by a metalloprotease (Oppong and Hooper, 1993). We have coined the term 'secretase' for this and other enzymes that release membrane proteins into a soluble form by proteolysis (for review see Hooper *et al.*, 1997b). ACE is cleaved by the secretase in the stalk region, the precise site of cleavage dependent on the isoform and species (Beldent *et al.*, 1993, 1995; Ehlers *et al.*, 1996). Subcellular fractionation, detergent solubilization, as well as pulse-chase and surface labelling experiments, have established that the secretase co-localizes with ACE in the plasma membrane (Beldent *et al.*, 1995; Oppong and Hooper, 1993; Ramchandran and Sen, 1995). The observation that the secretase can also be released from the membrane by trypsin (Parvathy *et al.*, 1997) suggests that, like its substrate ACE, it too may have a protease-susceptible stalk region. Recently, we have identified several small molecular weight, hydroxamic acid-based inhibitors of the secretase (Parvathy *et al.*, 1997), implying that it is also a zinc metalloenzyme. These inhibitors also block the release of a number of other cell-surface proteins, including tumour necrosis factor-α and its receptors, transforming growth factor-α, L-selectin, the amyloid precursor protein, the interleukin-6 receptor and the Fas ligand (Hooper *et al.*, 1997b). At the present time it is not clear whether there is a single secretase involved in the release of all these diverse proteins or whether there is a family of closely related secretases.

GPI-anchored peptidases are susceptible to release from the cell surface by proteases and by endogenous phospholipases. The fatty acids of the GPI anchor can be removed by the action of phospholipase C or D. Although a GPI-specific phospholipase D has been identified in serum and shown to be involved in the release of alkaline phosphatase (Solter and Hoffman, 1995), there is no evidence implicating it in the release of one of the GPI-anchored peptidases. Recently, we have identified MDP on the cell surface of differentiated 3T3-L1 adipocytes (S. Movahedi and N.M. Hooper, unpublished), and shown that, as well as being susceptible to cleavage by bacterial PI-PLC, it is released in response to insulin stimulation by an endogenous phospholipase C. In contrast, MDP in the pancreatic zymogen granule membrane appears to be released by the action of a phospholipase A (Hooper *et al.*, 1997a). Although phospholipase A will only cleave one of the two fatty acid chains from the GPI anchor, this would appear to be sufficient to promote release of the peptidase from the zymogen granule membrane. In support of this, is the observation that MDP exists in a soluble form in pancreatic secretions (N.M. Hooper and D. LeBel, unpublished observations).

3.6 Why different modes of membrane anchorage?

Why there should be three different modes of membrane anchorage for cell-surface peptidases is not readily apparent, but presumably the different anchors confer particular functions on the protein. This is most readily seen with the GPI anchor where unique properties, in addition to membrane anchorage, have been ascribed to this anchor. These include roles in intracellular sorting, in increased lateral mobility, in transmembrane signalling, in the novel endocytic process of potocytosis, in the selective release of the protein by phospholipases, and in the clustering of proteins in membrane microdomains (for reviews see Cross, 1990; Englund, 1993; Ferguson, 1992; McConville and Ferguson, 1993).

For example, GPI-anchored proteins have been reported to be more laterally mobile in the plane of the membrane because the lipids in the anchor interact only with the outer leaflet of the bilayer, although interactions of the ectodomain with other membrane components appear to be the major determinant of mobility (Zhang *et al.*, 1991). In contrast, a transmembrane anchor has the potential to interact with the cytoskeleton and other cytosolic proteins, which may cause such anchored proteins to be relatively immobile and/or to convey signals into the interior of the cell directly. Numerous studies have shown that a GPI anchor is a positive apical targeting signal in epithelial cells and a positive axonal targeting signal in neurons (Dotti *et al.*, 1991; Rodriguez-Boulan and Powell, 1992). However, a GPI anchor is not the only apical targeting signal as the majority of the transmembrane polypeptide-anchored peptidases are also targeted directly to this surface of the cell in detergent-insoluble glycolipid rafts alongside the GPI-anchored proteins (Danielsen, 1995, and Chapter 4). Probably one of the best characterized additional roles of a GPI anchor is the potential for cleavage by phospholipases. Several GPI-anchored proteins have now been shown to be cleaved by endogenous phospholipases to generate soluble forms, a process that is obviously restricted to this type of anchor. In contrast, both of the type I peptidases (ACE and EP-24.18) are susceptible to cleavage from the membrane by secretases or other proteases, whereas the type II peptidases are generally more resistant to proteolysis. Thus, it is possible that the use of different membrane anchors allows other properties to be conferred on a cell-surface peptidase, enabling it to carry out specific function(s). Answers to the question may come from detailed studies of the targeting, mobility, activity, cleavage, etc., of chimeric proteins having alternative anchoring domains.

Acknowledgements

I gratefully acknowledge the Medical Research Council, the Wellcome Trust and the Royal Society for financial support of my work on the anchorage of cell-surface peptidases.

References

Adachi H, Tawaragi Y, Inuzuka C, Kubota I, Tsujimoto M, Nishihara T, Nakazato H. (1990) Primary structure of human microsomal dipeptidase deduced from molecular cloning. *J. Biol. Chem.* **265:** 3992–3995.

Almenoff J, Teirstein AS, Thornton JC, Orlowski M. (1984) Identification of a thermolysin-like metallopeptidase in serum: activity in normal subjects and in patients with sarcoidosis. *J. Lab. Clin. Med.* **103:** 420–431.

Beldent V, Michaud A, Wei L, Chauvet M-T and Corvol P. (1993) Proteolytic release of human angiotensin converting enzyme. Localization of the cleavage site. *J. Biol. Chem.* **268:** 26428–26434.

Beldent V, Michaud A, Bonnefoy C, Chauvet M-T, Corvol P. (1995) Cell surface localization of proteolysis of human endothelial angiotensin I-converting enzyme. Effect of the amino-terminal domain in the solubilization process. *J. Biol. Chem.* **270:** 28962–28969.

Bordier C. (1981) Phase separation of integral membrane proteins in Triton X-114 solution. *J. Biol. Chem.* **256:** 1604–1607.

Brewis IA, Turner AJ, Hooper NM. (1994) Activation of the glycosyl-phosphatidylinositol-anchored membrane dipeptidase upon release from porcine kidney membranes by phospholipase C. *Biochem. J.* **303:** 633–638.

Brewis IA, Ferguson MAJ, Mehlert A, Turner AJ, Hooper NM. (1995) Structures of the glycosyl-phosphatidylinositol anchors of porcine and human membrane dipeptidase. Interspecies comparison of the glycan core structures and further structural studies on the porcine anchor. *J. Biol. Chem.* **270:** 22946–22956.

Broomfield SJ, Hooper NM. (1993) Characterization of an antibody to the cross-reacting determinant of the glycosyl-phosphatidylinositol anchor of human membrane dipeptidase. *Biochim. Biophys. Acta* **1145:** 212–218.

Brown DA. (1992) Interactions between GPI-anchored proteins and membrane lipids. *Trends Cell Biol.* **2:** 338–343.

Chang W-J, Ying Y, Rothberg KG, *et al.* (1994) Purification and characterization of smooth muscle cell caveolae. *J. Cell Biol.* **126:** 127–138.

Corbeil D, Gaudoux F, Wainwright S, Ingram J, Kenny AJ, Boileau G, Crine P. (1992) Molecular cloning of the α-subunit of rat endopeptidase-24.18 (endopeptidase-2) and co-localization with endopeptidase-24.11 in rat kidney by *in situ* hybridization. *FEBS Lett.* **309:** 203–208.

Cross GAM. (1990) Glycolipid anchoring of plasma membrane proteins. *Annu. Rev. Cell Biol.* **6:** 1–39.

Danielsen EM. (1995) Involvement of detergent-insoluble complexes in the intracellular transport of intestinal brush border enzymes. *Biochemistry* **34:** 1596–1605.

Delmas B, Gelfi J, Kut E, Sjostrom H, Noren O, Laude H. (1994) Determinants essential for the transmissible gastroenteritis virus–receptor interaction reside within a domain of aminopeptidase N that is distinct from the enzymatic site. *J. Virol.* **68:** 5216–5224.

Devault A, Lazure C, Nault C *et al.* (1987) Amino acid sequence of rabbit kidney neutral endopeptidase-24.11 (enkephalinase) deduced from a complementary DNA. *EMBO J.* **6:** 1317–1322.

Dotti CG, Parton RG, Simons K. (1991) Polarised sorting of glypiated proteins in hippocampal neurons. *Nature* **349:** 158–161.

Ehlers MRW, Scwager SLU, Scholle RR, Manji GA, Brandt WF, Riordan JF. (1996) Proteolytic release of membrane-bound angiotensin-converting enzyme: role of the juxtamembrane stalk sequence. *Biochemistry* **35:** 9549–9559.

Emoto N, Yanagisawa M. (1995) Endothelin-converting enzyme-2 is a membrane-bound, phosphoramidon-sensitive metalloprotease with acidic pH optimum. *J. Biol. Chem.* **250:** 15262–15268.

Englund PT. (1993) The structure and biosynthesis of glycosyl phosphatidylinositol protein anchors. *Annu. Rev. Biochem.* **62:** 121–138.

Erdos EG, Skidgel RA. (1987) The angiotensin I converting enzyme. *Lab. Invest.* **56:** 345–348.

Ferguson MAJ. (1992) Glycosyl-phosphatidylinositol membrane anchors: the tale of a tail. *Biochem. Soc. Trans.* **20:** 243–256.

Fulcher IS, Pappin DJC, Kenny AJ. (1986) The N-terminal amino acid sequence of pig kidney endopeptidase-24.11 shows homology with pro-sucrase-isomaltase. *Biochem. J.* **240:** 305–308.

Galasko GTF, Bao Y, Broomfield SJ, Hooper NM, Turner AJ, Larner J. (1995) Circulating factors and insulin resistance. I. A novel myoinositol 1,2-cyclic phosphate phosphoglycan insulin antagonist from human plasma is elevated in noninsulin-dependent diabetes mellitus. *J. Clin. Endocrinol. Metab.* **80:** 2419–2429.

Gorbea CM, Marchand P, Jiang W, Copeland NG, Gilbert DJ, Jenkins NA, Bond JS. (1993) Cloning, expression, and chromosomal localization of the mouse meprin β subunit. *J. Biol. Chem.* **268:** 21035–21043.

Guther MLS, Cardoso de Almeida ML, Rosenberry TL, Ferguson MAJ. (1994) The detection of phospholipase-resistant and -sensitive glycosyl-phosphatidylinositol membrane anchors by western blotting. *Analyt. Biochem.* **219:** 249–255.

Hesp JR, Hooper NM. (1997) Proteolytic fragmentation reveals the oligomeric and domain structure of porcine aminopeptidase A. *Biochemistry*, in press.

Hooper NM. (1992) Identification of a glycosyl-phosphatidylinositol anchor on membrane proteins. In: *Lipid Modification of Proteins: a Practical Approach* (eds N.M. Hooper and A.J. Turner), pp. 89–115. IRL Press, Oxford.

Hooper NM, Bashir A. (1991) Glycosyl-phosphatidylinositol-anchored membrane proteins can be distinguished from transmembrane polypeptide-anchored proteins by differential solubilization and temperature-induced phase separation in Triton X-114. *Biochem. J.* **280:** 745–751.

Hooper NM, Turner AJ. (1988a) Ectoenzymes of the kidney microvillar membrane. Aminopeptidase P is anchored by a glycosyl-phosphatidylinositol moiety. *FEBS Lett.* **229:** 340–344.

Hooper NM, Turner AJ. (1988b) Ectoenzymes of the kidney microvillar membrane. Differential solubilization by detergents can predict a glycosyl-phosphatidylinositol membrane anchor. *Biochem. J.* **250:** 865–869.

Hooper NM, Turner AJ. (1989) Ectoenzymes of the kidney microvillar membrane. Isolation and characterization of the amphipathic form of renal dipeptidase and hydrolysis of its glycosyl-phosphatidylinositol anchor by an activity in plasma. *Biochem. J.* **261:** 811–818.

Hooper NM, Keen J, Pappin DJC, Turner AJ. (1987a) Pig kidney angiotensin converting enzyme. Purification and characterization of amphipathic and hydrophilic forms of the enzyme establishes C-terminal anchorage to the plasma membrane. *Biochem. J.* **247:** 85–93.

Hooper NM, Low MG, Turner AJ. (1987b) Renal dipeptidase is one of the membrane proteins released by phosphatidylinositol-specific phospholipase C. *Biochem. J.* **244:** 465–469.

Hooper NM, Broomfield SJ, Turner AJ. (1991) Characterization of antibodies to the glycosyl-phosphatidylinositol membrane anchors of mammalian proteins. *Biochem. J.* **273:** 301–306.

Hooper NM, Cook S, Laine J, LeBel D. (1997a) Identification of membrane dipeptidase as a major protein of the pancreatic zymogen granule membrane, and evidence for cleavage of its GPI anchor by phospholipase A. *J. Biol. Chem.*, in press.

Hooper NM, Karran E, Turner AJ. (1997b) Membrane protein secretases. *Biochem. J.*, in press.

Hyde R, Hooper NM, Turner AJ. (1996) Molecular cloning and expression in Cos-1 cells of porcine kidney aminopeptidase P. *Biochem. J.* **319:** 197–200.

Ikezawa H, Yamanegi M, Taguchi R, Miyashita T, Ohyabu T. (1976) Studies on phosphatidylinositol phosphodiesterase (phospholipase C) of *Bacillus cereus*. I. Purification, properties and phosphatase-releasing activity. *Biochim. Biophys. Acta* **450:** 154–164.

Johnson AR, Coalson JJ, Ashton J, Larumbide M, Erdos EG. (1985) Neutral endopeptidase in serum samples from patients with adult respiratory distress syndrome. Comparison with angiotensin-converting enzyme. *Am. Rev. Respir. Dis.* **132:** 1262–1267.

Johnson GD, Hersh LB. (1992) Cloning a rat meprin cDNA reveals the enzymes is a heterodimer. *J. Biol. Chem.* **267:** 13505–13512.

Keller SR, Scott HM, Mastick CC, Aebersold R, Lienhard GE. (1995) Cloning and characterization of a novel insulin-regulated membrane aminopeptidase from Glut4 vesicles. *J. Biol. Chem.* **270:** 23612–23618.

Kenny AJ, Maroux S. (1982) Topology of microvillar membrane hydrolases of kidney and intestine. *Physiol. Rev.* **62:** 91–128.

Kenny AJ, Turner AJ. (1987) What are ectoenzymes? In: *Mammalian Ectoenzymes* (eds A.J. Kenny and A.J. Turner), pp.1–13. Elsevier, Amsterdam.

Kenny AJ, Fulcher IS, McGill KA, Kershaw D. (1983) Proteins of the kidney microvillar membrane. Reconstitution of endopeptidase in liposomes shows that it is a short stalked protein. *Biochem. J.* **211:** 755–762.

Laperche Y, Bulle F, Aissani T, Chobert M-N, Aggerbeck M, Hanoune J, Guellaen G. (1986) Molecular cloning and nucleotide sequence of rat kidney γ-glutamyl transpeptidase cDNA. *Proc. Natl Acad. Sci. USA* **83:** 937–941.

Lattion A-L, Soubrier F, Allegrini J, Hubert C, Corvol P, Alhenc-Gelas F. (1989) The testicular transcript of the angiotensin I-converting enzyme encodes for the ancestral, non-duplicated form of the enzyme. *FEBS Lett.* **252:** 99–104.

Li L, Wang J, Cooper MD. (1993) cDNA cloning and expression of human glutamyl aminopeptidase (aminopeptidase A). *Genomics* **17:** 657–664.

Lieberman J. (1975) Elevation of serum angiotensin converting enzyme (ACE) level in sarcoidosis. *Am. J. Med.* **59:** 365–372.

Louvard D, Maroux S, Vannier C, Desnuelle P. (1975) Topological studies on the hydrolases bound to the intestinal brush border membrane. Solubilization by papain and Triton X-100. *Biochim. Biophys. Acta* **375:** 236–248.

Low MG, Finean JB. (1977) Release of alkaline phosphatase from membranes by a phosphatidylinositol specific phospholipase C. *Biochem. J.* **167:** 281–284.

Malfroy B, Schofield PR, Kuang W-J, Seeburg PH, Mason AJ, Henzel WJ. (1987) Molecular cloning and amino acid sequence of rat enkephalinase. *Biochem. Biophys. Res. Commun.* **144:** 59–66.

Malfroy B, Kado-Fong H, Gros C, Giros B, Schwartz J-C, Hellmiss R. (1989) Molecular cloning and amino acid sequence of rat kidney aminopeptidase M: a member of a super family of zinc metallohydrolases. *Biochem. Biophys. Res. Commun.* **161:** 236–241.

Marchand P, Tang J, Bond JS. (1994) Membrane association and oligomeric organization of the α and β subunits of mouse meprin A. *J. Biol. Chem.* **269:** 15388–15393.

Marchand P, Tang J, Johnson GD, Bond JS. (1995) COOH-terminal proteolytic processing of secreted and membrane forms of the α subunit of the metalloprotease meprin A. Requirement of the I domain for processing in the endoplasmic reticulum. *J. Biol. Chem.* **270:** 5449–5456.

McConville MJ, Ferguson MAJ. (1993) The structure, biosynthesis and function of glycosylated phosphatidylinositols in the parasitic protozoa and higher eukaryotes. *Biochem. J.* **294:** 305–324.

Milhiet P-E, Corbeil D, Simon V, Kenny AJ, Crine P, Boileau G. (1994) Expression of rat endopeptidase-24.18 on Cos-1 cells: membrane topology and activity. *Biochem. J.* **300:** 37–43.

Ogata S, Misumi Y, Ikehara Y. (1989) Primary structure of rat liver dipeptidyl peptidase IV deduced from its cDNA and identification of the NH_2-terminal signal sequence as the membrane anchoring domain. *J. Biol. Chem.* **264:** 3596–3601.

Olsen J, Cowell GM, Konigshofer E, *et al.* (1988) Complete amino acid sequence of human intestinal aminopeptidase N as deduced from cloned cDNA. *FEBS Lett.* **238:** 307–314.

Oppong SY, Hooper NM. (1993) Characterization of a secretase activity which releases angiotensin-converting enzyme from the membrane. *Biochem. J.* **292:** 597–603.

Parkin ET, Turner AJ, Hooper NM. (1996) Isolation and characterization of two distinct low-density, Triton-insoluble complexes from porcine lung membranes. *Biochem. J.* **319**: 887–896.

Rached E, Hooper NM, James P, Semenza G, Turner AJ, Mantei N. (1990) cDNA cloning and expression in *Xenopus laevis* oocytes of pig renal dipeptidase, a glycosyl-phosphatidylinositol-anchored ectoenzyme. *Biochem. J.* **271:** 755–760.

Ramchandran R, Sen I. (1995) Cleavage processing of angiotensin-converting enzyme by a membrane-associated metalloprotease. *Biochemistry* **34:** 12645–12652.

Rodriguez-Boulan E, Powell SK. (1992) Polarity of epithelial and neuronal cells. *Annu. Rev. Cell Biol.* **8:** 395–427.

Rogi T, Tsujimoto M, Nakazato H, Mizutani S, Tomoda Y. (1996) Human placental leucine aminopeptidase/oxytocinase. A new member of type II membrane-spanning zinc metallopeptidase family. *J. Biol. Chem.* **271:** 56–61.

Schauder B, Schomburg L, Kohrle J, Bauer K. (1994) Cloning of a cDNA encoding an ectoenzyme that degrades thyrotropin-releasing hormone. *Proc. Natl Acad. Sci. USA* **91:** 9534–9538.

Scotto AW, Zakim D. (1985) Reconstitution of membrane proteins. Spontaneous association of integral membrane proteins with preformed unilamellar lipid bilayers. *Biochemistry* **24:** 4066–4075.

Semenza G. (1986) Anchoring and biosynthesis of stalked brush border membrane proteins: glycosidases and peptidases of enterocytes and renal tubuli. *Annu. Rev. Cell Biol.* **2:** 255–313.

Shimada K, Takahashi M, Tanzawa K. (1994) Cloning and functional expression of endothelin-converting enzyme from rat endothelial cells. *J. Biol. Chem.* **269:** 18275–18278.

Solter PF, Hoffman WE. (1995) Canine corticosteroid-induced alkaline phosphatase in serum was solubilized by phospholipase activity *in vivo. Am. J. Physiol.* **269:** G278-G286.

Soubrier F, Alhenc-Gelas F, Hubert C, Allegrini J, John M, Tregear G, Corvol P. (1988) Two putative active centers in human angiotensin I-converting enzyme revealed by molecular cloning. *Proc. Natl Acad. Sci. USA* **85:** 9386–9390.

Tan F, Chan S-J, Steiner DF, Schilling JW, Skidgel RA. (1989) Molecular cloning and sequencing of the cDNA for human membrane bound carboxypeptidase M. Comparison with carboxypeptidases A, B, H and N. *J. Biol. Chem.* **264:** 13165–13170.

Toutant J-P, Roberts WL, Murray NR, Rosenberry TL. (1989) Coversion of human erythrocyte acetylcholinesterase from an amphiphilic to a hydrophilic form by phosphatidylinostiol-specific phospholipase C and serum phospholipase D. *Eur. J. Biochem.* **180:** 503–508.

Udenfriend S, Kodukula K. (1995) How glycosyl-phosphatidylinositol-anchored membrane proteins are made. *Annu. Rev. Biochem.* **64:** 563–591.

Vogel M, Kowalewski H, Zimmermann H, Hooper NM, Turner AJ. (1992) Soluble low-K_m 5′-nucleotidase from electric-ray (*Torpedo marmorata*) electric organ and bovine cerebral cortex is derived from the glycosyl-phosphatidylinositol-anchored ectoenzyme by phospholipase C cleavage. *Biochem. J.* **284:** 621–624.

Wu Q, Lahti JM, Air GM, Burrows PD, Cooper MD. (1990) Molecular cloning of the murine BP-1/6C3 antigen: a member of the zinc-dependent metallopeptidase family. *Proc. Natl Acad. Sci. USA* **87:** 993–997.

Xu D, Emoto N, Giaid A, Slaughter C, Kaw S, deWit D, Yanagisawa M. (1994) ECE-1: a membrane-bound metalloprotease that catalyzes the proteolytic activation of big endothelin-1. *Cell* **78:** 473–485.

Zamze SE, Ferguson MAJ, Collins R, Dwek RA, Rademacher TW. (1988) Characterisation of the cross-reacting determinant (CRD) of the glycosyl-phosphatidylinositol membrane anchor of *Trypanosoma brucei* variant surface glycoprotein. *Eur. J. Biochem.* **176:** 527–534.

Zhang F, Crise B, Su B, Hou Y, Rose JK, Bothwell A, Jacobson K. (1991) Lateral diffusion of membrane-spanning and glycosylphosphatidylinositol-linked proteins: toward establishing rules governing the lateral mobility of membrane proteins. *J. Cell Biol.* **115:** 75–84.

Chapter 4

Trafficking and sorting of cell-surface peptidases in epithelial cells

E. Michael Danielsen, Lotte K. Vogel and Hans Sjöström

4.1 Introduction

From studies initially carried out on membrane proteins of enveloped viruses, it has long been established that membrane glycoproteins, destined for the cell surface, are part of the membrane flow out of the endoplasmic reticulum that was originally described for proteins bound for secretion (Palade, 1975). Thus, although by definition cell-surface peptidases reside at the plasmalemma, an ectopeptidase molecule will, during its lifespan, traverse other membraneous compartments at least once before reaching its final destination. This in turn implies that it has been subject to one or more sorting events, aimed to create for the molecule an environment of lipid and protein that exactly matches that of the cell-surface membrane. In this chapter, we will focus on biosynthetic membrane trafficking in epithelial cells where ectopeptidases invariably are expressed in a polarized fashion at the apical cell surface. For some time now, it has been evident that epithelial cells may employ at least two different targeting pathways to obtain this polarized steady-state distribution. However, in spite of an intense effort by many groups engaged in this field of research, progress in identifying structural motifs endowed with apical targeting information has been disappointingly slow and, as yet, we are nowhere near a full understanding of how, in molecular and mechanistic terms, apical sorting and targeting occurs. Below is an account of our current ideas and views on trafficking of cell-surface peptidases in epithelial cells. In addition, we present a kinetic model of membrane traffic in a polarized cell. The model, which is based on experimental data, shows how apical targeting can be achieved by different trafficking pathways and it may be useful in our efforts to obtain a comprehensive understanding of the ongoing membrane traffic in polarized cells.

4.2 Targeting to the apical cell surface: the direct versus the transcytotic route

It has long been realized that polarized cells may use two different routes for delivery of newly made apical cell-surface proteins; either a 'direct' one, leading from the trans-Golgi network (TGN) to the apical pole, or one involving two separate steps: an

Cell-Surface Peptidases in Health and Disease, edited by A.J. Kenny and C.M. Boustead.

initial exocytosis to the basolateral surface, followed by reinternalization into an endosomal compartment and transcytosis to the apical domain (Le Gall *et al.*, 1995; Matter and Mellman, 1994; Mellman *et al.*, 1993). The Madin–Darby canine kidney (MDCK) cell quickly became the archetypal cell using the direct route, and the hepatocyte, thought to lack this pathway, in turn became the paradigm of the latter mechanism (Bartles *et al.*, 1987; Simons and Fuller, 1985). To muddle the picture, subsequent studies of the biogenetic trafficking pathways in the intestinal enterocyte-like cell line Caco-2 convincingly showed that the two alternative pathways may operate within the same cell and that a given protein may use both to reach its ultimate apical destination (Le Bivic *et al.*, 1990; Matter *et al.*, 1990b). Very recently, this fact has been underlined by the observation that even non-polarized cells such as fibroblasts seem to possess both an apical and a basolateral cognate route from the TGN to the cell surface for delivery of newly synthesised proteins (Müsch *et al.*, 1996; Yoshimori *et al.*, 1996).

4.2.1 Direct targeting from the TGN to the apical cell surface

Early work by van Meer and Simons (1982, 1986) led to the realization that polarized cells not only generate and keep distinct apical and basolateral complements of plasma membrane proteins, but also maintain a polarized distribution of cell-surface membrane lipids in the ectoplasmic leaflet, with the apical cell surface being enriched in glycosphingolipids. This led the authors to propose the 'cluster' hypothesis (i.e. the existence of an integrated sorting mechanism which generates an apical microdomain in the TGN, including both lipids and proteins, that subsequently buds off to form an apically destined transport vesicle; Simons and van Meer, 1988). The major driving force in the formation of such membrane microdomains would be the ability of glycosphingolipids to self-associate by intermolecular hydrogen bonding, giving the individual lipid molecules less lateral mobility. Apically destined proteins would cluster within these viscous microdomains (often referred to as glycolipid 'rafts'), either by direct interactions with the glycosphingolipids, or with a putative raft-associated sorting protein (Simons and van Meer, 1988). Components with other subcellular destinations such as glycerolipids and basolateral membrane proteins would be excluded from the glycosphingolipid microdomains and thus be prevented from becoming part of apically targeted transport vesicles. Since this cluster model was first proposed, it has gained considerable experimental support; particularly concerning the apical targeting of proteins, attached to the membrane by a glycosyl phosphatidylinositol (GPI) anchor. More than 100 members of this class of membrane proteins have been identified (Brown and Stow, 1996; Low, 1989), including the ectopeptidases aminopeptidase P and renal dipeptidase (see Chapters 3 and 14). Generally, GPI-linked proteins are localized to the apical surface of epithelial cells where they typically resist solubilization at low temperature by non-ionic detergents with a low critical micellar concentration (CMC), such as Triton X-100 (Hooper and Turner, 1988). Several mutagenesis studies have shown that replacement of a transmembrane segment with a GPI anchor is sufficient to confer apical targeting in MDCK cells (Brown *et al.*, 1989; Lisanti *et al.*, 1989), qualifying it to be termed an apical sorting signal, the only such signal so far identified. However, notable exceptions are the Fischer rat thyroid cell line, where a transfected GPI-linked protein, gD1-DAF was predominantly basolateral, and a Con A-resistant MDCK cell line, where gD1-DAF was equally distributed between the two domains (Zurzolo *et al.*, 1994). In the first direct demonstration that glycosphingolipids and GPI-anchored proteins share a common biosynthetic pathway, Brown and Rose (1992) showed that newly made placental alkaline phosphatase in MDCK cells associates in the Golgi complex with detergent-insoluble complexes rich in glycolipids and cholesterol. Subsequent work by others has confirmed this biosynthetic association between glycosphingolipids and GPI-anchored proteins and has aimed to identify the protein components involved in microdomain formation (Dupree *et al.*, 1993; Fiedler *et al.*, 1993; Lisanti *et al.*, 1993; Sargiacomo *et al.*, 1993). One such component is caveolin/VIP21, a small hairpin-shaped integral membrane protein thought to constitute the cytoplasmic coat of *caveolae* which have long been recognized as distinct plasma membrane invaginations involved in a cellular internalization process termed *potocytosis* (Anderson *et al.*, 1992). Caveolin/VIP21 may well act both in the TGN and at the cell surface to recruit proteins to the glycolipid microdomains, possibly by formation of hetero-oligomeric complexes, but at present the exact relationship between detergent-insoluble glycolipid microdomains and caveolae is not fully understood (Parton and Simons, 1995). Another candidate component for a role in the apical sorting is VIP36, an integral membrane protein with sequence homology to plant lectins (Fiedler *et al.*, 1994), which has renewed the interest in the N-linked glycosylation as a possible apical sorting signal (Fiedler and Simons, 1995; Scheiffele *et al.*, 1995). A recent and surprising observation is that, contrary to many other intracellular trafficking events, the direct TGN to apical pathway in MDCK cells appears to operate independently of the 'normal' molecular machinery for vesicular transport, involving *N*-ethylmaleimide-sensitive factors, synaptosomal-associated proteins (NSF, SNAP, SNAREs) and rab proteins (Ikonen *et al.*, 1995). Instead, the same group has proposed that members of the annexin family of proteins might play a role in the direct vesicular transport between the TGN and the apical cell surface and they discovered and characterized a new member of the family, annexin XIIIb, which is exclusively expressed in kidney and intestine (Fiedler *et al.*, 1995). Annexin II also seems to be associated with detergent-insoluble complexes (Sargiacomo *et al.*, 1993), and was reported to be their major protein component in MDCK cells (Harder and Gerke, 1994). From a theoretical viewpoint, annexins would indeed seem able to perform some of the essential functions required for vesicular traffic to the apical cell surface: their ability to associate with phospholipids in a Ca^{2+}-dependent manner enables them to mediate an exocytic membrane fusion event (Creutz, 1992; Gruenberg and Emans, 1993), and the predominant localization of annexin II in the apical terminal web region of enterocytes and its interaction with the actin cytoskeleton places at least this member of the family at a strategic subcellular site for directing apical exocytic vesicles to their target membrane (Weinman *et al.*, 1994).

Although the details of the molecular machinery underlying the above 'cluster' hypothesis for sorting of apically destined membrane components remain to be fully worked out, it represents an attractively simple and plausible model for a direct apical delivery of newly synthesised glycosphingolipids and GPI-anchored proteins. However, most of the apical membrane proteins of polarized cells, including the majority of ectopeptidases, are anchored by a transmembrane segment of their polypeptide rather than by a glycophospholipid. Early work by Hooper and Turner (1988) showed that the transmembrane microvillar ectopeptidases of the kidney proximal tubule cell, unlike the GPI-linked proteins, are fully detergent-soluble. Studying ectoenzyme insolubility in Caco-2 cells, Garcia *et al.* (1993) found only one transmembrane enzyme, sucrase-isomaltase, to be partially insoluble (about 20%), whereas dipeptidyl peptidase IV (DPP-IV) was fully soluble. In contrast, we recently observed that several (but not all) of the major transmembrane hydrolases of the enterocyte brush border membrane [including aminopeptidases N and A (AP-N and -A) and DPP-IV] are Triton X-100-insoluble to a significant extent (35–50%), and that this property is conferred during intracellular transport (Danielsen, 1995). A similar, significant insolubility was also previously reported for the transmembrane influenza virus haemagglutinin (Fiedler *et al.*, 1993). From these reports, it does seem that transmembrane 'cargo' proteins generally have a weaker and more cell-type-dependent association with the glycolipid microdomains, as defined by detergent-extractability at low temperature. An explanation could be that this type of protein, unlike those attached by a GPI anchor, needs to associate with a sorting protein (possibly caveolin, VIP36 or an annexin) in order to cluster in the glycolipid 'rafts'. In this view, the extent to which apical transmembrane proteins make use of the direct pathway would reflect the level of expression of the particular sorting protein(s) in the given cell type.

4.2.2 Targeting via the transcytotic route

Transwell™ culture dishes, allowing the quantitative detection of newly synthesised proteins separately at the basolateral and apical cell surfaces of confluent monolayers of epithelial cells, have been widely used for studying membrane trafficking of ectopeptidases. This technique was used in studies showing that Caco-2 cells use both direct and transcytotic routes for the delivery of apical membrane proteins (Le Bivic *et al.*, 1990; Matter *et al.*, 1990b). Using essentially the same approach, a host of membrane and secretory proteins has been studied with the aim of identifying the sorting signals responsible for apical delivery. To tackle this problem thoroughly, the transport of a non-sorted membrane protein needs to be investigated as well, but unfortunately no such protein has been identified since the nature of the putative sorting signal is unknown. To circumvent this, several studies have compared the secretion of supposedly non-sorted secretory proteins with various transmembrane proteins made to be secretory by introducing a signal peptide cleavage site. Again for secretory proteins, no ideal marker for the signal-independent vesicular bulk transport has been found and we therefore rely on the accumulated evidence obtained from the analysis of various supposedly non-sorted proteins. In MDCK cells (strain II) the secretory granule proteins, oviduct lysozyme (Kondor-Kock *et al.,* 1985), growth hormone, gastric prochymosin, salivary gland α_{2u}-globulin (Gottlieb *et al.*, 1986), cystatin C (Vogel *et al.,* 1992) and pro-opiomelanocortin (Corbeil *et al.*, 1992) are secreted approximately 40% from the apical domain and 60% from the basolateral side. The immunoglobulin κ chain (Gottlieb *et al.*, 1986) is also secreted with a similar polarity. These results suggest that non-sorted secretory proteins in MDCK cells strain II are secreted in approximately equal quantities from both surface domains of the cell. Secretory forms of three transmembrane proteins, AP-N (Vogel *et al.*, 1992), endopeptidase-24.11 (E-24.11, Corbeil *et al.*, 1992) and DPP-IV (Weisz *et al.*, 1992) have also been investigated. AP-N was fused to the signal peptide, including the signal peptide cleavage site, of haemagglutinin (Vogel *et al.*, 1992), resulting in an enzymically active secretory form of the peptidase, predominantly secreted from the apical pole of MDCK cells. Corbeil and co-workers (1992) showed that a secretory form of E-24.11, consisting of the signal peptide of pro-opiomelanocortin and the ectodomain of E-24.11 is enzymically active and secreted from the apical pole of MDCK cells. Weisz *et al.* (1992) introduced a signal peptidase cleavage site into the DPP-IV cDNA and obtained MDCK cells expressing a secretory form of the enzyme that was shown to be secreted mainly from the apical surface. Taken together, and presuming that the secretory and transmembrane forms of the ectoenzymes interact with the cellular sorting machinery in a similar manner, these studies suggest that transport from the TGN to the apical plasma membrane is signal-mediated and that the sorting signal involved is located on the ectodomain of the transmembrane proteins. Unfortunately, this general conclusion could not be validated in cell lines known to use the indirect pathway to a greater extent than MDCK cells. Thus, the secretory form of E-24.11 was secreted about equally from both surfaces of LLC-PK_1 cells (Lanctôt *et al.*, 1995), as was the secretory form of AP-N when expressed in Caco-2 cells (Vogel *et al.*, 1995). At face value, these findings would suggest that the ectodomains of the two peptidases lack a sorting signal. However, the reason for this discrepancy could be that sorting along the transcytotic pathway has additional signal requirements, perhaps membrane attachment. Another explanation could be that secretory proteins trafficking via the transcytotic route are released from the basolateral surface before they get an opportunity to encounter the sorting machinery operating in the transcytotic pathway. At any rate, it must be concluded that the search by genetic engineering for an apical sorting signal for transmembrane proteins has yet to come up with a candidate for such a signal, be it a sequence or, perhaps more likely, a patch defined by higher order structure. High-resolution crystallography would probably be required to identify common structural motifs in the ectodomain of different cell-surface peptidases, but so far this approach has been hampered by the difficulties in obtaining suitable crystals of these large glycoproteins.

One common structural theme of several transmembrane ectoenzymes seems to be a short cytoplasmic tail most often combined with a type II membrane topology. Even in cells that preferably, or exclusively, use the transcytotic route to the apical surface, the short tail probably prevents them from being trapped at the basolateral surface by the mechanism described for the Na^+/K^+-ATPase, which involves selective association with the basolateral fodrin–ankyrin cytoskeleton (Nelson, 1992). Of course, this principle applies to the GPI-anchored ectoenzymes as well and would, in effect, promote apical localization of ectoenzymes that initially reach the basolateral cell surface ‘by default’.

Once arrived at their ultimate destination, do ectopeptidases also transcytose from the apical to the basolateral cell surface? Transcytosis in this direction is certainly a well-known mechanism used by the enterocyte of the neonate to take up maternal immunoglobulins from mothers’ milk (Rodewald, 1973). Whether this pathway is of functional importance in other epithelial cells remains to be established, but at least one apically located endogenous protein in MDCK cells has been reported to be selectively transcytosed bidirectionally between the apical and the basolateral plasma membrane (Brändli *et al.*, 1990). Recently, we have obtained evidence that AP-N also seems to be transcytosed bidirectionally between the two surface domains of MDCK cells (L.K. Vogel, G.H. Hansen, F. Moller, B. Delmas, O. Norén and H. Sjöström, unpublished results). Our results suggest that transcytosis from the apical to the basolateral surface only occurs at a relatively slow rate, but since AP-N has a long half-life in MDCK cells, a single enzyme molecule may shuttle back and forth between the two surface domains more than once in its lifetime. If our finding happens to reflect a more widespread phenomenon, it would imply a more complex membrane trafficking of ectopeptidases than was previously thought and, consequently, their predominant steady-state localization at the apical cell surface should be viewed more as a dynamic equilibrium between opposing trafficking pathways than as a static end-station. However, further studies are needed to elaborate this model and to find out if epithelial cells use this mechanism to regulate the apical expression of ectopeptidases. The transcytotic pathway seems indeed to be regulated, at least for some membrane proteins, as binding of IgA to the polymeric immunoglobulin receptor stimulates transcytosis by a phosphorylation/dephosphorylation mechanism (Cardone *et al.*, 1996). Otherwise, nothing much is known about the molecular machinery involved in transcytosis. A general vesicular transport factor TAP/p115 has been implicated (Barroso *et al.*, 1995), as well as rab17, a member of the family of small GTPases that is specifically expressed in epithelial cells of kidney and intestine (Lütcke *et al.*, 1993). The transcytotic route seems to pass through several compartments also involved in other types of transport, for example, transcytosis of the polymeric immunoglobulin receptor from the basolateral plasma membrane passes through both basolateral as well as apical endosomal compartments before reaching the apical cell surface (Apodaca *et al.*, 1994; Barroso and Sztul, 1994).

4.3 Cell-surface peptidases in endocytotic and lysosomal compartments

Several early immunogold electron microscopy studies on brush border ectoenzymes reported a lysosomal localization of AP-N and DPP-IV as well as glycosidases (Fransen *et al.*, 1985, 1988; Hansen *et al.*, 1987; Lorenzsonn *et al.*, 1987). A likely explanation for this localization would be that it represents the ongoing turnover of cell-surface proteins by endocytosis followed by lysosomal degradation. The relatively dense labelling could be explained by a high resistance to proteolysis, resulting in a high lysosomal steady-state concentration. However, from earlier morphological studies on the localization of biosynthetically labelled glycoproteins, it had been proposed that a fraction of newly synthesised cell-surface ectoenzymes is transported from the Golgi complex directly to the lysosomes instead of to the brush border membrane (Blok *et al.*, 1984). A direct experimental demonstration of this pathway is difficult to provide, but using a biochemical approach, comprising metabolic labelling and subcellular fractionation of Caco-2 cells, Matter *et al.* (1990a) were able to show that 7–9% of newly synthesised DPP-IV (and sucrase-isomaltase) appeared in the lysosomes within 7–8 h. This would be a long enough time for newly made proteins to reach the apical cell surface first and then become endocytosed, but the biphasic appearance of these proteins in the lysosomal fraction indicated that a significant amount had arrived directly, bypassing the brush border membrane. Endocytosis was shown to occur as well, but, interestingly, recycling back to the apical surface rather than delivery to the lysosomes seemed to be the major fate of the endocytosed ectoenzymes (Matter *et al.*, 1990a). The crinophagic pathway in Caco-2 cells has been further characterized recently by Baricault *et al.* (1995). They showed that within 1 h, exposure to the adenylate cyclase activator forskolin led to a decrease in the cell-surface expression of newly synthesised DPP-IV and caused an intracellular accumulation of this peptidase as well as other cell-surface proteins in an autophagic-like compartment (see Chapter 13). In line with this, we had previously observed that forskolin reduced the surface expression of newly synthesised AP-N in cultured intestinal explants (Danielsen *et al.*, 1987). Autophagy is considered to be a general mechanism whereby a cell may degrade its components in order to meet the metabolic needs under various stress conditions (Seglen and Bohley, 1992). Its sensitivity to forskolin in Caco-2 cells, as well as in enterocytes would suggest that the re-routing of newly made ectoenzymes to the lysosomes occurs in a regulated manner rather than by accidental missorting along the exocytic pathway. If this is the case, we still need to find out how this re-routing is achieved.

During fetal and neonatal development, the enterocyte gradually develops an elaborate apical endocytotic complex of tubulo-vesicular endosomes, multivesicular bodies and finally a giant lysosomal vacuole (Wilson *et al.*, 1991). This organellar system, which rapidly disappears after weaning, enables the neonate to assimilate and digest milk protein and, as mentioned above to obtain passive immunity by transcytosis of maternal immunoglobulins (Rodewald, 1973). We recently observed by immunogold electron microscopy, a dense labelling for AP-N in all these apical endocytotic structures in fetal pig enterocytes (Danielsen *et al.*, 1995). The significant intracellular localization was confirmed biochemically, and pulse-chase labelling indicated that the bulk of newly made AP-N was transported to the apical surface before appearing in the endocytotic structures. Interestingly, we observed fetal intestinal AP-N to reside in glycolipid rafts together with a GPI-linked iron-binding protein, melanotransferrin. The latter protein was clustered in flat microdomains and membrane invaginations between adjacent microvilli at the apical surface and, like AP-N, was seen in the apical endocytotic structures as well, indicating that glycolipid rafts are involved in endocytosis in the developing enterocyte (Danielsen and van Deurs, 1995).

In summary, it can be concluded that the ectopeptidases, which we normally think of as typical markers for the apical cell surface, do have a more widespread subcellular distribution in epithelial cells. Along these lines, we have already argued in this chapter that their localization in various compartments, including both basolateral and apical surface domains, should be viewed as reflecting a dynamic steady-state of a rather complex web of membrane trafficking pathways, operating between the involved compartments.

4.4 A kinetic model of membrane trafficking in an epithelial cell

We have developed a kinetic model (*Figure 4.1*) for the membrane trafficking of apical cell-surface proteins which aims to accomodate the view expressed in the foregoing

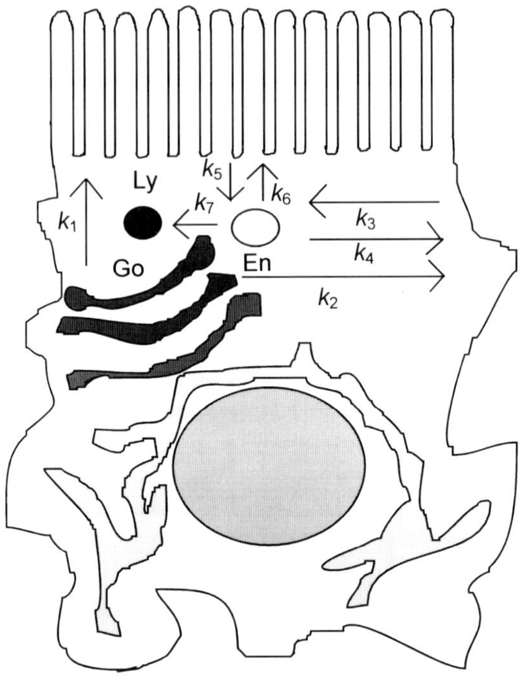

Figure 4.1. A kinetic model of membrane trafficking in an epithelial cell. The first-order rate constants of the various vesicular trafficking steps are marked $k_1 - k_7$. Ly, lysosome; En, endosome; Go, Golgi complex.

sections: that various compartments may not only be involved in the transport of these proteins at the time of biosynthesis but, to some extent, also during their 'adult' life. The model will be used to exemplify our considerations concerning the sorting mechanisms that operate in a given cell type to maintain the steady-state pool sizes of a membrane protein, such as an ectopeptidase, which, although predominantly localized at the apical cell surface, may also be found in other compartments.

The model assumes that the rate of vesicular transport between two compartments can be described by first-order kinetics, where the transport rate is proportional to the amount of a given protein in the donor compartment. Two parameters, the rates of vesicle loading and budding, combine to determine the size of the proportionality constant (the first-order rate constant) for a given transport step. Thus, a high rate constant reflects an effective recruitment of the protein into the budding vesicles combined with a high rate of budding. Furthermore, in a steady-state situation the synthesis and degradation rates are equal as, too, are the rates of transport to and from a compartment. Some of the first-order constants can be relatively easily extracted from experimental determinations of arrival and disappearance rates from the basolateral and apical membrane domains, using pulse-chase labelling in combination with surface biotinylation (L.K. Vogel, G.H. Hansen, B. Delmas, J. Møller, O. Norén and H. Sjöström, unpublished results). Using this approach for MDCK cells transfected with porcine AP-N, we obtained the following values: $k_1 = 1.0\ h^{-1}$, $k_2 = 0.3\ h^{-1}$, $k_3 = 0.4\ h^{-1}$ and $k_5 = 0.05\ h^{-1}$. In addition, the relative steady-state pool sizes of the compartments under consideration (Golgi, endosomal, basolateral and apical) according to the available biochemical and immunogold labelling data, suggest a ratio (G:E:B:A) of 1:9:90. Based on the above assumptions and experimental values, the remaining rate constants were then calculated (*Table 4.1*, data line 1). On the basis of this 'experimentally' obtained set of rate constants and relative compartment steady-state pool sizes, the model can be used for calculating how changes in the sets of rate constants affect the relative pool sizes of a given membrane protein in the Golgi, endosomal, basolateral and apical compartments. Data lines 2 and 3 show how changes of the rate contants in the transcytotic route drastically alter the basolateral:apical ratio. For instance, if no sorting occurs here ($k_3 = k_4$ and $k_5 = k_6$) a ratio of about 1:3 will be seen, a value well below the 1:9 ratio observed experimentally. Therefore, sorting in the TGN alone cannot account for the observed steady-state distribution of AP-N in MDCK cells; an additional sorting mechanism located either in the basolateral/endosomal or endosomal/ apical shuttle is required.

The G:E:B:A ratio derived from biochemical and immunogold labelling data (data line 1) may argue in favour of sorting in the endosomal/apical shuttle, where the relatively low value of k_5 would suggest that a major sorting occurs at the apical cell surface. Data line 4 of *Table 4.1* demonstrates that sorting in the apical/endosomal shuttle alone is fully capable of achieving the experimentally observed basolateral/apical ratio of about 1:9 ($k_1 = k_2$ and $k_5 = 0.05$; $k_6 = 0.6$). This even holds for cells which transport only minor amounts directly from the TGN to the apical plasma membrane ($k_1 = 0.05$; $k_2 = 0.8$; data line 5). Thus, different cell types (exemplified by MDCK, Caco-2 and hepatocytes) are able to remain roughly identical steady-state distributions of apically expressed membrane proteins despite greatly varying k_1/k_2 ratios. Finally, the last two lines of *Table 4.1* show that the turnover rate of a membrane protein (magnitude of transport rate to the lysosome, k_7) affects the intracellular (Golgi + endosomes) relative to the surface membrane (basolateral + apical) pool sizes, without significantly influencing the basolateral:apical ratio.

In summary, the model presented illustrates how a polarized distribution of a given membrane protein depends on many membrane trafficking pathways and it highlights

Table 4.1. Relationship between the rate constants (units: h^{-1}) and relative pool sizes in Golgi: endosomal: basolateral: apical compartments (G:E:B:A), based on the model shown in *Figure 4.1*[a]

k_1	k_2	k_3	k_4	k_5	k_6	k_7	G:E:B:A
1.0	0.3	0.4	0.4	0.05	0.4	0.15	1:9:9:90
1.0	0.3	0.4	0.4	0.05	**0.05**	0.15	1:9:9:29
1.0	0.3	0.4	**0.05**	0.05	**0.05**	0.15	1:9:2:29
0.5	**0.5**	0.4	0.4	0.05	**0.6**	0.15	1:7:8:90
0.05	**0.8**	0.4	0.4	0.05	**0.8**	0.15	1:6:8:92
1.0	0.3	0.4	0.4	0.05	0.4	**0.8**	1:2:2:33
1.0	0.3	0.4	0.4	0.05	0.4	**0.05**	1:2:27:228

[a] The set of values in data line 1 are derived from experimental data. Bold entries indicate changes in rate constants.

the potential importance of traffic regulation through the endosomal compartment. For simplicity, the model shows only one such compartment, but evidence suggests that the transcytotic route passes through separate basolateral and apical endosomes. Obviously, such a complexity further increases the ability of the cell to modulate the steady-state distribution of its membrane components. The future challenge to cell biologists will be to unravel the molecular mechanisms operating these membrane trafficking pathways.

References

Anderson RGW, Kamen BA, Rothberg KG, Lacey SW. (1992) Potocytosis: sequestration and transport of small molecules by caveolae. *Science* **255:** 410–411.

Apodaca G, Katz LA, Mostov KE. (1994) Receptor-mediated transcytosis of IgA in MDCK cells is via apical recycling endosomes. *J. Cell Biol.* **125**: 67–86.

Baricault L, Fransen JAM, Garcia M, Sapin C, Codogno P, Ginsel L, Trugnan G. (1995) Rapid sequestration of DPP IV/CD26 and other cell surface proteins in an autophagic-like compartment in Caco-2 cells treated with forskolin. *J. Cell Sci.* **108:** 2109–2121.

Barroso M, Sztul ES. (1994) Basolateral to apical transcytosis in polarized cells is indirect and involves BFA and trimeric G protein sensitive passage through the apical endosome. *J. Cell Biol.* **124:** 83–100.

Barroso M, Nelson D, Sztul E. (1995) Transcytosis-associated protein (TAP)/p115 is a general fusion factor required for binding of vesicles to acceptor membranes. *Proc. Natl Acad. Sci. USA* **92**: 527–531.

Bartles JR, Feracci HM, Stieger B, Hubbard AL. (1987) Biogenesis of the rat hepatocyte plasma membrane *in vivo:* comparison of the pathways taken by apical and basolateral proteins using subcellular fractionation. *J. Cell Biol.* **105**: 1241–1251.

Blok J, Fransen JAM, Ginsel LA. (1984) Turnover of brush border glycoproteins in human intestinal absorptive cells: do lysosomes have a regulatory function? *Cell Biol. Int. Rep.* **8:** 993–1014.

Brändli AW, Parton RG, Simons K. (1990) Transcytosis in MDCK cells: identification of glycoproteins transported bidirectionally between both plasma membrane domains. *J. Cell Biol.* **111**: 2909–2921.

Brown DA, Rose JK. (1992) Sorting of GPI-anchored proteins to glycolipid-enriched membrane subdomains during transport to the apical cell surface. *Cell* **68**: 533–544.

Brown D, Stow JL. (1996) Protein trafficking and polarity in kidney epithelium: from cell biology to physiology. *Physiol. Rev.* **76:** 245–297.

Brown DA, Crise B, Rose JK. (1989) Mechanism of membrane anchoring affects polarized expression of two proteins in MDCK cells. *Science (Wash.)* **245**: 1499–1501.

Cardone MH, Smith BL, Mennitt PA, Mochly-Rosen D, Silver RB, Mostov KE. (1996) Signal transduction by the polymeric immunoglobulin receptor suggests a role in regulation of receptor transcytosis. *J. Cell Biol.* **133:** 997–1005.

Corbeil D, Boileau G, Lemay G, Crine P. (1992) Expression and polarized apical secretion in Madin-Darby canine kidney cells of a recombinant soluble form of neutral endopeptidase lacking the cytosolic and transmembrane domains. *J. Biol. Chem.* **267:** 2798–2801.

Creutz CE. (1992) The annexins and exocytosis. *Science (Wash.)* **258:** 924–931.

Danielsen EM. (1995) Involvement of detergent-insoluble complexes in the intracellular transport of intestinal brush border enzymes. *Biochemistry* **34:** 1596–1605.

Danielsen EM, van Deurs B. (1995) A transferrin-like GPI-linked iron-binding protein in detergent-insoluble noncaveolar microdomains at the apical surface of fetal intestinal epithelial cells. *J. Cell Biol.* **131**: 939–950.

Danielsen EM, Hansen GH, Cowell GM. (1987) Biosynthesis of intestinal microvillar proteins. Forskolin reduces surface expression of aminopeptidase N. *Eur. J. Cell Biol.* **44:** 273–277.

Danielsen EM, Hansen GH, Niels-Christiansen, L. (1995) Localization and biosynthesis of aminopeptidase N in pig fetal small intestine. *Gastroenterology* **109:** 1039–1050.

Dupree P, Parton RG, Raposo G, Kurzchalia TV, Simons K. (1993) Caveolae and sorting in the *trans*-Golgi network of epithelial cells. *J. Cell Sci.* **12:** 1597–1605.

Fiedler K, Simons K. (1995) The role of N-glycans in the secretory pathway. *Cell* **81**: 309–312.

Fiedler K, Kobayashi T, Kurzchalia TV, Simons K. (1993) Glycosphingolipid-enriched detergent-insoluble complexes in protein sorting in epithelial cells. *Biochemistry* **32:** 6365–6373.

Fiedler K, Parton RG, Kellner R, Etzold T, Simons K. (1994) VIP36, a novel component of glycolipid rafts and exocytic carrier vesicles in epithelial cells. *EMBO J.* **13:** 1729–1740.

Fiedler K, Lafont F, Parton RG, Simons K. (1995) Annexin XIIIb: a novel epithelial specific annexin is implicated in vesicular traffic to the apical plasma membrane. *J. Cell Biol.* **128:** 1043–1053.

Fransen JA, Ginsel LA, Hauri HP, Sterchi E, Blok J. (1985) Immuno-electron microscopical localization of a microvillus membrane disaccharidase in the human small-intestinal epithelium with monoclonal antibodies. *Eur. J. Cell Biol.* **38**: 6–15.

Fransen JA, Ginsel LA, Cambier PH, Klumperman J, Oude Elferink RP, Tager JM. (1988) Immunocytochemical demonstration of the lysosomal enzyme alpha-glucosidase in the brush border of human intestinal epithelial cells. *Eur. J. Cell Biol.* **47**: 72–80.

Garcia M, Mirre C, Quaroni A, Reggio H, Le Bivic A. (1993) GPI-anchored proteins associate to form microdomains during their intracellular transport in Caco-2 cells. *J. Cell Sci.* **104:** 1281–1290.

Gottlieb TA, Beaudry G, Rizzolo L, Colman A, Rindler M, Adesnik M. Garoff H. (1986) Secretion of endogenous and exogenous proteins from polarized MDCK cell monolayers. *Proc. Natl Acad. Sci. USA* **83**: 2100–2104.

Gruenberg J, Emans N. (1993) Annexins in membrane traffic. *Trends Cell Biol.* **3:** 224–227.

Hansen GH, Sjöström H, Norén O, Dabelsteen E. (1987) Immunomicroscopic localization of aminopeptidase N in the pig enterocyte. Implications for the route of intracellular transport. *Eur. J. Cell Biol.* **43**: 253–259.

Harder T, Gerke V. (1994) The annexin II2p11(2) complex is the major protein component of the Triton X-100-insoluble low-density fraction prepared from MDCK cells in the presence of Ca^{2+}. *Biochim. Biophys. Acta* **1223:** 375–382.

Hooper NM, Turner AJ. (1988) Ectoenzymes of the kidney microvillar membrane. Differential solubilization by detergents can predict a glycosyl-phosphatidylinositol membrane anchor. *Biochem. J.* **250:** 865–869.

Ikonen E, Tagaya M, Ullrich O, Montecucco C, Simons K. (1995) Different requirements for NSF, SNAP, and Rab proteins in apical and basolateral transport in MDCK cells. *Cell* **81:** 571–580.

Kondor-Koch C, Bravo R, Fuller SD, Cutler D. Garoff H. (1985) Exocytotic pathways exist to both the apical and the basolateral cell surface of the polarized epithelial cell MDCK. *Cell* **43**: 297–306.

Lanctôt C, Fournier H, Howell S, Boileau G, Crine P. (1995) Direct targeting of neutral endopeptidase (EC 3.4.24.11) to the apical cell surface of transfected LLC-PK1 cells and unpolarized secretion of its soluble form. *Biochem. J.* **305**: 165–171.

Le Bivic A, Quaroni A, Nichols B, Rodriguez-Boulan E. (1990). Biogenetic pathways of plasma membrane proteins in Caco-2, a human intestinal epithelial cell line. *J. Cell Biol.* **111:** 1351–1361.

Le Gall AH, Yeaman C, Muesch A, Rodriguez-Boulan E. (1995) Epithelial cell polarity: new perspectives. *Sem. Nephrol.* **15:** 272–284.

Lisanti MP, Caras IW, Dovitz MA, Rodriguez-Boulan E. (1989) A glycophospholipid membrane anchor serves as an apical targeting signal in polarized epithelial cells. *J. Cell Biol.* **109:** 2145–2156.

Lisanti MP, Tang ZL, Sargiacomo M. (1993) Caveolin forms a hetero-oligomeric protein complex that interacts with an apical GPI-linked protein: implications for the biogenesis of caveolae. *J. Cell Biol.* **123:** 595–604.

Lorenzsonn V, Korsmo H, Olsen WA. (1987) Localization of sucrase-isomaltase in the rat enterocyte. *Gastroenterology* **92**: 98–105.

Low M. (1989) The glycosyl-phosphatidylinositol anchor of membrane proteins. *Biochim. Biophys. Acta* **988:** 427–454.

Lütcke A, Jansson S, Parton RG, Chavrier P, Valencia A, Huber LA, Lehtonen E, Zerial M. (1993) Rab17, a novel small GTPase, is specific for epithelial cells and is induced during cell polarization. *J. Cell Biol.* **121**: 553–564.

Matter K, Mellman I. (1994) Mechanisms of cell polarity: sorting and transport in epithelial cells. *Curr. Opin. Cell Biol.* **6:** 545–554.

Matter K, Stieger B, Klumperman J, Ginsel L, Hauri H-P. (1990a) Endocytosis, recycling, and lysosomal delivery of brush border hydrolases in cultured human intestinal epithelial cells (Caco-2). *J. Biol. Chem.* **265:** 3503–3512.

Matter K, Brauchbar M, Bucher K, Hauri H-P. (1990b) Sorting of endogeneous plasma membrane proteins occurs from two sites in cultured human intestinal epithelial cells (Caco-2). *Cell* **60:** 429–437.

Mellman I, Yamamoto E, Whitney JA, Kim M, Hunziker W, Matter K. (1993) Molecular sorting in polarized and non-polarized cells: common problems, common solutions. *J. Cell Sci.* **17:** 1–7.

Müsch A, Xu H, Shields D, Rodriguez-Boulan E. (1996) Transport of vesicular stomatitis virus G protein to the cell surface is signal mediated in polarized and nonpolarized cells. *J. Cell Biol.* **133**: 543–558.

Nelson WJ. (1992) Regulation of cell surface polarity from bacteria to mammals. *Science (Wash.)* **258**: 948–955.

Palade GE. (1975) Intracellular aspects of the process of protein secretion. *Science (Wash.)* **189**: 347–358.

Parton RG, Simons K. (1995) Digging into caveolae. *Science* **269**: 1398–1399.

Rodewald R. (1973) Intestinal transport of antibodies in the newborn rat. *J. Cell Biol.* **58:** 189–211.

Sargiacomo M, Sudol M, Tang ZL, Lisanti MP. (1993) Signal transducing molecules and glycosyl-phosphatidylinositol-linked proteins form a caveolin-rich insoluble complex in MDCK cells. *J. Cell Biol.* **122:** 789–807.

Scheiffele P, Peranen J, Simons K. (1995) N-glycans as apical sorting signals in epithelial cells. *Nature* **378**: 96–98.

Seglen PO, Bohley P. (1992) Autophagy and other vacuolar protein degradation mechanisms. *Experimentia* **48**: 158–172.

Simons K, Fuller SD. (1985) Cell surface polarity in epithelia. *Ann. Rev. Cell Biol.* **1:** 243–288.

Simons K, van Meer G. (1988) Lipid sorting in epithelial cells. *Biochemistry* **27:** 6197–6202.

van Meer G, Simons K. (1982) Viruses budding from either the apical or the basolateral plasma membrane domain of MDCK cells have unique phospholipid compositions. *J. Cell Sci.* **1:** 847–852.

van Meer G, Simons K. (1986) The function of tight junctions in maintaining differences in lipid composition between the apical and the basolateral cell surface domains of MDCK cells. *J. Cell Sci.* **5:** 1455–1464.

Vogel LK, Spiess M, Sjöström H, Norén O. (1992) Evidence for an apical sorting signal on the ectodomain of human aminopeptidase N. *J. Biol. Chem.* **267**: 2794–2797.

Vogel LK, Norén O, Sjöström H. (1995) Transcytosis of aminopeptidase N in Caco-2 cells is mediated by a non-cytoplasmic signal. *J. Biol. Chem.* **270**: 22933–22938.

Weinman JS, Feinberg JM, Rainteau DP, Gaspera BD, Weinman SJ. (1994) Annexins in rat enterocyte and hepatocyte: an immunogold electron-microscope study. *Cell Tissue Res.* **278:** 389–397.

Weisz OA, Machamer CE, Hubbard AL. (1992) Rat liver dipeptidylpeptidase IV contains competing apical and basolateral targeting information. *J. Biol. Chem.* **267**: 22282–22288.

Wilson JM, Whitney JA, Neutra MR. (1991) Biogenesis of the apical endosome-lysosome complex during differentiation of absorptive epithelial cells in rat ileum. *J. Cell Sci.* **100:** 133–143.

Yoshimori T, Keller P, Roth MG, Simons K. (1996) Different biosynthetic transport routes to the plasma membrane in BHK and CHO cells. *J. Cell Biol.* **133:** 247–256.

Zurzolo C, van't Hof W, van Meer G, Rodriguez-Boulan E. (1994) VIP21/caveolin, glycosphingolipid clusters and the sorting of glycosylphosphatidylinositol-anchored proteins in epithelial cells. *J. Cell Sci.* **13:** 42–53.

Chapter 5

The chemistry and pharmacology of cell-surface peptidase inhibitors

Ann Beaumont, Marie-Claude Fournié-Zaluski, Florence Noble, Rafael Maldonado and Bernard P. Roques

5.1 Introduction

Although the number of known cell-surface peptidases has not grown by alarming proportions over the last decade, and even if the rate of discovery of their potential substrates has slowed down to manageable levels, the need for potent selective inhibitors of each enzyme has not diminished. These inhibitors are required to establish the exact role of each enzyme in a given tissue, given that many have broad specificities, and to ascertain the consequences of inhibiting that enzyme, and thereby increasing endogenous peptide levels. An inhibitor should not only be potent and selective, but it should also have a long half-life *in vivo* and be able to reach its target enzyme. In addition to the need for selective inhibitors, the concept of 'mixed' inhibitors, capable of inhibiting two (dual) or even three (triple) enzymes, has become well-established over the last few years. Mixed inhibitors have applications where a peptide is metabolized by more than one peptidase or where it is thought desirable to change the endogenous levels of more than one peptide. This is a particularly challenging field: not only should an ideal mixed inhibitor be equipotent for the enzymes chosen, but selectivity should not be lost *vis-à-vis* other related enzymes. In addition to providing valuable insights into the roles of peptides and their metabolizing enzymes, some of these molecules are in clinical use as antihypertensives and antidiarrhoeal agents and there is a potential future for other inhibitors as novel analgesics and in the treatment of severe hypertension and heart failure.

In this chapter the synthesis and pharmacology of inhibitors of five cell-surface peptidases are considered, angiotensin-converting enzyme (ACE; EC 3.4.15.1), endopeptidase-24.11 (E-24.11; neprilysin; EC 3.4.24.11), endothelin-converting enzyme (ECE), aminopeptidase N (AP-N; EC 3.4.11.2) and aminopeptidase A (AP-A; EC 3.4.11.7). Some of the *in vivo* substrates of these enzymes and their cleavage sites are shown in *Figure 5.1*. In the central nervous system, the opioid peptides, Met and Leu enkephalins are rapidly metabolized by both AP-N and E-24.11, which cleave the pentapeptides between their Tyr1–Gly2 and Gly3–Phe4 bonds respectively (see Chapter 19). Inhibitors of these enzymes, by increasing the half-lives of the enkephalins induce significant 'morphine-like' effects without many of the side-effects of opiate drugs (Roques *et al.,* 1993). E-24.11 also inactivates atrial natriuretic peptide, a circulating

Cell-Surface Peptidases in Health and Disease, edited by A.J. Kenny and C.M. Boustead.

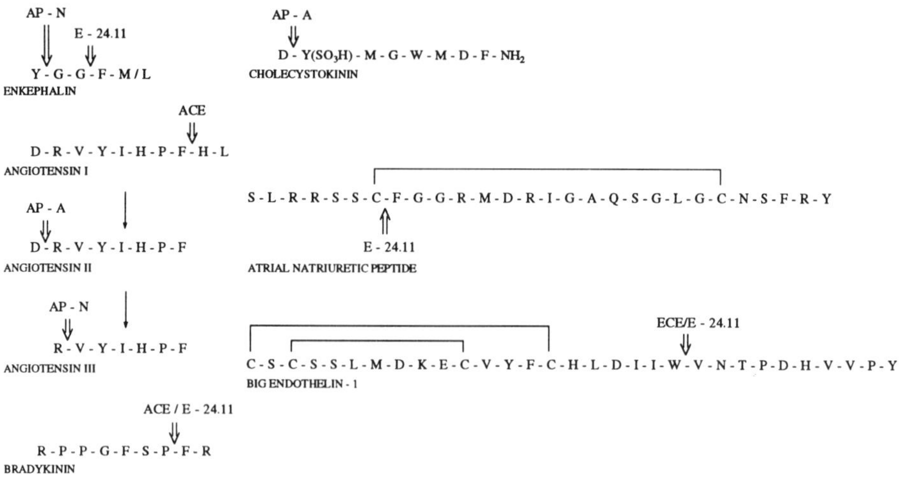

Figure 5.1. Hydrolysis of biologically active peptides involved in pain and cardiovascular control by cell-surface peptidases.

peptide which induces natriuresis and diuresis, by hydrolysing its Cys7–Phe8 bond (see Chapter 20). ACE removes the C-terminal dipeptide from the inactive peptide angiotensin I, releasing the hypertensive peptide angiotensin II (see Chapter 7), and both ACE and E-24.11 hydrolyse the hypotensive peptide bradykinin, mainly at Pro7–Phe8. ACE inhibitors have been in clinical use as antihypertensives for several years and it is hoped that combining ACE and E-24.11 inhibition will improve this form of treatment and extend it to heart failure (Roques and Beaumont, 1990). Inhibitors of ECE are also of potential interest in the cardiovascular field as this enzyme releases the vasoconstrictive peptide endothelin-1 (ET-1) from its inactive precursor big ET-1, by cleaving the Val21–Trp22 bond (see Chapter 9). E-24.11 may also have a role in the formation of endothelin (Lebel *et al.,* 1996) and interest is also being shown in the development of dual ECE/E-24.11 or triple ECE/E-24.11/ACE inhibitors. Selective inhibitors of AP-A, which have only recently become available, have shown that in the brain this enzyme hydrolyses the octapeptide cholecystokinin (CCK-8) (Migaud *et al.,* 1996) as well as converting angiotensin II to angiotensin III, which is in turn metabolized by AP-N (Zini *et al.,* 1996).

5.2 Inhibitors of zinc metallopeptidases

5.2.1 Inhibitor design

E-24.11, ACE, ECE, AP-N and AP-A are zinc metallopeptidases and the design of their inhibitors has often relied on structure–activity studies, based on active site models derived from structural data on thermolysin (EC 3.4.24.27) (see Roques and Fournié-Zaluski, 1986; Roques *et al.,* 1993; Thorsett and Wyvratt, 1987). Thermolysin has been crystallized with a variety of inhibitors (reviewed in Matthews, 1988) and its active site (*Figure 5.2*) contains a single essential zinc atom ligated by a water molecule and three residues, two of which, His142 and His146 form part of a zinc peptidase consensus sequence, His–Glu–Xaa–Xaa–His, whilst the third zinc-binding ligand, Glu166, forms part of a Glu–Xaa–Xaa–Xaa–Asp sequence. Glu143 acts as a general base in catalysis (Matthews, 1988) and Asp170 is involved in a hydrogen-bonding network important for catalytic activity (Marie-Claire *et al.,* 1997). The specificity of thermolysin is ensured by its S'_1 subsite, which accepts large hydrophobic residues such as Leu, Phe and Ile, and substrates and inhibitors can also interact with residues in the S_1, S_2 and S'_2 subsites. Other active site residues include Asn113 and Arg203, involved in substrate binding and His231, which stabilizes the transition state. Sequence data and site-directed mutagenesis studies (see Chapter 6) have confirmed the validity of using the thermolysin-based model for the mammalian enzymes. All of them contain the His–Glu–Xaa–Xaa–His consensus sequence and have a C-terminally situated Glu as the third zinc ligand which, in E-24.11, ECE and ACE also forms part of a Glu–Xaa–Xaa–Xaa–Asp sequence (Roques, 1993). The active site of E-24.11 appears to be very similar to that of thermolysin (see *Figure 5.2*) (Roques *et al.*, 1993 and

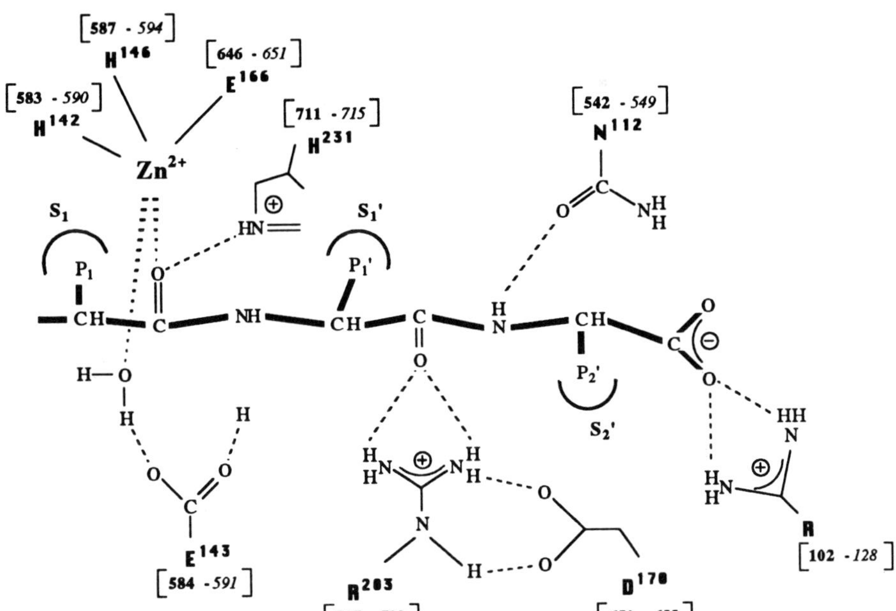

Figure 5.2. A schematic representation of the active site of thermolysin showing His142, His146 and Glu166 (zinc-binding), Glu143 (catalysis), His231 (transition-state binding), Asp170 (hydrogen bond network) and Asn112 and Arg230 (substrate-binding). Given in parentheses (bold) are the numbers of the residues shown by site-directed mutagenesis to be the equivalent residues of human E-24.11 (for references see Roques *et al.*, 1993; C. Marie-Claire, unpublished observations; see Chapter 6) and, in italics, the equivalent conserved residues of human ECE-1a.

Chapter 6). The enzymes have similar specificities and, in addition to the residues mentioned above, the active site of E-24.11 also contains residues corresponding to Asn113, Arg203 and His231. These residues are also conserved in ECE (*Figure 5.2* and Chapter 9). The only active site residue of E-24.11 so far identified with no thermolysin equivalent is Arg102, whose positioning is thought to confer the peptidyl dipeptidase activity that the mammalian enzyme displays with certain substrates, in contrast with thermolysin which is a strict endopeptidase.

Effective thermolysin inhibitors are molecules capable of interacting with at least the S'_1 subsite and which have a strong zinc-chelating group, such as a phosphoramidate, phosphonate, thiol, hydroxamate or carboxylate (reviewed in Matthews, 1988). These groups have been employed for inhibitors of the other zinc metallopeptidases, with their effectiveness varying from one enzyme to another. Thus thiol-containing inhibitors are generally more potent than carboxyl-containing inhibitors for E-24.11, but are equipotent for ACE and, in contrast to E-24.11, hydroxamate-containing inhibitors have low affinities for ACE (Roques *et al.*, 1993). Phosphonate or phosphoramidate-containing molecules, considered to be transition-state inhibitors (Matthews, 1988), give high potency against all these zinc metallopeptidases. Although the similarities in the active sites of the zinc metallopeptidases can make specificity difficult to achieve (currently a problem for ECE inhibition) this can be overcome by optimizing interactions with the relevant subsites and exploiting stereochemical preferences. The active site similarities have also led to the idea that mixed inhibitors can be developed by selecting a residue able to interact with at least one subsite of each enzyme, with the strong zinc-chelating group counterbalancing a less-than-perfect fit to other subsites (Fournié-Zaluski *et al.*, 1984, 1985).

5.2.2 ACE inhibitors

Potent dipeptide and tripeptide inhibitors of ACE have been synthesised with thiol, carboxylate or phosphate groups as zinc chelators. The salient features of these molecules are generally a small hydrophobic residue in the P'_1 position and a negative charge in P'_2, often borne by a proline residue, whose positioning is important for inhibition, in line with the peptidyl dipeptidase activity of the enzyme. The active site of ACE can also accommodate large constrained molecules in the S'_1–S'_2 subsites and lipophilic chains in the S_1 subsite. For reasons of space, the structures of only two of the most commonly used molecules are shown in *Table 5.1*, the mercapto-containing captopril (Ondetti *et al.*, 1977) and the carboxylate-containing enalaprilat (Patchett *et al.*, 1980). Fuller details can be found in Thorsett and Wyratt (1987) and Lawton *et al.* (1992).

5.2.3 E-24.11 inhibitors

As illustrated in *Table 5.1*, E-24.11 inhibitors have been synthesised with thiol, phosphonate, phosphonamide, hydroxamate and carboxylate groups as zinc-chelating agents. A hydrophobic group is normally incorporated to fit the S'_1 subsite, whilst the S'_2 subsite accommodates a wide variety of residues but, unlike ACE, proline is not

Table 5.1. Selective inhibitors of ACE and E-24.11

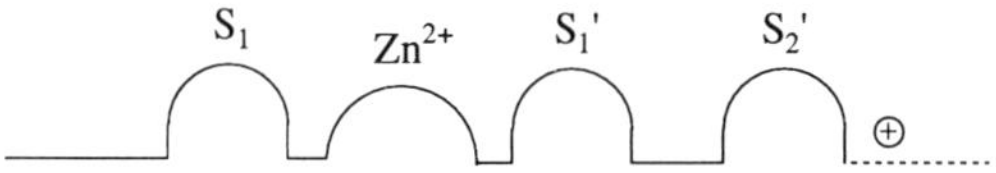

		K_i/IC_{50} (nM)	
		ACE	E-24.11
Captopril	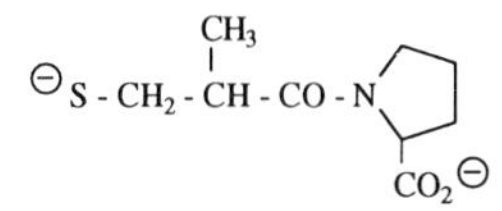	1.7	>10 000
Enalaprilat	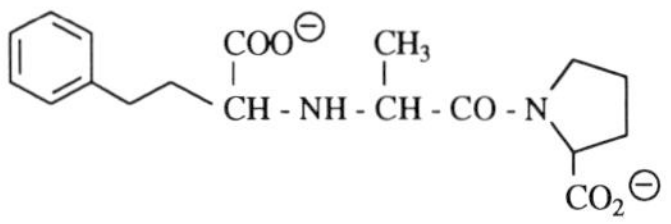	1.2	>10 000
Thiorphan	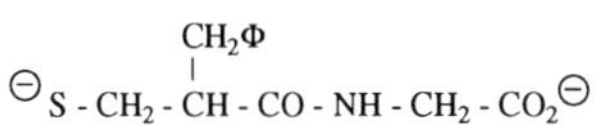	140	2.2
Retrothiorphan	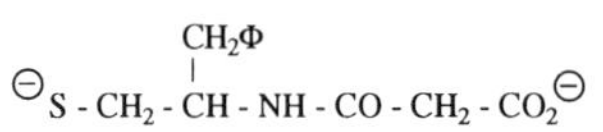	>10 000	2.3
RB 104	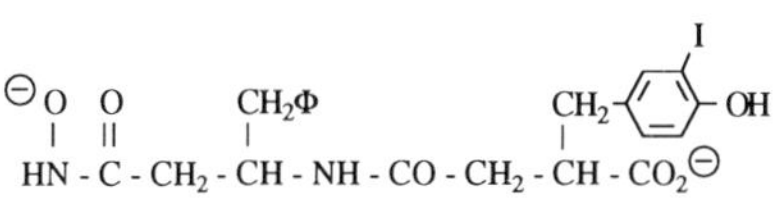	>10 000	0.03
UK 73 967	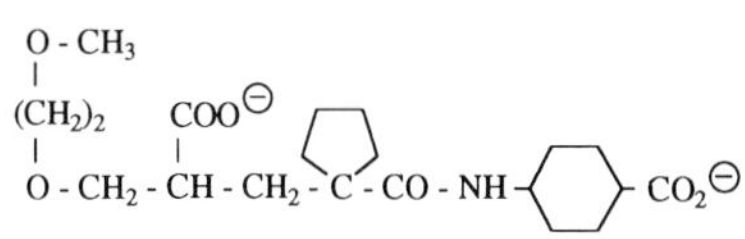	n.d.	14
CGS–24 592	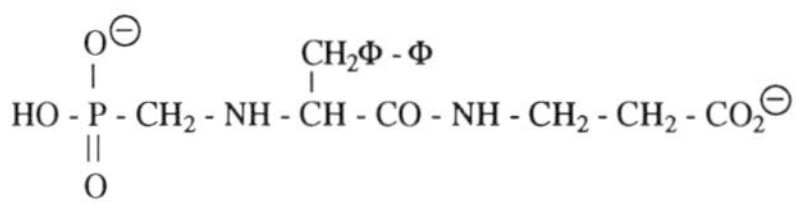	>100 000	2.6

n.d., not determined.

accepted. A free carboxyl group in P'_2 is advantageous for E-24.11 inhibition, but its positioning is not as critical as in ACE inhibition. Large molecules constrained in P'_1 and P'_2 are also accepted by the active site of E-24.11. Comprehensive reviews have dealt with the development of selective E-24.11 inhibitors (Roques and Fournié-Zaluski, 1986; Roques *et al.*, 1993; Thorsett and Wyvratt, 1987) and only a few key molecules are discussed here.

The thiol-containing inhibitor thiorphan (K_i = 2 nM; *Table 5.1*), which was the first synthetic inhibitor developed for E-24.11 (Roques *et al.*, 1980) is also a reasonably potent ACE inhibitor (K_i = 140 nM). However its analogue, retrothiorphan (R) (*Table 5.1*) retains a good affinity for E-24.11 but not for ACE (Fournié-Zaluski *et al.*, 1984) and illustrates how stereochemical preferences can be used to achieve selective inhibition. The thiol inhibitors are relatively hydrophilic and their bioavailability can be improved by protecting the thiol and carboxyl hydrophilic groups (as in acetorphan which is rapidly transformed to thiorphan by esterases in the brain and blood; Lecomte *et al.*, 1986).

N-terminal hydroxamate-containing molecules are also good E-24.11 inhibitors, with the advantage of having K_i values for ACE of more than 10 000 nM (Fournié-Zaluski *et al.*, 1985). One of these inhibitors, RB 104, is the most potent E-24.11 inhibitor described so far (*Table 5.1*; Fournié-Zaluski *et al.*, 1992c) and its radiolabelled derivative can be used as a sensitive probe for detecting the enzyme from different tissues (Solheilac *et al.*, 1992). Unlike ACE, carboxyl-containing inhibitors of E-24.11 have approximately 10-fold lower affinities than their corresponding thiol inhibitors, although several have been developed (such as UK 73, 967; Danilewicz *et al.*, 1989; *Table 5.1*) as prodrugs can be obtained by esterifying the carboxyl function with lipophilic alcohols (Chipkin *et al.*, 1988). The naturally occurring phosphoramidate inhibitor, phosphoramidon (*Table 5.2*), which has a K_i of 2 nM for E-24.11 (Kenny, 1977), was first described as a thermolysin inhibitor (K_i = 40 nM) and has been co-crystallized with the bacterial enzyme (Tronrud *et al.*, 1986). Potent phosphinic acid-containing inhibitors of E-24.11 have also been developed, such as CGS 24592 (*Table 5.1*; De Lombaert *et al.*, 1993).

5.2.4 Selective and mixed ECE inhibitors

ECE was first identified as a phosphoramidon-sensitive enzyme (Ikegawa *et al.*, 1990). However, phosphoramidon (*Table 5.2*) has K_i values of 1 μM for bovine adrenal ECE-1 (Emoto and Yanagisawa, 1995) and 2 nM for E-24.11 (Kenny, 1977) and to date there are no highly selective ECE inhibitors. In addition, widely different values are often reported for ECE inhibition, which may be partly due to the different assay systems used but also to the existence of isoenzymes, such as ECE-2, which may have an intracellular location and is also inhibited by phosphoramidon (K_i 4 nM; Emoto and Yanagisawa, 1995).

As both ECE and E-24.11 accept hydrophobic residues in S'_1, efforts to produce selective ECE inhibitors have concentrated mainly on the S'_2, S_1 and S_2 subsites. Phosphoramidon has been the starting point for most studies, and phosphinic acid derivatives have been favoured as they are less acid labile and more stable *in vivo* than phosphoramidates. Replacing the rhamnose moiety of phosphoramidon by hydrophobic groups such as naphthalene di-carboxyimidoethyl in the methylphosphinic acid inhibitor (no. 2 in *Table 5.2*) increased affinity (IC_{50} = 0.052 μM vs. IC_{50} phosphoramidon = 0.14 μM) and gave a 20-fold selectivity over E-24.11 inhibition (Keller *et al.*, 1996). In two other series of phosphinic acid derivatives, the most effective P'_2 residue was Trp (Chackalamannil *et al.*, 1996; Lloyd *et al.*, 1996), and Lys or Leu were found to be good P_2 residues (Chackalamannil *et al.*, 1996). Amongst these inhibitors no. 3 in *Table 5.2* has an IC_{50} of 25 nM for ECE (Chackalamannil *et al.*, 1996). The phosphinic acid-containing CGS 26303 (no. 4 in *Table 5.2*) has been described as a dual ECE/E-24.11 inhibitor, although with respective IC_{50} values of 1.1 μM and 0.9 nM it has more than a 1000-fold selectivity for the latter enzyme (De Lombaert *et al.*, 1994a). A prodrug of this molecule has been designated CGS 26393 (Trapani *et al.*, 1995). Another phosphinic acid inhibitor (no. 5 in *Table 5.2*) has been reported as a triple inhibitor of ECE, ACE and E-24.11, respectively, with nanomolar affinities for all three enzymes (McKittrick *et al.*, 1996).

Table 5.2. Selective and mixed inhibitors of ECE

		K_i/IC_{50} (nM)		
	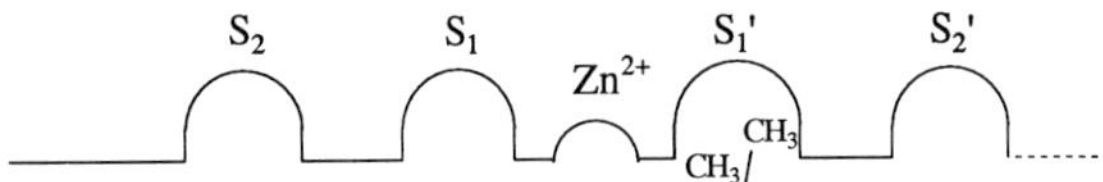	ECE	E-24.11	ACE
1 Phosphoramidon	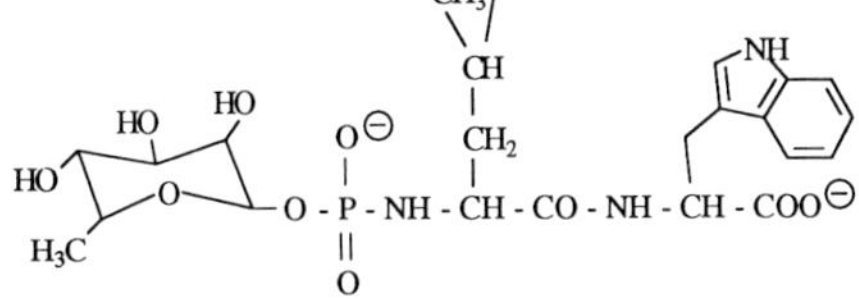	1000	2.0	>10 000
2	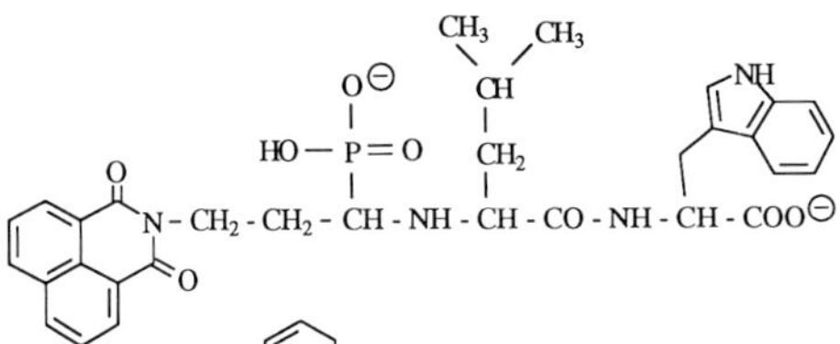	52	n.d.	n.d
3	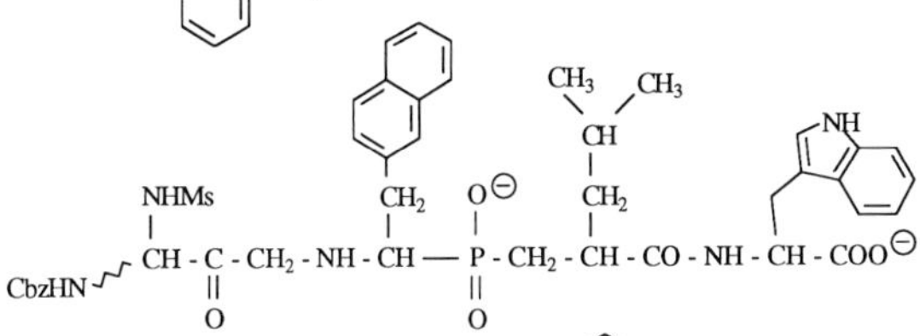	25	13% inhibition at 300 nM	n.d
4 CGS-26 303	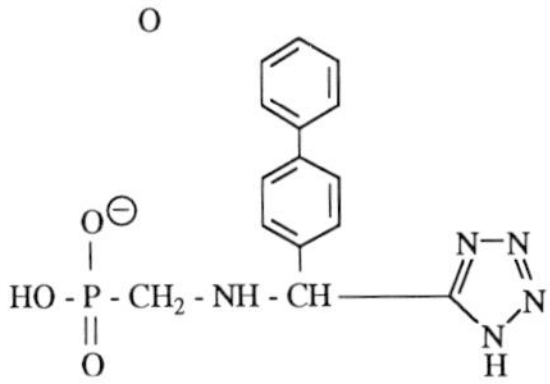	1000	0.9	n.d.
5	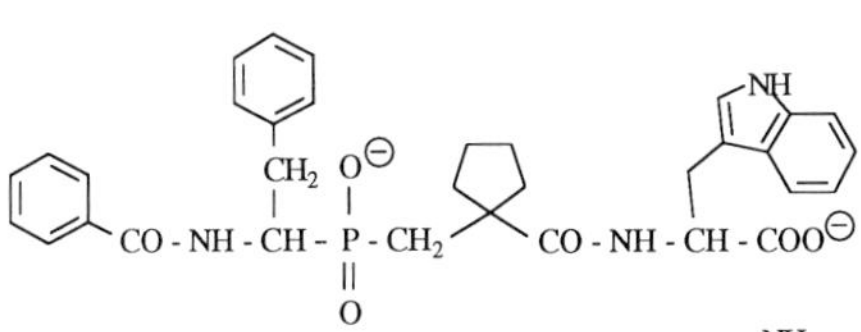	50	55	1.5
6 ST 55	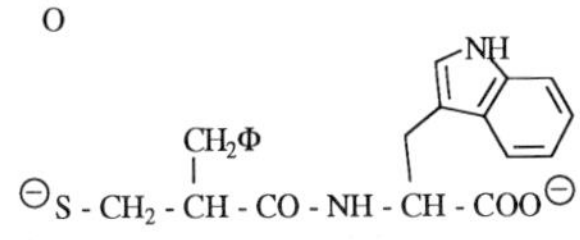	140	3.0	150

n.d., not determined.

Thiorphan (Emoto and Yanigasawa, 1995) and retrothiorphan (Kukkola *et al.,* 1996; *Table 5.1*) have K_i values of more than 100 μM for ECE-1, although several thiol inhibitors have been developed with affinities in the low micromolar range (Balwierczak *et al.*, 1995; Kukkola *et al.*, 1996). ST 55, a thiorphan derivative with Trp in P'_2 (6 in *Table 5.2*) has an IC_{50} for ECE of 140 nM and also good affinities for E-24.11 and ACE (M.-C. Fournié-Zaluski, S. Turcaud and B.P. Roques, unpublished observations), making it a useful lead compound for the futher development of selective ECE, or mixed inhibitors.

5.2.5 AP-N and AP-A inhibitors

AP-N preferentially removes N-terminal hydrophobic residues from substrates, its S_1 subsite optimally accommodating side chains of moderate bulk which have a certain degree of freedom (Fournié-Zaluski *et al.*, 1992a), whilst AP-A is more selective, cleaving N-terminal Glu and Asp residues. The exopeptidase specificity of both enzymes is thought to be ensured by the positioning of a negatively charged amino acid which interacts with the N-terminal α-amino group of the substrate (*Table 5.3*). Thiol groups have been shown to be more efficient zinc chelators for AP-N than carboxylate, hydroxamate or phosphonate groups (Fournié-Zaluski *et al.*, 1992a) and from structure–activity studies with β-amino-acid derivatives several molecules have been developed, such as Leu-thiol (*Table 5.3*; Chan, 1983) and PC 18 (*Table 5.3*; Fournié-Zaluski *et al.*, 1992a), which are more potent than the commonly used bestatin (K_i = 4 μM; Thorsett and Wyvratt, 1987).

Amastatin and Glu-thiol (*Table 5.3*), which have both been described as AP-A inhibitors, have similar or better potencies for AP-N (Wilk and Thurston, 1990). However, replacing the side-chain carboxylate group of Glu-thiol with a sulphonate group led to the development of EC 33 (*Table 5.3*) which has a 100-fold selectivity for AP-A over AP-N (Chauvel *et al.*, 1994).

5.2.6 Dual inhibitors of E-24.11 and AP-N

The first dual E-24.11 and AP-N inhibitors developed were hydroxamate-containing molecules such as kelatorphan, which also has a good affinity for a dipeptidyl aminopeptidase activity (*Table 5.3*; Fournié-Zaluski *et al.*, 1984). However, although these molecules gave promising results in numerous pharmacological tests (Roques *et al.*, 1993), they are too hydrophilic to cross the blood–brain barrier, precluding peripheral administration. Another strategy was therefore employed which led to the synthesis of RB101, whose disulphide bridge links the AP-N inhibitor PC 18 with the E-24.11 inhibitor ST 43 (*Table 5.3*). RB101, which is more hydrophobic than its constituent parts, crosses the blood–brain barrier and the disulphide bond, which is relatively stable in plasma, is rapidly reduced in the brain (Fournié-Zaluski *et al.*, 1992b). Recently phosphinic acid derivatives have been developed as mixed inhibitors (H. Chen, M.-C. Fournié-Zaluski and B.P. Roques, unpublished results) as exemplified by CH 23 (*Table 5.3*). These molecules can be considered to be the first true dual inhibitors of E-24.11 and AP-N, inasmuch as they incorporate, in a single molecule, groups designed to fit the important subsites of both the exopeptidase (S_1) and the endopeptidase (S'_1 and S'_2) and this is reflected in high affinities for both enzymes.

5.2.7 Dual inhibitors of ACE and E-24.11

For dual ACE and E-24.11 inhibition several constrained inhibitors have been developed. These thiol-containing molecules have provided potent, selective inhibitors for both enzymes and facilitate prodrug synthesis. Phosphonomethyl-containing dual inhibitors have also been reported (De Lombaert *et al.*, 1994b).

The mercaptoacyl dipeptide, S 21,402 (*Table 5.4*), previously known as RB 105, was

Table 5.3. Selective-inhibitors of AP-A and AP-N, and dual inhibitors of AP-N and E-24.11

	Structure	E-24.11 (K_i/IC_{50} (nM))	AP-N	AP-A
Leu-thiol	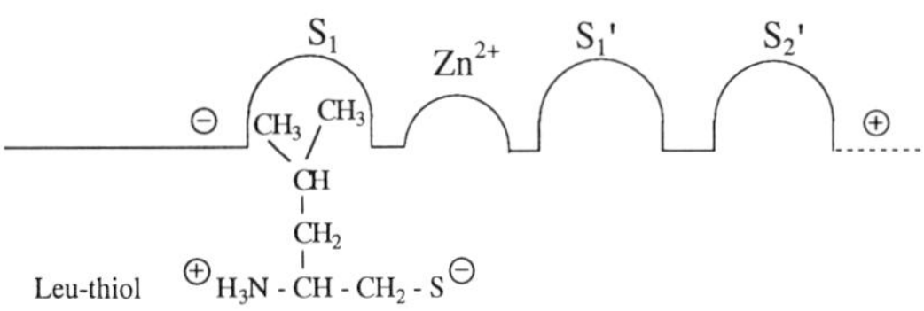	>10 000	22	>10 000
PC 18	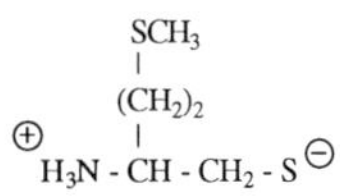	>10 000	11	>10 000
Glu-thiol	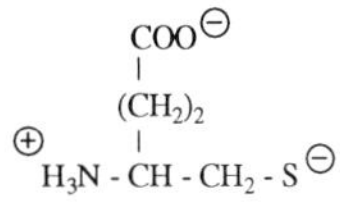	>10 000	120	140
EC 33	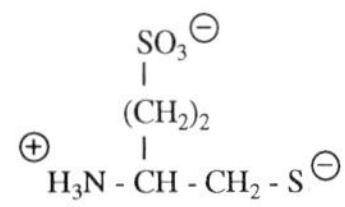	>10 000	25 000	290
Kelatorphan	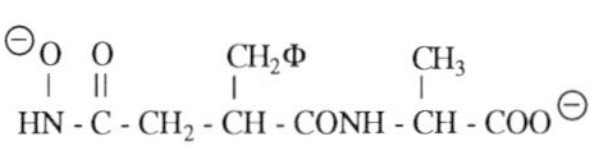	1.8	380	n.d.
RB 101	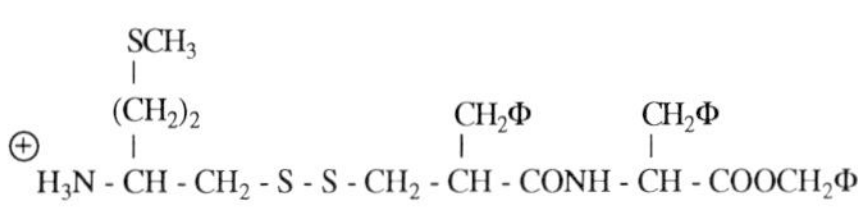	>10 000	>10 000	n.d.
ST 43	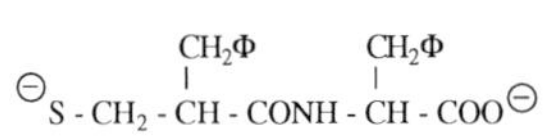	1.5	>1000	n.d.
CH 23	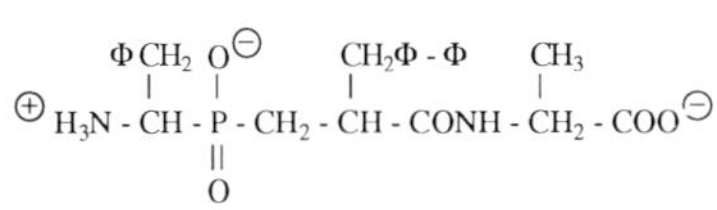	2.3	15	n.d.

n.d., not determined.

designed using an ACE pharmacore as a template (Fournié-Zaluski *et al.*, 1994a). Non-cyclic constraints were chosen, as such molecules are easier and cheaper to synthesise than cyclic compounds, and their residual flexibility aids target adaptation. In addition, it is easier to modulate their hydrophobic–hydrophilic balance than with more constrained molecules. S 21 402 has K_i values of 4.2 nM and 1.7 nM for ACE and E-24.11, respectively (Fournié-Zaluski *et al.*, 1994a). More recently, the same ACE pharmacore was used to design RB 106 (*Table 5.4*; Fournié-Zaluski *et al.*, 1996) which contains a glycine linked to a C-terminal 5-phenylproline and has K_i values of 0.35 nM for ACE and 1.6 nM for E-24.11. Molecular modelling studies show that RB 106 binds the S_1, S'_1 and S'_2 subsites of ACE, as illustrated in *Table 5.4*, and the S'_1 and S'_2 subsites of

Table 5.4. Dual inhibitors of ACE and E-24.11

Compound	Structure	K_i/IC_{50} (nM) ACE	K_i/IC_{50} (nM) E-24.11
1 S 21 402	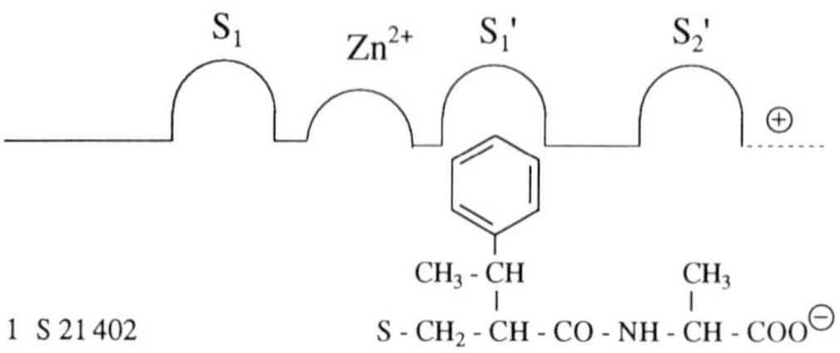	4.2	1.7
2 RB 106	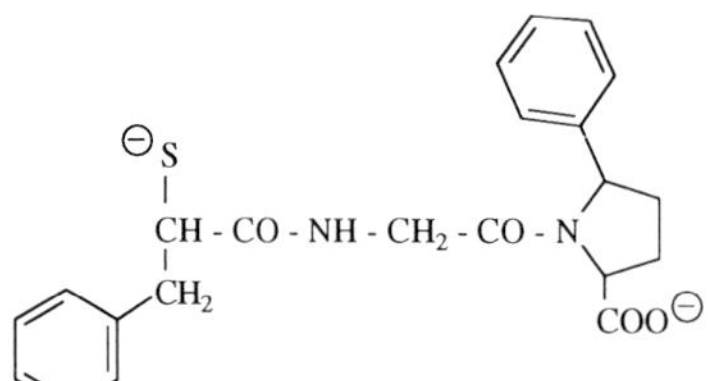	0.35	1.6
3 MDL 100 173	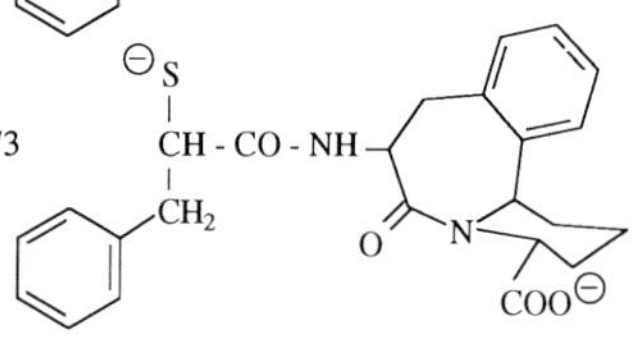	0.11[a] 2.5[b]	0.08[a] 2.2[b]
4 BMS 182 657	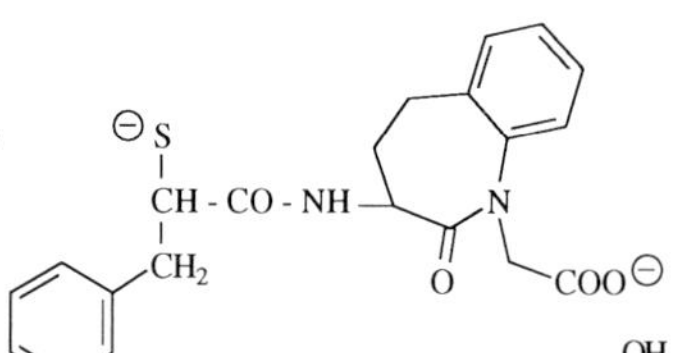	12.0	6
5	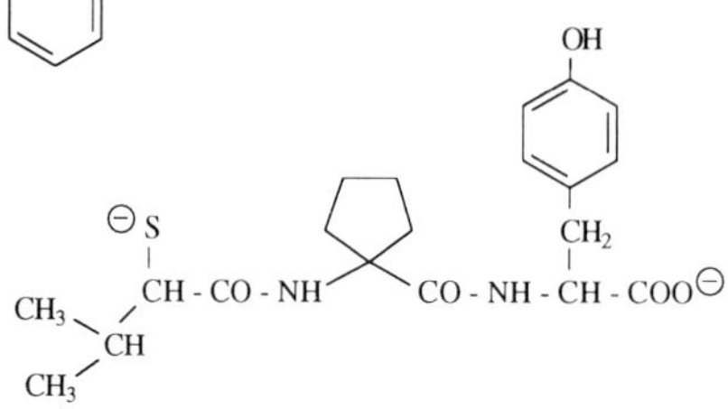	7.0	1.5

[a] Flynn *et al.* (1993); [b] Fournié-Zaluski *et al.* (1996).

E-24.11, with the 5-phenyl-proline moiety being outside the active site in the latter case (Fournié-Zaluski *et al.*, 1996).

MDL 100 173 (*Table 5.4*) is a highly constrained dipeptide derivative with good affinities for both enzymes (Flynn *et al.*, 1993) and a thioester prodrug has also been synthesised, designated MDL 100 240 (French *et al.*, 1994). Modifications of the Ala–Pro portion of a mercaptoacyl dipeptide ACE inhibitor led to the development of the benzazepine-based BMS 182 657 (*Table 5.4*) with IC_{50} values of 12 and 6 nM for ACE and E-24.11, respectively (Robl *et al.*, 1994; Trippodo *et al.*, 1995). In another series of thiol inhibitors the effect of incorporating a spiroalkyl amino acid in P'_1 was determined, leading to structure 5 (in *Table 5.4*) incorporating cycloLeu–Tyr (Fink *et al.*, 1995). A carboxylic acid ester-thioester-*O*-methyl tyrosine prodrug of this molecule has since been synthesised (Fink *et al.*, 1996).

5.3 Dual E-24.11 and AP-N inhibitors in pain management and opioid addiction

The pain suppression effect of morphine is related to the interaction of the alkaloid with μ and δ opioid receptors located in regions of the central nervous system involved in the regulation of nociceptive stimuli (e.g. spinal cord, periaqueductal grey matter and thalamus). The wide distribution of opioid receptors in the brain probably accounts for the multiplicity of pharmacological responses elicited by opiate administration as, in addition to its analgesic effects, morphine has anxiolytic and disinhibitory properties. Psychological dependence and respiratory depression, which are among the major side effects of narcotics, could be related to overstimulation of brain receptors involved in behavioural and bulbar respiratory control, respectively. The enkephalins, which also bind to μ and δ opioid receptors, are rapidly degraded by E-24.11 and AP-N *in vivo* (*Figure 5.1*) and the observation that the E-24.11 inhibitor thiorphan (*Table 5.1*) had antinociceptive properties in mice (Roques *et al.*, 1980) led to the idea that a more 'physiological' analgesia could be obtained by using inhibitors to increase the half-life of the peptides. This has recently been reinforced by the results obtained in mice deficient in pre-proenkephalin, confirming that enkephalins modulate responses to painful stimuli and participate in behavioural control (König *et al.*, 1996).

Although selective inhibitors of either enzyme alone have limited responses, their combined administration potentiates the effects of the endogenous enkephalins in a variety of pharmacological tests (Roques *et al.*, 1993). However, for clinical purposes these molecules not only have to cross the blood–brain barrier, allowing peripheral administration, but ideally they should have similar pharmacokinetic properties. It was thus proposed that one molecule, capable of inhibiting both enzymes would be advantageous, leading to the development of RB 101 (*Table 5.3*; Fournié-Zaluski *et al.*, 1992b).

RB 101 easily crosses the blood–brain barrier, as shown by the inhibition of cerebral E-24.11 following intravenous (i.v.) injection in mice (Noble *et al.*, 1992a), and by the increase in the extracellular levels of Met-enkephalin-like material in the nucleus accumbens of rats following intraperitoneal (i.p.) injection (Daugé *et al.*, 1996). The prodrug induces strong, dose-dependent antinociceptive responses in the hot-plate (ED_{50} = 9 mg kg^{-1}, i.v; 80 mg kg^{-1}, i.p.) and the writhing (ED_{50} = 3.25 mg kg^{-1}, i.v.) tests in mice and the tail-flick and the tail electric stimulation tests in rats, which are suppressed by the opiate antagonist naloxone (*Figure 5.3a*; Noble *et al.*, 1992a). Moreover, in a model of unilateral inflammatory pain and in the formalin test, which measures the response to a long-lasting nociceptive stimulus, peripheral administration of RB 101 also induced antinociceptive responses (Maldonado *et al.*, 1994; Noble *et al.*, 1995).

Chronic treatment with the inhibitor does not produce the severe physical withdrawal symptoms observed in mice treated with equivalent analgesic doses of morphine. In addition RB 101 does not induce tolerance or cross-tolerance with morphine

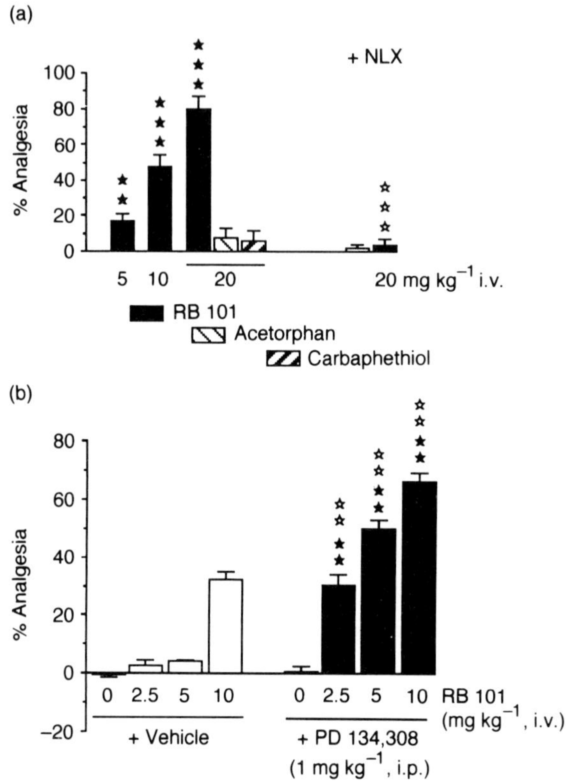

Figure 5.3. (a) A comparison of the analgesic potencies of i.v.-administered acetorphan (E-24.11 inhibitor), carbaphetiol (AP-N inhibitor) and RB 101 (Dual E-24.11/AP-N inhibitor) in the mouse hot-plate test, in the presence and absence of the opiate antagonist naloxone. (b) Antinociceptive effects of RB 101 (i.v.) in the rat tail-flick test and potentiation by the CCK-B antagonist PD-134 308.

(Noble *et al.*, 1992b) and its rewarding effects, measured in the place preference test, were significantly less than those produced by morphine (Noble *et al.*, 1992c; Valverde *et al.*, 1996). Side effects following chronic treatment with opiates are probably due to multiple cellular events involving several components of the cyclic AMP signal transduction cascade (Maldonado *et al.*, 1996). It has been suggested that the stimulation of opioid receptors by the endogenous opioids, prolonged by RB 101, is more specific than the ubiquitous stimulation by exogenously administered agonists, thus minimizing receptor desensitization or down-regulation. A limited opioid receptor occupation by the endogenous peptides is in agreement with *in vivo* binding studies, which demonstrated that the increase in tonically released endogenous enkephalins is too low to saturate opioid receptors (Ruiz-Gayo *et al.*, 1992a). In addition, because of their higher intrinsic efficacy, enkephalins need to occupy fewer opioid receptors than morphine to give the same pharmacological responses (Noble and Roques, 1995).

The craving for and self-administration of opiates is thought to be due to either a pre-existing deficit in the endogenous opioid system or a deficit induced by chronic administration. Dual inhibitors such as RB 101 reduce the severity of the withdrawal syndrome in rats (reviewed by Roques and Noble, 1995) and might therefore be of value in maintenance treatment. The most commonly used method for the study of physical dependence in animals is the precipitation of the withdrawal syndrome by administration of an opioid antagonist. However, in a clinical situation, this syndrome usually occurs when opiate addicts interrupt drug consumption. An experimental model of spontaneous morphine withdrawal and substitutive maintenance has therefore been developed in rats, in which RB 101 was as effective as methadone in inhibiting the expression of morphine-withdrawal symptoms (Ruiz *et al.*, 1996). The endogenous opioid system has also been linked to the aetiology of depression and 'anxious' behavioural responses triggered by conditioned suppression of motility and learned helplessness; these are attenuated by RB 101 (Baamonde *et al.*, 1992; Smadja *et al.*, 1995). Thus, in addition to reducing the severity of the withdrawal syndrome, the anti-depressant-like properties of dual AP-N and E-24.11 inhibitors might ameliorate the protracted abstinence syndrome, one of the main factors contributing to relapse.

CCK-8 (*Figure 5.1*), which has a similar distribution to the enkephalins in many pain-processing regions of the brain and spinal cord (Gall *et al.*, 1987), has been proposed to participate in an endogenous 'anti-opioid system'. In the central nervous system CCK interacts with two binding sites, CCK-A receptors and the predominant CCK-B receptors. Activation of CCK-A receptors potentiates the analgesic effects of both RB 101 and μ-opioid agonists, while activation of CCK-B receptors reduces their effects (Derrien *et al.*, 1993; Noble *et al.*, 1993), suggesting a regulatory mechanism between CCK and enkephalin systems in pain control (for a review, see Roques and Noble, 1996). The potentiation induced by stimulation of CCK-A receptors might be related to an increase in enkephalin release and/or to a direct improvement in the efficacy of transduction processes occurring at the μ sites. In contrast, CCK-B receptor activation appears to modulate negatively the opioidergic system. The *in vivo* binding of a CCK-B selective agonist is reduced by the administration of either RB 101 or a δ-selective agonist, suggesting that endogenous enkephalins increase extracellular CCK levels by activating δ-opioid receptors (Ruiz-Gayo *et al.*, 1992b). In addition the blockade of CCK-B-binding sites by selective antagonists significantly increases the antinociceptive responses induced by RB 101 in the rat tail-flick test and the mouse hot-plate test (Valverde *et al.*, 1994; *Figure 5.3b*). The RB 101-mediated reduction in the severity of the morphine-withdrawal syndrome, precipitated by naloxone is also potentiated by CCK-B antagonists (Maldonado *et al.*, 1995), although they have no effect on the spontaneous-withdrawal syndrome (Ruiz *et al.*, 1996). Moreover, the anti-depressant-type effects of RB 101 are increased by selective CCK-B antagonists in the suppression motility test in mice (Smadja *et al.*, 1995).

At the present time, no analgesics are available which fill the gap between opioid analgesics (morphine and surrogates) and antalgics (aspirin, paracetamol and non-steroidal anti-inflammatory drugs). Dual E-24.11/AP-N inhibitors, which produce naloxone-reversible analgesia in all tests where morphine is active, seem to be promising candidates. The therapeutic potential of these inhibitors, alone or in association with CCK-B antagonists, as analgesics or methadone substitutes, seems encouraging and preclinical trials are in progress.

In addition to their analgesic properties, opiates, such as the μ agonist loperamide, are also used to control the intestinal transit, although their use also leads to constipation. TIORFAN®, which is currently on the market as an antidiarrhoeal agent, is devoid of this side effect. This is thought to be due to the inhibitor increasing enkephalin levels which then exert an anti-secretory effect by stimulating δ receptors in the intestinal wall.

5.4 Dual ACE and E-24.11 inhibitors as therapeutic agents in cardiovascular disease

ACE forms part of the renin–angiotensin system, catalysing the release of the vasoconstrictive peptide angiotensin II (*Table 5.1*). Selective inhibitors of ACE, such as captopril and enalaprilat, which are used in the treatment of essential hypertension, are the most effective in cases with high renin activity and have no effect on hypertension resulting from sodium retention. E-24.11 inhibitors on the other hand, by increasing the circulating levels of atrial natriuretic peptide (*Table 5.1* and Chapter 20), promote its renal effects of natriuresis and diuresis. In addition, inhibitors of either enzyme increase the circulating levels of the hypotensive peptide bradykinin (*Table 5.1*), with ACE inhibitors having a mainly vascular action and E-24.11 a renal action (reviewed by Gonzales *et al.*, 1996). This led to the idea that dual inhibition of ACE and E-24.11 could have beneficial effects in the treatment of hypertension and heart failure (Roques and Beaumont, 1990).

Several rodent models have been developed to mimic the different forms of human hypertension such as the spontaneous hypertensive rat (SHR) and the renovascular hypotensive rat, characterized by normal and high renin levels, respectively, and the deoxycorticosterone acetate (DOCA-salt) hypertensive rat, a salt-dependent model, with ANP activation and low plasma renin levels. ACE inhibitors are active in the first two models but not in the DOCA-salt rat, whilst E-24.11 inhibitors are active only in the latter model. This is more fully discussed in Chapter 20, as are the advantages and disadvantages of the simultaneous inhibition of ACE and E-24.11 by the co-administration of selective inhibitors.

The pharmacological effects of S 21 402 (*Table 5.4*) are reviewed in Gonzales *et al.* (1996). The inhibitor simultaneously decreases circulating levels of angiotensin II and increases those of atrial natriuretic peptide and bradykinin. Oral administration (7.3 mg kg^{-1}) totally inhibited mouse lung ACE for 1 h and renal E-24.11 for up to 8 h while its ED_{50} values were calculated to be 5.75 mg kg^{-1} and 0.8 mg kg^{-1} for lung ACE and renal E-24.11, respectively (Fournié-Zaluski *et al.*, 1994a). S 21 402 also decreased blood pressure in the SHR, renovascular and DOCA-salt hypertensive rats to the same degree, despite different baseline plasma renin and atrial natriuretic peptide concentrations. Sodium excretion was also increased to the same extent in all three models. The more recently developed RB 106 has a long-lasting action *in vivo*, with E-24.11 and ACE inhibition being greater than 50% 12 h after oral administration of a single dose (25 mg kg^{-1}) and in the SHR oral administration (25 mg kg^{-1} day^{-1}) induced a prolonged hypotension, which was still significant 2 days after the end of the treatment (*Figure 5.4*). A strong hypotensive response and significant natriuresis was also obtained in the DOCA-salt hypertensive rat (Fournié-Zaluski *et al.*, 1996).

The prodrug of MDL 100 173, MDL 100 240 (3 in *Table 5.4*), at 30 mg kg^{-1} *per os* (p.o.) lowered systemic blood pressure in both SHR and DOCA/salt hypertensive rats with significant effects up to at least 5 h after administration (Flynn *et al.*, 1993; French *et al.*, 1994). BMS 182 657 (*Table 5.4*) also lowered mean arterial pressure in SHR rats (100 μmol kg^{-1} p.o.) and DOCA-salt hypertensive rats (100 μmol kg^{-1} i.v.), with responses greater than or equal to the effects of selective inhibitors (Trippodo *et al.*, 1995), and reduced blood pressure and stimulated natriuresis in conscious one-kidney one-clip hypertensive dogs, a model characterized by normal circulating levels of ANP and renin activity (Seymour *et al.*, 1995). Few pharmacological studies have been reported for compound 5 in *Table 5.4*, but this inhibitor (10 mg kg^{-1} p.o.) inhibited the angiotensin-I pressor response with a similar duration to captopril but with a smaller maximum response (Fink *et al.*, 1996). However, a carboxylic acid ester-thioester-O-methyl tyrosine prodrug had more potent responses and when orally administered (3 mg kg^{-1}) reduced blood pressure in a high-renin animal model and the SHR (Fink *et al.*, 1996).

A combination of selective ACE and E-24.11 inhibitors has been reported to have beneficial effects in experimental congestive heart failure (Margulies *et al.*, 1991; Seymour *et al.*, 1993; Trippodo *et al.*, 1993), which is characterized by sodium retention, high plasma concentrations of ANP, correlated with the severity of the disease, and a blunted renal response to exogenous and exogenous ANP. In catheterized conscious rats with a severe myocardial infarction S 21402 decreased mean blood pressure and increased plasma renin cyclic GMP levels, the latter correlating with increased plasma

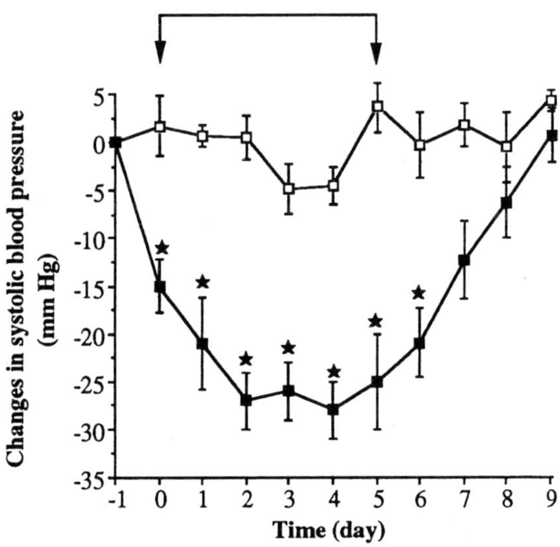

Figure 5.4. Variation in blood pressure measured during chronic oral administration of 25 mg kg^{-1} day of RB 106 for 5 days to SHR. ■, Animals receiving RB 106; □, control animals.

ANP levels. Natriuresis was also enhanced and urinary secretion of ANP and bradykinin levels increased in proportion with the size of the infarct (Gonzales *et al.*, 1996). BMS 182 657 (30 μmol kg^{-1} i.v.) reduced cardiac preload and afterload and augmented cardiac output in cardiomyopathic hamsters, an animal model of heart failure, where neither SQ 28 603 or captopril had significant effects (Trippodo *et al.*, 1995).

Mixed ACE/E-24.11 inhibitors thus appear to be living up to their original expectations as a new class of agents for the treatment of hypertension and chronic heart failure. As their actions are independent of the status of the renin–angiotensin system they have the potential to treat a broader range of patients than ACE inhibitors alone.

5.5 *In vivo* actions of ECE inhibitors

Due to the lack of selective ECE inhibitors few studies have been carried out to assess the role of the enzyme *in vivo* and the consequences of its inhibition, although phosphoramidon inhibits the pressor response to big-ET-1 in rats *in vivo* (McMahon *et al.*, 1991). Chronic subcutaneous administration of the dual ECE/E-24.11 inhibitor CGS 26303 (*Table 5.2*) has been found to reduce the mean arterial pressure in the SHR and was more effective than its E-24.11 selective analogue CGS 24592 (*Table 5.1;* De Lombaert *et al.*, 1994a). Oral administration of the prodrug of this molecule, CGS 26393, inhibited the pressor response induced by exogenous big ET-1 in rats, with inhibitory activity lasting up to 8 h (Trapani *et al.*, 1995).

5.6 *In vivo* actions of AP-A inhibitors

The selective AP-A inhibitor, EC 33 (*Table 5.3*), has been used to show the critical role of brain AP-A in the degradation of CCK-8, one of the most abundant peptides in the central nervous system which is thought to be involved in feeding, anxiety, memory and, as previously discussed, pain perception (Migaud *et al.*, 1996). Again in the central nervous system, the use of EC 33 has confirmed the role of AP-A in the conversion of angiotensin II to angiotensin III. Angiotensin III is in turn metabolized by AP-N and *in vivo* studies with EC 33 and an AP-N inhibitor have additionally established that angiotensin III, and not angiotensin II, is responsible for vasopressin release in the brain (Zini *et al.*, 1996).

References

Baamonde A, Daugé V, Ruiz-Gayo M, Fulga IG, Turcaud S, Fournié-Zaluski M-C, Roques BP. (1992) Antidepressant-type effects of endogenous enkephalins protected by systemic RB 101 are mediated by δ opioid and D_1 dopamine receptor stimulation. *Eur. J. Pharmacol.*, **216**: 157–166.

Balwierczak JL, Kukkola PJ, Savage P, Jeng AY. (1995) Effects of metalloprotease inhibitors on smooth muscle endothelin-converting enzyme activity. *Biochem. Pharmacol.* **49**: 291–296.

Chackalamannil S, Chung S, Stamford AW, McKittrick BA, Wang Y, Tsai H, Cleven R, Fawzi A, Czarniecki M. (1996) Highly potent and selective inhibitors of endothelin converting enzyme. *Bioorg. Med. Chem. Lett.* **6**: 1257–1260.

Chan WWCL. (1983) Leucine thiol. A potent inhibitor of leucine aminopeptidase. *Biochem. Biophys. Res. Commun.* **116**: 297–302.

Chauvel EN, Llorens-Cortes C, Coric P, Wilk S, Roques, BP, Fournié-Zaluski M-C. (1994) Investigation of the active site of aminopeptidase A and aminopeptidase N by new β-amino thiols. *J. Med. Chem.* **37**: 2950–2957.

Chipkin RE, Berger JG, Billard W, Iorio LC, Chapman R, Barnett A. (1988) Pharmacology of SCH-34,826, an orally active enkephalinase inhibitor analgesic. *J. Pharmacol. Exp. Ther.* **245**: 829–838.

Danilewicz, JC, Barclay PL, Barnish IT, Brown D, Campbell SF, James K, Samuels GMR, Terrett NK, Wythes MJ. (1989) UK-69. 578, a novel inhibitor of EC 3.4.24.11 which increases endogenous ANF levels and is natriuretic and diuretic. *Biochem. Biophys. Res. Commun.* **164**: 58–65.

Daugé V, Mauborge A, Cesselin F, Fournié-Zaluski M-C, Roques BP. (1996) The dual peptidase inhibitor RB 101 induces a long-lasting increase in the extracellular level of Met-enkephalin-like material in the nucleus acumbens of freely moving rats. *J. Neurochem.* **67**: 1301–1308.

De Lombaert S, Erion MD, Tan J, Blanchard L, El-Chehadi L, Ghai RD, Sakane Y, Berry C, Trapani AJ. (1993) *N*-Phosphonomethyl dipeptides and their phosphonate prodrugs, a new generation of neutral endopeptidase (NEP, EC 3.4.24.11) inhibitors. *J. Med. Chem.* **37**: 498–511.

De Lombaert S, Ghai RD, Jeng AY, Trapani AJ, Webb RL. (1994a) Pharmacological profile of a non-peptidic dual inhibitor of neutral endopeptidase 24.11 and endothelin-converting enzyme. *Biochem. Biophys. Res. Commun.* **204**: 407–412.

De Lombaert S, Tan J, Stamford LJ, Sakane Y, Berry C, Ghai RD. (1994b) Dual inhibition of neutral endopeptidase and angiotensin-converting enzyme by *N*-phosphonomethyl and *N*-carboxyalkyl dipeptides. *Bioorg. Med. Chem. Lett.* **4**: 2715–2720.

Derrien M, Noble F, Maldonado R, Roques BP. (1993) Cholecystokinin-A but not cholecystokinin-B receptor stimulation induces endogenous opioid-dependent antinociceptive effects in the hot plate test in mice. *Neurosci. Lett.* **160**: 193–196.

Emoto N, Yanagisawa M. (1995) Endothelin-converting enzyme-2 is a membrane-bound, phosphoramidon-sensitive metalloprotease with acidic pH optimum. *J. Biol. Chem.* **270**: 15262–15268.

Fink CA, Qiao Y, Berry CJ, Sakane Y, Ghai RD, Trapani AJ. (1995) New α-thiol inhibitors of angiotensin-I converting enzyme and neutral endopeptidase EC 3.4.24.11. *J. Med. Chem.* **38**: 5023–5030.

Fink CA, Carlson JE, McTaggart PA, Qiao Y, Webb R, Chatelain R, Jeng AY, Trapani AJ. (1996) Mercaptoacyl dipeptides as orally active dual inhibitors of angiotensin-converting enzyme. *J. Med. Chem.* **39**: 3158–3168.

Flynn GA, Beight DW, Mehdi S, Koehl JR, Giroux EL, French JF, Hahe PW, Dage RC. (1993) Application of a conformationally restricted Phe–Leu dipeptide mimetic to the design of a combined inhibitor of angiotensin I-converting enzyme and neutral-endopeptidase-24.11. *J. Med. Chem.* **36**: 2420–2423.

Fournié-Zaluski M-C, Chaillet P, Bouboutou R, Coulaud A, Chérot P, Waksman G, Costentin J, Roques BP. (1984) Analgesic effects of kelatorphan a new highly potent inhibitor of multiple enkephalin degrading enzymes. *Eur. J. Pharmacol.* **102**: 525–528.

Fournié-Zaluski M-C, Coulaud A, Bouboutou R, Chaillet P, Devine J, Waksman G, Costentin J, Roques B-P. (1985) New bidentates as full inhibitors of enkephalin degrading enzymes: synthesis and analgesic properties. *J. Med. Chem.* **28**: 1158–1169.

Fournié-Zaluski M-C, Coric P, Turcaud S, Bruetschy L, Lucas E, Noble F, Roques BP. (1992a) Potent and systemically active aminopeptidase N inhibitors designed from active site investigation. *J. Med. Chem.* **35**: 1259–1266.

Fournié-Zaluski M-C, Coric P, Turcaud S, Lucas E, Noble F, Maldonado R, Roques B-P. (1992b) Mixed-inhibitor-prodrug as a new approach towards systemically active inhibitors of enkephalin degrading enzymes. *J. Med. Chem.* **35**: 2474–2481.

Fournié-Zaluski M-C, Soleilhac J-M, Turcaud S, Lai-Kuen R, Crine P, Beaumont A, Roques B-P. (1992c) Development of [^{125}I]RB 104, a potent inhibitor of neutral endopeptidase-24.11 and its use in detecting nanogram quantities of the enzyme by 'Inhibitor Gel Electrophoresis'. *Proc. Natl Acad. Sci. USA* **89**: 6388–6392.

Fournié-Zaluski M-C, Coric P, Turcaud S *et al.* (1994a) New dual inhibitors of neutral endopeptidase and angiotensin-converting enzyme: rational design, bioavailability, and pharmacological responses in experimental hypertension. *J. Med. Chem.* **37**: 1070–1083.

Fournié-Zaluski M-C, Gonzalez W, Turcaud S, Pham I, Roques BP, Michel J-B. (1994b) Dual inhibition of angiotensin-converting enzyme and neutral endopeptidase by the orally active inhibitor Mixanpril 1: a new therapeutic approach in hypertension. *Proc. Natl Acad. Sci. USA* **91**: 4072–4076.

Fournié-Zaluski M-C, Coric P, Thery V, Gonzalez W, Meudal H, Turcaud S, Michel J-B, Roques BP. (1996) Design of orally active dual inhibitors of neutral endopeptidase and angiotensin-converting enzyme with long duration of action. *J. Med. Chem.* **39**: 2594–2608.

French JF, Flynn GA, Giroux EL, Mehdi S, Anderson B, Beach DC, Koehl JR, Dage RC. (1994) Characterization of a dual inhibitor of angiotensin I-converting enzyme and neutral endopeptidase. *J. Pharmacol. Exp. Ther.* **268**: 180–186

Gall C, Lauterborn J, Burks D, Serogy K. (1987) Co-localization of enkephalins and cholecystokinin in discrete areas of rat brain. *Brain Res.* **403**: 403–408.

Gonzalez W, Fournié-Zaluski M-C, Turcaud S, Roques BP, Michel, J-B. (1996) Pharmacology of S21402 (RB 105): a dual inhibitor of neutral endopeptidase and angiotensin converting enzyme. *Cardiovasc. Drug Rev.* **14**: 166–184.

Ikegawa R, Matsumura Y, Tsukahara Y, Takaoka M, Morimoto S. (1990) Phosphoramidon, a metalloproteinase inhibitor, suppresses the secretion of endothelin-1 from cultured endothelial cells by inhibiting a big endothelin-1 converting enzyme. *Biochem. Biophys. Res. Commun.* **171**: 669–675.

Keller PM, Lee C-P, Fenwick AE, Atkinson ST, Eliott JD, De Wolf WE Jr. (1996) Endothelin-converting enzyme: substrate specificity and inhibition by novel analogs of phosphoramidon. *Biochem. Biophys. Res. Commun.* **223**: 372–378.

Kenny AJ. (1977) Proteinases associated with cell membranes. In: *Proteinases in Mammalian Cells and Tissues* (ed. A.J. Barret), pp. 393–444. Elsevier/North-Holland Biomedical Press, Amsterdam.

König M, Zimmer AM, Steiner H, Holmes PV, Crawley JN, Browstein MJ, Zimmer A. (1996) Pain responses, anxiety and aggression in mice deficient in pre-proenkephalin. *Nature* **383**: 535–538.

Kukkola PJ, Bilci NA, Kozak WX, Savage P, Jeng AY. (1996) Optimization of retro-thiorphan for inhibition of endothelin converting enzyme. *Bioorg. Med. Chem. Lett.* **6**: 619–924.

Lawton G, Paciorek PM, Waterfall JF. (1992) The design and profile of ACE inhibitors. In: *Advances in Drug Research* (ed. B. Testa), pp.162–220. Academic Press, London.

Lebel N, D'Orleans-Juste P, Fournier A, Sirois P. (1996) Role of the neutral endopeptidase 24.11 in the conversion of big endothelins in guinea-pig lung parenchyma. *Br. J. Pharmacol.* **117**: 184–188.

Lecomte JM, Costentin J, Vlaiculescu A, Chaillet P, Marcais-Collado H, Llorens-Cortes C, Leboyer M, Schwartz J-C. (1986) Pharmacological properties of acetorphan, a parenterally active enkephalinase inhibitor. *J. Pharmacol. Exp. Ther.* **237**: 937–944.

Lloyd J, Schmidt JB, Hunt JT, Barrish JC, Little DK, Tymiak AA. (1996) Solid phase synthesis of phosphinic acid endothelin converting enzyme inhibitors. *Bioorg. Med. Chem. Lett.* **6**: 1323–1326.

Maldonado R, Valverde O, Turcaud S, Fournié-Zaluski M-C, Roques BP. (1994) Antinociceptive response induced by mixed inhibitors of enkephelin catabolism in peripheral inflammation. *Pain* **58**: 77–83.

Maldonado R, Valverde O, Ducos B, Blommaert A, Fournié-Zaluski M-C, Roques BP. (1995) Inhibition of morphine withdrawal by the association of RB 101, an inhibitor of enkephalin catabolism, and the CCKB antagonist PD-134, 308. *Br. J. Pharmacol.* **114**: 1031–1039.

Maldonado R, Blendy JA, Tzavara E, Gass P, Roques BP, Hanoune J, Schütz G. (1996) Reduction of morphine abstinence in mice with a mutation in the gene encoding CREB. *Science* **273**: 657–659.

Margulies KB, Perella MA, McKinley J, Burnett JC. (1991) Angiotensin inhibition potentiates the renal responses to neutral endopeptidase inhibition in dogs with congestive heart failure. *J. Clin. Invest.* **88**: 1636–1642.

Matthews BW. (1988) Structural basis of the action of thermolysin and related zinc peptidases. *Acc. Chem. Res.* **21**: 333–340.

McMahon EG, Paloma MA, Moore WM, McDonald JF, Stern MK. (1991) Phosphoramidon blocks the pressor activity of porcine big endothelin-1-(1–39) *in vivo* and conversion of big endothelin-1-(1–39) to endothelin-1-(1–21) *in vitro*. *Proc. Natl Acad. Sci. USA* **88**: 703–707.

McKittrick BA, Stamford AW, Weng X, Ma K, Chackalamanil S, Czarniecki M, Cleven RM, Fawzi AB. (1996) Design and synthesis of phosphinic acids that triply inhibit endothelin converting enzyme and neutral endopeptidase 24.11. *Bioorg. Chem. Lett.* **6**: 1629–1634.

Migaud, M, Durieux C, Viereck J, Soroca-Lucas E, Fournié-Zaluski M-C, Roques BP. (1996) The *in vivo* metabolism of cholecystokinin (CCK-8) is essentially ensured by aminopeptidase A. *Peptides*, **4**: 601–607.

Noble F, Roques BP. (1995) Assessment of endogenous enkephalins efficacy in the hot plate test in mice: comparative study with morphine. *Neurosci. Lett.* **185**: 75–78.

Noble F, Soleilhac JM, Soroca-Lucas E, Turcaud S, Fournié-Zaluski, M-C. Roques, BP. (1992a) Inhibition of the enkephalin-metabolizing enzymes by the first systemically active mixed inhibitor prodrug RB 101 induces potent analgesic responses in mice and rats. *J. Pharmacol. Exp. Ther.* **261**: 181–190.

Noble F, Coric P, Fournié-Zaluski M.-C, Roques BP. (1992b) Lack of physical dependence in mice after repeated systemic administration of the mixed inhibitor prodrug of enkephalin-degrading enzymes, RB 101. *Eur. J. Pharmacol.* **223**: 91–96.

Noble F, Turcaud S, Fournié-Zaluski M-C, Roques BP. (1992c) Repeated systemic administration of the mixed inhibitor of enkephalin-degrading enzymes, RB 101, does not induce either antinociceptive tolerance or cross-tolerance with morphine. *Eur. J. Pharmacol.* **223**: 83–89.

Noble F, Derrien M, Roques BP. (1993) Modulation of opioid antinociception by CCK at the supraspinal level: evidence of regulatory mechanisms between CCK and enkephalin systems in the control of pain. *Br. J. Pharmacol.* **109**: 1064–1070.

Noble F, Blommart A, Fournié-Zaluski M-C, Roques BP. (1995) A selective CCK_B receptor agonist potentiates μ-, but not δ-opioid receptor-mediated antinociception in the formalin test. *Eur. J. Pharmacol.* **273**: 145–151.

Ondetti MA, Rubin B, Cushman DW. (1977) Design of specific inhibitors of angiotensin-converting enzyme: a new class of orally active antihypertensive agents. *Science* **1996**: 441–444.

Patchett AA, Harris E, Tristram EW *et al.* (1980) A new class of angiotensin-converting enzyme inhibitors. *Nature* **288**: 280–283.

Robl JA, Simpkins LM, Stevenson J *et al.* (1994) Dual metalloprotease inhibitors I. Constrained peptidometics of mercaptoacyl dipeptides *BioMed. Chem. Lett.* **4**: 1789–1784.

Roques BP. (1993) Zinc metallopeptidases: active site structure and design of selective and mixed inhibitors – new approaches in the search for analgesics and antihypertensives. *Biochem. Soc. Transac.* **21**: 678–685.

Roques BP, Beaumont A. (1990) Neutral endopeptidase-24.11 inhibitors: from analgesics to antihypertensives? *Trends Pharmacol. Sci.* **11**: 245–249.

Roques BP, Fournié-Zaluski M-C. (1986) Enkephalin degrading enzyme inhibitors: a physiological way to new analgesics and psychoactive agents. *NIDA Res. Monogr. Ser.* **70**: 128–154.

Roques BP, Noble F. (1995) Dual inhibitors of enkephalin-degrading enzymes (neutral endopeptidase-24.11 and aminopeptidase-N) as potential new medications in the management of pain and opioid addiction. *NIDA Res. Monograph* **147**: 104–145.

Roques BP, Noble F. (1996) Association of enkephalin catabolism inhibitors and CCK-B antagonists: a potential use in the management of pain and opioid addiction. *Neurochem. Res.* **21**: 1395–1409.

Roques BP, Fournié-Zaluski MC, Soroca E, Lecomte JM, Malfroy B, Llorens C, Schwartz JC. (1980) The enkephalinase inhibitor thiorphan shows antinociceptive activity in mice. *Nature* **288**: 286–288.

Roques BP, Noble F, Daugé V, Fournié-Zaluski M-C, Beaumont A. (1993) Neutral endopeptidase 24.11. Structure, inhibition, and experimental and clinical pharmacology. *Pharmacol. Rev.* **45**: 87–146.

Ruiz F, Fournié-Zaluski M-C, Roques BP, Maldonado R. (1996) Similar decrease in spontaneous morphine abstinence by methadone and RB 101, an inhibitor of enkephalin catabolism. *Br. J. Pharmacol.* **119**: 174–182.

Ruiz-Gayo M, Baamonde A, Turcaud S, Fournié-Zaluski M-C, Roques BP. (1992a) *In vivo* occupation of mouse brain opioid receptors by endogenous enkephalins: blockade of enkephalin degrading enzymes by RB 101 inhibits [^{3}H]diprenorphine binding. *Brain Res.* **571**: 306–312.

Ruiz-Gayo M, Durieux C, Fournié-Zaluski M-C, Roques BP. (1992b) Stimulation of δ-opioid receptors reduces the *in vivo* binding of the CCK-B selective agonist [^{3}H]pBC264: evidence for a physiological regulation of CCKergic systems by endogenous enkephalins. *J. Neurochem.* **59**: 1805–1811.

Seymour AA, Asaad MM, Lanoce VM, Langenbacher KM, Fennell SA, Rogers WL. (1993) Systemic hemodynamics, renal function and hormonal levels during inhibition of neutral endopeptidase 3.4.24.11 and angiotensin-converting enzyme in dogs with pacing-induced heart failure. *J. Pharmacol. Exp. Ther.* **266**: 872–883.

Seymour AA, Abboa-Offei BE, Smith PL, Mathers PD, Assad MM, Rogers WL. (1995) Potentiation of natriuretic pepides by neutral endopeptidase inhibitors. *Clin. Exp. Pharm. Physiol.* **22**: 63–69.

Smadja C, Maldonado R, Turcaud S, Fournié-Zaluski M-C, Roques BP. (1995) Opposite role of CCK_A and CCK_B receptors in the modulation of endogenous enkephalin antidepressant-like effects. *Psychopharmacology* **120**: 400–408.

Soleilhac J-M, Lucas E, Beaumont A, Turcaud S, Michel J-B, Ficheux D, Fournié-Zaluski M-C, Roques B-P. (1992) A 94-kDa protein, identified as neutral endopeptidase-24.11, can inactivate atrial natriuretic peptide in the vascular endothelium. *Mol. Pharmacol.* **41**: 609–614.

Thorsett ED, Wyvratt MJ. (1987) Inhibition of zinc peptidases that hydrolyse neuropeptides. In: *Neuropeptides and their Peptidases* (ed. A.J. Turner), pp. 229–292. Horwood, Chichester.

Trapani AJ, De Lombaert S, Kuzmich S, Jeng AY. (1995) Inhibition of big-ET-induced pressor response by an orally active dual inhibitor of endothelin-converting enzyme and neutral endopeptidase 24.11. *J. Cardiovasc. Pharmacol.* **26**: S69-S71.

Trippodo NC, Fox M, Natarajan V, Panchal BC, Dorso CR, Asaad MM. (1993) Combined inhibition of neutral endopeptidase and angiotensin-converting enzyme in cardiomyopathic hamsters with compensated heart failure. *J. Pharmacol Exp. Ther.* **267**: 108–116.

Trippodo NC, Robl JA, Asaad MM, Bird JE, Panchal BC, Schaeffer TR, Giancarli MC, Hong SC. (1995) Cardiovascular effects of the novel dual inhibitor of neutral endopeptidase and angiotensin-converting enzyme BMS-182657 in experimental hypertension and heart failure. *J. Pharmacol. Exp. Ther.* **275**: 745–752.

Tronrud, DE, Monzingo AF, Matthews BW. (1986) Crystallographic structural analysis of phosphoramidates as inhibitors and transition-state analogs of thermolysin *Eur. J. Biochem.* **157**: 261–268.

Valverde O, Maldonado R, Fournié-Zaluski M-C, Roques BP. (1994) CCK-B antagonists strongly potentiate antinociception mediated by endogenous enkephalins. *J. Pharmacol. Exp. Ther.* **270**: 77–78.

Valverde O, Fournié-Zaluski M-C, Roques BP, Maldonado R. (1996) The CCK_B antagonist PD-134. 308 facilitates rewarding effects of endogenous enkephalins but does not induce place preference in rats. *Psychopharmacology.* **123**: 119–126.

Wilk S, Thurston LS. (1990) Inhibition of angiotensin III formation by thiol derivatives of acidic amino acids. *Neuropeptides* **16**: 163–168.

Zini S, Fournié-Zaluski M-C, Chauvel E, Roques BP, Corvol P, Llorens-Cortes C. (1996) Identification of metabolic pathways of brain angiotensin II and III using specific aminopeptidase inhibitors: predominant role of angiotensin III in the control of vasopressin release. *Proc. Natl Acad. Sci. USA* **93**: 11968–11973.

Chapter 6

Endopeptidase-24.11

Philippe Crine, Natalie Dion and Guy Boileau

6.1 Introduction

Endopeptidase-24.11 (EC 3.4.24.11; neprilysin, E-24.11) is an ectopeptidase belonging to the gluzincin family of zinc metallopeptidases (Hooper, 1994). This family of enzymes is characterized by the consensus sequence HEXXH where the two histidine residues are zinc ligands and the glutamic acid residue is involved in catalysis (see Section 6.2.1). In gluzincins, the third zinc ligand is a glutamic acid residue usually located towards the C terminus of the consensus sequence.

E-24.11 was first described by Kerr and Kenny (1974a,b) as a membrane-associated protein of the rabbit kidney brush border membrane. Quantitatively, E-24.11 represents 4–5% of the total protein mass of the renal brush border membrane, which lines the proximal convoluted tubule. When purified from kidney cortex, E-24.11 was associated with the microsomal fraction of the homogenate and could be released only after toluene–trypsin treatment, suggesting that this enzyme was an integral membrane protein (Kerr and Kenny, 1974a). The transmembrane nature of E-24.11 was confirmed by electron microscopic analysis of liposomes reconstituted in the presence of the enzyme (Kenny *et al.*, 1983). Furthermore, the reconstitution experiments showed that the bulk of the protein lies external to the membrane surface, projecting from it by a short (2 nm) stalk. The asymmetric association of E-24.11 with the cell membrane was established by labelling experiments using a photo-activated compound DIABS (3,5-di[^{125}I]iodo-4-azidobenzenesulphonate) (Booth and Kenny, 1980). The topology of the enzyme in the membrane thus suggests that the role of E-24.11 is confined to hydrolysis of susceptible substrates at the cell surface.

Since its purification from kidney brush border membranes, the enzyme has been shown to be present in several other tissues and cell types (for review, see Kenny *et al.*, 1987). An activity in brain, considered to be a putative peptidylcarboxypeptidase hydrolysing enkephalins ('enkephalinase') (Schwartz *et al.*, 1980), was shown to be E-24.11 (Fulcher *et al.*, 1982; Matsas *et al.*, 1983). In the central nervous system (see Chapter 19), it is present on a distinct population of neurons and on membranes surrounding the brain and spinal chord, as well as the choroid plexus, and is involved in the termination of the action of substance P and enkephalins. It is also present in peripheral nerves, where it is found on Schwann cells, rather than on neuronal elements (Barnes *et al.*, 1991). Peripherally, it plays an important role in the inactivation of natriuretic peptides (see Chapter 20). E-24.11 is also present at the surface of pre-B lymphocytes and

Cell-Surface Peptidases in Health and Disease, edited by A.J. Kenny and C.M. Boustead.

some leukaemic cells where it was shown to be identical to common acute lymphoblastic leukaemia antigen (CALLA, CD10; Letarte *et al.*, 1988; Shipp *et al.*, 1988; see Chapters 22 and 24). More recently, E-24.11 has also been found in endothelial cells from venous and arterial origins (Llorens-Cortes *et al.*, 1992).

E-24.11 is a integral membrane protein and circulating levels of the enzyme are very low. However, in patients with adult respiratory distress syndrome, the serum levels of E-24.11 can rise by 50- to 60-fold (Johnson *et al.*, 1985). Similarly, E-24.11 was detected in rat urine where it is the major kinin-degrading activity (Erdös and Skidgel, 1988). The precise structure of these soluble enzymes and the mechanisms by which they arise are yet to be characterized. However, one source of circulating enzyme could be the fibroblasts in lung or the epithelium of the airways and blood vessels.

E-24.11 has been purified from rabbit kidneys as a monomer of M_r 93 000 (Kerr and Kenny, 1974b). About 15% of the protein mass is made up by carbohydrate. In other species such as the pig, E-24.11 is isolated as a homodimer with the subunits having M_r values between 89 000 and 95 000. The monomers are not covalently linked by disulphide bridges. This difference in molecular weight observed between species and tissues is probably due to differences in the glycosylation pattern (Relton *et al.*, 1983). A similar phenomenon was observed in cultured cells expressing recombinant E-24.11 after transfection with viral vectors (Lemay *et al.*, 1990).

E-24.11 is a metalloenzyme with a zinc atom in its active site. The enzyme can be inhibited by general inhibitors of metalloenzymes such as ethylene diamine tetraacetic acid (EDTA), ethylene glycol bis-(β-aminoethylether) *N, N, N′, N′*-tetraacetic acid (EGTA) or 1,10-phenanthroline. Reactivation of the enzyme can be accomplished after dialysis of the inhibitor by the addition of metal divalent cations, zinc being the most efficient, restoring maximal activity at a concentration of 10^{-5} M (Kerr and Kenny, 1974b). The specificity of E-24.11 was first studied using the B-chain of insulin as substrate (Kerr and Kenny, 1974a). It was shown that E-24.11 prefers hydrophobic amino acids in the P_1' position in the substrate. In this regard, E-24.11 is similar to a group of microbial zinc metalloproteases that includes thermolysin (TLN), with which it also shares a common catalytic mechanism. Several lines of evidence suggest similarities in the active site of TLN and E-24.11 (Hersh and Morihara, 1986). Both enzymes cleave peptide bonds on the amino side of hydrophobic amino acid residues, indicating that their S_1' subsite accepts exclusively hydrophobic amino acid residues (Matsas *et al.*, 1984). E-24.11 shows a preference for another hydrophobic residue in the P_2' position (Quay *et al.*, 1994), but proline at this position decreases the interaction of substrates with both E-24.11 and TLN (Roques and Fournié-Zaluski, 1986). Furthermore an essential histidine residue present in the TLN active site was also found in E-24.11 (Bateman and Hersh, 1987; Beaumont and Roques, 1986). However, the use of extended synthetic peptides has revealed different requirements for subsite binding between E-24.11 and TLN (Almenoff and Orlowski, 1983; Matsas *et al.*, 1984; Orlowski and Wilk, 1981; Pozsgay *et al.*, 1986). Taking into account the similarities in the enzymic activity of E-24.11 and TLN, potent inhibitors of E-24.11 have been developed. Among these, thiorphan (Roques *et al.*, 1980) and retrothiorphan (Roques *et al.*, 1983) inhibit TLN and E-24.11 with the same stereochemical dependence. Like TLN,

E-24.11 is very sensitive to inhibition by phosphoramidon, a natural inhibitor of the bacterial enzyme (Kenny, 1977).

TLN (Coleman *et al.*, 1972; Holmes and Matthews, 1982) has been crystallized alone and in the presence of various inhibitors, and it provides a model for exploring the structure of this class of enzyme. The Zn^{2+} ion serves as a powerful electrophilic catalyst, probably by providing an activated water molecule and/or by contributing to the polarization of the carbonyl group of the scissile bond, since in both instances a lone pair of oxygen electrons of these moieties interact with the pentacoordinated zinc during the transition state (Vallée and Auld, 1990; Vallée and Galdes, 1984). A simplified model of the active sites of zinc metallopeptidases, derived from these crystallographic studies, was used by Ondetti *et al.* (1977) for the rational design of ACE inhibitors such as captopril.

The complete primary structure of E-24.11 has been deduced from the sequence of cDNA clones from human (Letarte *et al.*, 1988; Malfroy *et al.*, 1988), rabbit (Devault *et al.*, 1987), rat (Malfroy *et al.*, 1987) and mouse (Chen *et al.*, 1992). E-24.11 is a 749-amino-acid protein which has been predicted to consist of three distinct domains: a short amino-terminal cytosolic sequence (27 residues), a single transmembrane region (23 residues), and a large extracellular domain or ectodomain responsible for the catalytic activity of the enzyme. This ectodomain also contains asparagine residues that are part of consensus sequences for N-glycosylation. There are five potential glycosylation sites in the rabbit and mouse and six in the rat and human sequences. The ectodomain also contains 12 cysteine residues that may be involved in disulphide bridges stabilizing the conformation of the active enzyme.

Comparison of the sequence of E-24.11 with that of TLN, showed that the sequences of these two enzymes are very different. Despite these differences however, many of the amino acids involved in catalysis and binding of the zinc atom in TLN are found in two protein segments that have been conserved in E-24.11 (Devault *et al.*, 1987). In order to reveal additional hidden similarities between E-24.11 and TLN, comparison of the sequences was done by hydrophobic cluster analysis (Gaboriaud *et al.*, 1987). This method shows the occurrence of large similarities in the clustered sequences of TLN and E-24.11 (Benchetrit *et al.*, 1988). Based on these results, two possible sequence alignments have been proposed. The first model is characterized by a deletion of the Ca^{2+}-binding coil present in TLN and the second model by replacement of this coil by two α-helices. Both models suggest a significant structural analogy between TLN and E-24.11 at their active site.

6.2 The active site of E-24.11

6.2.1 Glutamic acid residue 584

The initial site-directed mutagenesis studies were designed to investigate the role of Glu584 in the HEXXH consensus sequence, which was proposed to correspond to Glu143 in TLN, an essential amino acid residue involved in catalysis. Changing Glu584 of E-24.11 for either Val or Asp completely abolished the activity of the recombinant mutant enzyme (Devault *et al.*, 1988a). However, the affinity of the mutated enzyme for the substrate-related tritiated inhibitor $[^3H]N$-[(2R,2S)-3-hydroxyamino-carbonyl-2-benzyl-1-oxopropyl]-glycine ($[^3H]$HACBO-Gly) remained unchanged. This last result strengthened the hypothesis that the mutation affected a key catalytic residue essential in peptide bond cleavage and was not due to alterations of the substrate binding site, either directly by changing a crucial residue involved in substrate binding, or indirectly by alteration of the active site tertiary structure.

6.2.2 Identification of the zinc ligands

X-ray diffraction studies of TLN (Coleman et al., 1972; Holmes and Matthews, 1982) have shown that the zinc ion is coordinated with three amino acid side chains and a water molecule. In TLN, His142 and His146 are two ligands of the zinc atom. These residues correspond to His583 and His587 in E-24.11. When these His residues were mutated into Phe, both mutations were found to essentially abolish all enzymic activity and HACBO-Gly binding (Devault et al., 1988b). As binding of the specific E-24.11 inhibitor is strictly dependent on the presence of a zinc atom in the active site, these results strongly suggest that His583 and His587 are two zinc ligands in E-24.11. In TLN the third zinc ligand is Glu166, located 23 amino acid residues on the C-terminal side of the catalytic glutamate residue. In order to find the equivalent residue in E-24.11, two glutamic acid residues located in position 616 and 646 on the C-terminal side of the active site consensus sequence were changed for either a valine or an aspartic acid residue (LeMoual et al., 1991). Mutations at position 616 had virtually no effect on the kinetic parameters of E-24.11. In contrast, the mutant where Glu646 had been replaced by Val was almost completely devoid of catalytic activity and unable to bind $[^3H]$HACBO-Gly. Replacing Glu for Asp at position 646 conserved the negative charge, and the mutant enzyme exhibited the same K_m value as the non-mutated enzyme but k_{cat} was decreased to less than 3% of the value of the non-mutated enzyme. The Asp646 mutant of E-24.11 showed a higher susceptibility to chelating agents than the non-mutated enzyme, but bound the tritiated inhibitor with the same affinity. Taken together, these observations strongly suggest that Glu646 of E-24.11 is the third zinc-coordinating residue and is equivalent to Glu166 in TLN.

6.2.3 Asp650 is crucial for catalytic activity of E-24.11

An extended zinc-coordination polyhedron has been structurally, but not functionally, characterized in TLN, carboxypeptidase A and in other crystallized zinc-dependent enzymes such as alcohol dehydrogenase and carbonic anhydrase II (Argos *et al.*, 1978; Christianson and Alexander, 1989, 1990). In these polyhedra, one of the zinc-coordinating histidine residues was found to bridge both the zinc ion and the carboxylate side chain of an Asp or Glu residue. Such an interaction could be important for protein folding (Christianson and Alexander, 1990; Higaki *et al.*, 1992). Alternatively, it could be critical for zinc binding by orientating the imidazole ring of the histidine residue and/or by enhancing the basicity of the histidine ligand (Argos *et al.*, 1978; Christianson, 1991; Higaki *et al.*, 1992). Moreover, the enhanced basic nature of the histidine zinc ligand could also modulate the electrophilic properties of the zinc ion (Christianson and Alexander, 1989). In TLN, His142 bridges the carboxylate side chain of Asp170 and the zinc ion. Alignment of the amino acid sequences of TLN and E-24.11 indicates that Asp650 in E-24.11 is possibly the equivalent of Asp170 in TLN. Site-directed mutagenesis was used to assess the functional role of this residue in E-24.11 (LeMoual *et al.*, 1994). Asp650 of E-24.11 was indeed found to be essential for hydrolytic activity. Substitution of Asp650 for either Glu, Asn or Ala resulted in mutant enzymes exhibiting drastic decreases in specific activity. Binding experiments with HACBO-Gly suggested that the zinc ion is present in the active site of these mutant enzymes.

6.2.4 Identification of a histidine residue at the active site

Crystallographic data have shown the presence of His231 at the active site of TLN (Kester and Matthews, 1977). According to Hangauer *et al.* (1984), this histidine residue participates in the stabilization of both the Michaelis complex and the tetrahedral intermediate. This residue might correspond to His637 in E-24.11, but replacing it by Phe had little effect on the catalytic and binding properties of E-24.11, suggesting that this histidine residue does not play any significant role in catalysis (Devault *et al.*, 1988b). Using photoaffinity labelling, Bateman *et al.* (1990) have shown that His711 of E-24.11 could be crosslinked to *N*-bromoacetyl-D-leucylglycine, an irreversible inhibitor of E-24.11 which mimics the transition state of the catalytic reaction. Hence, they proposed that residue His711 of E-24.11 was present at the active site and could explain the sensitivity of this enzyme to diethylpyrocarbonate (DEPC). However, mutations of this residue resulted in an inactive enzyme and no conclusion could be drawn as to its role in catalysis. A more recent study however (Dion *et al.*, 1993) created two mutants of E-24.11 where His711 was replaced by either Gln or Phe. Determination of the kinetic parameters showed that both mutants had K_m values very similar to that of the non-mutated enzyme but that their k_{cat} values were 25-fold lower. The calculated difference in free energy needed to form the transition state complex was increased by 2.2 kcal mol^{-1} for both mutants. These observations strongly suggest that His711 is involved in the stabilization of the transition state by forming a hydrogen bond with the oxyanion of the tetrahedral intermediate.

6.2.5 Two arginine residues are involved in substrate binding to the active site

Several groups have used the arginine-specific reagents, butanedione (Bateman *et al.*, 1989; Beaumont and Roques, 1986; Jackson and Hersh, 1986; Malfroy and Schwartz, 1982) and phenylglyoxal (Bateman *et al.*, 1989; Jackson and Hersh, 1986) to demonstrate the presence of an arginine residue at the active site of E-24.11. In this respect, the enzyme appears to differ from TLN, which is not affected by butanedione (Malfroy and Schwartz, 1982), even though Arg203 is known to be in the active site and to participate in substrate binding (Holmes and Matthews, 1982). From the computer-assisted alignment (Benchetrit *et al.*, 1988), four arginines Arg408, Arg409, Arg659, Arg747 have been suggested as possible active site residues of E-24.11. In addition, Arg102 has been identified as being irreversibly labelled after treatment of the enzyme with

[^{14}C]phenylglyoxal, and site-directed mutagenesis of this residue has shown that it could play an important role in substrate binding (Bateman *et al.*, 1989). The role of these five arginine residues was assessed using site-directed mutagenesis to replace each of them with a methionine residue (Beaumont *et al.*, 1991). The results indicate that two of the arginine residues (Arg747 and Arg102) may be involved in substrate binding. Arg747 appears to interact with the carbonyl amide group of the P_1' residue and Arg102 could interact, as suggested earlier (Bateman *et al.*, 1989), with the free carboxyl group of some substrates, such as the enkephalins. Whereas Arg747 can be modified when the enzyme is treated with either phenylglyoxal or butanedione, Arg102 is only modified with phenylglyoxal but not butanedione. It is not entirely clear why butanedione does not interact with Arg102 as it does with Arg747, but it is interesting to note that Arg102 is not located in the C-terminal region of E-24.11, where the other active site residues, so far identified, are found. This suggests a folding of the protein which brings Arg102 to the entrance of the active site cleft. As the complex formed between the guanidium group of arginine and butanedione is relatively unstable, it may be more readily reversible with a group in a hydrophilic environment than with a group in the more protected environment of the interior of the active site. Arg747 could be brought into the active site by a disulphide bridge involving Cys746. Taken together, these results suggest that Arg102 gives the enzyme its dipeptidylcarboxypeptidase activity with certain substrates (such as the enkephalins) in which the P_2' residue has a free carboxyl group. The arginine residue thus appears to play a role similar to Arg145 in carboxypeptidase A (EC 3.4.17.1) (Christianson and Lipscomb, 1985), although not entirely equivalent, as in the latter, Arg145 is positioned to block the binding of a C-terminally extended substrate, ensuring the strict exopeptidase activity of this enzyme.

Attempts to change enzyme specificity by charge polarity reversal have so far met with little success, probably due to a destabilization of the resulting ion pair in an environment naturally optimized for the inverted pair. This would not be the case in water where the energies of reverse ion pairs would be identical. Arg102, being probably located at the edge of the active site, may thus be in an environment favourable for polarity reversal. Based on this hypothesis, Arg102 was mutated to Glu, and data from inhibition studies were used to design a substrate with an optimally positioned C-terminal positive charge (Beaumont *et al.*, 1992). The results showed that, while mutating Arg102 to Glu reduces the specificity of a C-terminally negatively charged substrate by 16-fold, it increases that of a substrate with an optimally positioned positive charge by 29-fold. The results obtained thus confirm that Arg102 is situated at the edge of the active site of E-24.11. They also demonstrated for the first time, that it was possible in those conditions to change the specificity of a zinc metalloprotease.

6.2.6 Role of Asn542 in binding of the P_2' residue of substrates and inhibitors

Crystallization of TLN in the presence of the E-24.11 inhibitors, thiorphan and retrothiorphan, and determination of the complex structures by X-ray diffraction showed that Asn112 forms two hydrogen bonds with the P_2' residue of the inhibitors (Roderick *et al.*, 1989). Asn112 is part of a Asn111–Asn112–Ala113–Phe114–Trp115 sequence that makes a β-sheet which holds the substrate in the active site. Hydrophobic cluster analysis

(Benchetrit *et al.*, 1988) showed that the sequence Val541–Asn542–Ala543–Phe544–Tyr545 of E-24.11 has some homology with the Asn112-containing sequence of TLN and can also form a β-sheet. In order to assess the role of E-24.11 Asn542, site-directed mutagenesis was used to change it for either Gly or Gln. The effect of these mutations on substrate catalysis and inhibitor binding was examined with a series of thiorphan-like compounds containing various degrees of methylation at the P_2' residue (Dion *et al.*, 1995). For both mutated enzymes, determination of kinetic parameters with [D-Ala^2, Leu^5] enkephalin as substrate, showed that the large decrease in activity was attributable to an increase in K_m (14- to 16-fold), whereas k_{cat} values were only slightly affected (two- to three-fold decrease). This is in agreement with Asn542 being involved in substrate binding rather than directly in catalysis. Finally the IC_{50} values for thiorphan and substituted thiorphans strongly suggest that Asn542 of E-24.11 binds the substrate on the amino side of the P_2' residue by formation of a unique hydrogen bond. *Figure 6.1* presents a model of the active site of E-24.11, as established by these various site-directed mutation studies.

6.3 Topology of E-24.11 in the membrane

As reported for other kidney and intestinal hydrolases (Kenny and Maroux, 1982; Semenza, 1986), E-24.11 appears to be asymmetrically inserted into the lipid bilayer with a large extracytoplasmic domain and a very small anchoring tail (Booth and Kenny, 1980). Monoclonal antibodies were raised by immunizing mice with purified rabbit kidney brush border membranes (Aubry *et al.*, 1988). One monoclonal antibody

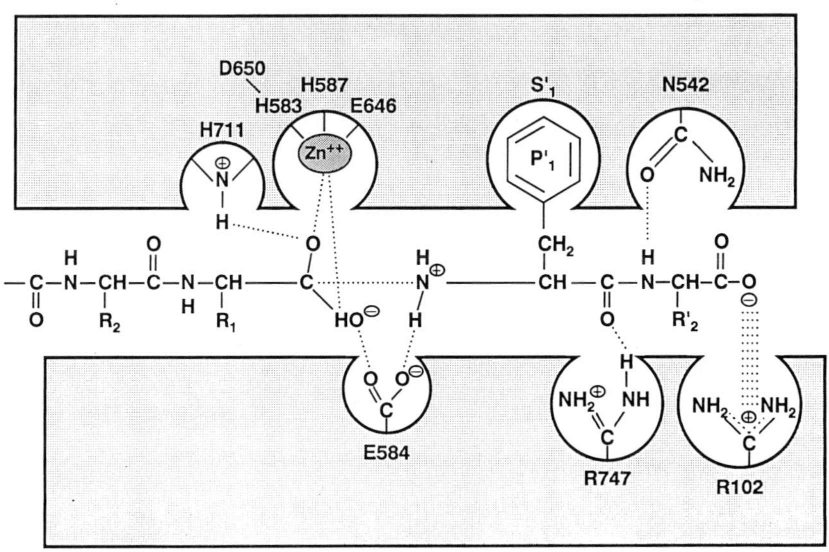

Figure 6.1. Model of the active site of endopeptidase 24.11. The interactions of the various amino acid residues identified by site-directed mutagenesis are indicated. The enzyme–substrate complex is represented in the tetrahedral intermediate form.

recognized a conformational epitope, which is destroyed after reduction and alkylation of the protein, and is located on the extracellular domain of the protein. Others bound to a linear epitope located on the cytosolic domain. These results demonstrated unambiguously the transmembrane nature of E-24.11.

The intracytoplasmic and membrane-spanning domains of E-24.11 were substituted for the signal peptide of a secretory protein (pro-opiomelanocortin) and the chimeric protein was expressed in COS-1 cells using an SV40-derived vector (Lemay *et al.*, 1989). The chimeric protein was found to be translocated into the endoplasmic reticulum (ER) lumen, where the heterologous signal peptide was cleaved. E-24.11, freed from its cytosolic and membrane-spanning domains, acquired terminal sugars in the Golgi and was found to be secreted as a soluble protein into the culture medium. This truncated form of E-24.11 was found to have the same catalytic properties as the membrane-bound enzyme extracted from kidney brush borders. These results therefore demonstrate that anchoring of the ectodomain in the plasma membrane is important neither for the catalytic activity of E-24.11, nor for its transport to the cell surface.

6.4 Glycosyl phosphatidylinositol-anchored constructs of E-24.11

A glycosyl phosphatidylinositol-anchored (GPI-anchored) form of E-24.11 has been constructed and expressed in COS-1 cells (Howell *et al.,* 1994a). The GPI-anchored form shows sensitivities to inhibitors similar to those of the soluble form and has a K_m for [D-Ala^2,Leu^5] enkephalin similar to that of the wild-type transmembrane-anchored form of E-24.11. Another chimeric form of E-24.11, containing a GPI anchor while retaining its transmembrane anchor, has also been constructed (Howell *et al.,* 1994b). This protein, expressed in COS-1 cells, is also enzymically active and is found to a large extent at the cell surface. Very little is known about the machinery involved in the biosynthesis of GPI-anchored proteins (see Chapter 3). It would seem that it is quite versatile, as it allows GPI anchor addition to a protein that is already anchored at its N terminus. One would imagine that double anchoring may only occur if the conformational strain imposed by both anchors is not too great, otherwise the protein may be distorted and not escape from the ER. The observation that a GPI anchor can be added to E-24.11 with retention of enzymic activity, would suggest that the C terminus lies at the protein surface within close enough proximity to the membrane to allow GPI anchoring. This observation also raises the possibility of naturally occurring proteins containing both transmembrane and GPI anchors (Hitt *et al.,* 1994).

The biosynthesis of the GPI-anchored forms of E-24.11 has also been compared with that of its transmembrane-anchored counterpart (Cailler *et al.,* 1995; Howell *et al.,* 1994a). The GPI-anchored form was delayed in reaching the Golgi apparatus compared with the transmembrane-anchored form, suggesting that folding events may delay its transport out of the ER. However, transport from the Golgi to the cell surface occurred at the same rate, regardless of the type of membrane anchor. Whether GPI- and peptide-anchored forms of E-24.11 use similar transport vesicles from the trans-Golgi network to the cell surface is not known. However, preliminary evidence suggests that in polarized Madin–Darby canine kidney cells, GPI-anchored forms of E-24.11, but not wild-type E-24.11 are associated with Triton X100-insoluble clusters recovered from low density fractions in sucrose gradients (Cailler *et al.,* 1995). These clusters, which are enriched in glycosphingolipids are thought to correspond to small apical transport vesicles carrying GPI-anchored proteins to the cell surface (see Chapter 4).

6.5 Role of glycosylation in transport and enzymic activity of E-24.11

Site-directed mutagenesis was used to alter each of the five consensus N-glycosylation sites in E-24.11 (Lafrance *et al.,* 1994). The recombinant proteins were expressed in COS-1 cells and transport of the glycosylation mutants to the cell surface was compared with that of the wild-type enzyme. Their enzymic activity was also determined. These studies were performed on the normal membrane-bound enzyme as well as on the secreted form (see Section 6.3). Immunoblotting of E-24.11 in COS-1 cell extracts and spent culture medium indicated that total expression of normal membrane-bound E-24.11 was not affected by mutation at glycosylation sites, whereas this expression level appeared to be strictly dependent on the number of glycosylation sites remaining on the soluble form. In both soluble and membrane-bound forms, progressive removal of glycosylation sites resulted in a concomitant increase in electrophoretic mobility, suggesting the use of all five potential glycosylation sites on the protein. Transport to the cell surface was also reduced by decreased glycosylation, but again, the phenomenon appeared more drastic in the case of the soluble form than for the membrane-bound enzyme. Enzyme activity was decreased by deglycosylation. However the presence of either of two crucial sites (Asn145, site 1 and Asn628, site 5) was sufficient to recover close to normal enzymic activities. More recently, a fully deglycosylated form of soluble E-24.11 was obtained by PNGase-F digestion of the recombinant protein produced in a baculovirus insect system (L. Gonneville, P. Crine and J. Sygusch, unpublished results). The oligosaccharide-free enzyme retained full enzymic activity and normal sensitivity to inhibitors. It thus appears that transport to the cell surface is much more dependent on a normal complement of oligosaccharide side chains than enzymic activity. This is most probably related to folding problems in the rough ER, as the amount of endo-H sensitive forms of the enzyme increased with the number of mutations. Restoration of near normal enzymic activity by maintaining either sites 1 or 5 on a partially deglycosylated enzyme produced *in vivo* suggests some local conformational constraints for enzyme activity. In that respect, it is noteworthy that glycosylation sites 1 and 5 are located closest to amino acid residues shown to be important for catalytic activity (Section 6.2). One explanation for these results could be that these sugar moieties participate in local folding of the polypeptide chain *in vivo.* If deglycosylation is performed *in vitro* on a fully folded protein, removal of glycans has little effect on both the conformation and enzymic activity of E-24.11.

6.6 Gene structure

The unique E-24.11 gene has been localized by *in situ* hybridization on chromosomes to region 3q21–27 of human chromosome 3 (Barker *et al.,* 1989; Tran-Paterson *et al.,*

```
                                1                      20                                             40                           60
E-24.11-h   .......... ....GK SESQMDITDI NTPKPKKKQR ....WTPLE. .......ISLS VLVLLLTIIA VTMIA....L YATYDDGICK SSDCIKSAAR
ECE-h       .STYKRATLD EEDLVDSLSE GDAYPNGLQV NFHSPRSGQR CWAARTQVE. ......KRLV VLVVLLAAGL VACLAALGIQ YQTRSPSVCL SEACVSVTSS
PEX-h       .......... .......... ....EAETGS SVETGKKANR ..GTRIALV. ......VFVG GTLVLGTILF LVSQGLLSLQ ...AKQEYCL KPECIEAAAA
KELL-h      .......... ...EGGDQSE EEPRERSQAG GMGTLWSQES TPEERLPVEG SRPWAVARRV LTAILILGLL LCFSVLLFYN FQNCGPRPCE TSVCLDLRDH
Consensus   ---------- ---------- ---------- ---------- ---------- ---------- ----L----- ---------- --------C- ---C------

                       80                        100                        120                        140                        160
E-24.11-h   LIQNMDATTE PCTDFFKYAC GGWLKRNVIP ETSSRYGNFD ILRDELEVVL KDVLQEPKT. .EDIVAVQKA KALYRSCINE SAIDSRGGEP LLKLLPD...
ECE-h       ILSSMDPTVD PCHDFFSYAC GGWIKANPVP DGHSRWGTFS NLWEHNQAII KHLLENSTA. .SVSEAERKA QVYYRACMNE TRIEELRAKP LMELIER...
PEX-h       ILSKVNLSVD PCDNFFRFAC DGWISNNPIP EDMPSYGVYP WLRHNVDLKL KELLEKSISR RRDTEAIQKA KILYSSCMNE KAIEKADAKP LLHILRHSPF
KELL-h      YLASGNTSVA PCTDFFSFAC G......... RAKETNNSFQ ELATKNKNRL RRILEVQNSW HPGS.GEEKA FQFYNSCMDT LAIEAAGTGP LRQVIEE...
Consensus   ---------- PC--FF--AC ---------- ---------- -L-------- ---L------ --------KA ---Y--C--- --I------P L---------

                                  180                        200                        220                        240
E-24.11-h   ...IYGWPVA TE..NWEQKY GASWTAEKAI AQLNSKYGKK VLINLFVGTD DKNSVNHVIH IDQPRLGLPS R.DYYECTGI YKEACTAYVD FMISVARLIR
ECE-h       ...LGGWNIT GP..WAKDNF ......QDTL QVVTAHYRTS PFFSVYVSAD SKNSNSNVIQ VDQSGLGLPS R.DYYLNKTE NEKVLTGYLN YMVQLGKLLG
PEX-h       RWPVLESNIG PEGVWSERKF S....LLQTL ATFRGQYSNS VFIRLYVSPD DKASNEHILK LDQATLSLAV REDYLDNSTE AKSYRDALYK FMVDTAVLLG
KELL-h      ...LGGWRIS GK........ WTSLNFNRTL RLLMSQYGHF PFFRAYLGPH PASPHTPVIQ IDQPEFDVPL KQD..QEQKI YAQIFREYLT YLNQLGTLLG
Consensus   ---------- ---------- ---------- ------Y--- ---------- ---------- -DQ------- --D------- ---------- -------L--

             260                   280                        300                        320                        340
E-24.11-h   QEERLPIDEN QLALEMNKVM ELEKEIANAT AKPEDRNDPM LLYNKMTLAQ IQNNFSLEIN GKPFSWLNFT NEI....MST VNISITNEED VVVYAPEYLT
ECE-h       GG.....DEE AIRPQMQQIL DFETALANIT IPQEKRRDEE LIYHKVTAAE LQ......TL APAINWLPFL NTI....FYP VEIN..ESEP IVVYDKEYLE
PEX-h       A......NSS RAEHDMKSVL RLEIKIAEIM IPHENRTSEA M.YNKMNISE L......SAM IPQFDWLGYI KKVIDTRLYP HLKDISPSEN VVVRVPQYFK
KELL-h      G......DPS KVQEHSSLSI SITSRLFQFL RPLEQRRAQG KLFQMVTIDQ LKE......M APAIDWLSCL QATFTP.... ..MSLSPSQS LVVHDVEYLK
Consensus   ---------- ---------- ---------- ---E-R---- ---------- ---------- -----WL--- ---------- ---------- -VV----Y--

                   360                        380                        400                        420                        440
E-24.11-h   KLKPILTK.. .YSARDLQNL MSWRFIMDLV SSLSRTYKES RNAFRKALYG TTSETA.... TWRR.CANYV NGNMENAVGR LYVEAAFAGE SKHVVEDLIA
ECE-h       QISTLINT.. .TDRCLLNNY MIWNLVRKTS SFLDQRFQDA DEKFMEVMYG TKKTCL.... PRWKFCVSDT ENNLGFALGP MFVKATFAED SKSIATEIIL
PEX-h       DLFRILGS.. .ERKKTIANY LVWRMVYSRI PNLSRRFQYR WLEFSRVIQG TT.TLL.... PQWDKCVNFI ESALPYVVGK MFVDVYFQED KKEMMEELVE
KELL-h      NMSQLVEEML LKQRDFLQSH MILGLVVTLS PALDSQFQEA RRKLSQKLRE LTEQPPMPAR PRWMKCVEET GTFFEPTLAA LFVREAFGPS TRSAAMKLFT
Consensus   ---------- ---------- ---------- --L------- ---------- ---------- ----C----- ---------- --V---F--- ----------
```

```
                                   460                   480                   500                   520                   540
E-24.11-h    QIREVFIQTL D.DLTWMDAE TKKRAEEKAL AIKERIGYPD DIVSNDNKLN NEYLELNYKE DEYFENIIQN LKFSQSKQLK KLREKVDKDE WISGAAVVNA
ECE-h        EIKKAFEESL S.TLKWMDEE TRKSAKEKAD AIYNMIGYP. NFIMDPKELD KVFNDYTAVP DLYFENAMRF FNFSWRVTAD QLRKAPNRDQ WSMTPPMVNA
PEX-h        GVRWAFIDML EKENEWMDAG TKRKAKEKAR AVLAKVGYP. EFIMNDTHVN EDLKAIKFSE ADYFGNVLQT RKYLAQSDFF WLRKAVPKTE WFTNPTTVNA
KELL-h       AIRDALITRL .RNLPWMNEE TQNMAQDKVA QLQVEMG.AS EWALKPELAR QEYNDIQLGS S.FLQSVLSC VRSLRARIVQ SFLQPHPQHR WKVSPWDVNA
Consensus    ---------L -----WM--- T---A--K-- ------G--- ---------- ---------- ---------- ---------- ---------- W------VNA

                                   560                   580                   600                   620                   640
E-24.11-h    FYSSGRNQIV FPAGILQPPF F.SAQQSNSL NYGGIGMVIG HEITHGFDDN GRNFNKDGDL VDWWTQQSAS NFKEQSQCMV YQYGNFSWDL AGGQHLNGIN
ECE-h        YYSPTKNEIV FPAGILQAPF Y.TRSSPKAL NFGGIGVVVG HELTHAFDDQ GREYDKDGNL RPWWKNSSVE AFKRQTECMV EQYSNYS... VNGEPVNGRH
PEX-h        FYSASTNQIR FPAGELQKPF FWGTEYPRSL SYGAIGVIVG HEFTHGFDNN GRKYDKNGNL DPWWSTESEE KFKEKTKCMI NQYSNYYWKK A.GLNVKGKR
KELL-h       YYSVSDHVVV FPAGLLQPPF FH.PGYPRAV NFGAAGSIMA HELLHIFYQL ..LLPGGCL .....ACDNH ALQEAHLCLK RHYAAF..PL PSRTSFNDSL
Consensus    -YS------- FPAG-LQ-PF ---------- --G--G---- HE--H-F--- -------G-L ---------- -------C-- --Y------- ----------

                                   660                   680                   700                   720
E-24.11-h    TLGENIADNG GLGQAYRAYQ NYI...KKNG EEKLLPGLDL NHKQLFFLNF AQVWCGTYRP EYAVNSIKTD VHSPGNFRII GTLQNSAEFS EAFHCRKNSY
ECE-h        TLGENIADNG GLKAAYRAYQ NWV...KKNG AEHSLPTLGL TNNQLFFLGF AQVWCSVRTP ESSHEGLITD PHSPSRFRVI GSLSNSKEFS EHFRCPPGSP
PEX-h        TLGENIADNG GLREAFRAYR KWINDRRQGL EEPLLPGITF TNNQLFFLSY AHVRCNSYRP EAAREQVQIG AHSPPQFRVN GAISNFEEFQ KAFNCPPNST
KELL-h       TFLENAADVG GLAIALQAYS KRLL...RHH GETVLPSLDL SPQQIFFRSY AQVMCRKPSP QDSH.....D THSPPHLRVH GPLSSTPAFA RYFRCARGAL
Consensus    T--EN-AD-G GL--A--AY- ---------- -E--LP---- ---Q-FF--- A-V-C----P ---------- -HSP---R-- G-------F- --F-C-----

            740
E-24.11-h    MNPE.KKCRV W
ECE-h        MNPP.HKCEV W
PEX-h        MNRGMDSCRL W
KELL-h       LNPS.SRCQL W
Consensus    -N-----C-- W
```

Figure 6.2. Computer alignment of sequences of zinc metallopeptidases of the E-24.11 family. The sequences of the human forms of E-24.11, endothelin-converting enzyme (ECE), the putative product of the *PEX* gene (PEX) and the erythrocyte Kell protein (KELL) are tabulated. The conserved residues are shown in the consensus line.

1989), close to two other type II membrane proteins, sucrase–isomaltase and the transferrin receptor. The gene covers at least 80 kb and the transcribed sequences are divided into 24 exons and 23 introns (D'Adamio *et al.,* 1989). Exon 3 contains the translation initiation codon, and the intracellular and transmembrane domains of the protein. Most of the ectodomain including the active site of the enzyme is encoded by exons 4–23. The last 32 amino acid residues of the protein and the 3′ untranslated region are encoded in exon 24. The 3′ untranslated region of E-24.11 mRNA shows five putative polyadenylation sites.

Exons 1 and 2 encode the 5′ untranslated region of the E-24.11 transcripts and appear to be responsible in part for the diversity of E-24.11 mRNAs observed. In humans, D'Adamio *et al.* (1989) have characterized three transcripts with differences in their 5′ untranslated region. Type 1 transcript is generated by splicing exon 1 directly to exon 3. The other two transcripts are obtained by splicing part of exon 2 to exon 3 (type 2a) or all of exon 2 to exon 3 (type 2b). The presence of two distinct donor splice sites 171 bp apart in exon 2 explains this differential splicing. In rodents, despite the presence of the two donor sites, transcript 2a is not found (Li *et al.*, 1995a). Other transcripts have been reported recently. Li *et al.* (1995b) have used reverse transcription coupled to polymerase chain reaction to show the presence of a third non-coding exon in the E-24.11 gene. This new non-coding exon located downstream of exon 2 is directly spliced to the first coding exon in type 3 transcripts. Such transcripts have been detected in human kidney and lung and also in rat tissues. Finally, Haouas *et al.* (1995) have detected in Nalm 6 cells, a human leukaemic cell line, the presence of a transcript containing sequences located downstream of exon 1. It has been suggested that the use of different transcription initiation sites determines the 5′ structure of E-24.11 transcripts (D'Adamio *et al.*, 1989; Li *et al.*, 1995b). These different transcripts appear to be expressed in a tissue-specific and probably developmentally regulated way. For example, in rat brain type 1 transcript is the major mRNA while type 2b is the most abundant in the kidney (Li *et al.*, 1995a). Furthermore, localized expression of the different transcripts has been observed in the brain: type 1 transcript is associated with neurons while type 3 mRNA is found in oligodendrocytes of the corpus callosum (Li *et al.*, 1995a). Similar tissue-specific distribution of the different 5′ transcripts is also observed in human tissues (Li *et al.*, 1995b). The complexity of the 5′ untranslated region and the presence of five putative polyadenylation sites in the 3′ untranslated region explain the diversity of the E-24.11 transcripts observed in different tissues and cell lines by Northern blot hybridization. Indeed, several species of E-24.11 mRNAs can be observed ranging in size from 2.7 to 5.7 kb, with major species at 3.7 and 5.7 kb.

6.7 Control of expression of E-24.11

E-24.11 is expressed in many tissues and cell types where it is probably involved in the inactivation of biologically active peptides. The topics of E-24.11 localization and functions are discussed in more detail in other sections of this book (see Chapters 19, 20 and 22) and will not be pursued further in this chapter. Expression of E-24.11 has been shown to be regulated by various modulators in several cell types. For example, E-24.11

expression has been shown to cycle in response to progesterone in human endometrium during the ovarian cycle (Casey *et al.,* 1991) and to be up-regulated by calcitonin and 1,25 dihydroxyvitamin D_3 in human osteoblasts (Howell *et al.,* 1993). Stimulation of neutrophils with C5a (Werfel *et al.,* 1991), f-MLP or substance P (Shipp *et al.,* 1991), GM-CSF (Connelly *et al.,* 1993) or lipopolysaccharides (Fagny *et al.,* 1995) also results in increased expression of E-24.11. The protein kinase C transduction pathway seems to be involved in regulating E-24.11 expression. Indeed, incubation of human neutrophils with phorbol 12-myristate 13-acetate (PMA) (Fagny *et al.,* 1995) or treatment of fibroblasts or transformed mammary epithelial cells with 12-*O*-tetradecanoylphorbol-13-acetate (Werb and Clark, 1989) results in a down-regulation of E-24.11 expression. In contrast, treatment of endothelial cells with thrombin or PMA, two activators of the protein kinase C pathway, provokes an up-regulation of E-24.11 expression (Graf *et al.,* 1995b). These different observations concerning the effects of the activation of protein kinase C may indicate an indirect role of this transduction pathway in regulating E-24.11 expression. In a manner similar to the protein kinase C pathway, activation of protein kinase A also stimulates expression of E-24.11 in endothelial cells (Graf *et al.,* 1995a). Finally, glucocorticoids were also shown to up-regulate E-24.11 expression in transformed human tracheal epithelial cells (Borson and Gruenert, 1991).

E-24.11 expression is developmentally regulated (Sunday *et al.,* 1992). In fetal human lung, E-24.11 is expressed at high levels early in development and appears to be involved in the growth of the undifferentiated airway epithelium in distal developing airways by controlling the activity of bombesin-like peptides. All these observations suggest that E-24.11 expression is controlled through several complex mechanisms.

The structure of the E-24.11 gene promoter region has been reported recently (Ishimaru and Shipp, 1995). Two sets of regulatory elements have been characterized in this region by transient expression assays. The first, called type 1 promoter, is located upstream of exon 1 and generates type 1 transcripts. The second, called type 2 promoter, is upstream of exon 2 and controls the production of type 2a and type 2b transcripts. Both promoters lack the classical TATA box and several transcription initiation sites have been determined for each promoter. The type 1 promoter is purine-rich and contains several putative PU.1 binding sites. PU.1 is a member of the *ets* family of transcription factors highly expressed in B lymphocytes and macrophages, where it controls the expression of several genes specific to these cells (Nelsen *et al.*, 1993; Pahl *et al.*, 1993; Pongubala *et al.*, 1992; Shin and Koshland, 1993; Zhang *et al.*, 1994). The abundance of type 1 transcripts found in fetal thymus and lymphoblastic leukaemia cell lines is consistent with a role of PU.1 in controlling E-24.11 expression in haematopoietic tissues. The type 1 regulatory region also contains several copies of the sequence GGAA, the core motif for binding of transcription factor PEA3. This factor is another member of the *ets* family of transcription factors. Interestingly, PEA3 is abundantly found in the brain, a tissue where type 1 transcript is the major E-24.11 mRNA (Ishimaru and Shipp, 1995).

The type 2 regulatory region is rich in guanine and cytosine and is characterized by the presence of putative SP1 binding sites. SP1 is a transcription factor believed to be necessary for initiation of transcription from TATA-less promoters (Pugh and Tjian,

1991). The type 2 promoter also contains an inverted CCAAT box which is often found in promoters of genes controlling initiation of the cell cycle (Knight *et al.*, 1987). This regulatory region is also characterized by the presence of a retinoblastoma control element (RCE). Interestingly, RCE sequences have been shown to be involved in up-regulating SP1-mediated transcription by binding the retinoblastoma protein (Kim *et al.*, 1992). Thus the E-24.11 gene promoter region contains several control elements that can explain the tissue-specific and developmental regulations observed for this cell-surface peptidase.

Regulation of E-24.11 expression is not only controlled at the transcriptional level. Recent work by Ishimaru *et al.* (1996) suggests that transcript stability might be involved. Indeed, treatment of REH lymphoblast cells with phorbol ester resulted in a decrease in the levels of E-24.11 transcripts. This decrease is not due to a slow down in transcriptional rate but rather to a destabilization and degradation of the E-24.11 mRNAs. It should be emphasized that exon 24 encoding the 3′-untranslated region of E-24.11 mRNA shows the presence of five putative polyadenylation sites which could be used to generate messages of different length and stability. Furthermore, the 3′-untranslated region contains several adenine/uridine rich elements which also could determine message stability and degradation (Shipp *et al.*, 1988). It thus appears that expression of E-24.11 in different cell types is regulated by several mechanisms.

6.8 E-24.11 as a model for other zinc metallopeptidases

It now appears that E-24.11 is part of a larger family of cell-surface metallopeptidases with similar amino acid sequences. One of these peptidases is endothelin-converting enzyme (ECE) which catalyses the conversion of the biosynthetic intermediate form ‘big endothelin’ to its biologically active product (for a recent review see: Turner and Murphy, 1996, and Chapter 9). Molecular cloning revealed the existence of various isoforms of ECE. ECE-1 has two isoforms, designated ECE-1a and ECE-1b. These two isoforms have different N-terminal amino acid sequences which are probably produced by alternative splicing. These isoforms show similar enzymic properties, tissue distribution and subcellular localization (Shimada *et al.,* 1995). More recently, another form of ECE (ECE-2), with a more acidic optimal pH has been cloned (Emoto and Yanagisawa, 1995). Both ECE-1 and ECE-2 (not shown) exhibit sequence similarities to E-24.11 (*Figure 6.2*). ECE-1 contains 14 cysteine residues, of which 10 are conserved in E-24.11. As disulphide bonds are probably crucial for maintaining the tertiary structure of this class of enzyme, this suggests that similarities also exist in the three-dimensional organizations of both enzymes. Unlike E-24.11, ECE-1 exists as a disulphide-linked dimer. Mutations introduced at the position of cysteine residue(s) showed that Cys412 in ECE was solely responsible for the formation of dimers (Shimada *et al.,* 1996). In addition to the conserved HEXXH motif, other residues essential for zinc ligation and substrate binding are conserved between E-24.11 and ECE. Arg102, involved in substrate binding in E-24.11, is conserved in ECE, but Arg747 is replaced in ECE by a glutamic acid residue. The third zinc ligand in E-24.11 (Glu646) is also present in ECE and is part of a conserved sequence ENXAD found also in all other members of the family. Asp650 in this sequence has been shown to be crucial for catalytic activity in E-24.11 (see Section 6.2.3). The functional role of these residues in ECE has recently been probed by site-directed mutagenesis (Shimada *et al.,* 1996). Asn542, involved in binding of the P_2' residue of substrate in E-24.11 is conserved throughout the members of the family. However its role in ECE has still to be verified experimentally.

The erythrocyte Kell protein (Lee *et al.*, 1991), retains a disulphide bonding pattern as well as zinc ligands and substrate binding residues equivalent to those in ECE and E-24.11. However, Arg102 and Arg747 are not conserved. No enzymic activity has been reported yet for Kell, and its function at the cell surface of erythrocytes is still a matter of debate.

PEX is a gene involved in X-linked hypophosphatemic rickets, a dominant disorder characterized by impaired phosphate uptake in the kidney (Francis *et al.,* 1995). This gene exhibits homology to the E-24.11 family of endopeptidases. The relation between *PEX* and phosphate reabsorption in the proximal tubules of the nephron is obscure. It is conceivable that PEX is a new endopeptidase acting on a still unknown peptide factor controlling phosphate transport in the kidney. As has been shown for ECE and Kell, there are substantial sequence similarities between E-24.11 and PEX, especially for cysteine residues and most of the important amino acids involved in catalysis, substrate binding and zinc chelation.

Considering the existence of several endopeptidases structurally related to E-24.11, one may wonder how many additional members of this family remain to be discovered. E-24.11 and ECE are the targets for the synthesis of inhibitors (see Chapters 5 and 20) that could be of benefit in a range of disorders, such as hypertension, congestive heart failure and stroke. Clearly, the existence of other members which could also be inhibited along with the intended target, should be of concern to minimize undesirable side effects of these drugs. Solving the atomic structure of E-24.11 by X-ray crystallography, which is well underway in our group, will hopefully lead to the structure of other members of the same family by homology modelling. Knowledge of these structures should help the rational design of specific inhibitors, which in turn will constitute invaluable tools for establishing the physiological function of the new members of the family.

Acknowledgement

Work in the laboratories of P.C. and G.B. was supported by grants from the Medical Research Council of Canada.

References

Almenoff J, Orlowski M. (1983) Membrane bound kidney neutral endopeptidase: interaction with synthetic substrates, natural peptides, and inhibitors. *Biochemistry* **22:** 590–599.

Argos P, Garavito RM, Eventoff W, Rossmann MG. (1978) Similarities in active center geometries of zinc-containing enzymes, proteases and dehydrogenases. *J. Mol. Biol.* **126:** 141–158.

Aubry M, Zollinger M, Fortin S, Vénien C, LeGrimellec C, Crine P. (1988) Monoclonal antibodies as probes for the transmembrane structure of neutral endopeptidase 24.11 ("enkephalinase"). *Biochim. Biophys. Acta* **967:** 56–64.

Barker PE, Shipp MA, D'Adamio L, Masteller EL, Reinherz EL. (1989) The common acute lymphoblastic leukemia antigen gene maps to chromosomal region 3 (q21–q27). *J. Immunol.* **142**, 283–287.

Barnes K, Bourne A, Cook PA, Turner AJ, Kenny AJ. (1991) Membrane peptidases in the peripheral nervous system of the pig: their localization by immunohistochemistry at light and electron microscopic levels. *Neuroscience* **44:** 245–261.

Bateman RC Jr, Hersh LB. (1987). Evidence for an essential histidine in neutral endopeptidase 24.11. *Biochemistry* **26:** 4237–4242.

Bateman RC Jr, Jackson D, Slaughter CA, Unnithan S, Chai YG, Moonaw C, Hersh LB. (1989) Identification of the active-site arginine in rat neutral endopeptidase 24.11 (enkephalinase) as arginine 102 and analysis of a glutamine 102 mutant. *J. Biol. Chem.* **264:** 6151–6157.

Bateman RC Jr, Kim Y-A, Slaughter C, Hersh LB. (1990) *N*-Bromoacetyl-D-leucylglycine. An affinity label for neutral endopeptidase 24.11. *J. Biol. Chem.* **265:** 8365–8368.

Beaumont A, Roques BP. (1986) Presence of a histidine at the active site of the neutral endopeptidase-24.11. *Biochem. Biophys. Res. Commun.* **139:** 733–739.

Beaumont A, Barbe B, LeMoual H, Boileau G, Crine P, Fournié-Zaluski M-C, Roques BP. (1992) Charge polarity reversal inverses the specificity of neutral endopeptidase-24.11. *J. Biol. Chem.* **267:** 2138–2141.

Beaumont A, LeMoual H, Boileau G, Crine P, Roques BP. (1991) Evidence that both arginine 102 and arginine 747 are involved in substrate binding to neutral endopeptidase (EC 3.4.24.11) *J. Biol. Chem.* **266:** 214–220.

Benchetrit T, Bissery V, Mornon JP, Devault A, Crine P, Roques BP. (1988) Primary structure homologies between two zinc metallopeptidases, the neutral endopeptidase 24.11 ("enkephalinase") and thermolysin, through clustering analysis. *Biochemistry* **27:** 592–596.

Booth AG, Kenny AJ. (1980) Proteins of the kidney microvillus membrane. Asymmetric labelling of the membrane by lactoperoxidase-catalysed radio-iodination and by photolysis of 3,5-di[^{125}I]iodo-4-azidobenzenesulphonate. *Biochem. J.* **187:** 31–44.

Borson DB, Gruenert DC. (1991) Glucocorticoids induce neutral endopeptidase in transformed human tracheal epithelial cells. *Am. J. Physiol. Lung Cell Mol. Physiol.* **260:** L83–L89.

Cailler F, Howell S, Crine P. (1995) Role of the glycosyl-phosphatidylinositol anchor in membrane sorting in MDCK cells. *Biochem. Soc. Trans.* **24:** 1865.

Casey ML, Smith JW, Nagai K, Hersh LB, MacDonald PC. (1991) Progesterone-regulated cyclic modulation of membrane metalloendopeptidase (enkephalinase) in human endometrium. *J. Biol. Chem.* **266:** 23041–23047.

Chen C-Y, Salles G, Seldin MF, Kister AE, Reinherz EL, Shipp MA. (1992) Murine common acute lymphoblastic leukemia antigen (CD10 neutral endopeptidase 24.11): molecular characterization, chromosomal localization, and modeling of the active site. *J. Immunol.* **148:** 2817–2825.

Christianson DW. (1991) Structural biology of zinc. *Adv. Prot. Chem.* **42:** 281–355.

Christianson DW, Alexander RS. (1989) Carboxylate-His zinc interactions in protein structure and function. *J. Am. Chem. Soc.* **111:** 6412–6419.

Christianson DW, Alexander RS. (1990) Another catalytic triad? *Nature* **346:** 225.

Christianson DW, Lipscomb WN. (1985) Binding of a possible transition state analogue to the active site of carboxypeptidase A. *Proc. Natl Acad. Sci. USA.* **82:** 6840–6844.

Coleman PM, Jansonius JN, Matthews BW. (1972) The structure of thermolysin: an electron density map at 2–3 Å resolution. *J. Mol. Biol.* **70:** 701–724.

Connelly JC, Chambless R, Holiday D, Chittenden K, Johnson AR. (1993) Up-regulation of neutral endopeptidase (CALLA) in human neutrophils by granulocyte-macrophage colony-stimulating factor. *J. Leuk. Biol.* **53:** 685–690.

D'Adamio L, Shipp MA, Masteller EL, Reinherz EL. (1989) Organization of the gene encoding common acute lymphoblastic leukemia antigen (neutral endopeptidase 24.11): multiple miniexons and separate 5′ untranslated regions. *Proc. Natl Acad. Sci. USA* **86:** 7103–7107.

Devault A, Lazure C, Nault C, *et al.* (1987) Amino acid sequence of rabbit kidney neutral endopeptidase 24.11 (enkephalinase) deduced from a complementary DNA. *EMBO J.* **6:** 1317–1322.

Devault A, Nault C, Zollinger M, Fournié-Zaluski M-C, Roques BP, Crine P, Boileau G. (1988a) Expression of neutral endopeptidase (enkephalinase) in heterologous COS-1 cells. Characterization of the recombinant enzyme and evidence for a glutamic acid residue at the active site. *J. Biol. Chem.* **263:** 4033–4040.

Devault A, Sales V, Nault C, Beaumont A, Roques BP, Crine P, Boileau G. (1988b) Exploration of the catalytic site of endopeptidase 24.11 by site-directed mutagenesis. Histidine residues 583 and 587 are essential for catalysis. *FEBS Lett.* **231:** 54–58.

Dion N, LeMoual H, Crine P, Boileau G. (1993) Kinetic evidence that His^{711} of neutral endopeptidase 24.11 is involved in stabilization of the transition state. *FEBS Lett.* **318:** 301–304.

Dion N, LeMoual H, Fournié-Zaluski M-C, Roques BP, Crine P, Boileau G. (1995) Evidence that Asn^{542} of neprilysin (EC 3.4.24.11) is involved in binding of the P_2' residue of substrates and inhibitors. *Biochem. J.* **311:** 623–627.

Emoto N, Yanagisawa M. (1995) Endothelin-converting enzyme-2 is a membrane-bound, phosphoramidon-sensitive metalloprotease with acidic pH optimum. *J. Biol. Chem.* **270:** 15262–15268.

Erdös EG, Skidgel RA. (1988) Human neutral endopeptidase 24.11 (NEP, enkephalinase); function, distribution and release. *Adv. Exp. Med. Biol.* **240:** 13–21.

Fagny C, Marchant A, De Prez E, Goldman M, Deschodt-Lanckman M. (1995) Lipopolysaccharide induces upregulation of neutral endopeptidase 24.11 on human neutrophils: involvement of the CD14 receptor. *Clin. Sci.* **89:** 83–89.

Francis F, Hennig S, Korn B, Reinhardt R, De Jong P, Poustka A, *et al. (HYP Consortium)* (1995) A gene (*PEX*) with homologies to endopeptidases is mutated in patients with X-linked hypophosphatemic rickets. *Nature Genetics* **11:** 130–136.

Fulcher IS, Matsas R, Turner AJ, Kenny AJ. (1982) Kidney neutral endopeptidase and the hydrolysis of enkephalin by synaptic membranes show similar sensitivity to inhibitors. *Biochem. J.* **203:** 519–522.

Gaboriaud C, Bissery V, Benchetrit T, Mornon JP. (1987) Hydrophobic cluster analysis: an efficient new way to compare and analyse amino acid sequences. *FEBS Lett.* **224:** 149–155.

Graf K, Koehne P, Gräfe M, Zhang M, Auch-Schwelk W, Fleck E. (1995a) Regulation and differential expression of neutral endopeptidase 24.11 in human endothelial cells. *Hypertension* **26:** 230–235.

Graf K, Kunkel K, Zhang MH, Gräfe M, Schultz K, Schudt C, Biros S, Fleck E, Kunkel G. (1995b) Activation of adenylate cyclase and phosphodiesterase inhibition enhance neutral endopeptidase activity in human endothelial cells. *Peptides* **16:** 1273–1278.

Hangauer DG, Monzingo AF, Matthews BW. (1984) An interactive computer graphics study of thermolysin-catalyzed peptide cleavage and inhibition by *N*-carboxymethyl dipeptides. *Biochemistry* **23:** 5730–5744.

Haouas H, Morello D, Lavenu A, Billard M, Jasmin C, Boucheix C. (1995) Characterization of the 5′ region of the CD10/neutral endopeptidase 24.11 gene. *Biochem. Biophys. Res. Commun.* **207:** 933–942.

Hersh LB, Morihara K. (1986) Comparison of the subsite specificity of the mammalian neutral endopeptidase 24.11 (enkephalinase) to the bacterial neutral endopeptidase thermolysin. *J. Biol. Chem.* **261:** 6433–6437.

Higaki JN, Fletterick RJ, Craik CS. (1992) Engineered metalloregulation in enzymes. *Trends Biochem. Sci.* **17:** 100–104.

Hitt AL, Lu TH, Luna EJ. (1994) Ponticulin is an atypical membrane protein. *J. Cell Biol.* **126:** 1421–1431.

Holmes MA, Matthews BW. (1982) Structure of thermolysin refined at 1.6 Å resolution. *J. Mol. Biol.* **160:** 623–639.

Hooper NM. (1994) Families of zinc metalloproteases. *FEBS Lett.* **354:** 1–6.

Howell S, Caswell AM, Kenny AJ, Turner AJ. (1993) Membrane peptidases on human osteoblast-like cells in culture: hydrolysis of calcitonin and hormonal regulation of endopeptidase-24.11. *Biochem. J.* **290:** 159–164.

Howell S, Lanctôt C, Boileau G, Crine P. (1994a) Expression of an enzymatically active glycosylphosphatidylinositol-anchored form of neutral endopeptidase (EC 3.4.24.11) in Cos-1 cells. *Biochem. J.* **299:** 171–176.

Howell S, Lanctôt C, Boileau G, Crine P. (1994b) A cleavable N-terminal signal peptide is not a prerequisite for the biosynthesis of glycosylphosphatidylinositol-anchored proteins. *J. Biol. Chem.* **269:** 16993–16996.

Ishimaru F, Shipp MA. (1995) Analysis of the human CD10/neutral endopeptidase 24.11 promoter region: two separate regulatory elements. *Blood* **85:** 3199–3207.

Ishimaru F, Potter NS, Shipp MA. (1996) Phorbol ester-mediated regulation of CD10/neutral endopeptidase transcripts in acute lymphoblastic leukemias. *Exp. Hematol.* **24:** 43–48.

Jackson DG, Hersh LB. (1986) Reaction of neutral endopeptidase 24.11 (enkephalinase) with arginine reagents. *J. Biol. Chem.* **261:** 8649–8654.

Johnson AR, Coalsopn JJ, Ashton J, Larumbide M, Erdös EG. (1985) Neutral endopeptidase in serum samples from patients with adult respiratory distress syndrome. Comparison with angiotensin-converting enzyme. *Am. Rev. Resp. Dis.* **132:** 1262–1267.

Kenny AJ. (1977) Proteinases associated with cell membranes. In: *Proteinases in Mammalian Cells and Tissues* (ed. A.J. Barrett), pp. 393–423. North-Holland, New York.

Kenny AJ, Maroux S. (1982) Topology of microvillar membrane hydrolases of kidney and intestine. *Physiol. Rev.* **62:** 91–128.

Kenny AJ, Fulcher IS, McGill KA, Kershaw D. (1983) Proteins of the kidney microvillar membrane. Reconstitution of endopeptidase in liposomes shows that it is a short-stalked protein. *Biochem. J.* **211:** 755–762.

Kenny AJ, Stephenson SL, Turner AJ. (1987) Cell-surface peptidases. In: *Mammalian Ectoenzymes* (eds A.J. Kenny and A.J. Turner), pp. 169–210. Elsevier, Amsterdam.

Kerr MA, Kenny AJ. (1974a) The purification and specificities of a neutral endopeptidase from rabbit kidney brush border. *Biochem. J.* **137:** 477–488.

Kerr MA, Kenny AJ. (1974b) The molecular weight and properties of a neutral endopeptidase from rabbit kidney brush border. *Biochem. J.* **137:** 489–495.

Kester WR, Matthews BW. (1977) Comparison of the structures of carboxypeptidase A and thermolysin. *J. Biol. Chem.* **252:** 7704–7710.

Kim SJ, Wagner S, Liu F, O'Reilly MA, Robbins PD, Green MR. (1992) Retinoblastoma gene product activates expression of the human TGF-beta 2 gene through transcription factor ATF-2. *Nature* **358:** 331–334.

Knight GB, Gudas JM, Pardee AB. (1987) Cell-cycle-specific interacton of nuclear DNA-binding proteins with a CCAAT sequence from the human thymidine kinases gene. *Proc. Natl Acad. Sci. USA* **84:** 8350–8354.

Lafrance M-H, Vézina C, Wang Q, Boileau G, Crine P, Lemay G. (1994) Role of glycosylation in transport and enzymic activity of neutral endopeptidase-24.11. *Biochem. J.* **302:** 451–454.

Lee S, Zambas ED, Marsh WL, Redman CM. (1991) Molecular cloning and primary structure of Kell blood group protein. *Proc. Natl Acad. Sci. USA.* **88:** 6353–6367.

Lemay G, Waksman G, Roques BP, Crine P, Boileau G. (1989) Fusion of a cleavable signal peptide to the ectodomain of neutral endopeptidase (EC 3.4.24.11) results in the secretion of an active enzyme in COS-1 cells. *J. Biol. Chem.* **264:** 15620–15623.

Lemay G, Zollinger M, Waksman G, Roques BP, Crine P, Boileau G. (1990) Recombinant neutral endopeptidase-24.11 expressed in mouse neuroblastoma cells is associated with neurite membranes. *Biochem. J.* **267:** 447–452.

LeMoual H, Devault A, Roques BP, Crine P, Boileau G. (1991) Identification of glutamic acid 646 as a zinc-coordinating residue in endopeptidase-24.11. *J. Biol. Chem.* **266:** 15670–15674.

LeMoual H, Dion N, Roques BP, Crine P, Boileau G. (1994) Asp^{650} is crucial for catalytic activity of neutral endopeptidase 24–11. *Eur. J. Biochem.* **221:** 475–480.

Letarte M, Vera S, Tran R, Addis JB, Onizuka RJ, Quackenbush EHJ, Jomgeneel CV, McInnes RR. (1988) Common acute lymphocytic leukemia antigen is identical to neutral endopeptidase. *J. Exp. Med.* **168:** 1247–1253.

Li C, Booze RM, Hersh LB. (1995a) Tissue-specific expression of rat neutral endopeptidase (neprilysin) mRNAs. *J. Biol. Chem.* **270:** 5723–5728.

Li C, Chen G, Gerard NP, Gerard C, Bozic CR, Hersh LB. (1995b) Comparison of the structure and expression of the human and rat neprilysin (endopeptidase 24.11)-encoding genes. *Gene* **164:** 363–366.

Llorens-Cortes C, Huang H, Vicart J, Gasc M, Paulin D, Corvol P. (1992) Identification and characterization of neutral endopeptidase in endothelial cells from venous or arterial origins. *J. Biol. Chem.* **267:** 14012–14018.

Malfroy B, Schwartz JC. (1982) Properties of "enkephalinase" from rat kidney: comparison of dipeptidyl-carboxypeptidase and endopeptidase activities. *Biochem. Biophys. Res. Commun.* **106:** 276–285.

Malfroy B, Schofield PR, Kuang WJ, Seeburg PH, Mason AJ, Henzel WJ. (1987) Molecular cloning and amino acid sequence of rat enkephalinase. *Biochem. Biophys. Res. Commun.* **144:** 59–66.

Malfroy B, Kuang WJ, Seeburg PH, Mason AJ, Schofield PR. (1988) Molecular cloning and amino acid sequence of human enkephalinase (neutral endopeptidase) *FEBS Lett.* **229:** 206–210.

Matsas R, Fulcher IS, Kenny AJ, Turner AJ. (1983) Substance P and [Leu]enkephalin are hydrolysed by an enzyme in pig caudate synaptic membranes that is identical with the endopeptidase of kidney microvilli. *Proc. Natl Acad. Sci. USA* **80:** 3111–3115.

Matsas R, Kenny AJ, Turner AJ. (1984) The metabolism of neuropeptides. The hydrolysis of peptides, including enkephalins, tachykinins and their analogues, by endopeptidase-24.11. *Biochem. J.* **223:** 433–440.

Nelsen B, Tian G, Erman B, Gregoire J, Maki R, Graves B, Sen R. (1993) Regulation of lymphoid-specific immunoglobulin mu heavy chain gene enhancer by ETS-domain proteins. *Science* **261:** 82–86.

Ondetti MA, Rubin B, Cushman DW. (1977) Design of specific inhibitors of angiotensin-converting enzyme: new class of orally active antihypertensive agents. *Science* **196:** 441–444.

Orlowski M, Wilk S. (1981) Purification and specificity of a membrane-bound metalloendopeptidase from bovine pituitaries. *Biochemistry* **20:** 4942–4950.

Pahl HL, Scheibe RJ, Zhang DE, Chen HM, Galson DL, Maki RA, Tenen DG. (1993) The proto-oncogene PU,1 regulates expression of the myeloid-specific CD11b promoter. *J. Biol. Chem.* **268:** 5014–5020.

Pongubala JM, Nagulapalli S, Klemsz MJ, McKercher SR, Maki RA, Atchison ML. (1992) PU,1 recruits a second nuclear factor to a site important for immunoglobulin kappa 3′ enhancer activity. *Mol. Cell Biol.* **12:** 368–378.

Pozsgay M, Michaud C, Liebman M, Orlowski M. (1986) Substrate and inhibitor studies of thermolysin-like neutral metalloendopeptidase from kidney membrane fractions. Comparison with bacterial thermolysin. *Biochemistry* **25:** 1292–1299.

Pugh BF, Tjian R. (1991) Transcription from a TATA-less promoter requires a multisubunit TFIID complex. *Genes Dev.* **5:** 1935–1945.

Quay T, Slaughter C, Davis TP, Merril BJ, Hersh LB. (1994) Positional effects in the neprilysin (neutral endopeptidase) reaction. *Arch. Biochem. Biophys.* **308:** 133–136.

Relton JM, Gee NS, Matsas R, Turner AJ, Kenny AJ. (1983) Purification of endopeptidase-24.11 ("enkephalinase") from pig brain by immunoadsorbent chromatography. *Biochem. J.* **215:** 519–523.

Roderick SL, Fournié-Zaluski M-C, Roques BP, Matthews BW. (1989) Thiorphan and retro-thiorphan display equivalent interactions when bound to crystalline thermolysin. *Biochemistry* **28:** 1493–1497.

Roques BP, Fournié-Zaluski M-C. (1986) Enkephalin degrading enzyme inhibitors: a physiological way to new analgesics and psychoactive agents. *NIDA Res. Monogr.* **70**: 128–154.

Roques BP, Fournié-Zaluski M-C, Soroca E, Lecomte J-M, Malfroy B, Llorens-Cortes C, Schwartz J-C. (1980) The enkephalinase inhibitor thiorphan shows antinociceptive activity in mice. *Nature* **288:** 286–288.

Roques BP, Lucas-Soroca E, Chaillet P, Costentin J, Fournié-Zaluski M-C. (1983) Complete differentiation between enkephalinase and angiotensin-converting enzyme inhibition by retro-thiorphan. *Proc. Natl Acad. Sci. USA* **80:** 3178–3182.

Schwartz JC, De La Baume S, Malfroy B, Patey G, Perdrisot R, Swerts JP, Fournié-Zaluski M-C, Gacel G, Roques BP. (1980) "Enkephalinase", a newly characterized dipeptidyl carboxypeptidase: properties and possible role in enkephalinergic transmission. *Int. J. Neurol.* **14:** 195–204.

Semenza G. (1986) Anchoring and biosynthesis of stalked brush border membrane proteins: glycosidases and peptidases of enterocytes and renal tubuli. *Annu. Rev. Cell Biol.* **2:** 255–313.

Shimada K, Takahashi M, Ikeda M, Tanzawa K. (1995) Identification and characterization of two isoforms of an endothelin-converting enzyme-1. *FEBS Lett.* **371:** 140–144.

Shimada K, Takahashi M, Turner AJ, Tanzawa K. (1996) Rat endothelin-converting enzyme-1 forms a dimer through Cys^{412} with a similar catalytic mechanism and a distinct substrate binding mechanism compared with neutral endopeptidase-24.11. *Biochem. J.* **315:** 863–867.

Shin MK, Koshland ME. (1993) Ets-related protein PU,1 regulates expression of the immunoglobulin J-chain gene through a novel Ets-binding element. *Genes Dev.* **7:** 2006–2015.

Shipp MA, Stefano GB, Switzer SN, Griffin JD, Reinherz EL. (1991) CD10 (CALLA)/neutral endopeptidase 24.11 modulates inflammatory peptide-induced changes in neutrophil morphology, migration, and adhesion proteins and is itself regulated by neutrophil activation. *Blood* **78:** 1834–1841.

Skipp MA, Richardson NE, Sayre PH, Brown NR, Masteller EL, Clayton LK, Ritz J, Reinherz EL. (1988) Molecular cloning of the common acute lymphoblastic leukemia antigen (CALLA) identifies a type II integral membrane protein. *Proc. Natl Acad. Sci. USA* **85:** 4819–4823.

Sunday ME, Hua J, Torday JS, Reyes B, Shipp MA. (1992) CD10/neutral endopeptidase 24.11 in developing human fetal lung. Patterns of expression and modulation of peptide-mediated proliferation. *J. Clin. Invest.* **90:** 2517–2525.

Tran-Paterson R, Willard HF, Letarte M. (1989) The common acute lymphoblastic leukemia antigen (neutral endopeptidase-3.4.24.11) gene is located on human chromosome 3. *Cancer Genet. Cytogenet.* **42:** 129–134.

Turner AJ, Murphy LJ. (1996) Molecular pharmacology of endothelin converting enzymes. *Biochem. Pharmacol.* **51:** 91–102.

Vallée BL, Auld DS. (1990) Zinc coordination, function and structure of zinc enzymes and other proteins. *Biochemistry* **29:** 5647–5659.

Vallée BL, Galdes A. (1984) The metallobiochemistry of zinc enzymes. *Adv. Enzymol. Relat. Areas Mol. Biol.* **56:** 283–430.

Werb Z, Clark EJ. (1989) Phorbol diesters regulate expression of the membrane neutral metalloendopeptidase (EC 3.4.24.11) in rabbit synovial fibroblasts and mammary epithelial cells. *J. Biol. Chem.* **264:** 9111–9113.

Werfel T, Sonntag G, Weber MH, Götze O. (1991) Rapid increases in the membrane expression of neutral endopeptidase (CD10), aminopeptidase N (CD13), tyrosine phosphatase (CD45), and Fcγ-RIII (CD16) upon stimulation of human peripheral leukocytes with human C5a. *J. Immunol.* **147:** 3909–3914.

Zhang DE, Fujioka K, Hetherington CJ, Shapiro LH, Chen HM, Look AT, Tenen DG. (1994) Identification of a region which directs the monocytic activity of the colony-stimulating factor 1 (macrophage colony-stimulating factor) receptor promoter and binds PEBP2/CBF (AMIL1) *Mol. Cell Biol.* **14:** 8085–8095.

Chapter 7

Biochemical and molecular aspects of angiotensin I-converting enzyme

Pierre Corvol and Tracy A. Williams

7.1 Introduction

Angiotensin I-converting enzyme (ACE; peptidyl dipeptidase A; EC 3.4.15.1) is a zinc metallopeptidase which cleaves the C-terminal dipeptide from angiotensin I (AI) to produce the potent vasopressor octapeptide angiotensin II (AII) (Skeggs *et al.*, 1956) and inactivates bradykinin (BK) by the sequential removal of two C-terminal dipeptides (Yang *et al.*, 1970). In addition to these two main physiological substrates which are involved in blood pressure regulation and water and salt metabolism, ACE cleaves C-terminal dipeptides from various oligopeptides with a free C terminus. ACE is also able to act as an endopeptidase on certain substrates which are amidated at the C terminus by cleaving a C-terminal dipeptide amide. Clearly, ACE displays a wide substrate specificity and none of the names assigned to this enzyme adequately describe its varied actions: ACE has been implicated in a range of other physiological processes unrelated to blood pressure regulation such as immunity, reproduction and neuropeptide regulation due to its localization and/or the *in vitro* cleavage of a range of biologically active peptides.

The role of ACE in blood pressure control and water and salt metabolism has been defined mainly by the use of highly specific ACE inhibitors which have been developed during the last decade (Waeber *et al.*, 1990). These drugs clearly show their efficacy in the treatment of hypertension, congestive heart failure and diabetic nephropathy. Although the physiological importance of ACE is undebated, it is only relatively recently that its structure and function have been studied in detail by the application of molecular biological techniques.

Several reviews have been published on the biochemistry of ACE (Ehlers and Riordan, 1989; Erdös and Skidgel, 1987; Hooper, 1991), on its main enzymic characteristics and on ACE inhibitors (Lawton *et al.*, 1992). The purpose of the present review is to discuss new aspects of the structure and function of ACE as revealed by the recent discoveries made on the molecular biology of ACE .

Cell-Surface Peptidases in Health and Disease, edited by A.J. Kenny and C.M. Boustead.

7.2 Distribution

ACE is an ectoenzyme anchored to the plasma membrane, with the bulk of its mass exposed at the extracellular surface of the cell. There are two ACE isoforms: a somatic form of around 150–180 kDa found in endothelial, epithelial and neuronal cells and a smaller isoform (90–110 kDa) present in germinal cells. The structural relationship between these two isoenzymes has been elucidated by molecular cloning of the corresponding cDNAs and of the ACE gene (see Section 7.4).

7.2.1 Somatic ACE

ACE is found in the plasma membrane of vascular endothelial cells with high levels present at the vascular endothelial surface of the lung, such that the active sites of ACE are poised to metabolize circulating substrates (Oparil and Haber, 1974a,b). Indeed, early experiments suggested that the predominant conversion of AI to AII takes place at the endothelial surface of the lungs (Ng and Vane, 1967, 1968). However, more recently this concept has been challenged in view of the discovery of renin–angiotensin systems in other tissues, which may be the major sites of AI conversion. In addition to the endothelial location of ACE, the enzyme is also expressed in the brush borders of absorptive epithelia of the small intestine and the kidney proximal convoluted tubule (Bruneval *et al.*, 1986). ACE is also found in mononuclear cells, such as monocytes after macrophage differentiation and T-lymphocytes, and in fibroblasts (Costerousse *et al.*, 1993; Friedland *et al.*, 1978).

In vitro autoradiography, employing radiolabelled specific ACE inhibitors, and immunohistochemical studies have mapped the principal locations of ACE in brain (Barnes *et al.*, 1988; Defendini *et al.*, 1983; Strittmatter *et al.*, 1984). ACE was found primarily in the choroid plexus, which may be the source of ACE in cerebrospinal fluid, ependyma, subfornical organ, basal ganglia (caudate-putamen and globus pallidus), substantia nigra and pituitary. The co-localization of ACE and the mammalian tachykinin peptide, substance P, to the striatonigral neuronal pathway led to suggestions of a physiological role for ACE in the hydrolysis of substance P in brain (Strittmatter and Snyder, 1987; Strittmatter *et al.*, 1985). However, another mammalian ectoenzyme, endopeptidase-24.11, also present in the same neuronal pathway, is more likely to be involved in the *in vivo* degradation of this neuropeptide (Barnes *et al.*, 1988; Matsas *et al.*, 1984; Oblin *et al.*, 1988). Neuronal ACE is smaller in size than the more widely distributed form of the enzyme, a difference which has been attributed to differences in the extent of N-glycosylation (Hooper and Turner, 1987). Neuronal ACE and the endothelial form of the enzyme display the same neuropeptide specificities and do not represent distinct isoenzymes (Williams *et al.*, 1991).

7.2.2 Soluble ACE

A soluble form of ACE is present in many biological fluids such as serum (Das *et al.*, 1977), seminal fluid (El-Dorry *et al.*, 1983), amniotic fluid (Yasui *et al.*, 1984) and cerebrospinal fluid (Schweisfurth and Schioberg-Schiegnitz, 1984). The soluble form of ACE appears to be derived from the membrane-bound form of the enzyme in endothelial cells (see Section 7.8).

7.2.3 Germinal ACE

A smaller (90–100 kDa) isoenzyme of ACE (germinal ACE), is found exclusively in the testes (El-Dorry *et al.*, 1982; Velletri, 1985). The expression of germinal ACE is specifically regulated, being found in spermatids and not in immature forms of germinal cells (Berg *et al.*, 1986; Brentjens *et al.*, 1986). The maximum expression of ACE occurs during the acrosome phase in murine species. ACE is exclusively produced in haploid germ cells and belongs to the family of proteins whose expression during definite maturation steps of spermiogenesis appears to be correlated with the unique process of germ cell differentiation (Sibony *et al.*, 1994). The inactivation of the ACE gene by homologous recombination leads to homozygous male mice with markedly reduced blood pressure, severe renal abnormalities and the male homozygous mice display significantly reduced fertility (Esther *et al.*, 1996; Krege *et al.*, 1995). The effect on male fertility is attributable to the inactivation of the germinal form of ACE, although the physiological substrates of this ACE isoform have not yet been identified.

7.3 Catalytic properties

7.3.1 ACE is a zinc metallopeptidase

The metal dependency of ACE was inferred, when the enzyme was first identified from its inhibition by the metal-chelating agent, ethylene diamine tetraacetic acid (EDTA; Skeggs *et al.*, 1956) and subsequent studies, using atomic absorption spectroscopy, revealed the associated metal to be zinc (Bünning and Riordan, 1985; Ehlers and Riordan, 1991; Williams *et al.*, 1992). Zinc is essential for the catalytic activity of ACE and functions exclusively in the catalytic step of peptide hydrolysis (Bünning and Riordan, 1985). The role of the zinc ion in ACE catalysis is probably analogous to the active site zinc ion in carboxypeptidase A, thermolysin and endopeptidase-24.11, where the zinc ion functions directly in the catalytic step of peptide hydrolysis by ionizing or polarizing the zinc-bound water molecule which thus becomes activated to initiate the nucleophilic attack on the substrate carbonyl scissile bond (Vallee and Auld, 1990).

7.3.2 Substrate specificity

The primary specificity of ACE is to cleave C-terminal dipeptides from oligopeptide substrates with a free C terminus and the absence of a penultimate proline residue (Hooper, 1991), by which action ACE hydrolyses both AI and BK. BK is, in fact, the more favourable substrate for ACE (k_{cat}/K_m 3900–5000 sec^{-1} mM^{-1}), with a k_{cat}/K_m considerably higher than that of AI hydrolysis (147–189 sec^{-1} mM^{-1}; Ehlers and Riordan, 1990). ACE hydrolyses *in vitro* a wide range of substrates, in addition to AI and BK, by virtue of its classical specificity described above. These substrates include neurotensin (Skidgel *et al.*, 1984), [Met^5] enkephalin, [Met^5] enkephalin–Arg^6–Gly^7–Leu^8, β-neoendorphin, dynorphins

(Skidgel and Erdös, 1987) and insulin B chain (Igic *et al.*, 1972) . The specificity of ACE is not restricted to its peptidyl dipeptidase action: it has been shown to act as an endopeptidase on certain substrates which are blocked at the C terminus. In this respect ACE is able to release the C-terminal tripeptide amides from substance P (Skidgel *et al.*, 1984; Yokosawa *et al.*, 1983), luteinizing hormone-releasing hormone (LHRH; Skidgel and Erdös, 1985), the C-terminal tripeptide from des-Arg^9-BK, which contains a penultimate proline residue (Inokuchi and Nagamatsu, 1981) and the C-terminal dipeptide amides from cholecystokinin-8 and various gastrin analogues (Dubreuil *et al.*, 1989). In addition, and perhaps more surprisingly, ACE is also able to release the N-terminal tripeptide from LHRH as well as the blocked C-terminal tripeptide (Skidgel and Erdös, 1985). However, it should be noted that, while the hydrolysis of all these peptides has been demonstrated *in vitro,* there is no convincing evidence that they are so hydrolysed *in vivo* (with the exception of AI and BK).

ACE has recently been implicated in the *in vitro* hydrolysis of the tetrapeptide *N*-acetyl-seryl-aspartyl-lysyl-proline (AcSDKP), which is involved in the control of haematopoietic stem cell proliferation by preventing their recruitment into S phase (Rieger *et al.*, 1993). The acute administration of captopril, an ACE inhibitor, produces a seven-fold increase in the plasma concentration of AcSDKP in normal volunteers, demonstrating the *in vivo* role of the enzyme in the metabolism of this substrate (Azizi *et al.*, 1996). The increase in AcSDKP which results from chronic ACE inhibitor treatment is particularly marked during renal failure (M. Azizi, unpublished results) and may account for the rare cases of anaemia and/or leukopenia which have been observed in these circumstances.

7.3.3 Anion activation

An interesting feature of the catalytic activity of ACE is the activating effect of monovalent anions. The requirement of ACE for chloride for the hydrolysis of AI has long been recognized (Skeggs *et al.*, 1956), ACE being unable to hydrolyse this substrate in the absence of chloride (Bünning and Riordan, 1983). However, the presence of chloride is not essential for the hydrolysis of all substrates, for example, BK is hydrolysed in the absence of chloride with a maximal rate of hydrolysis occurring at only 20 mM chloride (Bünning and Riordan, 1983; Dorer *et al.*, 1974). The kinetics of chloride activation are rather complex, being both substrate- and pH-dependent (Ehlers and Kirsch, 1988; Shapiro *et al.*, 1983). An attractive hypothesis has been proposed for the mechanism of chloride activation in which chloride binding induces a subtle change in the conformation of the active site which facilitates substrate binding (Ehlers and Riordan, 1990).

It is conceivable that chloride activation may be a mechanism for regulating ACE activity under different physiological conditions of pH and chloride concentration. This is particularly interesting in view of more recent studies, discussed later, which demonstrate the different chloride dependence of the two active ACE sites, which may display different substrate specificities. Thus, different conditions of chloride and pH could alter the relative activities of the two active sites of somatic ACE and, as such, determine the substrate specificity of the enzyme. The identity of the amino acid residue of

ACE involved in the chloride activation is still unknown. Reductive methylation of rabbit pulmonary ACE by formaldehyde and sodium cyanoborohydride has implicated a single lysine residue, located at or near the active site of ACE, as the chloride-binding site (Shapiro and Riordan, 1983). However, the involvement of this residue has not been confirmed and the location of the putative chloride-binding lysine requires further study.

7.4 Molecular structure of ACE

cDNA cloning of the somatic form of ACE expressed in endothelial cells revealed that the enzyme is composed of two highly homologous domains, indicating a gene duplication event in the course of evolution. This hypothesis was supported by the elucidation of the structure of the germinal form of ACE expressed in the testes which contains only one of the two homologous domains of somatic ACE and therefore corresponds to the putative ancestral form of the ACE gene. The identification of two transcription initiation sites and the determination of the structure of the ACE gene enabled the determination of the mechanism of transcription of the two ACE mRNAs.

7.4.1 Somatic ACE

Complete cDNAs for the somatic form of ACE were first isolated from human endothelial and from mouse kidney cDNA libraries (Bernstein *et al.*, 1988; Soubrier *et al.*, 1988). Human endothelial ACE mRNA is 4.3 kb in length and in rabbit the pulmonary ACE mRNA is 5.0 kb (Soubrier *et al.*, 1988; Thekkumkara *et al.*, 1992). In mice, two ACE mRNAs have been detected in kidney and in lung of 4.9 and 4.15 kb, respectively (Bernstein *et al.*, 1989). A high level of sequence similarity, 80–90% in both the nucleotide and amino acid sequences, has been found between the different somatic mammalian ACE cDNA sequences cloned from human, mouse, rabbit, bovine and rat ACE (Bernstein *et al.*, 1989; Koike *et al.*, 1994; Shai *et al.*, 1992; Soubrier *et al.*, 1988; Thekkumkara *et al.*, 1992). Human endothelial ACE comprises 1306 amino acid residues, of which 14 are cysteine residues, and contains 17 potential N-linked glycosylation sites but no Ser/Thr-rich region indicative of O-glycosylation (Soubrier *et al.*, 1988). The primary sequence of ACE reveals two hydrophobic regions: a sequence of 29 amino acids at the N terminus is absent in the mature enzyme and therefore constitutes the cleaved signal peptide, and a second hydrophobic sequence of 17 amino acids is located near the C terminus of the enzyme which comprises the membrane anchor (*Figure 7.1*).

Molecular cloning of the somatic form of ACE also demonstrated that the enzyme is composed of two highly homologous domains, called thereafter N and C domains, indicative of the duplication of an ancestral gene (Bernstein *et al.*, 1989; Soubrier *et al.*, 1988). The overall sequence homology in human ACE is 60% in both the nucleotide and amino acid sequences but increases to 89% in a sequence of 40 amino acids in each domain containing essential residues of the active site, hence each domain of ACE contains a putative catalytic site. Each domain contains the zinc-binding motif His–Glu–Xaa–Xaa–His identified in many zinc metalloproteases including thermolysin

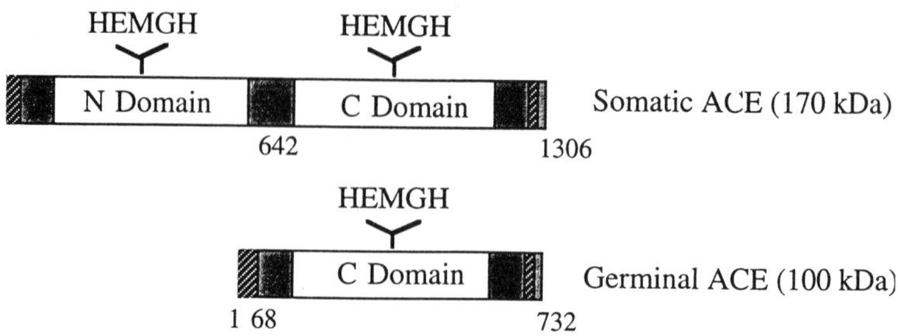

Figure 7.1. Schematic representation of somatic and germinal ACE. Somatic ACE comprises two highly similar domains, called N and C domains, each of which comprises the consensus sequence for zinc binding (HEXXH). A hydrophobic sequence in the C-terminal region (shaded box on the right) comprises the membrane-spanning domain. Germinal ACE comprises a single catalytically active domain identical to the C-domain of human ACE except for an N-terminal 67-residue germinal ACE-specific sequence.

and endopeptidase-24.11 (Vallee and Auld, 1990). In this consensus sequence, the two histidine residues provide two of the three zinc co-ordinating ligands and the glutamate residue is the base donor in the catalytic reaction. The third zinc ligand has been identified in both thermolysin and endopeptidase-24.11, as a glutamate residue which is separated, on the C-terminal side, from the second histidine zinc-binding ligand by 19 and 58 amino acid residues, respectively (Kester and Mathews, 1977; LeMoual *et al.*, 1991). The putative glutamate third zinc co-ordinating residues in ACE have been proposed as Glu389 and Glu987 in the N and C domains of ACE (Soubrier *et al.*, 1988) and have been confirmed by site-directed mutagenesis (see Section 7.5).

The ACE gene appears to be duplicated in all mammalian species from which ACE has been cloned to date. Furthermore, a peptidyl dipeptidase was purified by affinity chromatography from the electric organ of *Torpedo marmorata* (Turner *et al.*, 1987). The molecular mass of the *Torpedo* ACE was 190 kDa, compared to 170 kDa of the human somatic form of ACE, and therefore it would appear that *Torpedo* ACE is also transcribed from a duplicated gene.

An ACE-like enzyme has also been characterized in the housefly, *Musca domestica* (Lamango and Isaac, 1993). This enzyme is able to hydrolyse the synthetic ACE substrate, Hip–His–Leu, and is inhibited by the specific ACE inhibitor captopril with an IC_{50} of 50 nM. In addition, a cDNA encoding an enzyme with a high degree of sequence similarity to ACE has also been cloned in *Drosophila melanogaster* (Cornell *et al.*, 1995) and was named AnCE. The peptide sequence between residues 311 and 395 of AnCE, which includes the region containing the zinc-binding motif, displays 72 and 76% identity with the corresponding regions in the N and C domains of human ACE. Interestingly, the ACE-like enzyme from *D. melanogaster* is not duplicated, since AnCE contains a single active site and has a molecular mass of 74 kDa. In addition, there is no hydrophobic sequence in the C-terminal region for membrane anchorage and the recombinant enzyme is fully secreted (Cornell *et al.*, 1995). AnCE has been recently expressed on a large scale in *Pichia pastoris* in order to crystallize the enzyme and solve its tertiary structure (*Figure 7.2*). The unglycosylated mutant obtained by

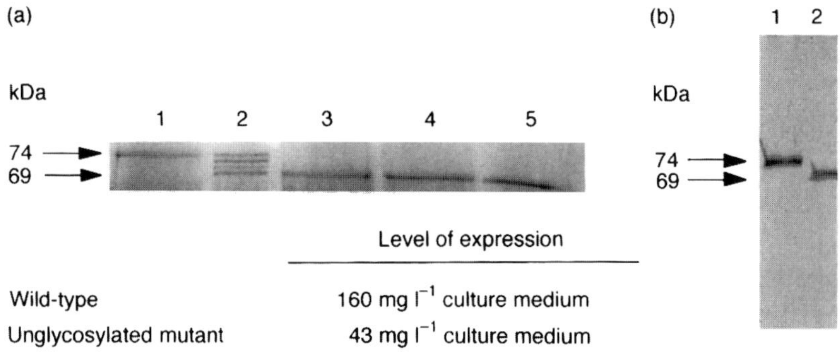

Figure 7.2. Expression of *Drosophila melanogaster* ACE (AnCE) and an unglycosylated AnCE mutant in the yeast *Pichia pastoris.* (a) The AnCE primary structure reveals three consensus sequences for N-linked glycosylation. Partial deglycosylation of AnCE by endoglycosidase F resolves AnCE to four discrete bands (lane 2) indicating that all three consensus sequences are occupied by oligosaccharide chains. Deglycosylated AnCE has a molecular mass of 69 kDa (lane 3) which co-migrates with an unglycosylated AnCE mutant (lane 4) whose size is unaltered by endoglycosidase F digestion (lane 5). (b) Wild-type AnCE (lane 1) and unglycosylated mutant AnCE (lane 2) are expressed to high levels in *Pichia pastoris* constituting the only visible proteins secreted into the culture medium on silver-stained SDS-PAGE. Modified from Williams *et al.* (1996).

suppressing three N-glycosylation sites by site-directed mutagenesis was fully active, but was less thermostable than the wild-type enzyme (Williams *et al.*, 1996).

7.4.2 Germinal ACE

The primary structure of the germinal form of ACE found in testes has been deduced from the cDNA sequence from human (Ehlers *et al.*, 1989; Lattion *et al.*, 1989), rabbit (Kumar *et al.*, 1989) and mouse (Howard *et al.*, 1990). Germinal ACE is comprised of 732 amino acids, compared with 1306 in somatic ACE, and corresponds to the C domain of somatic ACE. It therefore contains a single active site (*Figure 7.1*). In fact, amino acid residues 68–732 of human germinal ACE are identical to residues 642–1306 of human somatic ACE, hence germinal ACE contains the same hydrophobic membrane-anchoring domain located near the C terminus of somatic ACE. The N-terminal 67-amino-acid sequence of germinal ACE is specific to this form of the enzyme and contains the signal peptide and a Ser/Thr-rich region for O-glycosylation. Germinal ACE is indeed heavily O-glycosylated with 90% of the O-linked carbohydrates added to a region covering 36 amino acids in the N-terminal germinal ACE-specific sequence (Ehlers *et al.*, 1992). The function of the O-linked carbohydrates has not been determined, but it has been demonstrated that there is no relation between the O-linked glycosylation and enzyme activity or stability (Ehlers *et al.*, 1992).

7.5 Identification of active site residues

Several amino acid residues have been proposed to play an important structural/functional role in ACE as a result of site-directed mutagenesis and chemical modification of the protein with specific functional reagents. However, the precise organization of the catalytic domains of ACE must await the resolution of the three-dimensional structure.

The highly homologous N and C domains of ACE both contain the zinc-binding motif found in the active sites of many zinc metalloproteases and therefore contain a putative active site. In order to establish whether both putative active sites in somatic ACE are functional, a series of ACE mutants containing only a single intact domain were constructed (Wei *et al.*, 1991a). These mutants contained point mutations of putative critical active site residues in either the N or C domain, or contained a deletion of either the N or C domain to form a truncated mutant. Both the N and C domains displayed catalytic activity, hydrolysing the synthetic ACE substrate Hip–His–Leu and the physiological substrate angiotensin I. Each domain displayed a similar K_m for substrate binding, but the N domain exhibited a nine- and three-fold lower k_{cat} compared to the C domain for the hydrolysis of Hip–His–Leu and AI respectively. In addition, both the N- and C-domain active sites possessed an absolute zinc requirement for activity and appeared to function independently. Thus, this study demonstrated conclusively that both putative active sites in ACE were functional. In agreement with these findings, the zinc content of somatic ACE has been determined as two atoms of zinc bound per molecule of enzyme and each domain binds a specific ACE inhibitor (Ehlers and Riordan, 1991; Perich *et al.*, 1992; Williams *et al.*, 1992).

Sequence alignment of each ACE domain with other zinc metalloproteases indicated a glutamate residue, which could represent the third zinc ligand, and an aspartate residue which might be indirectly involved in zinc interaction. This hypothesis was confirmed by site-directed mutagenesis of the ACE C domain which showed that Glu987 constituted the third zinc ligand, as its substitution by a valine residue completely abolished the activity of this ACE domain, and its replacement by an aspartyl residue resulted in a 300-fold decrease in k_{cat} (Williams *et al.*, 1994). In the same report, Asp991 was found to be important for catalytic activity probably playing a structural role in positioning the first histidine zinc ligand (*Figure 7.3*).

A tyrosine and a lysine residue have been identified as essential active site residues in the C domain of rabbit pulmonary (the somatic form) and also in rabbit testicular (germinal form) ACE (Bünning *et al.*, 1990; Chen and Riordan, 1990). These residues were identified by a chemical modification using 1-fluoro-2,4-dinitrobenzene (Dnp-F) which inactivated both forms of the enzyme. Amino acid analysis and sequencing of tryptic digests of the modified enzymes defined the amino acid positions of the lysine and tyrosine residues as Lys118 and Tyr200, respectively (Chen and Riordan, 1990). The functional role of the tyrosine residue in human testicular ACE was investigated by mutating this residue to a phenylalanine (Chen *et al.*, 1992). The mutant enzyme displayed a decreased specific activity for the hydrolysis of the synthetic substrate furanacryloyl–Phe–Gly–Gly and AI. The change in the K_m for the binding of these substrates was quite minor and the K_a for chloride activation was similar to that of the wild-type enzyme. In contrast, the mutant displayed a 100-fold decrease in affinity for the specific ACE inhibitor lisinopril, which is believed to act partly as a transition state analogue (Patchett and Cordes, 1985). Thus, these results indicated that the tyrosine

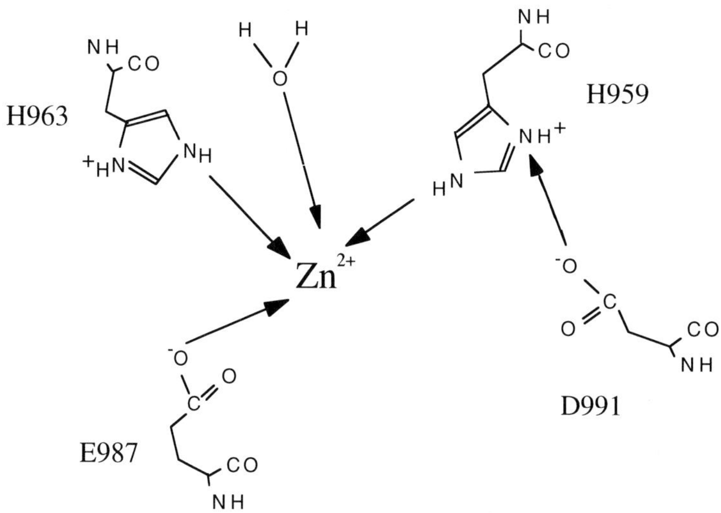

Figure 7.3. Zinc ligands in the C-domain active site of ACE. The amino acid residues which coordinate the active site zinc ion are shown with their corresponding amino acid positions. The water molecule provides the fourth zinc ligand. The aspartate residue (D991) which may function in the precise positioning of the first histidine zinc ligand is also shown. Arrows represent zinc coordinating bands and the dashed arrow indicates a salt link. This is a hypothetical model based on the known active site of thermolysin (Colman *et al.*, 1972). Modified from Williams *et al.* (1994) with permission from *The Journal of Biological Chemistry.*

residue was not involved in substrate or chloride binding and was not crucial for catalysis but could influence the rate of catalysis by stabilizing the transition state complex, in a manner analogous to Tyr198 in carboxypeptidase A (Gardell *et al.*, 1987). However, a subsequent study, on rabbit testicular ACE, produced quite different results. In this study the lysine and tyrosine residues previously identified by chemical labelling with Dnp-F were mutated to produce two single mutants (Sen *et al.*, 1993). These mutated enzymes exhibited the same enzymic properties as the wild-type enzyme, indicating that the single replacement of either residue can be tolerated without disturbing the usual functional properties of the enzyme.

7.6 Differences in the characteristics of the N and C domains

The presence of two active sites in somatic ACE raises the question of differences in the properties of each domain, and of a specific function (substrate) attributable to the N and C domains.

7.6.1 Chloride activation

As discussed previously, ACE is activated by chloride. However, the enzymic activity of the N and C domains is affected differently by this anion (Wei *et al.*, 1991a). The N domain appears to be less chloride-dependent than the C domain, such that catalytic activity is observed in the absence of chloride and optimal activity is achieved for the hydrolysis of AI at 10 mM chloride. In contrast, the C domain is essentially inactive in the absence of chloride and higher chloride concentrations are required for maximal activity (30 mM for AI hydrolysis). The chloride-dependence of the C domain is particularly marked for a synthetic substrate, such as Hip–His–Leu, which requires 800 mM chloride for maximum enzymic activity.

7.6.2 Substrate specificity

Some enzymes, for example, the brush-border hydrolases sucrase-isomaltase and lactase-phlorizin hydrolase, contain two active sites and are truly bifunctional, since they exhibit distinct functions. For these enzymes, the duplication of an ancestral gene occurred in the course of evolution to form two active sites, but a subsequent divergent evolution resulted in a distinct catalytic activity for each active site. For ACE to be classed as a bifunctional enzyme it must be demonstrated that the N domain is able to hydrolyse a substrate with more favourable kinetic parameters compared with the C domain. Somatic ACE has been reported to perform the endopeptidase cleavage from the amino terminus of LHRH (Trp3–Ser4) at a faster rate than testicular ACE, thus implying that it is the N-domain active site which is primarily responsible for this cleavage *in vitro* (Ehlers and Riordan, 1991). In support of this observation, a subsequent study using the ACE mutants described above containing only a single intact domain, demonstrated that the N domain active site of ACE was able to perform the cleavage of the Trp3–Ser4 bond of LHRH 30 times faster than the C domain active site (Jaspard *et al.*, 1993). This same study also reported that both active sites in ACE exhibit the same kinetic parameters for the hydrolysis of BK, although only the C-domain activity is stimulated by chloride. In addition, both domains of ACE hydrolysed substance P to produce the carboxyl-terminal dipeptide and tripeptide amides. For the hydrolysis of this substrate the C-domain active site was more readily activated by chloride and hydrolysed substance P at a faster rate than the N domain.

The peptide AcSDKP is the first highly specific substrate for the N-active site of ACE with kinetic constants in the range of physiological substrates. AcSDKP hydrolysis by ACE was studied using wild-type recombinant ACE and two full-length mutants containing a single functional site. Both the N- and C-domain active sites of ACE exhibit peptidyl dipeptidase activity towards AcSDKP, with K_m values of 31 and 39 μM, respectively. However, the N-domain active site hydrolyses the peptide 50 times faster than the C-domain active site, with K_{cat}/K_m values of 0.5 and 0.01 sec^{-1} μM^{-1}, respectively. The predominant role of the N-domain active site in AcSDKP hydrolysis was confirmed by the inhibition of hydrolysis using a monoclonal antibody specifically directed against the N-domain active site (Rousseau *et al.*, 1995). This suggests that ACE might be involved via its N-terminal active site in the *in vivo* regulation of the local concentration of this haemoregulatory peptide.

7.6.3 ACE inhibitors

The binding of various ACE inhibitors (see Chapter 5) has been used to probe possible structural differences existing between the N- and C-domain active sites (Perich *et al.*,

1992). The binding of ACE inhibitors to purified rat lung and testis ACE was studied and the binding parameters for the two inhibitor-binding sites in lung ACE were determined from the displacement of the ACE inhibitor [^{125}I]-Ro 31–8472, with either Ro 31–8472 or 351A, an analogue of lisinopril. Similar K_d values for Ro31–8472 were determined for the two binding sites in lung ACE, however, with 351A, the K_d for the N domain was more than 100-fold higher than that for the C domain. Since the main structural difference between these two inhibitors is the length of the inhibitor side chain which interacts with the active site zinc atom in each domain of ACE, it was proposed that the N domain of ACE contains a deeply recessed active site and therefore the N domain active site zinc atom is less accessible to the inhibitor 351A than that of the C domain.

Wei *et al.* (1992) reported that both ACE domains contain a single high-affinity binding site for the highly potent inhibitor trandolaprilat, and that chloride enhances inhibitor potency essentially by stabilizing the enzyme–inhibitor complex and slowing the dissociation rate. Interestingly, the effect of chloride on the action of inhibitors was more marked for the C domain than the N domain. Furthermore, the order of potency of the specific ACE inhibitors captopril, enalaprilat and lisinopril differed for the N and C domains such that the potency for the C domain, in decreasing order, was lisinopril > enalaprilat > captopril, whereas this order was reversed for the N domain.

A recent study (A. Michaud, T.A. Williams, M.-T. Chauvet and P. Corvol, unpublished results) evaluated the substrate dependence of ACE inhibition. They studied the inhibition of AI and AcSDKP hydrolysis, as well as synthetic substrates, by wild-type ACE and the N and C domains, using a range of specific ACE inhibitors. They showed that the K_i of captopril was 16-fold lower with AcSDKP than with AI as substrate, when using wild-type ACE. This raises the interesting possibility of using a specific N domain ACE inhibitor for enhancing plasma AcSDKP levels during chemotherapy to specifically protect haemopoietic stem cells. These results highlight structural and functional differences between the two ACE active sites, despite their high degree of sequence homology. AcSDKP is a highly specific substrate for the N domain but a specific substrate for the C domain, which might be important for male reproduction, has not yet been identified.

7.7 Structure of the ACE gene

The complete intron–exon structure of the human ACE gene was determined from the restriction mapping of genomic clones and by sequencing the intron–exon boundaries (Hubert *et al.*, 1991; see also Chapter 8). The human ACE gene contains 26 exons: the somatic ACE mRNA is transcribed from exon 1 to 26, but exon 13 is removed by splicing from the primary RNA transcript; the germinal ACE mRNA is transcribed from exon 13–26 (*Figure 7.4*). The structure of the human ACE gene provides further support for the hypothesis of the duplication of an ancestral gene. Exons 4–11 and 17–24, which encode the two homologous domains of the enzyme, are highly similar in size and in sequence. In contrast, the intron sizes are not conserved.

The human somatic and germinal forms of ACE mRNA are transcribed from a single

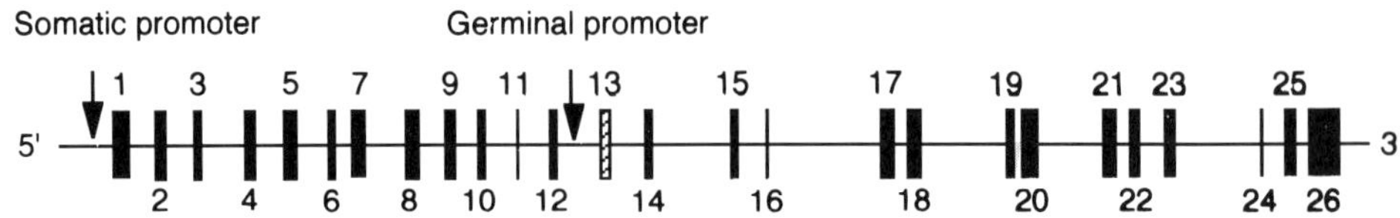

Figure 7.4. Organization of the human ACE gene. The relative locations of the 26 exons (vertical bars) are shown. Exon 13 (hatched bar) is specific to the testicular ACE mRNA. The two promoters are indicated by vertical arrows. Modified from Hubert *et al.* (1991).

gene (Hubert *et al.*, 1991), thus the two species of ACE mRNA could be generated by either differential splicing of a common primary RNA transcript or by the initiation of transcription from alternative start sites under the control of separate promoters. Several studies have demonstrated the presence of two functional promoters in the ACE gene. The somatic ACE promoter is located on the 5′ side of the first exon of the gene. The fusion of fragments of various sizes of the somatic ACE promoter to a reporter gene and the subsequent expression of the reporter gene demonstrated the presence of positive regulatory elements inside the 132 bp region upstream from the transcription start site and also indicated the presence of negative regulatory elements between positions –132 and –343 and between –472 and –754 from the transcription start site (Testut *et al.*, 1993).

The germinal ACE promoter is located on the 5′ side of the 5′ end of the germinal ACE mRNA. Primer extension and ribonuclease (RNase) protection assays on mouse, rabbit and human germinal RNA demonstrated that the transcription initiation site was located inside the ACE gene (Howard *et al.*, 1990; Hubert *et al.*, 1991; Kumar *et al.*, 1991). Thus, intron 12, which corresponds to the genomic sequence 5′ to the germinal specific exon 13, as deduced from the complete analysis of the ACE gene in humans, was proposed as the putative germinal ACE promoter. Unequivocal evidence for the promoter function of this sequence was obtained by using intron 12 to drive the transcription of a reporter gene in a germinal-specific fashion, in elongating spermatozoa (Langford *et al.*, 1991). In another series of transgenic mice, a 91 bp fragment of intron 12 of the ACE gene was used as the promoter and was sufficient to confer a germinal cell-restricted pattern of transcription to the transgene (Howard *et al.*, 1993). Further mapping of the elements controlling transcription was achieved by deoxyribonuclease (DNase) footprinting and gel mobility shift assays which demonstrated that a sequence between positions –42 and –62 specifically binds nuclear factors from testes extracts and contains a consensus cAMP-responsive element (CRE; Howard *et al.*, 1993). The two alternative promoters of the ACE gene exhibit highly contrasting cell specificities since the somatic ACE promoter is active in several cell types in contrast to the germinal ACE promoter, which is only active in a stage-specific manner in male germinal cells (Nadaud *et al.*, 1992).

7.8 Mechanism of ACE anchorage and solubilization

The mode of anchorage of ACE to the plasma membrane was first studied by Hooper *et al.* (1987) who showed that ACE was inserted into the membrane via a sequence located at the C terminus of the enzyme, since the N termini of the enzyme released by trypsin digestion or by solubilization with detergent were identical. Further evidence for this observation was provided by the primary structure of ACE, which contains a highly hydrophobic sequence located near the C-terminal end of the enzyme (Soubrier *et al.*, 1988). This sequence constitutes the membrane anchor since it is the only region of high hydrophobicity capable of spanning the lipid bilayer.

Further evidence for a C-terminal anchor was obtained by site-directed mutagenesis studies on human endothelial ACE and human testicular ACE (Wei *et al.*, 1991a). Transfection of Chinese hamster ovary (CHO) cells with the wild-type recombinant ACE cDNA resulted in 95% of the newly synthesised enzyme becoming bound to the cell-surface and the remaining 5% being secreted into the culture medium. However, when a stop codon was introduced on the 5′ side of the sequence encoding the putative membrane anchor, all newly synthesised ACE was liberated into the culture medium. Similarly, the expression of mutants of testicular ACE with a deletion of the putative membrane anchor also results in the production of a secreted form of the enzyme (Ehlers *et al.*, 1991). In addition, the soluble form of the enzyme purified from human plasma is not recognized by an antibody raised against the C-terminal cytoplasmic domain of the enzyme (Wei *et al.*, 1991b). Therefore, the secreted, circulating form of ACE would seem to be derived from the membrane-bound form by a post-translational proteolytic cleavage occurring in the C-terminal region of the membrane-bound form of ACE.

Amino acid C-terminal micro-sequencing of the spontaneously secreted recombinant human somatic ACE from CHO cells established that the post-translational cleavage occured at the Arg1137–Leu1138 bond (Beldent *et al.*, 1993; *Figure 7.5*). In addition, sequencing of the C terminus of the circulating form of ACE in human plasma determined the same cleavage site as above and it would appear that a similar mechanism for ACE secretion exists in human vascular cells and in CHO cells. Transfection of mouse epithelial cells with rabbit testicular ACE cDNA demonstrated that the cleavage process is enhanced by phorbol esters (Ramchandran *et al.*, 1994). In addition, this same study showed that the cleavage occurs exclusively at the cell surface and the cleavage site was identified as Arg663–Ser664 (rabbit testicular ACE numbering). The enzyme involved in the secretion of ACE from pig kidney microvillar membranes has been partially characterized (Oppong and Hooper, 1993; see also Chapter 3). The ACE-secreting enzyme is tightly associated with the membrane and is EDTA-sensitive, with the activity being substantially restored by the addition of Mg^{2+} and, to a lesser extent, by Zn^{2+} and Mn^{2+}. Ramchandran and Sen (1995) recently found that both somatic and testicular rabbit ACE were solubilized by an integral membrane metalloprotease which was specifically inhibited by a synthetic hydroxamic acid derivative.

The subcellular localization of ACE proteolysis was recently established by pulse-chase experiments, cell surface immunolabelling, and biotinylation of radiolabelled mature proteins (Beldent *et al.*, 1995). The proteolysis of ACE takes place primarily at the plasma membrane. The solubilization of ACE is less than 2% within 1 h, is increased 2.4-fold by phorbol esters, but is not influenced by ionophores. An ACE mutant, lacking the transmembrane domain and the cytosolic part, is secreted at a faster rate without a carboxyl-terminal cleavage, and phorbol esters or ionophores have no effect on its rate of secretion into the medium. The proteolysis of ACE is therefore

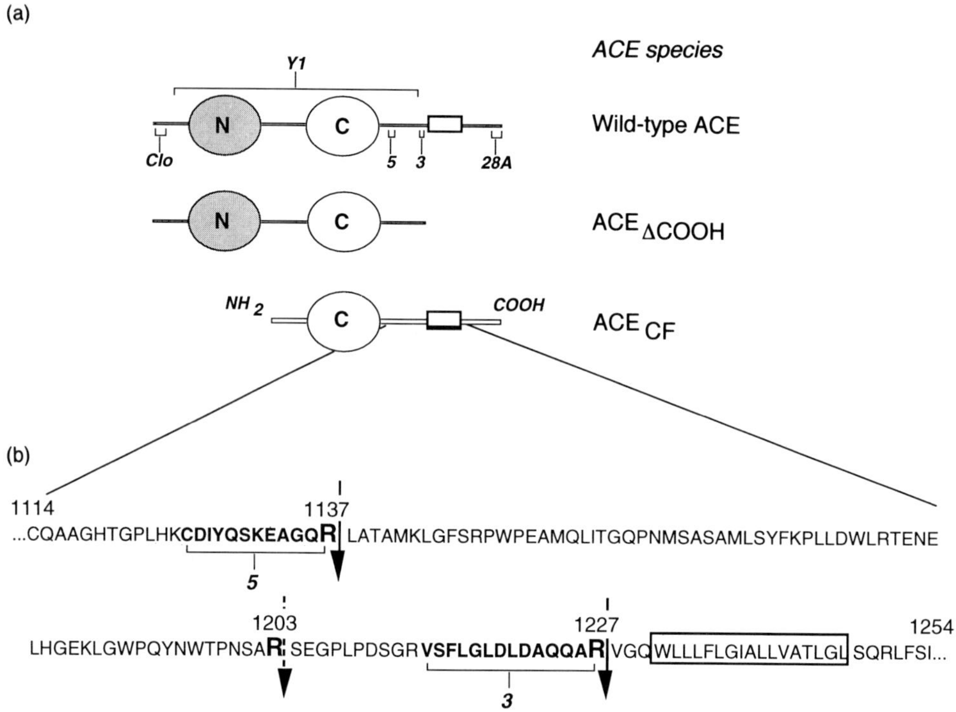

Figure 7.5. Wild-type ACE, C- and N-terminally truncated mutants. (a) Diagram of human ACEs encoded by the cDNA constructions. Human wild-type ACE, encoded by the full-length cDNA, is composed of 1277 residues, comprising the N and C homologous domains (egg-shaped boxes) in the extracellular region and the hydrophobic segment of 17 amino acids near the C terminus (squared boxes). Antisera (Y1, Clo and 28A) used for immunoprecipitation or Western blot analysis are also shown. The C-terminally truncated ACE ($ACE_{\Delta COOH}$) comprises 1230 residues. The 47 amino acids deleted from the C terminus correspond to the transmembrane and the cytosolic regions. The N-terminally truncated ACE (ACE_{CF}) comprises residues 1–4 of the N terminus followed by residues 572–1277 of mature ACE. (b) C-terminal sequence of ACE from Cys1114 to Ile1254. Cleavage sites of human ACE_{CF} are shown by an arrow. Corresponding regions of the two antibodies 5 and 3 are indicated. The hydrophobic domain is boxed. Reproduced from Beldent *et al.* (1995) with permission from *The Journal of Biological Chemistry.*

dependent on the presence of the membrane anchor and the present data suggest that the secretase(s) involved is also membrane-associated. An ACE mutant lacking the N domain (ACE_{CF}) is secreted 10-fold faster than wild-type ACE. The solubilization of ACE_{CF} also occurs at the plasma membrane and is stimulated 2.7-fold by phorbol esters. The cleavage site occurs between Arg1227 and Val1228. The N domain of ACE slows down the proteolysis and seems to act as a 'conformational inhibitor' of the proteolytic cleavage, possibly via interactions with the 'stalk' of ACE and the secretase(s) itself. It will be of considerable interest to elucidate the nature of the secretase involved in the process of ACE solubilization because it may involve a category of enzyme(s) implicated in the solubilization of other ectoproteins, such as transforming growth factor-α receptor, β-amyloid precursor protein and colony-stimulating factor.

References

Azizi M, Rousseau A, Ezan E, Guyene T-T, Michelet S, Grognet J-M, Lenfant M, Corvol P, Ménard J. (1996) Acute angiotensin-converting enzyme inhibition increases the plasma level of the natural stem cell regulator *N*-acetyl-seryl-aspartyl-lysyl-proline. *J. Clin. Invest.* **97**: 839–844.

Barnes K, Matsas R, Hooper NM, Turner AJ, Kenny AJ. (1988) Endopeptidase-24.11 is striosomally ordered in pig brain and, in contrast to aminopeptidase N and dipeptidyl dipeptidase, an angiotensin converting enzyme, is a marker for a set of striatal efferent fibers. *Neuroscience* **27**: 799–817.

Beldent V, Michaud A, Wei L, Chauvet M-T, Corvol P. (1993) Proteolytic release of human angiotensin-converting enzyme. Localization of the cleavage site. *J. Biol. Chem.* **268**: 26428–26433.

Beldent V, Michaud A, Bonnefoy C, Chauvet M-T, Corvol P. (1995) Cell surface localization of proteolysis of human endothelial angiotensin I-converting enzyme. *J. Biol. Chem.* **270:** 28962–28969.

Berg T, Sulner J, Lai CY, Soffer RL. (1986) Immunohistochemical localization of two angiotensin I-converting isoenzymes in the reproductive tract of the male rabbit. *J. Histochem. Cytochem.* **34**: 753–760.

Bernstein KE, Martin BM, Bernstein EA, Linton J, Striker L, Striker G. (1988) The isolation of angiotensin converting enzyme cDNA. *J. Biol. Chem.* **263**: 11021–11024.

Bernstein KE, Martin BM, Edwards AS, Bernstein EA. (1989) Mouse angiotensin I-converting enzyme is a protein composed of two homologous domains. *J. Biol. Chem.* **264**: 11945–11951.

Brentjens JR, Matsuo S, Andres GA, Caldwell PRB, Zamboni L. (1986) Gametes contain angiotensin converting enzyme (kininase II). *Experientia* **42**: 399–402.

Bruneval P, Hinglais N, Alhenc-Gelas F, Tricottet V, Corvol P, Ménard J, Camilleri J-P, Bariety J. (1986) Angiotensin I converting enzyme in human intestine and kidney. Ultrastructural immunohistochemical localization. *Histochemistry* **85**: 73–80.

Bünning P, Riordan JF. (1983) Activation of angiotensin converting enzyme by monovalent anions. *Biochemistry* **22**: 110–116.

Bünning P, Riordan JF. (1985) The functional role of zinc in angiotensin converting enzyme. Implication for the enzyme mechanism. *J. Inorg. Biochem.* **24**: 183–198.

Bünning P, Kleemann SG, Riordan JF. (1990) Essential residues in angiotensin converting enzyme: modification with 1-fluoro-2,4-dinitrobenzene. *Biochemistry* **29**: 10488–10492.

Chen YNP, Riordan JF. (1990) Identification of essential tyrosine and lysine residues in angiotensin converting enzyme: evidence for a single active site. *Biochemistry* **29**: 10493–10498.

Chen YNP, Ehlers MRW, Riordan JF. (1992) The functional role of tyrosine-200 in human testis angiotensin-converting enzyme. *Biochem. Biophys. Res. Commun.* **184:** 306–309.

Colman PM Jansonius JN, Matthews BW. (1972) The structure of thermolysin: an electron density map at 2.3Å resolution. *J. Mol. Biol.* **70**: 701–724.

Cornell MJ, Williams TA, Lamango NS, Coates D, Corvol P, Soubrier F, Hoheisel J, Lehrach H, Isaac RE. (1995) Cloning and expression of an evolutionary conserved single-domain angiotensin converting enzyme from *Drosophila melanogaster*. *J. Biol. Chem.* **270**: 13613–13619.

Costerousse O, Allegrini J, Lopez M, Alhenc-Gelas F. (1993) Angiotensin I-converting enzyme in human circulatory mononuclear cells. Genetic polymorphism in T-lymphocytes. *Biochem. J.* **290**: 33–40.

Das M, Hartley JL, Soffer RL. (1977) Serum angiotensin-converting enzyme. Isolation and relationship to the pulmonary enzyme. *J. Biol. Chem.* **252**: 1316–1319.

Defendini R, Zimmerman EA, Weare JA, Alhenc-Gelas F, Erdös EG. (1983) Angiotensin-converting enzyme in epithelial and neuroepithelial cells. *Neuroendocrinology* **37**: 32–40.

Dorer FE, Kahn JR, Lentz KE, Levine M, Skeggs LT. (1974) Hydrolysis of bradykinin by angiotensin-converting enzyme. *Circ. Res.* **34**: 824–827.

Dubreuil P, Fulcrand P, Rodriguez M, Fulcrand H, Laur J, Martinez J. (1989) Novel activity of angiotensin-converting enzyme hydrolysis of cholecystokinin and gastrin analogues with release of the amidated carboxy-terminal dipeptide. *Biochem. J.* **262**: 125–130.

Ehlers MRW, Kirsch RE. (1988) Catalysis of angiotensin I hydrolysis by human angiotensin-converting enzyme: effect of chloride and pH. *Biochemistry* **27**: 5538–5544.

Ehlers MRW, Riordan JF. (1989) Angiotensin-converting enzyme: new concepts concerning its biological role. *Biochemistry* **28**: 5311–5318.

Ehlers MRW, Riordan JF. (1990) Angiotensin-converting enzyme. Biochemistry and molecular biology. In: *Hypertension Pathophysiology, Diagnosis and Management* (eds J.H. Laragh and B.M. Brenner), pp. 1217–1231. Raven Press, New York.

Ehlers MRW, Riordan JF. (1991) Angiotensin-converting enzyme: zinc- and inhibitor binding stoichiometries of the somatic and testis isozymes. *Biochemistry* **30**: 7118–7126.

Ehlers MRW, Fox EA, Strydom DJ, Riordan JF. (1989) Molecular cloning of human testicular angiotensin-converting enzyme: the testis isozyme is identical to the C-terminal half of endothelial angiotensin-converting enzyme. *Proc. Natl Acad. Sci. USA* **86**: 7741–7745.

Ehlers MRW, Chen YNP, Riordan JF. (1991) Spontaneous solubilization of membrane-bound human testis angiotensin-converting enzyme expressed in chinese hamster ovary cells,. *Proc. Natl Acad. Sci. USA* **88**: 1009–1013.

Ehlers MRW, Chen YNP, Riordan JF. (1992) The unique N-terminal sequence of testis angiotensin-converting enzyme is heavily O-glycosylated and unessential for activity or stability. *Biochem. Biophys. Res. Commun.* **183**: 199–205.

El-Dorry H, Iwata K, Thornberg NA, Cordes EH, Soffer RL. (1982) Molecular and catalytic properties of rabbit testicular dipeptidyl carboxypeptidase. *J. Biol. Chem.* **257**: 14128–14133.

El-Dorry HA, MacGregor JS, Soffer RL. (1983) Dipeptidyl carboxypeptidase from seminal fluid resembles the pulmonary rather than the testicular isoenzyme. *Biochem. Biophys. Res. Commun.* **115**: 1096–1100.

Erdös EG, Skidgel RA. (1987) The angiotensin I-converting enzyme. *Lab. Invest.* **56**: 345–348.

Esther CR, Howard TE, Marino EM, Goaddard JM, Capecchi MR, Bernstein KE. (1996) Mice lacking angiotensin-converting enzyme have low blood pressure, renal pathology and reduced male fertility. *Lab. Invest.* **74**: 953–965.

Friedland J, Setton C, Silverstein E. (1978) Induction of angiotensin-converting enzyme in human monocytes in culture. *Biochem. Biophys. Res. Commun.* **83**: 843–849.

Gardell SJ, Hilvert D, Barnett J, Kaiser ET, Ruter WJ. (1987) Use of directed mutagenesis to probe the role of tyrosine 198 in the catalytic mechanism of carboxypeptidase A. *J. Biol. Chem.* **262**: 576–582.

Hooper NM. (1991) Angiotensin-converting enzyme. Implications from molecular biology for its physiological functions. *Int. J. Biochem.* **23**: 641–647.

Hooper NM, Turner AJ. (1987) Isolation of two differentially glycosylated forms of peptidyl-dipeptidase A (angiotensin-converting enzyme) from pig brain: a re-evaluation of their role in neuropeptide metabolism. *Biochem. J.* **241**: 625–633.

Hooper NM, Keen J, Pappin DJC, Turner AJ. (1987) Pig kidney angiotensin-converting enzyme. Purification and characterization of amphiphatic and hydrophilic forms of the enzyme establishes C-terminal anchorage to the plasma membrane. *Biochem. J.* **247**: 85–93.

Howard TE, Shai SY, Langford KG, Martin BM, Bernstein KE. (1990) Transcription of testicular angiotensin-converting enzyme (ACE) is initiated within the 12th intron of the somatic ACE gene. *Mol. Cell. Biol.* **10:** 4294–4302.

Howard T, Balogh R, Overbeek P, Bernstein KE. (1993) Sperm-specific expression of angiotensin-converting enzyme (ACE) is mediated by a 91-base-pair promoter containing a CRE-like element. *Mol. Cell. Biol.* **13**: 1–27.

Hubert C, Houot A-M, Corvol P, Soubrier F. (1991) Structure of the angiotensin I-converting enzyme gene. Two alternate promoters correspond to evolutionary steps of a duplicated gene. *J. Biol. Chem.* **266**: 15377–15383.

Igic R, Erdös EG, Yen HSG, Sorrells K, Nakajima T. (1972) Angiotensin I converting enzyme of the lung. *Circ. Res.* **31** (Suppl. 2): 51–61.

Inokuchi JJ, Nagamatsu A. (1981) Tripeptidyl carboxypeptidase activity of kininase II (angiotensin converting enzyme). *Biochim. Biophys. Acta* **662**: 300–307.

Jaspard E, Wei L, Alhenc-Gelas F. (1993) Differences in properties and enzymatic specificities between the two active sites of angiotensin I-converting enzyme. Studies with bradykinin and other natural peptides. *J. Biol. Chem.* **268**: 9496–9503.

Kester WR, Matthews BW. (1977) Crystallographic study of the binding of dipeptide inhibitors to thermolysin: implications for the mechanism of catalysis. *Biochemistry* **16**: 2506–2516.

Koike G, Krieger JE, Jacob HJ, Mukoyama M, Pratt RE, Dzau VJ. (1994) Angiotensin-converting enzyme and genetic hypertension: cloning of rat cDNAs and characterization of the enzyme. *Biochem. Biophys. Res. Commun.* **198**: 380–386.

Krege JH, John SWM, Laugenbach LL, Hodgin JB, Hagaman JR, Bachman ES, Jennette JC, O'Brien DA, Smithies O. (1995) Male–female differences in fertility and blood pressure in ACE-deficient mice. *Nature* **375**: 146–148.

Kumar RS, Kusari J, Roy SN, Soffer RL, Sen GC. (1989) Structure of testicular angiotensin-converting enzyme. A segmental mosaic isozyme. *J. Biol. Chem.* **264**: 16754–16758.

Kumar RS, Thekumkara TJ, Sen G. (1991) The mRNAs encoding the two angiotensin-converting isozymes are transcribed from the same gene by a tissue-specific choice of alternative transcription initiation sites. *J. Biol. Chem.* **266**: 3854–3862.

Lamango N, Isaac RE. (1993) Identification of an ACE-like peptidyl dipeptidase activity in the housefly, *Musca domestica. Biochem. Soc. Trans.* **21**: 245S.

Langford KG, Shai SY, Howard TE, Kovac MJ, Overbeek PA, Bernstein KE. (1991) Transgenic mice demonstrate a testis-specific promoter for angiotensin-converting enzyme. *J. Biol. Chem.* **266**: 15559–15562.

Lattion AL, Soubrier F, Allegrini J, Hubert C, Corvol P, Alhenc-Gelas F. (1989) The testicular transcript of the angiotensin I-converting enzyme encodes for the ancestral, non-duplicated form of the enzyme. *FEBS Lett.* **252**: 99–104.

Lawton G, Paciorek PH, Waterfall JF. (1992) The design and biological profile of ACE inhibitors. In: *Advances in Drug Research*, Vol. 23, pp.161–220. Academic Press, New York.

LeMoual H, Devault A, Roques BP, Crine P, Boileau G. (1991) Identification of glutamic acid 616 as a zinc-coordinating residue in endopeptidase 24.11. *J. Biol. Chem.* **266:** 15670–15674.

Matsas R, Kenny AJ, Turner AJ. (1984) The metabolism of neuropeptides. The hydrolysis of peptides including enkephalins, tachykinins and their analogues by endopeptidase 24.11 (E.C. 3.4.24.11). *Biochem. J.* **223**: 433–440.

Nadaud S, Houot AM, Hubert C, Corvol P, Soubrier F. (1992) Functional study of the germinal angiotensin I-converting enzyme promoter. *Biochem. Biophys. Res. Commun.* **189**: 134–140.

Ng KKF, Vane JR. (1967) Conversion of angiotensin I to angiotensin II. *Nature* **216**: 762–766.

Ng KKF, Vane JR. (1968) Fate of angiotensin I in the circulation. *Nature* **218**: 144–150.

Oblin A, Danse MJ, Zivkovic B. (1988) Degradation of substance P by membrane peptidases in the rat substantia nigra: effect of selective inhibitors. *Neurosci. Lett.* **84**: 91–96.

Oparil S, Haber E. (1974a) The renin angiotensin system. *New Engl. J. Med.* **291**: 389–401.

Oparil S, Haber E. (1974b) The renin angiotensin system. *New Engl. J. Med.* **291**: 446–457.

Oppong SY, Hooper NM. (1993) Characterization of a secretase activity which releases angiotensin-converting enzyme from the membrane. *Biochem. J.* **292**: 597–603.

Patchett AA, Cordes EH. (1985) The design annd properties of *N*-carboxyalkyldipeptide inhibitors of angiotensin converting enzyme. *Adv. Enzymol.* **57:** 1–84.

Perich RB, Jackson B, Rogerson F, Mendelsohn FAO, Paxton D, Johnston CI. (1992) Two binding sites on angiotensin-converting enzyme: evidence from radioligand binding studies. *Mol. Pharmacol.* **42**: 286–293.

Ramchandran R, Sen I. (1995) Cleavage processing of angiotensin-converting enzyme by a membrane-associated metalloprotease. *Biochemistry* **34**: 12645–12652.

Ramchandran R, Sen GC, Misono K, Sen I. (1994) Regulated cleavage-secretion of the membrane-bound angiotensin-converting enzyme. *J. Biol. Chem.* **269:** 2125–2130.

Rieger KJ, Saez-Servent N, Papet MP, Wdzieczak-Bakala J, Morgat JL, Thierry J, Woelter W, Lenfant M. (1993) Involvements of human plasma angiotensin converting enzyme in the degradation of the hemoregulatory peptide *N*-acetyl-seryl-aspartyl-lysyl-proline. *Biochem. J.* **296:** 373–378.

Rousseau A, Michaud A, Chauvet M-T, Lenfant M, Corvol P. (1995) The hemoregulatory peptide *N*-acetyl-Ser-Asp-Lys-Pro is a natural and specific substrate of the N-terminal active site of human angiotensin-converting enzyme. *J. Biol. Chem.* **270:** 3656–3661.

Schweisfurth H, Schioberg-Schiegnitz S. (1984) Assay and biochemical characterization of angiotensin I-converting enzyme in cerebrospinal fluid. *Enzyme* **32**: 12–19.

Sen I, Kasturi S, Jabbar MA, Sen GC. (1993) Mutations in two specific residues of testicular angiotensin-converting enzyme change its catalytic properties. *J. Biol. Chem.* **268**: 25748–25754.

Shai SY, Fishel RS, Martin BM, Berk BC, Bernstein KB. (1992) Bovine angiotensin converting enzyme cDNA cloning and regulation. Increased expression during endothelial cell growth arrest. *Circ. Res.* **70**: 1274–1281.

Shapiro R, Riordan JF. (1983) Critical lysine residue at the chloride binding site of angiotensin converting enzyme. *Biochemistry* **22**: 5315–5321.

Shapiro R, Holmquist B, Riordan JF. (1983) Anion activation of angiotensin-converting enzyme: dependence on nature of substrate. *Biochemistry* **22**: 3850–3857.

Sibony M, Segretain D, Gasc J-M. (1994) Angiotensin-converting enzyme in murine testis: Step-specific expression of the germinal isoform during spermiogenesis. *Biol. Reprod.* **50:** 1015–1026.

Skeggs LT, Kahn JR, Shumway NP. (1956) The preparation and function of the hypertensin-converting enzyme. *J. Exp. Med.* **103**: 295–299.

Skidgel RA, Erdös EG. (1985) Novel activity of human angiotensin I converting enzyme: release of the NH_2- and COOH-terminal tripeptides from the luteinizing hormone-releasing hormone. *Proc. Natl Acad. Sci. USA* **82**: 1025–1029.

Skidgel RA, Erdös EG. (1987) The broad substrate specificity of human angiotensin I converting enzyme. *Clin. Exp. Hypertens.* **A9**: 243–259.

Skidgel RA, Engelbrecht S, Johnson AR, Erdös EG. (1984) Hydrolysis of substance P and neurotensin by converting enzyme and neutral endopeptidase. *Peptides* **5**: 769–776.

Soubrier F, Alhenc-Gelas F, Hubert C, Allegrini J, John M, Tregear G, Corvol P. (1988) Two putative active centers in human angiotensin I-converting enzyme revealed by molecular cloning. *Proc. Natl Acad. Sci. USA* **85**: 9386–9390.

Strittmatter SM, Snyder SH. (1987) Angiotensin-converting enzyme immuno-histochemistry in rat brain and pituitary gland: correlation of isozyme type with cellular localization. *Neuroscience* **21**: 407–420.

Strittmatter SM, Lo MMS, Javitch JA, Snyder SH. (1984) Autoradiographic visualization of angiotensin-converting enzyme in rat brain with [^{3}H]captopril: localization to a striatonigral pathway. *Proc. Natl Acad. Sci. USA* **81**: 1599–1603.

Strittmatter SM, Thiele EA, Kapiloff MS, Snyder SH. (1985) A rat brain isozyme of angiotensin-converting enzyme. *J. Biol. Chem.* **260**: 9825–9832.

Testut P, Corvol P, Hubert C. (1993) Functional analysis of the somatic angiotensin I converting enzyme promoter. *Biochem. J.* **293**: 843–848.

Thekkumkara TJ, Livingston IW, Kumar RS, Sen GC. (1992) Use of alternative polyadenylation sites for tissue-specific transcription of two angiotensin-converting enzyme mRNAs. *Nucleic Acids Res.* **20**: 683–687.

Turner AJ, Hryszko J, Hooper NM, Dowdall MJ. (1987) Purification and characterization of a peptidyl dipeptidase resembling angiotensin converting enzyme from the electric organ of *Torpedo marmorata. J. Neurochem.* **48**: 910–916.

Vallee BL, Auld DS. (1990) Active-site zinc-ligands and activated H_2O of zinc enzymes. *Proc. Natl Acad. Sci. USA* **87:** 220–224.

Velletri PA. (1985) Testicular angiotensin I-converting enzyme (E.C. 3.4.15.1). *Life Sci.* **36**: 1597–1608.

Waeber B, Nussberger J, Brunner HS. (1990) Angiotensin-converting enzyme inhibitors in hypertension. In: *Hypertension: Pathophysiology, Diagnosis and Management* (eds J.H. Laragh and B.M. Brenner), pp. 2209–2232. Raven Press, New York.

Wei L, Alhenc-Gelas F, Corvol P, Clauser E. (1991a) The two homologous domains of the human angiotensin I-converting enzyme are both catalytically active. *J. Biol. Chem.* **266**: 9002–9008.

Wei L, Alhenc-Gelas F, Soubrier F, Michaud A, Corvol P, Clauser E. (1991b) Expression and characterization of recombinant human angiotensin I-converting enzyme. Evidence for a C-terminal transmembrane anchor and for a proteolytic processing of the secreted recombinant and plasma enzymes. *J. Biol. Chem.* **266:** 5540–5546.

Wei L, Clauser E, Alhenc-Gelas F, Corvol P. (1992) The two homologous domains of human angiotensin I-converting enzyme interact differently with competitive inhibitors. *J. Biol. Chem.* **267**: 13398–13405.

Williams TA, Hooper NM, Turner AJ. (1991) Characterization of neuronal and endothelial forms of angiotensin converting enzyme in pig brain. *J. Neurochem.* **57**: 193–199.

Williams TA, Barnes K, Kenny AJ, Turner AJ, Hooper NM. (1992) A comparison of the zinc content and substrate specificity of the endothelial and testicular forms of porcine angiotensin converting enzyme, and the isolation of isoenzyme specific antisera. *Biochem. J.* **288**: 878–881.

Williams TA, Corvol P, Soubrier F. (1994) Identification of two active site residues in human angiotensin I-converting enzyme. *J. Biol. Chem.* **269**: 29430–29434.

Williams TA, Michaud A, Houard X, Chauvet M-T, Soubrier F, Corvol P. (1996) *Drosophila melanogaster* angiotensin I-converting enzyme expressed in *Pichia pastoris* resembles the C domain of the mammalian homologue and does not require glycosylation for secretion and enzymic activity. *Biochem. J.* **318:** 125–131.

Yang HYT, Erdös EG, Levin Y. (1970) A dipeptidyl carboxypeptidase that converts angiotensin I and inactivates bradykinin. *Biochim. Biophys. Acta* **214**: 374–376.

Yasui T, Alhenc-Gelas F, Corvol P, Menard J. (1984) Angiotensin I-converting enzyme in amniotic fluid. *J. Lab. Clin. Med.* **104**: 741–751.

Yokosawa H, Endo S, Ogura Y, Ishii S. (1983) A new feature of angiotensin-converting enzyme in the brain: hydrolysis of substance P. *Biochem. Biophys. Res. Commun.* **116**: 735–742.

Chapter 8

Genetic and physiological aspects of angiotensin I-converting enzyme

François Alhenc-Gelas, Olivier Costerousse and Sergei Danilov

8.1 Introduction

Angiotensin I-converting enzyme (peptidyl dipeptidase A, kininase II, EC 3.4.15.1, ACE) is a transmembrane ectopeptidase of vascular cells, also present as a soluble form in plasma (for a detailed account of the biochemical and molecular aspects of ACE, see Chapter 7). This enzyme plays an important role in cardiovascular homeostasis through its action on angiotensin I (AI) and bradykinin (Erdös, 1990; Skeggs *et al.*, 1956). AI is converted into the vasopressor and aldosterone-stimulating peptide, angiotensin II (AII), by removal of a single carboxyl-terminal dipeptide. ACE also cleaves sequentially two carboxyl-terminal dipeptides from bradykinin suppressing the vasodepressor action of this peptide. Because ACE is a relatively non-specific peptidase *in vitro*, with two active sites, it may have other physiological functions not yet identified (Azizi *et al.*, 1996; Jaspard *et al.*, 1993, 1995; Wei *et al.*, 1991). Numerous studies using competitive ACE inhibitors indicate that ACE participates in the control of vascular tone and maintains vasoconstriction, especially in physiological situations where renin secretion, and therefore AI production, are stimulated (Cushman and Ondetti, 1980). These inhibitors have become widely used as therapeutic agents to reduce blood pressure in hypertension, or decrease vascular resistance in cardiac insufficiency. ACE inhibition seems also to protect against impairment of renal function in type I diabetes with incipient diabetic nephropathy and in other types of renal disease (Lewis *et al.*, 1993; Marre *et al.*, 1988; Maschio *et al.*, 1996).

The ACE gene is constitutively expressed in several types of vascular cells, endothelial cells, smooth muscle cells (at least in the rat) and fibroblasts, as well as in some subtypes of circulating mononuclear cells (Arnal *et al.*, 1994; Caldwell *et al.*, 1976; Costerousse *et al.*, 1993a). ACE gene expression also occurs in extravascular tissues where the functions of the enzyme remain largely unknown. This is the case of the absorptive epithelial cells in the intestine, kidney tubule, placenta and choroid plexus, where ACE is one of the most abundant brush border peptidases (Erdös, 1990). ACE is also synthesised in neurones and is abundant in certain areas of the brain and medulla (Turner *et al.*, 1987). ACE is also abundant in the testis, where a different form of the

Cell-Surface Peptidases in Health and Disease, edited by A.J. Kenny and C.M. Boustead.

enzyme is synthesised in germinal cells, encoded by a shorter mRNA transcribed under the influence of an intragenic promoter.

The soluble form of ACE circulates in plasma at the relatively high concentration of 1–6 nM, where it is considered to be less important for the processing of peptides in the circulation than the membrane-bound endothelial form (Caldwell *et al.,* 1976; Vane, 1972). Plasma ACE, and probably also the soluble forms in cerebrospinal fluid and amniotic fluid, is derived from cellular ACE by proteolysis (see Chapter 7).

ACE gene expression can be followed in cells and tissues by mRNA quantification using somatic or germinal ACE cDNA and cRNA probes, which are available for man and several animal species (Bernstein *et al.,* 1989; Costerousse *et al.,* 1994; Ehlers *et al.,* 1989; Koike *et al.,* 1994; Kumar *et al.,* 1991; Lattion *et al.,* 1989; Shai *et al.,* 1992; Soubrier *et al.,* 1988), or by polymerase chain reaction (PCR) quantification of ACE mRNAs (Lanzillo *et al.,* 1994; Paul *et al.,* 1993). ACE levels in plasma and tissue can be studied by a number of enzymic or immunological assays (Alhenc-Gelas *et al.,* 1983; Cushman and Cheung, 1971; Danilov *et al.,* 1994, 1996; Friedland *et al.,* 1978). In man, plasma ACE levels have been extensively studied in healthy subjects and in disease. Recent work has established that the inter-individual variability of human plasma and cellular ACE levels is large, under the influence of at least one major gene. A genomic polymorphism has been identified in the ACE gene, which relates to ACE levels in plasma and cells. Because of the role of ACE in vasoactive peptide metabolism, the polymorphism of ACE has been studied in relation to the genetic susceptibility to cardiovascular and renal diseases, for example, in several degenerative vascular diseases, especially myocardial infarction and diabetic nephropathy. ACE may also be involved in the progression of several types of glomerular disease.

This chapter will address the questions of the distribution and regulation of the expression of the ACE gene in tissues, especially in the cardiovascular system (CVS) and kidney. The genetic polymorphism of human ACE levels, the hypotheses concerning its molecular basis and the possible physiopathological consequences of the level of ACE gene expression in the CVS and kidney will also be covered.

8.2 Regulation of ACE gene expression in somatic and germinal cells

The ACE gene contains two alternative promoters (see also Chapter 7). The promoter located 5′ of the gene initiates transcription of the duplicated form of the enzyme with two active sites (the form synthesised in all somatic tissues). The intragenic promoter located in intron 12 initiates transcription of a shorter form of the enzyme containing only the carboxyl-terminal domain and active site of somatic ACE, preceded by a short, heavily glycosylated, testis-specific sequence encoded by exon 13. This intragenic promoter seems to be activated only in mature male germinal cells, triggered during the late stages of differentiation and maturation of spermatocytes into spermatozoons (El Dorry *et al.,* 1982; Langford *et al.,* 1991; Sibony *et al.,* 1993). Because of the ontogeny of ACE gene expression in germinal cells, ACE mRNA and activity are very low in the testis before puberty, when they display a dramatic increase (Costerousse *et al.,* 1994; Cushman and Cheung, 1971). Conversely, in the adult rat, the disappearance of ACE from the testis after hypophysectomy (but not from the genital tract where the somatic form of ACE is synthesised) reflects the interruption of the spermatogenesis induced by the disappearance of testosterone secretion (Strittmatter *et al.*, 1985).

ACE is located, as a membrane-bound enzyme, mainly in the acrosome and intermediate zone of mature spermatozoa (Sibony *et al.*, 1993; Velletri *et al.*, 1985). The function of ACE in the testis was unknown, partly because most ACE inhibitors do not readily cross the blood–testis barrier and cannot help in studies of the function of testicular ACE (Jackson *et al.*, 1988). However, it has recently been shown that inactivation of the ACE gene by homologous recombination in embryonic stem cells induces hypofertility in male mice homozygous for the deletion (Esther *et al.*, 1996; Krege *et al.*, 1995). Preliminary observations in these animals suggest that the spermatogenesis proceeds apparently normally until the last stages. ACE may, however, play an important role in the capacitation process of spermatozoa, their migration and penetration of the oöcyte. The molecular basis of this effect remains to be established, as does the role of ACE in fertility in species other than the mouse.

The somatic form of the enzyme is synthesised in all other ACE-containing tissues where the transcription of the ACE gene initiated by the 5′ promoter occurs constitutively, activated in a tissue-specific manner during development (Costerousse *et al.*, 1994; Yosipiv *et al.*, 1994). The regulation of ACE synthesis has been studied in cultured endothelial cells, but results should be extrapolated with caution to the *in vivo* situations. Thus, although *in vitro* corticosteroids and thyroid hormones have been shown to increase ACE synthesis and secretion in endothelial cells and in smooth muscle cells (Fishel *et al.*, 1995a; Krulewitz *et al.*, 1984; Mendelsohn *et al.*, 1982), these hormones do not seem to regulate ACE gene expression *in vivo* in normal physiological situations. The increase in plasma ACE levels observed in hyperthyroid patients (Nakamura *et al.*, 1982; Yotsumoto *et al.*, 1982), could be due to an activation of the solubilization process of ACE, or to a reduction in its removal from the circulation, rather than to an increased synthesis of the enzyme, as suggested by observations made in euthyroid and hypothyroid rats during post-natal development, when plasma ACE levels increase at the time of thyroid maturation (Costerousse *et al.*, 1994). This effect of thyroid secretion probably explains the transient rise of plasma ACE levels observed during childhood (Lieberman, 1976). It has been suggested that infusion of AII can down-regulate ACE gene expression in the lung and testis, but observations made with AII antagonists or during renin inhibition do not support a negative feedback effect of AII on ACE gene expression (Schunkert *et al.*, 1993; O. Costerousse *et al.*, unpublished observations). Other factors have been reported to alter ACE synthesis in cultured endothelial cells, including fibroblast growth factor, calcium ionophores and drugs altering cellular cAMP levels, but the physiological significance of these observations remains unknown (Dasarathy and Fanburg, 1990; Fishel *et al.*, 1995b; Lloyd *et al.*, 1987; Okabe *et al.*, 1987).

8.3 Induction of ACE gene expression in physiological and pathological situations

Induction of ACE gene expression occurs normally in somatic and germinal cells during their differentiation and maturation process. It is also observed at tissue sites, during several physiological or pathological processes and also, interestingly, during ACE inhibition. In several cell types, normally expressing low levels of the enzyme, phenotypic alterations induced by haemodynamic or humoral mechanisms are accompanied by a dramatic increase in ACE gene expression: this is especially the case in monocytes, vascular smooth muscle cells, and fibroblasts. *In vitro* differentiation of monocytes into activated macrophages is accompanied by a large increase in ACE gene expression and ACE secretion (Costerousse *et al.,* 1993a; Friedland *et al.,* 1978). *In vivo*, activated macrophages seem to contain more ACE than resident macrophages. On the other hand, mild activation of T-lymphocytes, which in the resting state synthesise more ACE than monocytes, is not accompanied by a detectable increase in ACE synthesis. The physiological role of ACE in the immune system, especially in macrophages, remains unknown, it may be related to the metabolism of peptides involved in the inflammatory response, or to the control of blood flow at inflammation sites. It has also been suggested that ACE can participate, together with other peptidases, in the antigen recognition process, but this remains hypothetical (for review, see Costerousse *et al.,* 1993b). No apparent alterations of immune function have been reported in patients treated with ACE inhibitors. In the disease sarcoidosis, the granuloma cells derived from activated macrophages, as well as the isolated macrophages infiltrating tissues, contain high levels of ACE, and the increase in plasma ACE observed in this disease probably reflects secretion by these cells. ACE does not seem to play a pathogenic role in the disease, but measurements of plasma ACE levels are considered helpful for the diagnosis of the disease and for the follow-up care of patients (Lieberman, 1976).

ACE induction occurs also in the CVS during remodelling. The heart normally contains very little ACE, present in coronary endothelial cells and valvular endocardium (Johnston, 1994). However, when the ventricle become hypertrophied as a consequence of pressure overload, ACE mRNA and ACE activity increase (Schunkert *et al.,* 1990). The increased ACE synthesis in the hypertrophic heart is, however, not due to an induction of ACE gene expression in myocardial cells, but is rather related to the so-called 'interstitial activation' occurring under these circumstances, where the interstitium is infiltrated by activated macrophages and fibroblasts, as in an inflammatory response (Challah *et al.,* 1995). These activated cells, as well as the vascular cells in hypertrophied vessels and a expanded capillary network, produce, and perhaps secrete, high levels of ACE. The same phenomenon of ACE induction in the cardiac interstitium and vessels is observed during experimental ischaemic necrosis (Falkenhahn *et al.,* 1995; Passier *et al.,* 1995). Increased levels of ACE mRNA have been measured in human failing hearts (Studer *et al.,* 1994). Overproduction of ACE is also observed in the hypertrophied vessels, proximal tubule and activated interstitium of the hypertensive rat kidney (Vio and Alhenc-Gelas, 1995). Increased ACE gene expression is also observed in the hypertensive aortic smooth muscle in the rat, although in this case the induction of ACE gene expression occurs in vascular cells, especially in the hypertrophied smooth muscle cells (Arnal *et al.,* 1994; Shiota *et al.,* 1992). Likewise, a large induction of ACE gene expression is observed in proliferative smooth muscle cells at the site of an experimental endothelial injury, indicating that stimulation of smooth muscle cells by both proliferative or hypertrophic stimuli is accompanied by an activation of the ACE gene by unknown mechanisms (Rakugi *et al.,* 1994). Interestingly ACE inhibitors have been shown to prevent neointima formation after balloon-induced endothelial injury in the rat, although it is not clear whether this effect is due to the lowering of blood pressure or to a direct antiproliferative effect at the lesion site, or to both (Powell *et al.,* 1989).

A generalized induction of ACE gene expression in vascular and extravascular tissues, accompanied by an increase in plasma ACE levels, occurs during prolonged ACE inhibition *in vivo* (Forslund *et al.,* 1982; O. Costerousse *et al.,* unpublished observations) and has also been reported to occur *in vitro* in cultured endothelial cells (King and Oparil, 1992). The mechanism of this effect, which is not related to AII suppression or to haemodynamic alterations, remains unknown, but this observation suggests that other substrates or products of ACE can regulate, perhaps indirectly, its synthesis.

8.4 Genetic polymorphism of ACE levels in man

The inter-individual variation of plasma ACE levels is very high in healthy subjects, levels differing up to six-fold among individuals (Alhenc-Gelas *et al.,* 1991). When measured repeatedly in a given adult subject plasma, ACE levels remain remarkably constant. No association with candidate hormonal or environmental parameters was found to explain this large inter-individual variability. To test the hypothesis that the inter-individual variability in ACE levels was genetically determined, a study of plasma ACE levels has been conducted in nuclear families. Intrafamilial correlation between genetically related members was indeed observed, and the analysis of several transmission models suggested that a major gene effect was responsible for a large part of the inter-individual variability in ACE levels (Cambien *et al.,* 1988).

As the ACE gene was a logical candidate for this effect, polymorphism of this gene was studied after cloning of the ACE cDNA. A major, frequent polymorphism was recognized, corresponding to the presence or absence of a 287-base-pair *Alu*-type sequence located in intron 16, and this polymorphism was shown to be associated with the concentration of ACE in plasma. Subjects homozygous for the insertion have lower levels than those homozygotes without the insertion, while heterozygotes display intermediate levels (Rigat *et al.,* 1990). These findings have been extended to the membrane-bound form of ACE by studying circulating mononuclear cells, where ACE was found to be synthesised mainly in T lymphocytes. The membrane-bound ACE levels in T lymphocytes, like the levels of soluble ACE in plasma, are highly variable from one subject to another and are associated with the ACE gene polymorphism (Costerousse *et al.,* 1993a). A similar observation has been made in normal human hearts, where cardiac ACE levels were also found to be associated with the ACE gene insertion/deletion (I/D) polymorphism (Danser *et al.,* 1995). These observations suggest that in most, if not all tissues synthesising the enzyme, the inter-individual variability in ACE level is large and in part genetically determined.

The mechanisms linking the I/D polymorphism to ACE levels remain unknown. The insertion is an *Alu* repeat type sequence, located in intron 16 and not found in the ACE gene of other primates. In the 3′ part of this intron two *Alu*-type sequences in tandem are found in all individuals and a third *Alu*-type sequence, oriented in an opposite direction and located 3′ of the others, is present in some subjects only. The allele carrying this third *Alu* sequence has been designated as the I allele, and the other allele without it as the D allele (*Figure 8.1*). This sequence is unlikely to directly affect the level of transcription of the ACE gene promoter, but it may possibly affect the transcription and maturation of the ACE pre-mRNA. This remains, however, hypothetical. The I/D polymorphism may also be only a neutral marker in linkage disequilibrium with a causal variant (Rigat *et al.,* 1990; Tiret *et al.,* 1992). Although other polymorphisms have been found in the ACE gene, all in non-coding sequences, the putative variant affecting the transcription of the promoter has not yet been identified (Doria *et al.,* 1994; Villard *et al.,* 1996). In any case, the I/D polymorphism is associated with a large part of the variance of plasma ACE levels in populations (*Figure 8.2*). In European Caucasian populations where the frequency of the I allele (the allele associated with lower ACE levels) is about 46%, this part has been estimated to be between 20% and 47%, in different studies (Cambien *et al.,* 1994; Danilov *et al.,* 1996; Rigat *et al.,* 1990). Recent segregation studies have suggested that in African-Caribbean subjects two polymorphic genes are involved in the determinism of the interindividual variability of plasma ACE levels, one being the ACE gene, and the other an unknown, unrelated gene (McKenzie *et al.,* 1995). Candidates for this second gene logically include the gene coding for the unknown membrane-bound ACE-cleaving enzyme.

The ACE gene I/D polymorphism, initially identified by restriction fragment length polymorphism (RFLP) analysis, can be more conveniently studied in large series of subjects by PCR amplification of intron 16, preferably using flanking primers for primary amplification and an insertion, or allele-specific oligonucleotide for secondary amplification, or for dot-blot hybridization (Ludwig *et al.,* 1995; Marre *et al.,* 1996; Rigat *et al.,* 1992; Shanmugam *et al.,* 1993). The I/D polymorphism has been studied in many populations. The allele frequencies in Caucasian and African or African-American populations are close, with a slightly lower frequency of the I allele in Africans. However, in Asian populations there is a much higher frequency of the I allele (Barley *et al.,* 1996; Batzer *et al.,* 1994; Ishigami *et al.,* 1995; Lee, 1994; Nakai *et al.,* 1994; Rotimi *et al.,* 1996). A comprehensive study of the allelic distribution of the ACE

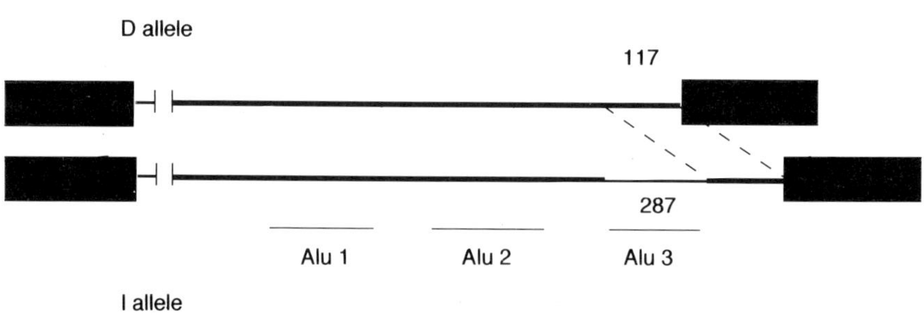

Figure 8.1. Schematic representation of intron 16 of the human ACE gene with its two allelic forms. Black boxes represent exon 16 (on the left) and 17, (on the right). The polymorphic *Alu* sequence is indicated by a thinner line. Numbers indicate size (in bases) of corresponding portions of DNA.

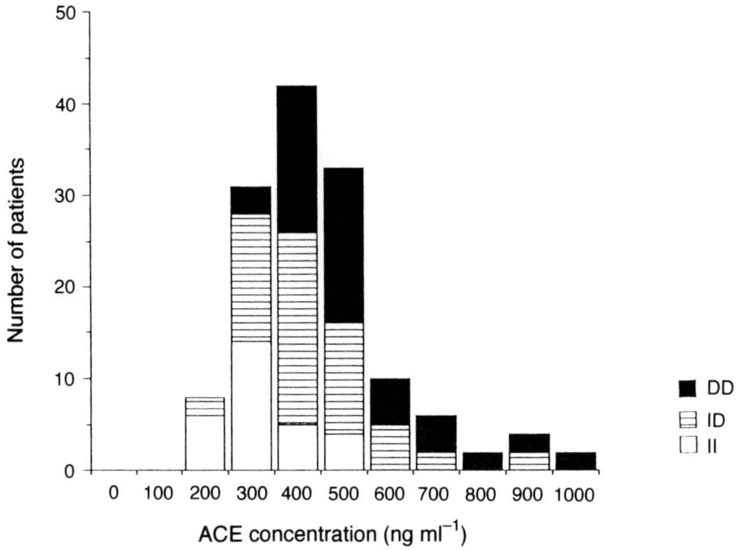

Figure 8.2. Distribution of plasma ACE levels in a healthy human population, according to the three I/D genotypes. ACE concentration was determined by ELISA. Reproduced from Danilov *et al.* (1996) with permission from the Gower Publishing Co., Aldershot, UK.

gene polymorphism in different populations (together with that of several other human-specific polymorphic *Alu* sequences), has been reported recently. Considering that the D allele is the ancestral allele, this study suggests that the insertion appeared in Africa, early during human evolution (Batzer *et al.*, 1994).

A rare genetic abnormality consisting of a familial elevation of plasma ACE (sometimes more than 10 times above usual values) with an autosomal dominant mode of transmission has been reported in a few isolated families (Okabe *et al.*, 1987). The molecular basis of this abnormality is unknown.

8.4.1 Association of the genetic polymorphism of ACE levels with diseases

Because of its involvement in cardiovascular and renal regulations, the ACE gene is a candidate gene for cardiovascular and renal diseases, and for hypertension. The ACE gene polymorphism and plasma ACE levels have therefore been studied with this possible association in view. Most studies did not find an association between the ACE gene polymorphism and blood pressure (Cambien *et al.*, 1992; Harrap *et al.*, 1993; Schmidt *et al.*, 1993), except perhaps in Black subjects (Barley *et al.*, 1996; Duru *et al.*, 1994; Rotimi *et al.*, 1996) and the ACE gene locus was not found to be linked to hypertension in a segregation analysis of Utah kindreds (Jeunemaître *et al.*, 1992). Although studies performed in crosses of a normotensive Wistar-Kyoto rat strain, and a derived hypertensive rat strain, suggested that a gene located at or near the ACE gene locus was a major determinant of high blood pressure in these animals (Hilbert *et al.*, 1991; Jacob *et al.*, 1991), further studies did not confirm the involvement of the ACE gene in this genetic form of rat hypertension (Kreutz *et al.*, 1995).

Other studies have suggested a role for ACE in the susceptibility to several types of cardiovascular diseases. In a large multicentre study of patients having suffered from a myocardial infarction compared to appropriate controls (the ECTIM study), an abnormal proportion of subjects of the DD genotype, and of young subjects of all genotypes with high plasma ACE levels, was observed among cases, independent of blood pressure level, or of other classical risk factors (Cambien *et al.,* 1992, 1994). An abnormal proportion of subjects carrying the D allele was also observed among subjects with a family history of myocardial infarction (Tiret *et al.,* 1993). This study suggests that ACE is a risk factor for myocardial infarction. Many other studies have been subsequently carried out in this area. Some were confirmatory (e.g. Arbustini *et al.,* 1995; Badenhop *et al.,* 1995; Bohn *et al.,* 1993; Ludwig *et al.,* 1995; Mattu *et al.,* 1995; Nakai *et al.,* 1994; Oike *et al.,* 1995; Ruiz *et al.,* 1994), others gave conflicting results (Friedl *et al.,* 1995; Lindpaintner *et al.,* 1995; Miettenen *et al.,* 1994), but there seems to be growing evidence in favour of an association of the ACE gene polymorphism with myocardial infarction in both Caucasian and Japanese populations (Samani *et al.,* 1996). However, these reports were case–control studies or, in one case (Lindpaintner *et al.,* 1995), a *post hoc* analysis of a therapeutic intervention trial in highly selected populations. Such studies can be subject to bias, and the role of ACE as a cardiovascular risk factor needs to be definitively established in prospective studies, several of which are now in progress.

Although early studies also suggested a role of the ACE gene polymorphism in the development of several types of cardiomyopathies, or in the risk of myointimal proliferation and re-stenosis after coronary angioplasty, these hypotheses are considered to be very controversial (Hamon *et al.,* 1995; Iwai *et al.,* 1994; Kupari *et al.,* 1994; Lechin *et al.,* 1995; Lindpaintner *et al.,* 1995; Marian *et al.,* 1993; Raynolds *et al.,* 1993; Samani *et al.,* 1995; Schunkert *et al.,* 1994; Van Bockxmeer, 1995). I/D polymorphism and plasma ACE levels, have also been reported to be associated with vascular wall thickness in unselected populations (Boniton-Koop *et al.,* 1994; Castellano *et al.,* 1995) and, recently, a possible relationship between the ACE polymorphism and the risk for ischaemic cerebrovascular disease has also been discussed (Catto *et al.,* 1996; Margaglione *et al.,* 1996; Markus *et al.,* 1995; Pullicino *et al.,* 1996).

The genetic polymorphism of ACE levels was also associated with diabetic nephropathy in type I diabetes and subsequently to the progression of other types of glomerular diseases, suggesting that ACE can have a deleterious effect on the renal as well as the coronary circulations. In type I diabetic patients, the risk of diabetic nephropathy, often leading to death from coronary artery disease, is probably in part genetically transmitted (Krolewski *et al.,* 1987). Vasoactive peptide systems affecting glomerular haemodynamics may be involved in the development of diabetic nephropathy.

In a case–control study of type I diabetic patients exposed to the risk of nephropathy because of poor glycaemic control, and all with proliferative retinopathy, an abnormal proportion of subjects carrying the D allele, or of subjects with high plasma ACE levels, was observed in nephropathic subjects, compared with paired non-albuminuric diabetic patients (Marre *et al.,* 1994). A similar, independent study also found an association between an ACE gene polymorphism and diabetic nephropathy in North American type I diabetic patients (Doria *et al.*, 1994). Since then, several studies have addressed the question in both insulin and non-insulin-dependent diabetic patients, again with confirmatory or different results (Doi *et al.*, 1996; Dudley *et al.*, 1995; Fogarty *et al.*, 1994; Keavney *et al.*, 1995; Mizuri *et al.*, 1995; Ohno *et al.*, 1996; Parving *et al.*, 1996; Pfeufer *et al.*, 1996; Schmidt *et al.*, 1995; Tarnow *et al.*, 1995). Recently a large multicentre case–control study of type I diabetic patients, selected for high risk of nephropathy because of long-term poor glycaemic control, did confirm the association and also suggested a dominant effect of the D allele on the progression of the disease (Marre *et al.*, 1996). Association between the ACE D allele and the severity of renal dysfunction, or the rate of decline in renal function, was also observed in studies of several glomerular diseases, especially IgA nephropathy (Harden *et al.*, 1995; Parving *et al.*, 1996; Van Essen *et al.*, 1996; Yorioka *et al.*, 1995; Yoshida *et al.*, 1996). These observations suggest that constitutive ACE levels are critical for the susceptibility of the glomerular circulation to external injury, whatever the nature of the injury. However, these studies, concerning diabetic nephropathy and non-diabetic glomerular diseases, are case–control studies that need to be confirmed in follow-up studies of renal function in exposed patients. Interestingly, it has been reported that the D allele is associated with longevity in a study of candidate genes in centenarians (Schachter *et al.*, 1994). Because of the physiological role of the renin–angiotensin system this hypothesis should not be considered as definitely conflicting with that of ACE being a cardiovascular risk factor.

There is a theoretical possibility that the polymorphisms of the ACE gene are in linkage disequilibrium with polymorphisms of another unknown, but closely located gene, partly responsible for the susceptibility to diseases. However, the association of the ACE gene I/D polymorphism with ACE levels in plasma and tissue, together with the strong association between these levels and disease, suggest a pathogenetic role for ACE (Cambien *et al.*, 1994; Marre *et al.*, 1994), a role that may be mediated by angiotensin (AII) production and kinin inactivation.

8.5 Old and new hypotheses concerning the physiological role of ACE

The conversion of AI is generally not considered as a rate-limiting step in AII formation because of the high abundance of ACE in the lung and the large capacity of the pulmonary circulation for AI conversion. Two studies have assessed the conversion of intravenously injected AI in the pulmonary, large arterial and forearm circulations, in relation to ACE gene polymorphism. They provided different results, perhaps because of different design and basal state of activation of the renin–angiotensin system (Lachurie *et al.*, 1995; Ueda *et al.*, 1995). The fact that the D allele does not seem to be associated with higher blood pressure in most studies, could suggest that the inter-individual variability in ACE levels has only minimal consequences for the overall rate of AI conversion through the lung. However, AI is produced in arteries and in peripheral vascular beds, as well as in interstitial tissues. Studies in man or pig indicate that a large part of the total production of AI in the body occurs in kidney and heart (Admiraal *et al.*, 1993; Danser *et al.*, 1992). The conversion of AI may be a critical, rate-limiting step in the production of AII in these organs, especially in the kidney (Vane, 1972), where renin and AI are high, while ACE in renal vessels is very low, at least in the human. The low conversion of AI in the kidney, especially in the glomeruli, may depend mostly on circulating ACE (Alhenc-Gelas *et al.,* 1989).

Bradykinin is another ACE substrate, produced in the kidney and able to participate in the control of renal circulation (Roman *et al.,* 1988). Kinins may also perhaps be produced in the heart and have a protective effect in experimental cardiac ischaemia (Linz *et al.,* 1996). As ACE is the major kinin-inactivating enzyme in the circulation, and ACE gene expression is up-regulated in the heart and kidney during ischaemia and remodelling, the enzyme may also play a pathogenetic role through kinin inactivation, or perhaps though hydrolysis of other physiological substrates yet to be identified. In any case it will be of considerable interest to establish the role of ACE levels in plasma and tissues in cardiovascular and renal homeostasis.

Acknowledgements

We thank Nathalie Coustou for typing the manuscript. Our research on ACE is supported by INSERM and by a grant from the Bristol Myers–Squibb Institute for Medical Research (Princeton, NJ, USA).

References

Admiraal PJ, Danser AH, Jong MS, Pieterman H, Derkx FH, Shalekamp MA. (1993) Regional angiotensin II production in essential hypertension and renal artery stenosis. *Hypertension* **21**: 173–174.

Alhenc-Gelas F, Weare J, Johnson R, Erdös E. (1983) Measurements of human converting enzyme level by direct radioimmunoassay. *J. Lab. Clin. Med.* **101**: 83–96.

Alhenc-Gelas F, Baussant T, Hubert C, Soubrier F, Corvol P. (1989) The angiotensin I-converting-enzyme in the kidney. *J. Hypertens.* **7**: S9–S13.

Alhenc-Gelas F, Richard J, Courbon D, Warnet J, Corvol P. (1991) Distribution of plasma angiotensin I-converting-enzyme levels in healthy men: relationship to environmental and hormonal parameters. *J. Lab. Clin. Med.* **117**: 33–39.

Arbustini E, Grasso M, Fasani R *et al.* (1995) Angiotensin converting enzyme gene deletion allele is independently and strongly associated with coronary atherosclerosis and myocardial infarction. *Br. Heart J.* **74**: 584–591.

Arnal JF, Battle T, Rasetti C, Challah M, Costerousse O, Vicaut E, Michel JB, Alhenc-Gelas F. (1994) ACE in the tunicae of rat aorta: expression in smooth muscle and effect of renovascular hypertension. *Am. J. Physiol.* **267**: H1777–H1784.

Azizi M, Rousseau A, Elan E, Guyenne TT, Michelet S, Grognet JM, Lenfant M, Corvol P, Menard J. (1996) Acute ACE inhibition increases the plasma level of the natural stem cell regulator *N*-acetyl-seryl-aspartyl-lysyl-proline. *J. Clin. Invest.* **97**: 588–592.

Badenhop RF, Wang XL, Wilcken DE. (1995) Angiotensin-converting enzyme genotype in children and coronary events in their grandparents. *Circulation* **91**: 1655–1658.

Barley J, Blackwood A, Miller M, Markandu ND, Carter ND, Jeffrey S, Cappucio FP, MacGregor GA, Sagnella GA. (1996) Angiotensin converting enzyme insertion/deletion polymorphism, blood pressure and the renin–angiotensin system in caucasian and afro Caribbean people. *J. Hypertens.* **10**: 31–35.

Batzer MA, Stoneking M, Alegria-Hartman M *et al.* (1994) African origin of human-specific polymorphic Alu insertions. *Proc. Natl Acad. Sci. USA* **91**: 12288–12292.

Bernstein KE, Martin BM, Edwards AS, Bernstein EA. (1989) Mouse angiotensin-converting-enzyme is a protein composed of two homologous domains. *J. Biol. Chem.* **264**: 11945–11951.

Bohn M, Berge KE, Bakken A, Erikssent J, Berg K. (1993) Insertion/deletion (I/D) polymorphism at the locus for angiotensin I-converting enzyme and parental history of myocardial infarction. *Clin. Genet.* **44**: 298–301.

Boniton-Koop, Ducimetière P, Touboul PJ, Feve JM, Billaud E, Courbon D, Heraud V. (1994) Plasma angiotensin-converting-enzyme activity and carotid wall thickening. *Circulation* **89**: 952–954.

Caldwell PR, Seegal BC, Hsu KC, Das M, Soffer RL. (1976) Angiotensin-converting-enzyme: vascular endothelial localization. *Science* **191**: 1050–1051.

Cambien F, Alhenc-Gelas F, Herbeth B, Andre JL, Rakotovao R, Gonzalez MF, Allegrini J, Bloch C. (1988) Familial resemblance of plasma angiotensin-converting-enzyme level: the Nancy Study. *Am. J. Hum. Genet.* **43**: 774–780.

Cambien F, Poirier O, Lecerf L *et al.* (1992) Deletion polymorphism in the gene for angiotensin-converting-enzyme is a potent risk factor for myocardial infarction. *Nature* **359**: 641–644.

Cambien F, Costerousse O, Tiret L *et al.* (1994) Plasma level and gene polymorphism of angiotensin-converting-enzyme in relation to myocardial infarction. *Circulation* **90**: 669–676.

Castellano M, Muiesan ML, Rizzoni D *et al.* (1995) Angiotensin-converting enzyme I/D polymorphism and arterial wall thickness in a general population. The Vobarno Study. *Circulation* **91**: 2721–2724.

Catto A, Carter AM, Barrett JH, Stickland M, Bamford J, Davies JA, Grant PJ. (1996) Angiotensin-converting enzyme insertion/deletion polymorphism and cerebrovascular disease. *Stroke* **27**: 435–440.

Challah M, Nicoletti A, Arnal JF, Philippe M, Laboulandine I, Allegrini J, Alhenc-Gelas F, Danilov S, Michel JB. (1995) Cardiac angiotensin-converting-enzyme overproduction indicates interstitial activation in renovascular hypertension. *Cardiovasc. Res.* **30**: 231–239.

Costerousse O, Allegrini J, Lopez M, Alhenc-Gelas F. (1993a) Angiotensin I-converting enzyme in human circulating mononuclear cells: genetic polymorphism of expression in T-lymphocytes. *Biochem. J.* **290**: 33–40.

Costerousse O, Jaspard E, Alhenc-Gelas F. (1993b) Molecular and genetic aspects of dipeptidylcarboxypeptidase I (the angiotensin I converting-enzyme). Expression in the immune system. *Adv. Neuroimmunol.* **3**: 217–224.

Costerousse O, Allegrini J, Huang H, Bouhnik J, Alhenc-Gelas F. (1994) Regulation of ACE gene expression and plasma levels during rat postnatal development. *Am. J. Physiol.* **267**: E745–E753.

Cushman DW, Cheung HS. (1971) Concentrations of angiotensin-converting-enzyme in tissues of the rat. *Biochim. Biophys. Acta* **250**: 261–265.

Cushman DW, Ondetti MA. (1980) Inhibitors of angiotensin-converting-enzyme. *Prog. Med. Chem.* **17**: 42–104.

Danilov S, Jaspard E, Churakova T, Towbin H, Savoie F, Wei L, Alhenc-Gelas F. (1994) Structure–function analysis of angiotensin I-converting-enzyme using monoclonal antibodies. Selective inhibition of the amino-terminal active site. *J. Biol. Chem.* **269**: 26806–26814.

Danilov S, Savoie F, Lenoir B, Jeunemaître X, Azizi M, Tarnow L, Alhenc-Gelas F. (1996) Development of enzyme linked immunoassays for human angiotensin I-converting-enzyme suitable for large studies. *J. Hyperten.* **14**: 709–717.

Danser AH, Koning MM, Admiraal PJ, Sassen LM, Derkx FH, Verdouw PD, Shalekamp ADH. (1992) Production of angiotensin I and II at tissue sites in intact pigs. *Am. J. Physiol.* **263**: H429–H437.

Danser AH, Schalekamp MA, Bax WA, Van Den Brink AM, Satema PR, Riegger GA, Schunkert H. (1995) Angiotensin-converting enzyme in the human heart. Effect of the deletion/insertion polymorphism. *Circulation* **92**: 1387–1388.

Dasarathy Y, Fanburg BL. (1990) Elevation of bovine endothelial cell angiotensin-converting-enzyme by cationophores and inhibition by ouabain. *Biochim. Biophys. Acta* **1051**: 14–20.

Doi Y, Yoshigumi H, Lino K, Yamamoto M, Ichikawa K, Iwase M, Fujishima M. (1996) Association between a polymorphism in the angiotensin-converting-enzyme gene and microvascular complications in Japanese patients with NIDDM. *Diabetologia* **39**: 97–102.

Doria A, Warram JH, Krolewski AS. (1994) Genetic predisposition to diabetic nephropathy: evidence for a role of angiotensin-I-converting-enzyme gene. *Diabetes* **43**: 690–695.

Dudley CRK, Keavney B, Stratton IM, Turner RC, Ratcliffe PJ. (1995) U.K prospective diabetes study XV: relationship of renin–angiotensin system gene polymorphisms with microalbuminuria in NIDDM. *Kidney Int.* **48**: 1907–1911.

Duru K, Farrow S, Wang JM, Lockette W, Kurtz T. (1994) Frequency of a deletion polymorphism in the gene for angiotensin-converting-enzyme is increased in African-Americans with hypertension. *Am. J. Hypertens.* **7**: 759–762.

Ehlers M, Fox E, Strydom D, Riordan J. (1989) Molecular cloning of human testicular angiotensin-converting-enzyme: the testis isozyme is identical to the C-terminal half of endothelial angiotensin-converting-enzyme. *Proc. Natl Acad. Sci.* USA **86**: 7741–7745.

El-Dorry HA, Pickett CB, MacGregor JS, Soffer RL. (1982) Tissue-specific expression of mRNAs for dipeptidyl carboxypeptidase isoenzymes. *Proc. Natl Acad. Sci.* USA **79**: 4295–4297.

Erdos EG. (1990) Angiotensin I converting enzyme and the changes in our concepts through the years. Lewis K. Dahl memorial lecture. *Hypertension* **16**: 363–370.

Esther CR Jr, Howard TE, Marino EM, Goddard JM, Capecchi M, Bernstein KE. (1996) Mice lacking angiotensin-converting-enzyme have low blood pressure, renal pathology, and reduced male fertility. *Lab. Invest.* **74:** 953–965.

Falkenhahn M, Franke F, Bohle RM, Zhu YC, Stauss HM, Bachmann S, Danilov S, Unger T. (1995) Cellular distribution of angiotensin-converting-enzyme after myocardial infarction. *Hypertension* **25**: 219–226.

Fishel RS, Eisenberg S, Shai SY, Redden RA, Bernstein KE, Berk BC. (1995a) Glucocorticoids induce angiotensin-converting-enzyme expression in vascular smooth muscle. *Hypertension* **25**: 343–349.

Fishel RS, Thourani V, Eisenberg SJ. (1995b) Fibroblast growth factor stimulates angiotensin-converting-enzyme expression in vascular smooth muscle cells. Possible mediator of the response to vascular injury. *J. Clin. Invest.* **95**: 377–387.

Fogarty DG, Maxwell AP, Hughes A, Nevin NC, Doherthy CC. (1994) A deletion polymorphism in the angiotensin-converting-enzyme gene is a risk factor for diabetic nephropathy. *EDTA Proceedings of the XXI SF Congress,* p. 117.

Forslund T, Fyhrquist F, Gronhagen-Riska C. (1982) Induction of angiotensin-converting-enzyme with the ACE inhibitory compound MK-421 in rat lung. *Eur. J. Pharmacol.* **80**: 121–125.

Friedl W, Krempler F, Paulweber B, Tickkanen I. (1995) A deletion polymorphism in the angiotensin-converting-enzyme gene is not associated with coronary heart disease in an Austrian population. *Atherosclerosis* **112:** 137–143.

Friedland J, Setton C, Silverstein E. (1978) Induction of angiotensin-converting-enzyme in human monocytes in culture. *Biochem. Biophys. Res. Commun.* **83**: 843–849.

Hamon M, Bauters C, Amant C, McFadden EP, Helbecque N, Lablanche JM, Bertrand ME, Amouyel P. (1995) Relation between the deletion polymorphism of the angiotensin-converting enzyme gene and late luminal narrowing after coronary angioplasty. *Circulation* **92**: 296–299.

Harden PN, Geddes C, Rowe PA *et al.* (1995) Polymorphisms in angiotensin-converting-enzyme gene and progression of IgA nephropathy. *Lancet* **345**: 1540–1542.

Harrap SB, Davidson HR, Connor JM, Soubrier F, Corvol P, Fraser R, Foy CJ, Watt GC. (1993) The angiotensin I-converting-enzyme gene and predisposition to high blood pressure. *Hypertension* **21**: 455–460.

Hilbert P, Lindpaintner K, Beckmann JS *et al.* (1991) Chromosomal mapping of two genetic loci associated with blood-pressure regulation in hereditary hypertensive rats. *Nature.* **353:** 521–529.

Ishigami T, Iwamoto T, Tamura K, Yamaguchi S, Iwasawa K, Uchino K, Umemura S, Ishii M. (1995) Angiotensin I converting enzyme (ACE) gene polymorphism and essential hypertension in Japan. Ethnic difference of ACE genotype. *Am. J. Hypertens.* **8**: 95–97.

Iwai N, Ohmichi N, Nakamura Y. (1994) DD genotype of the angiotensin-converting-enzyme gene is a risk factor for left ventricular hypertrophy. *Circulation* **90**: 2622–2628.

Jackson B, Cubela RB, Sakaguchi K, Johnston CI. (1988) Characterization of angiotensin-converting-enzyme (ACE) in the testis and assessment of the *in vivo* effects of the ACE inhibitor perindopril. *Endocrinology* **123:** 50–55.

Jacob HJ, Lindpaintner K, Lincoln SE, Kusumi K, Bunker RK, Mao YP, Ganten D, Dzau VJ, Lander ES. (1991) Genetic mapping of a gene causing hypertension in the stroke-prone spontaneously hypertensive rat. *Cell* **67:** 213–224.

Jaspard E, Alhenc-Gelas F. (1995) Catalytic properties of the two active sites of angiotensin I-converting enzyme on the cell surface. *Biochem. Biophys. Res. Commun.* **211**: 528–34.

Jaspard E, Wei L, Alhenc-Gelas F. (1993) Differences in the properties and enzymatic specificities of the two active sites of angiotensin I-converting-enzyme (kininase II). Studies with bradykinin and other natural peptides. *J. Biol. Chem.* **268**: 9496–9503.

Jeunemaître X, Lifton RP, Hunt SC, Williams RR, Lalouel JM. (1992). Absence of linkage between the angiotensin-converting-enzyme locus and human essential hypertension. *Nature Genetics* **1**: 72–75.

Johnston CI. (1994) Tissue angiotensin-converting-enzyme in cardiac and vascular hypertrophy, repair, and remodelling. *Hypertension* **23**: 258–268.

Keavney BD, Dudley CR, Stratton IM, Holman RR, Mathews DR, Ratcliffe PJ, Turner RC. (1995) UK prospective diabetes study (UKPDS) 14: association of angiotensin- converting enzyme insertion/deletion polymorphism with myocardial infraction in NIDDM. *Diabetologia* **38**: 948–952.

King SJ, Oparil S. (1992) Converting-enzyme inhibitors increase converting-enzyme mRNA and activity in endothelial cells. *Am. J. Physiol.* **263:** C743–C749.

Koike G, Krieger JE, Jacob HJ, Muroyama M, Pratt RE, Dzau V. (1994) Angiotensin converting enzyme and genetic hypertension: cloning of rat cDNAs and characterization of the enzyme. *Biochem. Biophys. Res. Commun.* **198:** 380–386.

Krege JH, John SW, Langenbach LL, Hodgin JB, Hagaman JR, Bachman ES, Jennette JC, O'Brien DA, Smithies O. (1995) Male–female differences in fertility and blood pressure in ACE-deficient mice. *Nature* **375:** 146–148.

Kreutz R, Hubner N, James MR, Bihoreau MT, Gaughier D, Lathrop GM, Ganten D, Lindpaintner K. (1995) Dissection of a quantitative trait locus for genetic hypertension on rat chromosome 10. *Proc. Natl Acad. Sci. USA* **92**: 8778–8782.

Krolewski AJ, Warram JH, Rand LI, Kahn CR. (1987) Epidemiologic approach to the etiology of type I diabetes mellitus and its complications. *New Engl. J. Med.* **317:** 1390–1398.

Krulewitz AH, Baur WE, Fanburg BL. (1984) Hormonal influence on endothelial cell angiotensin I-converting-enzyme activity. *Am. J. Physiol.* **247** (*Cell Physiol.* 16): C163–C168.

Kumar RS, Thekkumkara TJ, Sen GC. (1991) The mRNAs encoding the two angiotensin-converting isozymes are transcribed from the same gene by a tissue-specific choice of alternative transcription initiation sites. *J. Biol. Chem.* **266**: 3854–3862.

Kupari M, Perola M, Koskinen P. (1994) Left ventricular size, mass, and function in relation to angiotensin-converting enzyme gene polymorphism in humans. *Am. J. Physiol.* **267**: H1107–H1111.

Lachurie ML, Azizi M, Guyene TT, Alhenc-Gelas F, Menard J. (1995) Angiotensin-converting enzyme gene polymorphism has no influence on the circulating renin–angiotensin-aldosterone system or blood pressure in normotensive subjects. *Circulation* **91**: 2933–2942.

Langford KG, Shai SY, Howard TE. (1991) Transgenic mice demonstrate a testis-specific promoter for angiotensin-converting enzyme. *J. Biol. Chem.* **266**: 15559–15562.

Lanzillo JJ, Kong XJ, Fanburg BL. (1994) A competitive deletion mutant quantitative PCR assay for angiotensin-converting enzyme mRNA in smooth muscle cells. *PCR Methods Appl.* **4**: 167–171.

Lattion AL, Soubrier F, Allegrini J Hubert C, Corvol P, Alhenc-Gelas F. (1989) The testicular transcript of the angiotensin I-converting-enzyme encodes for the ancestral, non-duplicated form of the enzyme. *FEBS Lett.* **252**: 99–104.

Lechin M, Quinones MA, Omran A *et al.* (1995) Angiotensin-I converting enzyme genotypes and left ventricular hypertrophy in patients with hypertrophic cardiomyopathy. *Circulation* **92**: 1808–1812.

Lee EJ. (1994) Population genetics of angiotensin II-converting enzyme in chinese. *Br. J. Clin. Pharmacol.* **37**: 212–214.

Lewis EJ, Hunsiker LG, Bain RP, Rohde RD, for the Collaborative Study Group. (1993) The effect of angiotensin converting-enzyme inhibition on diabetic nephropathy. *New Engl. J. Med.* **329**: 1456–1462.

Lieberman J. (1976) The specificity and nature of serum-angiotensin-converting-enzyme (serum ACE) elevations in sarcoidosis. *Ann. NY Acad. Sci.* **278**: 488–497.

Lindpaintner K, Pfeffer MA, Kreutz R, Stampfer MJ, Grodstein F, Lamotte F, Buring J, Mennekens CH. (1995) A prospective evaluation of an angiotensin-converting-enzyme: gene polymorphism and the risk of ischemic heart disease. *New Engl. J. Med.* **332**: 706–11.

Lindpaintner K, Lee M, Larson MG *et al.* (1996) Absence of association or genetic linkage between the angiotensin-converting-enzyme gene and left ventricular mass. *New Engl. J. Med.* **334**: 1023–1028.

Linz W, Wiemer G, Scholkens BA. (1996) Role of kinins in the pathophysiology of myocardial ischemia. *In vitro* and *in vivo* studies. *Diabetes* **45** (Suppl. **1**:) S51–S58.

Lloyd CJ, Cary DA, Mendelsohn FA. (1987) Angiotensin converting enzyme induction by cyclic AMP and analogues in cultured endothelial cells. *Mol. Cell. Endocrinol.* **52**: 219–225.

Ludwig E, Corneli PS, Anderson JL, Marshall HW, Lalouel JM, Ward RH. (1995) Angiotensin-converting enzyme gene polymorphism is associated with myocardial infarction but not with development of coronary stenosis. *Circulation* **91**: 2120–2124.

Margaglione M, Celentano E, Grandone E *et al.* (1996) Deletion polymorphism in the angiotensin-converting-enzyme gene in patients with a history of ischemic stroke. *Arterioscler. Thromb. Vasc. Biol.* **16**: 304–309.

Marian AJ, Yu QT, Workman R, Greve G, Roberts R. (1993) Angiotensin-converting enzyme polymorphism in hypertrophic cardiomyopathy and sudden cardiac death. *Lancet* **342**: 1085–1086.

Markus HS, Barley J, Lunt R, Bland JM, Jeffery S, Carter ND, Brown MM. (1995) Angiotensin-converting enzyme gene deletion polymorphism. A new risk factor for lacunar stroke but not carotid atheroma. *Stroke* **26**: 1329–1333.

Marre M, Chatellier G, Leblanc H, Guyenne TT, Menard J, Passa P. (1988) Prevention of diabetic nephropathy with enalapril in normotensive diabetics subjects with persistent microalbuminuria. *Br. Med. J.* **297:** 1092–1095.

Marre M, Bernadet P, Gallois Y *et al.* (1994) Relationships between angiotensin I-converting-enzyme gene polymorphism, plasma levels, and diabetic retinal and renal complications. *Diabetes* **43**: 384–388.

Marre M, Jeunemaître J, Gallois Y *et al.* for the Genediab Study Group. (1997) Contributions of genetic polymorphisms of the renin-angiotensin-system to renal complications in insulin dependent diabetes. *J. Clin. Invest.* **99:** 1585–1595.

Maschio G, Alberti D, Janin G, Locatelli F, Mann JFE, Motolese M, Ponticelli C, Ritz E, Ziuchelli P. (1996) ACE inhibition in Progressive Renal Insufficiency Study Group: effect of the ACE inhibitor benazepril on the progression of chronic renal insufficiency. *New Engl. J. Med.* **334**: 939–945.

Mattu RK, Needham EW, Galton DJ, Frangos E, Clark AJ, Caufield M. (1995) A DNA variant at the angiotensin-converting-enzyme gene locus associates with coronary artery disease in the Caerphilly Heart Study. *Circulation* **91**: 270–274.

McKenzie CA, Julier C, Forrester T, McFarlane-Anderson N, Keavney B, Lathrop GM, Ratcliffe PJ, Farall M. (1995) Segregation and linkage analysis of serum angiotensin I-converting-enzyme levels: evidence for two quantitative-trait loci. *Am. J. Hum. Genet.* **57**: 1426–1435.

Mendelsohn FA, Lloyd CJ, Kachel C, Funder JW. (1982) Induction by glucocorticoids of angiotensin-converting-enzyme production from bovine endothelial cells in culture and rat lung *in vivo*. *J. Clin. Invest.* **70**: 684–692.

Miettinen HE, Korpela K, Hamalainen L, Kontula K. (1994) Polymorphisms of the apolipoprotein and angiotensin-converting-enzyme genes in young North Karelian patients with coronary heart disease. *Hum. Genet.* **94**: 189–192.

Mizuiri S, Hemmi H, Inoue A *et al.* (1995) Angiotensin-converting enzyme polymorphism and development of diabetic nephropathy in non-insulin-dependent diabetes mellitus. *Nephron* **70**: 455–459.

Morris BJ, Zee RY, Schrader AP. (1994) Different frequencies of angiotensin-converting-enzyme genotypes in older hypertensive individuals. *J. Clin. Invest.* **94**: 1085–1089.

Nakai K, Itoh C, Miura Y, Nakai K, Syo T, Musya T, Hiramori K. (1994). Deletion polymorphism of the angiotensin I-converting-enzyme gene is associated with serum ACE concentration and increased risk for CAD in the Japanese. *Circulation* **90**: 2199–2202.

Nakamura Y, Takeda T, Ishii M, Nishiyama K, Yamakada M, Hirata Y, Kimura K, Murao S. (1982) Elevation of serum angiotensin-converting-enzyme activity in patients with hyperthyroidism. *J. Clin. Endocrinol. Metab.* **55**: 931–934.

Ohno T, Kawazu S, Tomono S. (1996) Association analyses of the polymorphisms of angiotensin-converting-enzyme and angiotensinogen genes with diabetic nephropathy in Japanese non-insulin-dependent diabetics. *Metabolism* **45**: 218–222.

Oike Y, Hata A, Ogata Y, Numata Y, Shido K, Kondo K. (1995) Angiotensin converting enzyme as a genetic risk factor for coronary artery spasm. Implication in the pathogenesis of myocardial infarction. *J. Clin. Invest.* **96**: 2975–2979.

Okabe T, Yamagata K, Fujisawa M, Watanabe J, Takaku F, Lanzillo JJ, Fanburg BL. (1985) Increased angiotensin-converting-enzyme in peripheral blood monocytes from patients with sarcoidosis. *J. Clin. Invest.* **75**: 911–914.

Okabe T, Fujisawa M, Watanabe J, Yotsumoto H, Takaku F. (1987) Familial hyper-angiotensin-converting-enzyme (ACE)-emia: increased production of ACE by monocyte-macrophage. *Jpn. J. Med.* **26**: 140–146.

Parving HH, Jacobsen P, Tarnow L, Rossing P, Lecerf L, Poirier O, Cambien F. (1996) Effect of a deletion polymorphism of angiotensin-converting-enzyme gene on progression of diabetic nephropathy during inhibition of angiotensin-converting-enzyme, observational follow up study. *Br. Med. J.* **313:** 591–594.

Passier RC, Smits JF, Verluyten MJ, Studer R, Drexler H, Daemen MJ. (1995) Activation of angiotensin-converting-enzyme expression in infarct zone following myocardial infarction. *Am. J. Physiol.* **269:** H1268–H1276.

Paul M, Wagner J, Dzau VJ. (1993) Gene expression of the renin–angiotensin system in human tissues. Quantitative analysis by the polymerase chain reaction. *J. Clin. Invest.* **91**: 2058–2064.

Pfeufer A, Osterziel KJ, Urata H, Borck G, Schuster H, Wienker T, Dietz R, Luft FC. (1996) Angiotensin-converting enzyme and heart chymase gene polymorphisms in hypertrophic cardiomyopathy. *Am. J. Cardiol.* **78**: 362–364.

Powell JS, Clozel JP, Muller RK, Kuhn H, Hefti F, Baumgartner HR. (1989) Inhibitors of angiotensin-converting-enzyme prevent myointimal proliferation after vascular injury. *Science* **245**: 186–188.

Pullicino P, Kwen PL, Greenberg S, Becker AL, Glenister N. (1996) Angiotensin-converting enzyme gene and lacunar stroke. *Stroke* **27**: 569–570.

Rakugi H, Kim DK, Krieger JE, Wang DS, Dzau VJ, Pratt RE. (1994) Induction of angiotensin-converting-enzyme in the neointima after vascular injury. Possible role in restenosis. *J. Clin. Invest.* **93**: 339–346.

Raynolds MV, Bristow MR, Bush EW, Abraham WT, Lowes BD, Zisman LS, Taft CS, Perryman MB. (1993) Angiotensin-converting enzyme DD genotype in patients with ischaemic or idiopathic dilated cardiomyopathy. *Lancet* **342**: 1073–1075.

Rigat B, Hubert C, Alhenc-Gelas F, Cambien F, Corvol P, Soubrier F. (1990) An insertion/deletion polymorphism in the angiotensin I-converting-enzyme gene accounting for half the variance of serum enzyme levels. *J. Clin. Invest.* **86**: 1343–1346.

Rigat B, Hubert C, Corvol P, Soubrier F. (1992) PCR detection of the insertion/deletion polymorphism of the human angiotensin-converting-enzyme gene (DCP1) (dipeptidyl carboxypeptidase 1). *Nucleic Acids Res.* **20:** 1433.

Roman RJ, Kaldunski ML, Scicli AG, Carretero OA. (1988) Influence of kinins and angiotensin II on the regulation of papillary blood flow. *Am. J. Physiol.* **255**: 801–804.

Rotimi C, Puras A, Cooper R, McFarlane-Anderson N, Forrester T, Ogunbiyi O, Morrison L, Ward R. (1996) Polymorphisms of renin-angiotensin genes among Nigerians, Jamaicans, and African Americans. *Hypertension.* **27**: 558–563.

Ruiz J, Blanche H, Cohen N, Velho G, Cambien F, Cohen D, Passa P, Froguel P. (1994). Insertion/deletion polymorphism of the angiotensin-converting-enzyme gene is strongly associated with coronary heart disease in non-insulin-dependent diabetes mellitus. *Proc. Natl Acad. Sci.* USA **91**: 3662–3665.

Samami NJ, Martin DS, Brack M *et al.* (1995) Insertion/deletion polymorphism in the angiotensin-converting-enzyme gene and risk of restenosis after coronary angioplasty. *Lancet* **345**: 1013–1016.

Samami NJ, Thompson JR, O'Toole L, Channer K, Woods KL. (1996) A meta analysis of the association of the deletion allele of the angiotensin-converting-enzyme gene with myocardial infarction. *Circulation* **94**: 708–712.

Schachter F, Faure-Delanef L, Guenot F, Rouger H, Froguel P, Lesueur-Ginot L, Cohen D. (1994) Genetic associations with human longevity at the APOE and ACE loci. *Nature Genetics* **6**: 29–32.

Schmidt S, Van Hooft IM, Grobbee DE, Ganten D, Ritz E. (1993) Polymorphism of the angiotensin I-converting-enzyme gene is apparently not related to high blood pressure: Dutch Hypertension and Offspring Study. *J. Hypertens.* **11**: 345–348.

Schmidt S, Schone N, Ritz E. (1995) Association of ACE gene polymorphism and diabetic nephropathy? The Diabetic Nephropathy Study Group. *Kidney Int.* **47**: 1176–1181.

Schunkert H, Dzau VJ, Tang SS, Hirsch AT, Apstein CS, Lorell BH. (1990) Increased rat cardiac angiotensin-converting-enzyme activity and mRNA expression in pressure overload left ventricular hypertrophy. Effects on coronary resistance, contractility, and relaxation. *J. Clin. Invest.* **86**: 1913–1920.

Schunkert H, Ingelfinger JR, Hirsch AT, Pinto Y, Remme WJ, Jacob H, Dzau VJ. (1993) Feedback regulation of angiotensin-converting-enzyme activity and mRNA levels by angiotensin II. *Circul. Res.* **72:** 312–318.

Schunkert H, Hense HW, Holmer SR. (1994) Association between a deletion polymorphism of the angiotensin-converting-enzyme gene and left ventricular hypertrophy. *New Engl. J. Med.* **330**: 1634–1638.

Shai SY, Fishel RS, Martin BM, Berk BC, Bernstein KE. (1992) Bovine angiotensin-converting-enzyme cDNA cloning and regulation. Increased expression during endothelial cell growth arrest. *Circul. Res.* **70**: 1274–1281.

Shanmugam V, Sell KW, Saha BK. (1993) Mistyping ACE heterozygotes. *PCR Meth. Applic.* **3**: 120–121.

Shiota N, Miyazaki M, Okunishi H. (1992) Increase of angiotensin-converting-enzyme gene expression in the hypertensive aorta. *Hypertension* **20**: 168–174.

Sibony M, Gasc JM, Soubrier F, Alhenc-Gelas F, Corvol P. (1993) Gene expression and tissue localization of the two isoforms of angiotensin I-converting-enzyme. *Hypertension* **21**: 827–835.

Skeggs LT, Kahn JR, Shumway NP. (1956) The preparation and function of the hypertensin converting enzyme. *J. Exp. Med.* **103**: 295–299.

Soubrier F, Alhenc-Gelas F, Hubert C, Allegrini J, John V, Treggear G, Corvol P. (1988). Two putative active centers in human angiotensin I-converting-enzyme revealed by molecular cloning. *Proc. Natl Acad. Sci.* USA **85**: 9386–9390.

Strittmatter SM, Thiele EA, De Souza EB, Snyder SH. (1985) Angiotensin-converting enzyme in the testis and epididymis: differential development and pituitary regulation of isozymes. *Endocrinology* **117**: 1374–1379.

Studer R, Reinecke H, Muller B, Holtz J, Just H, Drexler H. (1994) Increased angiotensin-I converting enzyme gene expression in the failing human heart. Quantification by competitive RNA polymerase chain reaction. *J. Clin. Invest.* **94:** 301–310.

Tarnow L, Cambien F, Rossing P Nielsen FS, Hansen BV, Lecerf L, Poirier O, Danilov S, Parving HH. (1995) Lack of relationship between an insertion/deletion polymorphism in the angiotensin I-converting-enzyme gene and diabetic nephropathy and proliferative retinopathy in IDDM patients. *Diabetes* **44**: 489–494.

Tiret L, Rigat B, Visvikis S, Breda C, Corvol P, Cambien F, Soubrier F. (1992) Evidence, from combined segregation and linkage analysis, that a variant of the angiotensin I-converting-enzyme (ACE) gene controls plasma ACE levels. *Am. J. Hum. Genet.* **51**: 197.

Tiret L, Kee F, Poirier O *et al.* (1993) Deletion polymorphism in angiotensin-converting-enzyme gene associated with parental history of myocardial infarction. *Lancet* **341**: 991–992.

Turner AJ, Hooper NM, Kenny AJ. (1987) Metabolism of neuropeptides. In: *Mammalian Ectoenzymes* (eds A.J. Kenny and A.J. Turner), pp. 211–248. Elsevier, Amsterdam.

Ueda S, Elliott HL, Morton JJ, Connell JM. (1995) Enhanced pressor response to angiotensin I in normotensive men with the deletion genotype (DD) for angiotensin-converting-enzyme. *Hypertension* **25**: 1266–1269.

Van Bockxmeer FM, Mamotte CD, Gibbons FA, Burke V, Taylor RR. (1995) Angiotensin-converting enzyme and apolipoprotein E genotypes and restenosis after coronary angioplasty. *Circulation* **92:** 2066–2071.

Van Essen GG, Rensma PL, De Zeeuw D, Sluiter WJ, Scheffer H, Aperloo AJ, De Jong PE. (1996) Association between angiotensin-converting-enzyme gene polymorphism and failure of renoprotective therapy. *Lancet* **347**: 94–95.

Vane GR. (1972) Sites of conversion of angiotensin I. In: *Hypertension* (eds J. Genest and E. Koine), pp. 523–532. Springer Verlag, Berlin.

Velletri PA, Aquilano DR, Bruckwick E, Tsai-Morris CH, Dufau ML, Lovenberg W. (1985) Endocrinological control and cellular localization of rat testicular angiotensin-converting-enzyme (EC 3.4.15.1). *Endocrinology* **116**: 2516–2522.

Villard E, Tiret L, Visvikis S, Rakotovao R, Cambien F, Soubrier F. (1996) Identification of new polymorphisms of the angiotensin I-converting-enzyme (ACE) gene, and study of their relationship to plasma ACE levels by two-QTL segregation-linkage analysis. *Am. J. Hum. Genet.* **58**: 1268–1278.

Vio CP, Alhenc-Gelas F. (1995) Differential distribution of angiotensin I-converting-enzyme in normal and hypertensive rat kidneys. *Hypertension* **25**: 18.

Wei L, Alhenc-Gelas F, Corvol P, Clauser E. (1991) The two homologous domains of human angiotensin I-converting-enzyme are both catalytically active. *J. Biol. Chem.* **266**: 9002–9006.

Yorioka T, Suehiro T, Yasuoka N, Hashimoto K, Kawada M. (1995) Polymorphism of the angiotensin-converting-enzyme gene and clinical aspects of IgA nephropathy. *Clin. Nephrol.* **44**: 80–85.

Yoshida H, Mitarai T, Kawamura T *et al.* (1996) Role of the deletion of polymorphism of the angiotensin-converting-enzyme gene in the progression and therapeutic responsiveness of IgA nephropathy. *J. Clin. Invest.* **96**: 2162–2169.

Yosipiv IV, Dipp S, El-Dahr SS. (1994) Ontogeny of somatic angiotensin-converting-enzyme. *Hypertension* **23**: 369–374.

Yotsumoto H, Imai Y, Kuzuya N, Uchimura H, Matsuzaki F. (1982) Increased levels of serum angiotensin-converting-enzyme activity in hyperthyroidism. *Ann. Intern. Med.* **96**: 326–328.

Chapter 9

Endothelin-converting enzymes

Anthony J. Turner

9.1 Introduction

The discovery of the endothelin family of peptides followed on from that of endothelial-derived relaxing factor (Furchgott and Zawadczi, 1980) which was later recognized to be nitric oxide. At about this time work was also in progress on an endothelial-derived vasoconstrictor, whose presence was first demonstrated by Hickey *et al.* (1985). The nature of that factor was clarified by O'Brien *et al.* (1987) who reported that cultured bovine endothelial cells of aortic and pulmonary origin released a vasoconstrictive peptide into the culture medium. It was, however, the pioneering work of Yanagisawa *et al.* (1988) which succeeded in isolating and characterizing this vasoconstrictor from the conditioned media of cultured porcine aortic endothelial cells. The compound turned out to be a 21-amino-acid peptide with two intramolecular disulphide bonds, which was named endothelin (ET). Subsequently, three distinct ET genes were identified encoding three closely related peptides (ET-1, ET-2 and ET-3; Inoue *et al.*, 1989) (*Figure 9.1*). All three ET peptides are structurally related to sarafotoxins, vasoactive peptides which are found in the venom of the snake *Atractaspis engaddensis* (Kloog *et al.*, 1988) (*Figure 9.1*). Endothelins have now been found in a variety of extravascular tissues, including the lung, pancreas, spleen and brain. Synthetic ET-1 proved to be a powerful constrictor of coronary arteries *in vitro* and a potent hypertensive agent *in vivo,* producing a long-lasting pressor response when injected intravenously (Yanagisawa *et al.*, 1988). Because of its unique pharmacological properties, the peptide may play an important role in the maintenance of vascular tone and blood pressure (Masaki and Yanagisawa, 1992; Masaki *et al.*, 1991). ETs have not only been implicated in the pathophysiology of cardiovascular diseases but also in renal and respiratory diseases (Masaki and Yanagisawa, 1992). ET-1 may participate in vascular modelling and regulation of cell proliferation through its mitogenic activity, for example mediating cardiac growth and hypertrophy (Shubeita *et al.*, 1990). More recently it has been shown, through targeted gene disruption, that the ET peptides may also play a role as morphogens in a variety of developmental events (Baynash *et al.*, 1994; Kurihara *et al.*, 1994).

9.2 Endothelin biosynthesis

9.2.1 Identification of the relevant 'endothelin-converting enzyme'

The biosynthesis of ET-1 occurs predominantly in vascular endothelial or smooth muscle

Cell-Surface Peptidases in Health and Disease, edited by A.J. Kenny and C.M. Boustead.

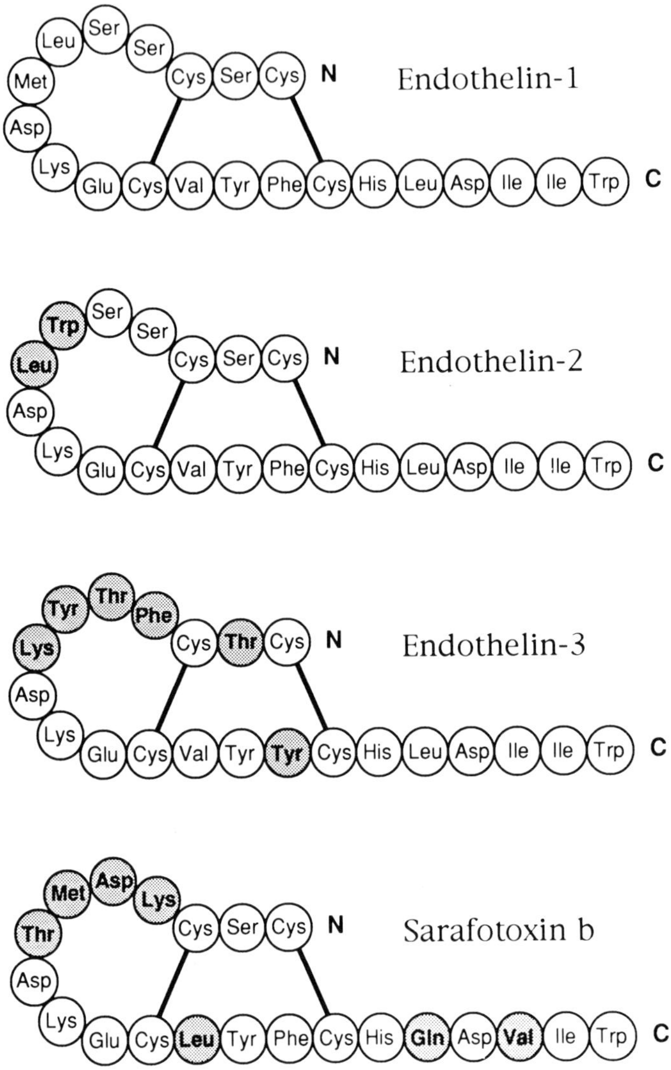

Figure 9.1. Amino acid sequences of human endothelins-1, -2 and -3 and the homologous snake venom peptide, sarafotoxin b. Residues that differ from those in ET-1 are shown shaded.

cells through the constitutive secretory pathway. The processing events involved in production of ET-1 were indicated from the cloned sequence of the precursor (Yanagisawa *et al.*, 1988). The initial precursor (prepro-ET-1) is of approximately 200 amino acids. Sandwiched between pairs of basic residues, which serve as recognition signals for a prohormone convertase, is a sequence of 38 or 39 amino acids depending on species, referred to as 'big ET-1' which is biologically inactive. Conversion of big ET-1 to the fully active ET-1 involves a further and unique protein processing event catalysed by an activity termed 'endothelin-converting enzyme' (ECE) by Yanagisawa *et al.* (1988) (*Figure 9.2*).

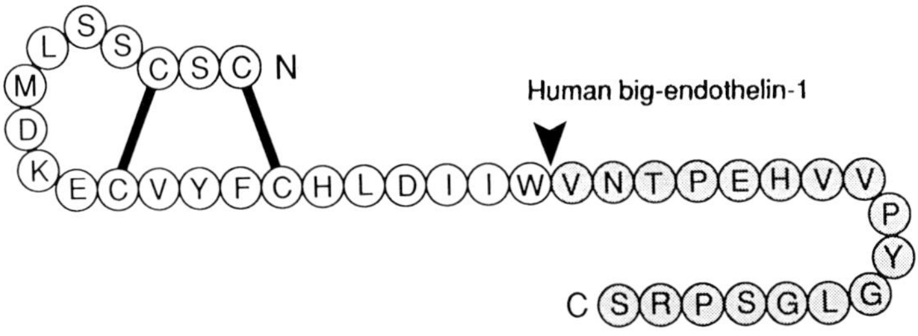

Figure 9.2. Structure of human big endothelin-1 and the reaction catalysed by endothelin-converting enzyme. The Trp21–Val22 bond cleaved by ECE is indicated by the arrow. The products of the reaction are ET-1 (open circles) and the C-terminal fragment (shaded circles). No further degradation of these products occurs with ECE.

This conversion involves hydrolysis of the Trp21–Val22 bond in big ET-1, and an identical cleavage is involved in the conversion of big ET-2 to ET-2. In big ET-3, however, the scissile bond is Trp–Ile. The recognition that the ECE reaction is specific to ET processing and that its inhibition could selectively attenuate a powerful vasoconstrictor has, not surprisingly, attracted much attention in the pharmaceutical industry. Identification of the relevant ECE and its subsequent isolation and cloning has, however, occupied 6 years since the first isolation of the ET peptide.

The search for the physiologically relevant ECE has embraced all classes of proteases: serine, thiol, metallo- and aspartic. The primary specificity of ECE was initially suspected to be the tryptophan residue in the P_1 position which led Yanagisawa *et al.* (1988) to suggest that ECE may be a chymotrypsin-like activity and it was shown that chymotrypsin could indeed convert big ET to products that produced similar responses to ET both *in vitro* and *in vivo* (McMahon *et al.*, 1989). However, chymotrypsin initially hydrolysed the Tyr31–Gly32 bond of big ET followed by the correct Trp21–Val22 but then further degraded the mature ET peptide (Takaoka *et al.*, 1990). Subsequently, Wypij *et al.* (1992) identified a chymostatin-sensitive peptidase in rat lung able to process extracellular big ET-1 to ET-1. This activity was identified as mast cell chymase I but there is no evidence to date that this, or any related serine protease, is responsible for big ET processing *in vivo* in endothelial cells. Parallel studies had begun to focus on aspartyl proteases as processing activities following the observation that pepsin could hydrolyse big ET-1 to ET-1 and its C-terminal fragment (Takaoka *et al.*, 1990) and an aspartyl activity with the correct processing requirements was identified in porcine aortic endothelial cells (Matsumura *et al.*, 1990a). Others also identified similar activities which were inhibited by pepstatin A and exhibited an acidic pH optimum. Both cathepsin D and cathepsin E as well as thiol proteases were briefly considered as candidate ECE species but interest in these activities was superseded by the discovery that the conversion of big ET-1 to ET-1 both *in vivo* and *in vitro* was markedly sensitive to phosphoramidon (Okada *et al.*, 1990), a potent inhibitor of the zinc metallopeptidase, endopeptidase-24.11 (E-24.11; see Chapter 6). Most subsequent work on ECE has therefore focused on a phosphoramidon-sensitive metalloenzyme. A more detailed discussion of the preliminary work leading up to the identification of the physiologically relevant ECE is to be found in Opgenorth *et al.* (1992).

Several groups reported, more or less simultaneously, the presence in endothelial cells derived from different sources of a phosphoramidon-sensitive ECE active at neutral pH (unlike the aspartyl protease activities) and which was not inhibited by another E-24.11 inhibitor, thiorphan (Matsumura *et al.*, 1991; Takada *et al.*, 1991). Phosphoramidon at micromolar concentrations was shown to prevent the release of ET-1 from endothelial cells and to alter the ET-1 to big ET ratio in media and within cells in a manner consistent with blocking ET-1 formation by inhibition of ECE (Sawamura *et al.*, 1991; Shields *et al.*, 1991). Phosphoramidon has also been shown to abolish the haemodynamic effects of intravenously administered big ET-1 (McMahon *et al.*, 1991; Pollock and Opgenorth, 1991) again suggesting blockade of ECE *in vivo*. Since phosphoramidon does not readily enter cells, it was likely that this effect was achieved by inhibiting an activity on the surface of endothelial cells (i.e. an ectoenzyme) and that the final stage in endothelin biosynthesis occurred as a post-secretory processing event. This enzyme would appear to be distinct from, but possibly related to, E-24.11 on the basis of inhibition by phosphoramidon at micromolar rather than nanomolar concentrations and the insensitivity to thiorphan.

9.2.2 E-24.11 as a model for ECE

Although E-24.11 and ECE may be distinct proteins, certain similarities suggested that E-24.11 may serve as a useful model for ECE and may itself be able to effect the conversion of big ET-1 to ET-1. Although the specificity of E-24.11, cleaving on the amino side of hydrophobic residues (see Chapter 6), would be consistent with it cleaving the Trp–Val bond of big ET-1, a number of studies have shown that purified E-24.11 can degrade ET-1 *in vitro*, consistent with the presence of other hydrophobic residues in the sequence (Sokolovsky *et al.*, 1990; Vijayaraghavan *et al.*, 1990). Furthermore, renal E-24.11 has been implicated in the physiological inactivation of circulating ET-1 (Abassi *et al.*, 1992). Two studies on the possible conversion of big ET-1 to ET-1 by E-24.11 have arrived at differing conclusions. Abassi *et al.* (1993), using recombinant human E-24.11, failed to observe the formation of ET-1 from big ET-1 by high performance liquid chromatography (HPLC). A more detailed study employing a combination of radioimmunoassay and mass spectrometry of resolved peptide products was able to detect the formation of significant quantities of both ET-1 and its C-terminal fragment by purified pig kidney E-24.11 (Murphy *et al.*, 1994). However the levels of ET-1 generated were seen to depend critically on the enzyme to substrate (big ET-1) ratio as the generated ET-1 is rapidly degraded to the ET-1 (17–21) and ET-1 (19–21) fragments, although steady-state levels of ET-1 remain detectable. The comparative kinetics of big ET-1 and ET-1 hydrolysis by E-24.11 confirm that degradation of the ET-1 is favoured (Murphy *et al.*, 1994) consistent with the preference of the enzyme for smaller peptides. These observations tend to discount a role for E-24.11 as a physiological endothelin-converting activity but imply that assays for ECE in tissue preparations which also contain E-24.11 could be compromised unless E-24.11 activity was first inhibited with thiorphan. The failure of Abassi *et al.* (1993) to detect ET-1 production from big ET-1 is likely to reflect the low sensitivity of their detection system coupled with relatively poor resolution of substrate and product peptides by HPLC. Ultimately, full characterization of a metallopeptidase ECE in endothelial cells that was distinct from E-24.11 would depend on purification of the relevant activity.

9.3 Isolation of ECE

It took 5 years from the discovery of ET-1 to the first purifications of ECE reported at about the same time by Takahashi *et al.* (1993) and Ohnaka *et al.* (1993). Two major difficulties hindered attempts at purification. Unlike E-24.11, which is found in high abundance in the renal brush border membrane, no rich source of ECE activity has been found which could be used as a starting material. Also, no rapid and sensitive assay for the enzyme had been devised since, as confirmed subsequently, ECE is highly specific for its natural substrate(s) and attempts to develop synthetic fluorogenic peptide substrates had proved unsuccessful. Most of the studies on ECE have relied on a radioimmunoassay or enzyme-linked immunosorbent assay (ELISA) to detect the ET-1 product but this is time-consuming and inconvenient for enzyme purification work.

The purification of ECE by Takahashi *et al.* (1993) used a tissue preparation (rat lung) as starting material and devised a rapid scintillation proximity assay for detection of the ET-1 product. The enzyme was purified from a membrane preparation in a three-step procedure employing lectin chromatography on wheat germ agglutinin–Sepharose, followed by zinc chelating-Sepharose and Blue B–agarose. An overall purification of 6300-fold was achieved from the starting material with a yield of 42%. The final product was homogeneous on reducing sodium dodecyl sulphate polyacrylamide gel electrophoresis (SDS-PAGE) with an apparent M_r of 130 000. Ohnaka *et al.* (1993) isolated ECE from porcine aortic endothelium requiring an overall purification of 12 000-fold, the final product migrating as a polypeptide of 120 kDa on SDS-PAGE. Other properties of the two preparations of ECE were similar. The enzyme exclusively cleaved the Trp21–Val22 bond of big ET-1 and no further degradation of the product was detected. Activity was inhibited by chelating agents and by phosphoramidon (IC_{50} = 0.35 μM; Takahashi *et al.*, 1993) but thiorphan and captopril were ineffective. In addition to big ET-1, purified ECE cleaved big ET-2 and big ET-3 correctly but at substantially lower rates than for big ET-1. A K_m of 0.2 μM was reported for the rat lung enzyme using big ET-1 as substrate. These two studies therefore confirmed that ECE was a novel endothelial, membrane-associated metalloprotease, possibly related to E-24.11, and set the scene for the cloning of the protein. Two further purifications of ECE have subsequently been reported with broadly similar characteristics. Xu *et al.* (1994) observed that bovine adrenal cortex was a relatively good source of ECE activity and the enzyme was purified approximately 3000-fold to homogeneity as a glycoprotein of 126 kDa on reducing SDS-PAGE, by adopting a purification procedure similar to that described by Takahashi *et al.* (1993). Schmidt *et al.* (1994) purified ECE from a membrane preparation of a permanent bovine endothelial cell line. These authors took advantage of a novel and unexpected observation that pre-incubation of these cells in the presence of 10^{-4} M phosphoramidon for 48 h increased the specific activity of ECE 10-fold (see Section 9.8), considerably aiding purification.

9.4 Cloning and expression of ECE-1

Within 12 months of the purification of ECE, the first cloning of the enzyme had been described, with two reports appearing almost simultaneously. Shimada *et al.* (1994) cloned ECE (now named ECE-1) from a transformed rat vascular endothelial cell line, TRLEC-03. The 4487-nucleotide clone encoded a 754-amino-acid sequence containing a zinc consensus sequence (HELTH) consistent with the protein being a zinc metallopeptidase. The overall topology of the predicted protein was that of a type II integral membrane protein with a short N-terminal domain (52 amino acids), a hydrophobic membrane-spanning region (residues 53–73) and a large C-terminal domain (681 residues) containing 10 possible consensus sequences for N-linked glycosylation. Most significantly, the amino acid sequence of rat ECE-1 shared 37% identity with rat E-24.11 (Malfroy *et al.*, 1987) and 31% with human erythrocyte Kell protein, a putative neutral endopeptidase (Lee *et al.*, 1991). Other features conserved between ECE-1, E-24.11 and Kell are 10 conserved cysteine residues, four of which are clustered in a 32-amino-acid sequence close to the membrane-anchoring region, as well as several of the substrate-binding and catalytic residues (see below). The sequences of ECE-1 and E-24.11 are compared schematically in *Figure 9.3*. The similarity of ECE-1 to E-24.11 suggested that it may exist as an ectoenzyme. Subsequent immunohistochemical studies using a monoclonal antibody to rat ECE-1 have confirmed this topology where ECE-1 is seen to cluster in patches on the surface of endothelial cells (Takahashi *et al.*, 1995). The bovine adrenal enzyme (Xu *et al.*, 1994) was seen to have similar characteristics to the rat lung enzyme except that it differed substantially in sequence in its N-terminal region. Cloning of ECE-1 from a number of different species (Ikura *et al.*, 1994; Schmidt *et al.*, 1994; Shimada *et al.*, 1995a,b; Yorimitsu *et al.*, 1995) has now confirmed that ECE-1 exists in isoforms. The predicted protein sequences for ECE-1 range

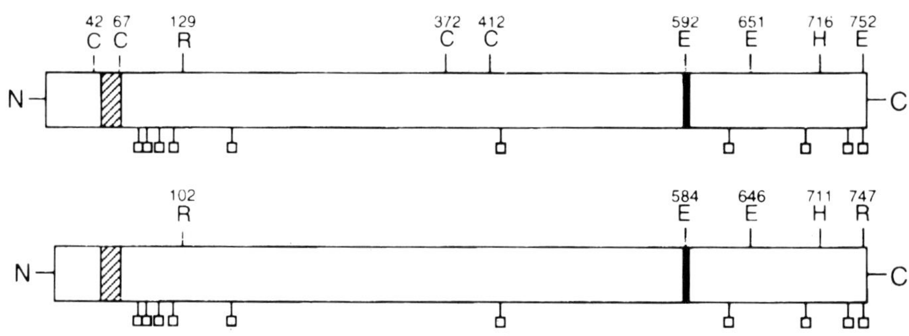

Figure 9.3. Schematic representation of the primary structures of ECE-1 and E-24.11. Single letter codes indicate residues that have been mutated in ECE-1 (upper panel) and the corresponding residues in E-24.11 (lower panel). Cysteine residues conserved in both ECE-1 and E-24.11 are indicated by open squares. The closed box represents the zinc-binding motif (HELTH), and the hatched box indicates the putative transmembrane region. Reproduced from Shimada *et al.* (1996) by permission of The Biochemical Society and Portland Press Ltd.

in length from 753 to 770 residues. All are strikingly similar in their transmembrane hydrophobic and putative extracellular domains where they show more than 95% identity in amino acid sequence. The N-terminal cytoplasmic domain of bovine ECE-1 (Schmidt *et al.*, 1994) is highly similar (>92%) to those reported for the rat and human sequences by Shimada *et al.* (1994) and Schmidt *et al.* (1994). However, the N-terminal sequences of two other reported bovine ECE-1 clones (Ikura *et al.*, 1994; Xu *et al.*, 1994) differ especially in the first 32 amino acids from the N terminus (*Figure 9.4*). This led to the suggestion that distinct subunits of ECE-1 exist, differing substantially only in their cytoplasmic domains, perhaps as a result of alternative splicing. This concept has been confirmed by additional cloned sequences (Shimada *et al.*, 1995b; Yorimitsu *et al.*, 1995) which are compared in *Figure 9.4*. The 'short' form of ECE-1 (51–52 residues in the cytoplasmic domain) originally reported by Shimada *et al.* (1994) and Schmidt *et al.* (1994) we and others have designated the α-subunit (Shimada *et al.*, 1995b; Turner and Murphy, 1996) and the longer form (56–60 cytoplasmic residues) the β-subunit. Confusingly, Valdenaire *et al.* (1995) have used an inverse nomenclature (i.e. they refer to the α-subunit as the b-subunit and vice versa). It is conceivable that the α and β isoforms of ECE-1 may differ in targeting, localization and some other properties. However, Shimada *et al.* (1995b) report that human ECE-1α and ECE-1β are both localized to the plasma membrane and show no significant differences in specific activity or substrate specificity. The significance of these two isoforms is therefore unclear at present. In all tissues examined, the major isoform appears to be ECE-1α.

ECE-1 protein has been localized immunohistochemically to endothelial cells in aorta, lung, kidney, liver and heart and is also present in some secretory cells (e.g. β cells in pancreatic islets and chromaffin cells in the adrenal glands; Takahashi *et al.*, 1995). Testis also contains high levels of immunoreactive ECE-1 protein where it is restricted to interstitial cells. Recently, ECE-1 immunoreactivity has been localized in rat hippocampus, in primary cultured astrocytes and in both neuroblastoma and glial cell lines (Barnes *et al.*, 1997). ECE-1 is not detectable in the renal brush-border membranes and therefore does not co-localize with E-24.11. Both *in situ* hybridization and Northern blot analysis also reveal a broad distribution of ECE-1 with mRNA detectable in lung, pancreas, placenta, adrenal gland, ovary, testis and brain (Barnes *et al.*, 1997; Ikura *et al.*, 1994; Schmidt *et al.*, 1994; Shimada *et al.*, 1994; Xu *et al.*, 1994). Aortic smooth muscle cells do not contain detectable ECE-1, suggesting that the endothelin-converting activity in this tissue must represent a different gene product.

Study of the structure of the human ECE-1 gene (Valdenaire *et al.*, 1995) has provided clues to the origin of the isoforms of ECE-1. ECE-1α and ECE-1β (referred to as ECE-1b and -1a in that paper, respectively) are encoded by the same gene, which is composed of 19 exons spanning more than 68 kb and which has been mapped to the 1p36 band of the human genome (35, 36). The precursor mRNAs of ECE-1α and -1β are transcribed from two distinct start sites, upstream of exon 1 and exon 3, respectively. The two isoform mRNAs display different tissue distributions but no direct comparison has been made between α and β forms. Of two promoters separated by approximately 11 kb, the region surrounding the ECE-1α initiation point has the features of a housekeeping gene

```
Rat α            1 MM SSYKRATLDE EDLVDSLSEG DVYPNGLQVN FRSPRSGQRC WAARTSVEKR  52
Bovine α         1 ** *T***P**** *********S *****H**** **G**N**** *****P****  52
Human α           1 * *T******** ********** *A******** *H******** *****Q****  51

Rat β      1 MGSLRPPQGL GLQWSSFFLG KKGPGLTVSL PLLASSLQVN FRSPRSGQRC WAARTSVEKR  60
Bovine β         1 MS*R*Q D*LR*PLL** SEA****S*P FR*PP***** **G**N**** *****P****  56
Human β          1 M*L*** G**RNP*LQ* *R*****S*P ***PP***** *H******** *****Q****  56
```

Figure 9.4. Comparison of the N-terminal sequences of human, rat and bovine ECE-1. The predicted amino acid sequences of the putative cytoplasmic domains of human, rat and bovine ECE α- and β-subunits are compared. The one-letter code for amino acids is used and the numbers indicate the position in the sequence. Amino acids identical to the rat α- and β-subunits are indicated by asterisks. The amino acid sequences are obtained from: rat-α (Shimada *et al.*, 1994), bovine-α (Schmidt *et al.*, 1994), human-α (Schmidt *et al.*, 1994), rat-β (Shimada *et al.*, 1995b), bovine-β (Ikura *et al.*, 1994; Xu *et al.*, 1994), human-β (Shimada *et al.*, 1995a). The designation α and β for the two subunits is from Turner and Murphy (1996) and Shimada *et al.* (1995b). However, Valdenaire *et al.* (1995) have used the terminology a and b for the β–and α-subunits respectively. The sequence for human ECE reported by Yorimitsu *et al.* (1995) is not included here as it is identical to that described by Schmidt *et al.* (1994) except that it contains an additional 17 residues at the N terminus (not shown).

promoter. In contrast, the ECE-1β transcription start site shows different transcriptional regulation and contains potential binding sites for glucocorticoid receptors and a number of other transcription factors suggesting that it may be induced in the pathological state.

9.5 Site-directed mutagenesis studies of ECE-1

The sequence similarities between E-24.11 and ECE-1 have suggested that the equivalent residues in ECE-1 may be critical for enzyme activity and preliminary modelling of the active site of ECE supports this contention (Sansom *et al.*, 1995). In order to confirm the similarities between these two enzymes, we have conducted a series of site-directed mutagenesis studies on rat ECE-1 and have expressed and characterized the wild-type and mutants in COS-1 cells (Shimada *et al.*, 1996). The sequences and some of the conserved residues in E-24.11 and ECE are compared in *Figure 9.3*. The active site and mechanism of E-24.11 is discussed in Chapter 6. Mutation of the glutamate (E592) in the HEXXH zinc consensus motif results in a complete loss in specific activity of ECE-1 although protein is expressed at normal levels at the cell surface. Thus, by analogy with other zinc peptidases, the two histidines act as zinc ligands and the glutamate is essential for catalytic activity. The third ligand in E-24.11 is E646 and the comparable residue in rat ECE-1 is E651. Mutation of this residue to glutamine also resulted in a complete loss of enzyme activity. H716 in ECE-1 is likely to be involved in stabilization of the tetrahedral intermediate during the transition state, comparable with H711 in E-24.11. The H716Q mutant of ECE-1 was expressed at wild-type levels at the cell surface but was also inactive. These data confirm that the structures of the catalytic site and the zinc-binding domains of E-24.11 are fully conserved in ECE-1 implying a similar proteolytic mechanism for these two endopeptidases.

Differences are seen, however, when residues in ECE-1 equivalent to those important for substrate binding in E-24.11 are changed. Two arginine residues (R102 and R747) have been identified as playing important roles in substrate binding and specificity of E-24.11 (Bateman *et al.*, 1989; Beaumont *et al.*, 1991). R102 is able to bind the free carboxyl group in the P2′ position of certain substrates such as the enkephalins, endowing the enzyme with its peptidyldipeptidase character. The equivalent residue in ECE-1 is also an arginine (R129) but its mutation to glutamine produced only a small change in specific activity of the expressed enzyme compared with wild-type and little change in k_{cat}/K_m or sensitivity to phosphoramidon. Thus, R129 plays little if any role in substrate binding, consistent with the lack of detectable peptidyldipeptidase activity in ECE-1 and the lack of a free carboxyl group in the P2′ position of the substrate, big ET. Arg747 of E-24.11 is replaced by a residue of opposite charge in ECE-1 (Glu752). Mutation of this glutamate to glutamine, aspartate or arginine again produced no significant change in properties of the expressed mutants. In particular, changing the glutamate to arginine did not endow the mutant ECE-1 with 'E-24.11' properties, for example sensitivity to thiorphan or ability to hydrolyse E-24.11 substrates. Thus, it would appear that the residues important for substrate binding in ECE-1 are distinct from those in E-24.11 and remain to be identified. These differences between the two enzymes should be exploitable to allow development of ECE-selective inhibitors. The mutagenesis data are summarized in *Table 9.1*.

Table 9.1. Kinetic parameters of wild-type and some mutant forms of rat ECE-1

Enzyme	Specific activity (nmol h^{-1} mg^{-1})	k_{cat}/K_m (min^{-1} μM^{-1})	IC_{50} (μM)
Wild-type	156 ± 6.3	2.73	0.69 ± 0.07
R129Q	264 ± 9.5	4.24	1.34 ± 0.12
E752Q	134 ± 6.7	2.34	1.28 ± 0.25
E752D	110 ± 1.1	2.23	1.55 ± 0.35
E752R	92 ± 9.5	1.45	1.08 ± 0.30
C412S	32 ± 1.1	0.33	3.50

The data are derived from Shimada *et al.* (1996). Specific activities were determined at 0.1 μM big ET-1 substrate concentration. The inhibitor for IC_{50} determinations was phosphoramidon. The following mutants were expressed fully at the cell surface and exhibited the same M_r as wild-type ECE-1 but were inactive: E592Q, E651Q, H716Q, see text for details.

One important structural difference between E-24.11 and ECE-1 has emerged. In most species, E-24.11 exists as a non-covalently associated homodimer (Fulcher and Kenny, 1983). However, when ECE-1 was subjected to SDS-PAGE under non-reducing conditions, the protein migrated as a polypeptide of approximately 300 kDa rather than 130 kDa, suggesting that it occurred as a covalent dimer (Schmidt *et al.*, 1994; Takahashi *et al.*, 1995). Occurrence as a covalent dimer is relatively rare among the brush border peptidases, the only other known examples being membrane dipeptidase, aminopeptidase A and endopeptidase-24.18 (meprin). The number and location of the interchain disulphide bonds in membrane dipeptidase and endopeptidase-24.18 have recently been reported (Chevallier *et al.*, 1996; Keynan *et al.*, 1996). In order to identify the cysteine involved in subunit dimerization in ECE-1 we have again conducted a site-directed mutagenesis approach (Shimada *et al.*, 1996). Comparison of the sequences of ECE-1 and E-24.11 revealed four non-conserved cysteines in ECE-1. One of these is located in the putative cytoplasmic domain (C42) and another in the transmembrane domain (C67) and neither is likely to fulfil this role. The two remaining cysteines (C372 and C412) are present in the extracellular domain and are likely candidates. Mutation of each of these cysteines in turn to serine, followed by expression and analysis by SDS-PAGE under non-reducing conditions, has established that it is C412 alone in rat ECE-1 that mediates dimer formation. A similar study with human ECE-1 has confirmed that it is the equivalent residue (C416) that is involved (V.M. Hoang and A.J. Turner, unpublished observations). The kinetic parameters of monomer and dimer rat ECE-1 were compared. The k_{cat}/K_m for the dimer was eight-fold greater than for the monomer, suggesting that dimerization in some way favours substrate conversion. The C412S mutant was also found to be three-fold less sensitive to phosphoramidon. Since two isoforms of ECE-1 subunits exist (α and β), it is possible that heterodimers (αβ) occur as well as the homodimers (αα and ββ) if both subunits are normally expressed in the same cell. A model for the structure of rat ECE-1 based on these mutagenesis studies is shown in *Figure 9.5*.

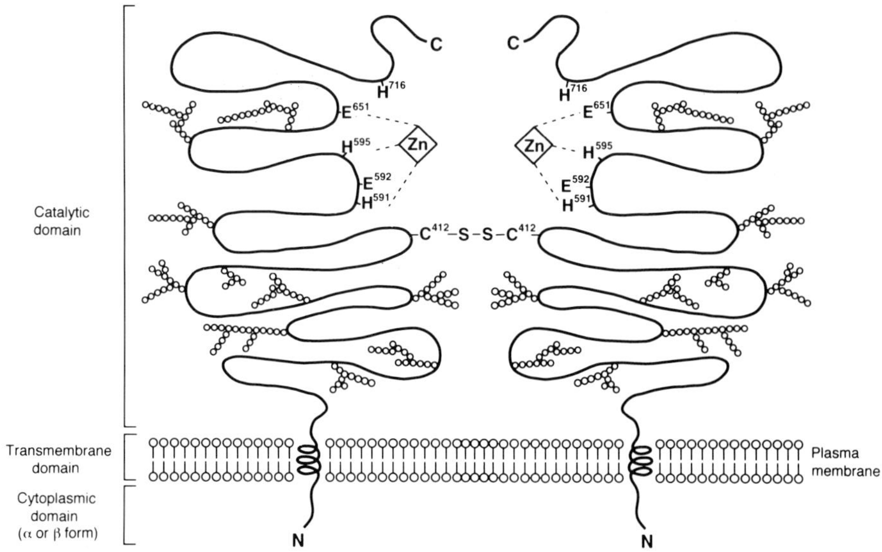

Figure 9.5. Topology and dimeric structure of rat ECE-1. Rat ECE-1 exists as an ectoenzyme on endothelial cells (Takahashi *et al.*, 1995). The large catalytic domain projects outside the plasma membrane with the small N-terminal domain within the cytoplasm. The enzyme forms a disulphide-linked dimer via Cys412 and each subunit contains a zinc ion co-ordinated to His591, His595 and Glu651. Glu592 has a catalytic role and His716 contributes to the stabilization of the intermediate during the transition state. There are 10 putative N-linked glycosylation sites in the extracellular domain. ECE-1 occurs in two isoforms (α and β) which differ only in part of their cytoplasmic domains (see *Figure 9.4*). Reproduced from Shimada *et al.* (1996) by permission of The Biochemical Society and Portland Press Ltd.

9.6 Substrate specificity and inhibition of ECE-1

A major difference between E-24.11 and ECE-1 is in their respective substrate specificities. E-24.11 has a broad specificity, the primary recognition site being a hydrophobic residue of the substrate located in the P1′ subsite of the enzyme. Substrates ranging in size from the tripeptide fMet–Leu–Phe up to approximately 40 amino acids in length can be cleaved efficiently. However for ECE-1, the big ETs and very closely related peptides are the only substrates yet identified. Furthermore the enzyme converts big ET-1 to ET-1 by hydrolysis of the Trp21–Val22 bond but no further hydrolysis of the product occurs. Even among the big ETs differences in the relative rates of conversion are seen. For example, most of the cloned and expressed ECE-1 proteins convert big ET-1 > big ET-2 ≫ big ET-3 (Schmidt *et al.*, 1994; Shimada *et al.*, 1994) with the exception of the cloned bovine enzyme in which the order of conversion was big ET-1 ≫ big ET-3 > big ET-2 (Xu *et al.*, 1994). The enzyme purified from the transformed endothelial cell line, EA.hy926, also processes big ET-1 at least 10 times more efficiently than the other precursors (Waxman *et al.*, 1994). It is likely that an isoform of ECE yet to be discovered serves as a distinct big ET-3 convertase. This may reflect the fact that the scissile bond in big ET-3 is W–I rather than W–V and that the C-terminal eight amino acids of big ET-3 differ completely from big ET-1 and big ET-2. The C terminus is also amidated in big ET3. Although an extended sequence C-terminal to the scissile bond is essential for efficient conversion of the big ETs, the N-terminal region of big ET actually hinders conversion since truncated forms have substantially higher specific activities (Okada *et al.*, 1993). The minimal sequence of the substrate that is required on the N-terminal side of the scissile bond is only IIW although DIIW produces higher specific activity (Okada *et al.*, 1993). The Trp21 residue of big ET-1 is important for its efficient processing and secretion (Ergul and Puett, 1995). However, replacement of Val22 (P1′ position in substrate) by Phe produces almost a 10-fold enhancement in specific activity and thus the synthetic peptide: $DIIWFNTPEHVVPYGLG_{amide}$ provides a useful substrate for rapid and sensitive HPLC assay of ECE in cell lines and during purification (V.M. Hoang, K. Barnes and A.J. Turner, unpublished observations).

Both random screening of fermentation broths and rational peptide synthesis based on known E-24.11 inhibitors have been used to try to develop potent and specific ECE inhibitors. In general, however, synthetic ECE inhibitors have tended to inhibit E-24.11 more potently. For example, the compound CGS 26303 [(S)-2-biphenyl-4-yl-1-(1H-tetrazol-5-yl)-ethylamino-methyl phosphonic acid)] based on a single amino acid substituent rather than the di- or tripeptide framework typical of E-24.11 inhibitors is one of the most potent synthetic inhibitors of ECE reported to date (IC_{50} = 0.8 μM) (De Lombaert *et al.*, 1994). However, it is still 20-fold more potent against E-24.11 (IC_{50} = 40 nM). A specific pharmacophore for selective ECE inhibitors has not yet been found but random screening of a culture broth of *Saccharothrix* has recently identified a novel ECE inhibitor (compound WS75624; Tsurumi *et al.*, 1995a) that is 30 times more potent against ECE than E-24.11 and which may prove a useful lead compound. Dual inhibitors of E-24.11 and ECE may, however, themselves be valuable therapeutic agents simultaneously protecting atrial natriuretic peptide from degradation and preventing the formation of ET-1.

9.7 Endothelin-converting enzyme-2 (ECE-2)

Since some cell types that produce mature ET-1 do not contain detectable ECE-1, Emoto and Yanagisawa (1995) screened for additional ECE species using highly degenerate oligonucleotide primers based on a 25-amino-acid sequence showing high similarity between ECE-1 and E-24.11. A clone was isolated from a bovine adrenal cortex cDNA library that differed somewhat in sequence from ECE-1 and the product of this clone was named ECE-2. ECE-2 cDNA encodes a novel 787-amino-acid polypeptide which shares a number of structural features with both ECE-1 and E-24.11. ECE-2 is a type II integral membrane protein possessing a much longer cytoplasmic tail (82 residues) than ECE-1 or E-24.11, a transmembrane segment and a large catalytic domain (682 residues) possessing an HEXXH motif and with conservation of a number of the critical substrate-binding and catalytic residues of ECE-1. The two pairs of cysteine residues close to the transmembrane region of ECE-1 and E-24.11 are also conserved. ECE-2 is highly glycosylated, with 10 putative N-linked glycosylation sites and, on immunoblot analysis, the protein appears as a polypeptide of M_r 130 000 under reducing conditions. C412 which constitutes the disulphide link in ECE-1 is conserved as a cysteine, suggesting that ECE-2 is also a disulphide-linked dimer although this has not been directly tested. Highest levels of expression of ECE-2 were seen in the nervous system although, even here, the levels were significantly less than for ECE-1. In cultured endothelial cells the level of expression of ECE-2 mRNA was only 1–2% of that of ECE-1 mRNA (Emoto and Yanagisawa, 1995)

ECE-2 resembles ECE-1 in that it converts big ET-1 more efficiently than big ET-2 or big ET-3 and it is inhibited by phosphoramidon but not by thiorphan or captopril. Two striking differences were observed, however, between ECE-1 and ECE-2 when the respective cDNAs were expressed in CHO cells. ECE-2 exhibits an acidic pH optimum (pH 5.5) rather than the neutral pH optimum of ECE-1 or E-24.11. This suggests that it may not be an ectoenzyme but, rather, located in an acidic intracellular compartment where it could effect intracellular processing of big ETs. Secondly, ECE-2 is 250-fold more sensitive to phosphoramidon than ECE-1 (IC_{50} values 1 μM and 4 nM for ECE-1 and ECE-2 respectively). Another ECE inhibitor [FR901533; WS79089B; 1,6,9,14-tetrahydroxy-3-(2-hydroxypropyl)-7-methoxy-8,13-dioxo-5,6,8,13-tetrahydrobenzo-[α] naphthacene-2-carboxylate-Na; Tsurumi *et al.*, 1995b], however, was equipotent against ECE-1 and ECE-2 (IC_{50} values 2 μM and 3 μM for ECE-1 and ECE-2, respectively). Further studies are needed on the cellular and subcellular locations of ECE-2 as well as on its precise role in the physiological production of the ETs. Since neither ECE-1 nor ECE-2 converts big ET-3 efficiently, it is likely that there are additional ECE species yet to be identified.

9.8 Cell biology of endothelin processing

The cell biology of ET biosynthesis has proven controversial with evidence for both intracellular processing to mature ET as well as post-secretory processing at the cell surface. Thus, inhibition of the conversion of big ET-1 to ET-1 *in vivo* by phosphoramidon has been interpreted as evidence for an ectoenzyme topology for ECE (Corder and Vane, 1995; Matsumura *et al.*, 1990b; McMahon *et al.*, 1991; Pollock and Opgenorth, 1991) and is consistent with the similar location of the closely related E-24.11. However, some studies using cultured endothelial cells have demonstrated that ET-1 rather than big ET-1 is the major released product implicating intracellular conversion (Ikegawa *et al.*, 1990; Sawamura *et al.*, 1991), consistent with the detection of both big ET-1 and ET-1 in a vesicle preparation isolated from bovine aortic endothelial cells (Harrison *et al.*, 1995). The existence of multiple isoforms of ECE may help to explain these anomalies. ECE-2 has an acidic pH optimum and could be located within secretory vesicles. Immunohistochemical studies have located ECE-1 predominantly at the cell surface (Shimada *et al.*, 1995b; Takahashi *et al.*, 1995) although the up-regulation of ECE activity that occurs on pre-treatment of cells with phosphoramidon (Barnes *et al.*, 1996; Schmidt *et al.*, 1994) appears to result in a dramatic relocation of ECE-1 immunoreactivity to an intracellular compartment, probably the Golgi (Barnes *et al.*, 1996). No change in the level of ECE-1α mRNA was detected suggesting the changes may reflect decreased turnover of ECE protein or induction of another isoform of ECE. More studies are needed to elucidate the regulation of ECE levels and localization of the various isoforms under normal and pathological conditions.

9.9 Conclusions

During the last 3 years substantial characterization of ECE has occurred at the molecular and cellular levels. Long-standing knowledge on the properties of the homologue E-24.11 has considerably aided this progress. The detection of multiple forms of ECE both through alternative splicing to generate ECE-1α and ECE-1β, and the existence of distinct gene products (ECE-1 and ECE-2), was unexpected but has helped to resolve some of the anomalies in the cellular locations of endothelin processing. ECE inhibitors have yet to have proven therapeutic value in cardiovascular or renal medicine but there are high hopes in this direction. The recent reports that ECE-1 may be involved in vascular injury and human coronary atherosclerosis (Minamino *et al.*, 1996; Wang *et al.*, 1996) have extended interest in ECE still further and we may expect more information on the pathophysiological roles of ECE in the near future.

Acknowledgements

I would like to thank the British Heart Foundation for their generous support of my group's work on ECE.

References

Abassi ZA, Tate J, Golomb E, Keiser H. (1992) Role of neutral endopeptidase in the metabolism of endothelin. *Hypertension* **20:** 89–95.

Abassi ZA, Golomb E, Bridenbaugh R, Keiser HR. (1993) Metabolism of endothelin-1 and big endothelin-1 by recombinant neutral endopeptidase EC 3.4.24.11. *Br. J. Pharmacol.* **109:** 1024–1028.

Barnes K, Shimada K, Takahashi M, Tanzawa K, Turner AJ. (1996) Metallopeptidase inhibitors induce an upregulation of endothelin-converting enzyme levels and its redistribution from the plasma membrane to an intracellular compartment. *J. Cell Sci.* **109:** 919–928.

Barnes K, Walkden BJ, Wilkinson TC, Turner AJ. (1997) Expression of endothelin-converting enzyme in both neuroblastoma and glial cell lines and its localization in rat hippocampus. *J. Neurochem.*, in press.

Bateman RC Jr, Jackson D, Slaughter CA, Unnithan S, Chai YG, Moomaw C, Hersh LB. (1989) Identification of the active-site arginine in rat neutral endopeptidase 24.11 (enkephalinase) as arginine 102 and analysis of a glutamine 102 mutant. *J. Biol. Chem.* **264:** 6151–6157.

Baynash AG, Hosoda K, Giaid A, Richardson JA, Emoto N, Hammer RE, Yanagisawa M. (1994) Interaction of endothelin-3 with endothelin-B receptor is essential for development of epidermal melanocytes and enteric neurons. *Cell* **79:** 1277–1285.

Beaumont A, Le Moual H, Boileau G, Crine P, Roques BP. (1991) Evidence that both arginine 102 and arginine 747 are involved in substrate binding to neutral endopeptidase (EC 3.4.24.11). *J. Biol. Chem.* **266:** 214–220.

Chevallier S, Ahn J, Boileau G, Crine P. (1996) Identification of the cysteine residues implicated in the formation of α2 and α/β dimers of rat meprin. *Biochem. J.* **317:** 731–738.

Corder R, Vane JR. (1995) Radioimmunoassay evidence that the pressor effect of big endothelin-1 is due to local conversion to endothelin-1. *Biochem. Pharmacol.* **49:** 375–380.

De Lombaert S, Ghai RD, Jeng AY, Trapani AJ, Webb RL. (1994) Pharmacological profile of a non-peptidic dual inhibitor of neutral endopeptidase-24.11 and endothelin-converting enzyme. *Biochem. Biophys. Res. Commun.* **204:** 407–412.

Emoto N, Yanagisawa M. (1995) Endothelin-converting enzyme-2 is a membrane-bound, phosphoramidon-sensitive metalloprotease with acidic pH optimum. *J. Biol. Chem.* **270**: 15262–15268.

Ergul A, Puett D. (1995) Trp-21 is important in the processing and secretion of big endothelin-1. *Mol. Cell. Endocrinol.* **110:** 89–94.

Fulcher IS, Kenny AJ. (1983) Proteins of the kidney microvillar membrane. The amphipathic forms of endopeptidase purified from pig kidney. *Biochem. J.* **211:** 743–753.

Furchgott RF, Zawadzki JV. (1980) The obligatory role of endothelial cells in the relaxation of arterial smooth muscle by acetylcholine. *Nature* **288:** 373–376.

Harrison VJ, Barnes K, Turner AJ, Wood E, Corder R, Vane JR. (1995) Identification of endothelin-1 and big endothelin-1 in secretory vesicles isolated from bovine aortic endothelial cells. *Proc. Natl Acad. Sci. USA* **92:** 6344–6348.

Hickey KA, Rubanyi G, Paul RJ, Highsmith RF. (1985) Characterization of a coronary vasoconstrictor produced by cultured endothelial cells. *Am. J. Physiol.* **248:** C550-C556.

Ikegawa R, Matsumura Y, Tsukuhara Y, Takaoka M, Morimoto S. (1990) Phosphoramidon, a metalloproteinase inhibitor, suppresses the secretion of endothelin-1 from cultured endothelial cells by inhibiting a big endothelin-1 converting enzyme. *Biochem. Biophys. Res. Commun.* **171:** 669–675.

Ikura T, Sawamura T, Shiraki T, *et al.* (1994) cDNA cloning and expression of bovine endothelin converting enzyme. *Biochem. Biophys. Res. Commun.* **203:** 1417–1422.

Inoue A, Yanagisawa M, Kimura S, Tomobe Y, Kobayashi M, Mitsui Y, Yazaki Y, Goto K, Masaki, T. (1989) The human endothelin family: three structurally and pharmacologically distinct isopeptides predicted by three separate genes. *Proc. Natl Acad. Sci. USA* **86:** 2863–2867.

Keynan S, Habgood NT, Hooper NM, Turner AJ. (1996) Site-directed mutagenesis of conserved cysteine residues in porcine membrane dipeptidase. Cys 361 alone is involved in disulfide-linked dimerisation. *Biochemistry* **35**: 12511–12517.

Kloog Y, Ambar I, Sokolovsky M, Kochva E, Wollberg Z, Bdolah A. (1988) Sarafotoxin, a novel vasoconstrictor peptide: phosphoinositide hydrolysis in rat heart and brain. *Science* **242:** 268–270.

Kurihara Y, Kurihara H, Suzuki H, *et al.* (1994) Elevated blood pressure and craniofacial abnormalities in mice deficient in endothelin-1. *Nature* **368:** 703–710.

Lee S, Zambas ED, Marsh WL, Redman CM. (1991) Molecular cloning and primary structure of Kell blood group protein. *Proc. Natl Acad. Sci. USA* **88:** 6353–6357.

Malfroy B, Schofield PR, Kuang WJ, Seeburg PH, Mason AJ, Henzel WJ. (1987) Molecular cloning and amino acid sequence of rat enkephalinase. *Biochem. Biophys. Res. Commun.* **144:** 59–66.

Masaki T, Yanagisawa M. (1992) Endothelins. *Essays Biochem.* **27:** 79–89.

Masaki T, Kimura S, Yanagisawa M, Goto K. (1991) Molecular and cellular mechanism of endothelin regulation. Implications for vascular function. *Circulation* **84:** 1457–1468.

Matsumura Y, Ikegawa R, Takaoka M, Morimoto S. (1990a) Conversion of porcine big endothelin to endothelin by an extract from porcine aortic endothelial cell. *Biochem. Biophys. Res. Commun.* **167:** 203–210.

Matsumura Y, Hisaki K, Takaoka M, Morimoto S. (1990b) Phosphoramidon, a metalloproteinase inhibitor, suppresses the hypertensive effect of big endothelin-1. *Eur. J. Pharmacol.* **185:** 103–106.

Matsumura Y, Ikegawa R, Tsukahara Y, Takaoka M, Morimoto S. (1991) *N*-ethylmaleimide differentiates endothelin converting activity by two types of metalloproteinases derived from vascular endothelial cells. *Biochem. Biophys. Res. Commun.* **178:** 531–538.

McMahon EG, Fok KF, Moore WM, Smith CE, Siegel NR, Trapani AJ. (1989) *In vitro* and *in vivo* activity of chymotrypsin-activated big endothelin (porcine 1–40). *Biochem. Biophys. Res. Commun.* **161:** 406–413.

McMahon EG, Palomo MA, Moore WM, McDonald JF, Stern MK. (1991) Phosphoramidon blocks the pressor activity of porcine big endothelin-1-(1–39) *in vivo* and conversion of big endothelin-1-(1–39) to endothelin-1-(1–21) *in vitro*. *Proc. Natl Acad. Sci. USA* **88:** 703–707.

Minamino T, Kurihara H, Takahashi M, *et al.* (1996) Endothelin converting enzyme (ECE-1) expression in the rat vascular injury model and human coronary atherosclerosis. *Circulation*, in press.

Murphy LJ, Corder R, Mallet AI, Turner AJ. (1994) Generation by the phosphoramidon-sensitive peptidases, endopeptidase-24.11 and thermolysin, of endothelin-1 and C-terminal fragment from big endothelin-1. *Br. J. Pharmacol.* **113:** 137–142.

O'Brien RF, Robbins RJ, McMurtry IF. (1987) Endothelial cells in culture produce a vasoconstrictor substance. *J. Cell. Physiol.* **132:** 263–270.

Ohnaka K, Takayanagi R, Nishikawa M, Haji M, Nawata H. (1993) Purification and characterization of a phosphoramidon-sensitive endothelin-converting enzyme in porcine aortic endothelium. *J. Biol. Chem.* **268:** 26759–26766.

Okada K, Miyazaki Y, Takada J, Matsuyama K, Yamaki T, Yano M. (1990) Conversion of big endothelin-1 by membrane-bound metalloendopeptidase in cultured bovine endothelial cells. *Biochem. Biophys. Res. Commun.* **171:** 1192–1198.

Okada K, Arai Y, Hata M, Matsuyama K, Yano M. (1993) Big endothelin-1 structure important for specific processing by endothelin-converting enzyme of bovine endothelial cells. *Eur. J. Biochem.* **218:** 493–498.

Opgenorth TJ, Wu-Wong JR, Shiosaki K. (1992) Endothelin-converting enzyme. *FASEB J.* **6:** 2653–2659.

Pollock DM, Opgenorth TJ. (1991) Evidence for metalloprotease involvement in the *in vivo* effects of big endothelin-1. *Am. J. Physiol.* **261:** R257–R263.

Sansom CE, Hoang VM, Turner AJ. (1995) Molecular modeling of the active site of endothelin-converting enzyme. *J. Cardiovasc. Pharmacol.* **26** (Suppl. 3): S75–S77.

Sawamura T, Kasuya Y, Matsushita Y, *et al.* (1991) Phosphoramidon inhibits the intracellular conversion of big endothelin-1 to endothelin-1 in cultured endothelial cells. *Biochem. Biophys. Res. Commun.* **174:** 779–784.

Schmidt M, Kröger B, Jacob E, Seulberger H, Subkowski T, Otter R, Meyer T, Schmalzing G, Hillen H. (1994) Molecular characterization of human and bovine endothelin converting enzyme (ECE-1). *FEBS Lett.* **356:** 238–243.

Shields PP, Gonzales TA, Charles D, Gilligan JP, Stern W. (1991) Accumulation of pepstatin in cultured endothelial cells and its effect on endothelial processing. *Biochem. Biophys. Res. Commun.* **177:** 1006–1012.

Shimada K, Takahashi M, Tanzawa K. (1994) Cloning and functional expression of endothelin-converting enzyme from rat endothelial cells. *J. Biol. Chem.* **269:** 18275–18278.

Shimada K, Matsushita Y, Wakabayashi K, Takahashi M, Matsubara A, Iijima Y, Tanzawa K. (1995a) Cloning and functional expression of human endothelin converting enzyme cDNA. *Biochem. Biophys. Res. Commun.* **207:** 807–812.

Shimada K, Takahashi M, Ikeda M, Tanzawa, K. (1995b) Identification and characterization of two isoforms of an endothelin converting enzyme-1. *FEBS Lett.* **371:** 140–144.

Shimada K, Takahashi M, Turner AJ, Tanzawa K. (1996) Rat endothelin converting enzyme-1 forms a dimer through Cys^{412} with a similar catalytic mechanism and a distinct substrate binding mechanism compared with neutral endopeptidase-24.11. *Biochem. J.* **315:** 863–867.

Shubeita HE, McDonough PM, Harris AN, Knowlton KU, Glembotski CC, Brown JH, Chien KR. (1990) Endothelin induction of inositol phospholipid hydrolysis, sarcomere assembly and cardiac gene expression in ventricular myocytes. Paracrine mechanism for myocardial cell hypertrophy. *J. Biol. Chem.* **265:** 20555–20562.

Sokolovsky M, Galron R, Kloog Y, Bdolah A, Indig FE, Blumberg S, Fleminger G. (1990) Endothelins are more sensitive than sarafotoxins to neutral endopeptidase: possible physiological significance. *Proc. Natl Acad. Sci. USA* **87:** 4702–4706.

Takada J, Okada K, Ikenaga T, Matsuyema K, Yano M. (1991) Phosphoramidon-sensitive endothelin converting enzyme in the cytosol of cultured bovine endothelial cells. *Biochem. Biophys. Res. Commun.* **174**: 860–865.

Takahashi M, Matsushita Y, Iijima Y, Tanzawa K. (1993) Purification and characterization of endothelin-converting enzyme from rat lung. *J. Biol. Chem.* **268:** 21394–21398.

Takahashi M, Fukuda K, Shimada K, Barnes K, Turner AJ, Ikeda M, Koike M, Yamamoto Y, Tanzawa K. (1995) Localization of rat endothelin-converting enzyme to vascular endothelial cells and some secretory cells. *Biochem. J.* **311:** 657–665.

Takaoka M, Miyata Y, Takenobu Y, Ikegawa R, Matsumura Y, Morimoto S. (1990) Mode of cleavage of pig big endothelin-1 by chymotrypsin. Production and degradation of mature endothelin-1. *Biochem. J.* **270:** 541–544.

Tsurumi Y, Fujie K, Nishikawa M, Kiyoto S, Okuhara M. (1995a) Biological and pharmacological properties of highly selective new endothelin converting enzyme inhibitor WS79089B isolated from *Streptosporangium roseum* No. 79089. *J. Antibiot.* **48:** 169–174.

Tsurumi Y, Ueda H, Hayashi K, Takase S, Nishikawa M, Kiyoto S, Okuhara M. (1995b) WS75624 A and B, new endothelin converting enzyme inhibitors isolated from *Saccharothrix* sp. no. 75624. I. Taxonomy, fermentation, isolation, physico-chemical properties and biological activities. *J. Antibiot.* **48:** 1066–1072.

Turner AJ, Murphy LJ. (1996) Molecular pharmacology of endothelin converting enzymes. *Biochem. Pharmacol.* **51:** 91–102.

Valdenaire O, Rohrbacher E, Mattei M-G. (1995) Organisation of the gene encoding the human endothelin-converting enzyme (ECE-1). *J. Biol. Chem.* **270:** 29794–29798.

Vijayaraghavan J, Scicli AG, Carretero OA, Slaughter C, Moomaw C, Hersh LB. (1990) The hydrolysis of endothelins by neutral endopeptidase 24.11 (enkephalinase). *J. Biol. Chem.* **265:** 14150–14155.

Wang X, Douglas SA, Louden C, Vickery-Clark LM, Feuerstein GZ, Ohlstein EH. (1996) Expression of endothelin-1, endothelin-3, endothelin-converting enzyme, and endothelin-A and endothelin-B receptor mRNA after angioplasty-induced neointimal formation in the rat. *Circ. Res.* **78:** 322–328.

Waxman L, Doshi KP, Gaul SL, Wang S, Bednar RA, Stern AM. (1994) Identification and characterisation of endothelin converting activity from EAHY 926 cells: evidence for the physiologically relevant human enzyme. *Arch. Biochem. Biophys.* **308:** 240–253.

Wypij DM, Nichols JS, Novak PJ, Stacy DL, Berman J, Wiseman JS. (1992) Role of mast cell chymase in the extracellular processing of big endothelin-1 to endothelin-1 in the perfused rat lung. *Biochem. Pharmacol.* **43:** 845–853.

Xu D, Emoto N, Giaid A, Slaughter C, Kaw S, deWit D, Yanagisawa M. (1994) ECE-1: a membrane-bound metalloprotease that catalyzes the proteolytic activation of big endothelin-1. *Cell* **78:** 473–485.

Yanagisawa M, Kurihara H, Kimura S, Tomobe Y, Kobayishi Y, Mitsui Y, Yazaki Y, Goto K, Masaki T. (1988) A novel potent, vasoconstrictor peptide produced by vascular endothelial cells. *Nature* **332:** 411–415.

Yorimitsu K, Moroi K, Inagaki N, Saito T, Matsuda Y, Masaki T, Sein, S, Kimura S. (1995) Cloning and sequencing of a human endothelin converting enzyme in renal adenocarcinoma (ACHN) cells producing endothelin-2. *Biochem. Biophys. Res. Commun.* **208:** 721–727.

Chapter 10

Endopeptidase-24.18

Erwin E. Sterchi, Dagmar Hahn and Daniel Lottaz

10.1 Introduction

The microvillus membranes of epithelial cells from the small intestine and the proximal tubules of the kidney are rich in exopeptidases and endopeptidases. Endopeptidase-24.18 (E-24.18) was discovered in the early 1980s. Kenny *et al.* (1981) showed that approximately 50% of the hydrolytic activity of rat kidney microvillus membranes against [^{125}I]-insulin B-chain was insensitive to phosphoramidon, a characteristic inhibitor of E-24.11 (EC 3.4.24.11, neprilysin). Subsequently the two enzymes were chromatographically resolved and the phosphoramidon-insensitive endopeptidase peak (endopeptidase-2) was biochemically characterized (Kenny and Ingram, 1987). Another research group (Beynon *et al.*, 1981) reported independently on the purification and partial characterization of a metallo-endopeptidase from mouse kidney. Both enzymes isolated from rat and mouse tissue were effective in hydrolysing insulin B-chain and azocasein. The mouse enzyme has become known under the acronym 'meprin' (metallo endopeptidase from renal tissue) (Bond *et al.*, 1983). At about the same time another membrane-bound endopeptidase from human and rat small intestine was investigated in our laboratory (Sterchi *et al.*, 1982, 1983). The enzyme hydrolysed *N*-benzoyl-L-tyrosyl-*p*-aminobenzoic acid (paba-peptide), a substrate used in the clinical evaluation of exocrine pancreatic function. This enzyme, referred to as paba-peptide hydrolase (PPH) is localized to the microvillus membrane of intestinal epithelial cells of the small intestine (Sterchi *et al.*, 1988b). PPH cleaves a number of biologically active peptides, including insulin B-chain. The membrane-bound rodent and human endopeptidases were later characterized and turned out to have very similar biochemical properties (Butler *et al.*, 1987; Kenny and Ingram, 1987; Sterchi *et al.*, 1988b). They were found to form disulphide-linked homo- or hetero-oligomers (Butler and Bond, 1988; Butler *et al.*, 1987; Kenny and Ingram, 1987; Sterchi *et al.*, 1988b), and two types of subunit (α and β) with molecular weights of between 90 and 110 kDa were identified. In mice, the expression and assembly of the two subunits were found to be genetically determined. Randomly bred mice and some inbred mice express both α- and β-subunits in an α_4 and $\alpha_2\beta_2$ constellation (meprin A). On the other hand, some inbred mouse strains, such as C3H/He, express only β-subunits as β_4 (meprin B) in which the peptidase activity is latent (Butler and Bond, 1988; Gorbea *et al.*, 1991; Kounnas *et al.*, 1991). As meprin A and meprin B show somewhat different substrate specificities the Enzyme

Cell-Surface Peptidases in Health and Disease, edited by A.J. Kenny and C.M. Boustead.

Nomenclature Committee of the International Union of Biochemistry has assigned EC no. 3.4.24.18 to meprin A and EC no. 3.4.24.63 to meprin B. No tetrameric β_4 meprin B has been found in rat and human.

10.2 Molecular structure

Dumermuth *et al.* (1991) reported on the partial cloning of human intestinal PPH and mouse kidney meprin A. The two mature enzymes showed 82% identity in the N-terminal amino acid sequence (192 residues). Furthermore, comparison of the deduced sequences with protein databases revealed strong similarities with a digestive protease from freshwater crayfish (*Astacus astacus*) and an N-terminal domain of a human bone morphogenetic protein (BMP-1). Multiple alignment analysis indicated that human PPH and mouse meprin A had approximately 30% identity with astacin and BMP-1. Thirty-seven amino acid residues, including three cysteines, were found to be strictly conserved in all four proteins in a domain equivalent to the complete 200-residue sequence of the *Astacus* protease. The four proteins were further shown to contain a zinc-binding motif (HEXXH) within an extended sequence of HEXXHXXGFXHE, which distinguished them from other known metallo-endopeptidases. In a 24-amino-acid sequence that included the active site, the four proteins were 54% identical. A fifth protein, UVS-2, a developmentally regulated protein from frog, was also shown to have sequence identity with the metallo-endopeptidases. These data provided strong evidence for an evolutionary relationship among these proteins and the authors proposed the name 'astacins' for this family of metallo-endopeptidases.

10.2.1 E-24.18 is a multidomain protein

In 1992 the full-length sequences of the α-subunit of rat and mouse kidney E-24.18 were reported (Corbeil *et al.*, 1992; Jiang *et al.*, 1992) and these were followed by the sequence for the human α-subunit (Dumermuth *et al.*, 1993). The human enzyme showed 75% identity to mouse and 78% identity to rat α-subunit, respectively. In 1992 the β-subunit of rat E-24.18 was cloned, followed in 1993 by the mouse β-subunit (Gorbea *et al.*, 1993; Johnson and Hersh, 1992) and in 1994 the full-length sequence of the human β-subunit was submitted to the nucleic acid and protein databases (GenBank AC 81333, paper submitted).

The general molecular structure of E-24.18, with its two subunits α and β, is depicted in *Figure 10.1*. Both subunits have a similar multidomain structure, containing an N-terminal signal peptide sequence, followed by a propeptide sequence preceding the astacin protease domain. The catalytic protease domain is followed by a MAM or adhesive domain, an intervening sequence, an epithelial growth factor (EGF)-like domain, a transmembrane domain and a short cytosolic domain. Initially, the β-subunit of rat E-24.18 was thought to be quite different from the α-subunit (Johnson and Hersh, 1992) because a sequencing error led to the N-terminal signal sequence being interpreted as a non-cleavable signal peptide thus serving as a membrane anchor. The error was subsequently amended (Johnson and Hersh, 1993). The sequences of the two subunits are over 40%

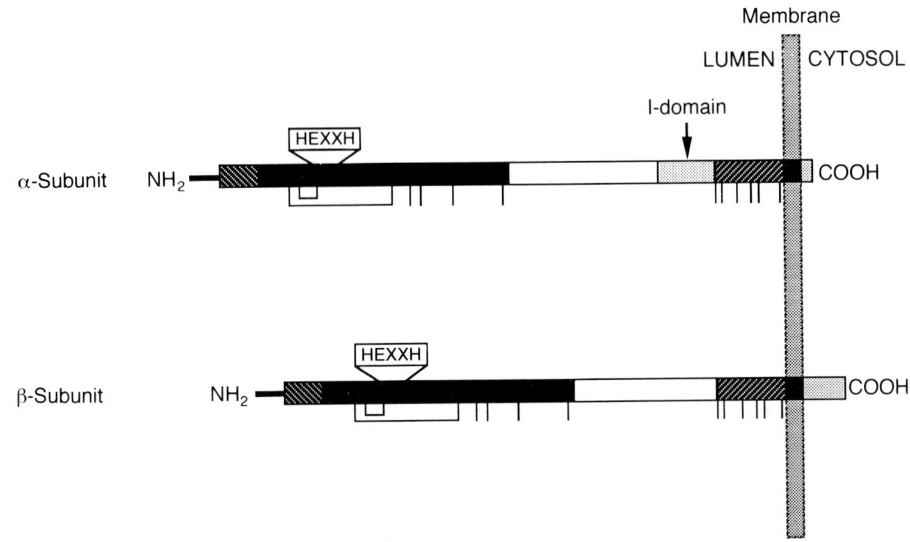

Figure 10.1. Domain structure of α-and β-subunits of E-24.18.
Signal peptide, propeptide, astacin (protease) domain, MAM domain, intervening sequence, inserted domain, EGF-like domain, transmembrane domain. cytosolic domain

identical and about 60% similar. A higher degree of homology exists in the N-terminal half of the molecules, the highest homology being in the catalytic domain. Similarity in the C-terminal region is considerably less, with the α-subunit having an additional domain immediately before the EGF-like domain, termed the I-domain, which is missing in the β-subunit. This additional domain appears to be important in determining C-terminal proteolysis, essential for transport and secretion of the α-subunit. The cytosolic domain of the β-subunit is 24 amino acids long as compared with six amino acids in the α-subunit and contains a consensus sequence for phosphorylation by protein kinase C. It has not been shown whether this subunit is actually phosphorylated. The C-terminal regions of the β-subunits from mouse, rat and human also show considerable variability and this may be important in species-specific processing of this subunit.

Protease domain. The X-ray crystal structure of astacin has recently been determined to a resolution of 1.8 Å (Bode *et al.*, 1992; Gomis-Rüth *et al.*, 1993). The enzyme is kidney-shaped, consisting of an N-terminal and a C-terminal domain separated by an active site cleft. The zinc ion is located at the bottom of this cleft and is penta-coordinated in a trigonal bipyramidal arrangement by three histidines, a tyrosine and a glutamic acid residue via a water molecule bound to its carboxyl side chain. The N-terminal domain consists predominantly of two long α-helices and a five-stranded β-pleated sheet. The amino terminus folds back into a water-filled cavity of the C domain and forms a salt bridge with a glutamic acid residue. This is achieved only after proteolytic removal of the propeptide sequence. Although no inactive zymogen has been isolated from crayfish, nucleic acid sequence analysis of astacin cDNA suggests that proforms of this enzyme do exist (Stöcker *et al.*, 1993). Based on this astacin structure, the proteolytic domain of the α-subunit of mouse E-24.18 has been modelled (Stöcker *et al.*, 1993). For a detailed discussion of the three-dimensional structure of astacin and its implications for E-24.18, see Chapter 2.

MAM domain. The MAM domain immediately following the protease domain was first described by Beckmann and Bork (1993). It is present in a number of proteins, such as protein tyrosine phosphatase μ, A5 protein, and the α- and β-subunits of E-24.18. Recently, it has been shown that this domain is also shared by enteropeptidase, the intestinal protease responsible for the initiation of the activation cascade for pancreatic proteases (Kitamoto *et al.*, 1994). The exact function of the MAM domain is not known, but the four conserved cysteine residues in this domain suggests that it may serve a common adhesive function in these proteins. In rat and mouse E-24.18, Cys309 and Cys320 have very recently been shown to be involved in dimerization of the subunits (Chevallier *et al.*, 1996; Marchand *et al.*, 1996). Expression of mutated enzymes resulted in the synthesis of monomers, which were more susceptible to proteolytic degradation and had a diminished enzymic activity towards protein substrates.

EGF-like domain. Another prominent feature in both subunits of E-24.18 is an EGF-like domain comprising approximately 40 amino acids and including six conserved cysteine residues (Appella *et al.*, 1988). EGF-like domains have been found in a number of proteins with diverse functions (Davis, 1990). Many of these are growth factors or membrane receptors. Some proteases have also been found to contain EGF-like domains and include blood coagulation factors involved in proteolytic activation of zymogens or precursor proteins (Clarke *et al.*, 1992; Zhong *et al.*, 1994). Some EGF-like domains are clearly involved in calcium binding while others are not (Campell and Bork, 1993; Davis, 1990). Their function is thought to be primarily in regulatory mechanisms.

10.2.2 E-24.18 is a member of the astacin family of metallo-endopeptidases

As indicated above, the astacin family of metallo-endopeptidases was recognized in the 1990s when mammalian E-24.18 was cloned and sequenced (Dumermuth *et al.*, 1991). The family has since grown in size to over 20 members. In the classification of peptidases, astacins are now grouped in family 12A which belongs to clan MB of the metallo-endopeptidases (N. Rawlings, Classification of peptidases, SWISS-PROT Protein Sequence Data Bank; http: //expasy.hcuge.ch/cgi-bin/lists?peptidas.txt). The X-ray crystal structures of members of clan MB have revealed a strong overall topological similarity, having an almost identical extended HEXXHXXGXXH zinc-binding motif and, in addition, a loop or turn containing a methionine (‘Met-turn’), which provides the hydrophobic basis for the zinc and the three histidine residues that bind this metal ion. The term *metzincins* is sometimes used for this superfamily of zinc endopeptidases which, in addition to astacins, includes adamalysins (snake venoms), serralysin (*Serratia* proteases) and matrixins (collagenases) (Bode *et al.*, 1993; Stöcker and Bode, 1995; Stöcker and Zwilling, 1995, and

Chapter 2). Astacin proteases have been found in diverse species among eukaryotes and recently also in prokaryotes. In addition to astacin and E-24.18, they include bone morphogenetic protein-1 from human, mouse, frog, sea urchin and fruit fly, CAM-1 from the Japanese quail, LCE and HCE from fish, UVS-2 from frog, HMP-1 from hydra, BP10 and SpAN from sea urchin, and tolloid from human, mouse and fruit fly. There are also examples from crab-beetle and worm and a prokaryote member, flavastacin. For a list of members together with their accession numbers, see SWISS-PROT and EMBL/GENBANK, as well as references given in *Table 10.1*.

Most astacins show an assembly of different domains, the organization of which is complex and conveys to the individual members characteristic structural and functional features (*Figure 10.2*). Most astacins are secreted from the cells and are thus synthesised as pre-pro-enzymes which require post-translational proteolytic processing to acquire

Table 10.1. List of astacin family members with respective accession numbers for SWISS-PROT and EMBL/GENBANK respectively and references to original reports.

Protein	SWISS-PROT, EMBL/ GENBANK	Reference (first author)
Astacin	P07584 / X95684	Titani (1987)
h E-24.18		
α-subunit	M82962	Dumermuth *et al.* (1993)
β-subunit	X81333	Unpublished
m E-24.18		
α-subunit	P28825 / M74897	Jiang *et al.* (1992)
β-subunit	L15193	Gorbea *et al.* (1993)
r E-24.18		
α-subunit	S43408	Corbeil *et al.* (1992)
β-subunit	P28826 / M88601	Johnson and Hersh (1992)
h BMP-1	P13497 / L35279	Wozney (1988)
m BMP-1	P98063 / L35280	Fukagawa (1994)
x BMP-1	P98070 / L12249	Maeno (1993)
s BMP-1	P98069 / L23838	Hwang (1994)
CAM-1	P42662 / U12642	Elaroussi and DeLuca (1994)
LCE	P31579 / M96169	Yasumasu (1992)
HCE-1	P31580 / M96170	Yasumasu (1992)
HCE-2	P31581 / M96171	Yasumasu (1992)
UVS-2	P42669 / M27162	Sato (1990)
YNT5-CAEEL	P98060 / U00036	Unpublished
YPZ8-CAEEL	P98061 / U10414	Unpublished
HMP-1	U22380	Yan *et al.* (1995)
BP10	P42674 / X56224	Lepage (1992)
SpAN	P98068 / M84144	Reynolds (1992)
h Tolloid	L35279	Takahara (1994)
d Tolloid	P25723 / M76976	Shimell (1991)
	U04239	Finelli (1994)
dTir-1/Tolkin	U12634	Nguyen (1994)
COG4_CHIOP	P34156	Unpublished
Flavastacin	L37784	Tarentino (1995)

d, *Drosophila*; h, human; m, mouse; r, rat; s, sea urchin; x, *Xenopus*.

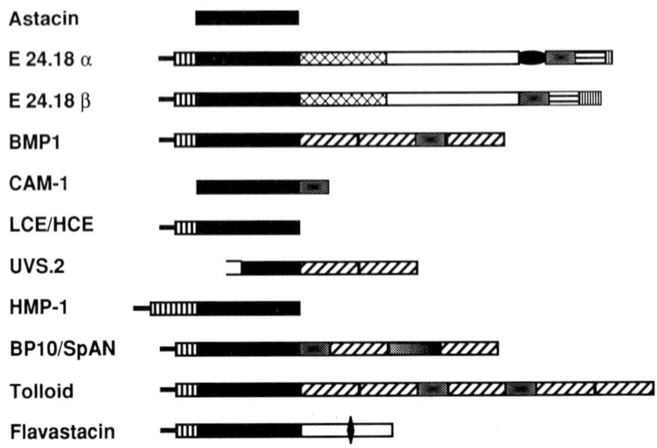

Figure 10.2. Schematic representation of the domain structure of the astacin family members. Signal peptide, propeptide, astacin (protease) domain, MAM domain, intervening sequence, inserted domain, EGF-like domain, transmembrane domain, cytosolic domain, CUB domain. The diamond shape represents a glycosylation site (O-glycosylation).

catalytic activity. CAM-1 represents the exception to the rule in that it does not appear to contain signal and propeptide sequences (Elaroussi and DeLuca, 1994). The mature forms of astacin, HMP-1 and LCE/HCE contain only the protease domain. Evidently, this is sufficient for proteolytic cleavage of peptides and proteins. E-24.18 α- and β-subunits are the only members identified so far that have transmembrane domains. They are also the only members for which the cleavage of the propeptide associated with enzymic activation has been experimentally demonstrated.

A common functional feature of some astacins appears to be the involvement in embryonic development which in turn necessitates either the proteolytic modification of growth factors or tissue remodelling through interactions with extracellular matrix proteins. The different domains probably mediate contact with cells or extracellular matrix or may be involved in the regulation of these proteases.

10.3 Substrates and inhibitors

Rodent E-24.18 activity was initially detected using insulin B-chain and azocasein as substrates (Beynon *et al.*, 1981; Kenny *et al.*, 1981). The human enzyme was initially assayed using *N*-benzoyl-L-tyrosyl-*p*-aminobenzoic acid (paba-peptide), a substrate for chymotrypsin (Sterchi *et al.*, 1982, 1983). Subsequently, E-24.18 was analysed using a wide selection of peptide and peptide-like substrates (Butler *et al.*, 1987; Kenny and Ingram, 1987; Sterchi *et al.*, 1988b). Small peptides and peptide analogues are generally poor substrates for E-24.18. This is also the case for paba-peptide (K_m 16 mM for the human and 4.7 mM for the rat enzyme). Enkephalins (pentapeptides) are not cleaved, angiotensin I (octapeptide) and II (decapeptide) and bradykinin (nonapeptide) are efficiently hydrolysed. Bradykinin and angiotensins are specifically cleaved after an aromatic residue (Butler *et al.*, 1987; Stephenson and Kenny, 1988; Sterchi *et al.*, 1988b; Wolz and Bond, 1990). Insulin B-chain on the other hand appears to be cleaved rather less specifically (Butler *et al.*, 1987). Wolz and Bond (1990) have described the use of a chromogenic substrate, Phe_5(4-nitro)-bradykinin, in kinetic studies. The use of a series of bradykinin analogues enabled them to partially map the active site of the mouse enzyme (Wolz *et al.*, 1991). The authors concluded that multiple subsites are involved in the binding and catalytic reaction of E-24.18. In a recent kinetic study meprin A and astacin were compared and found to have several substrates in common, indicating some similarity among different members of the astacin family (Wolz, 1994).

E-24.18 has also been reported to degrade extracellular matrix proteins (Kaushal *et al.*, 1994). In hydra, HMP-1 has also been suggested to function as a modifier of extracellular matrix proteins (Yan *et al.*, 1995) and in a recent report the C-proteinase that processes procollagen to fibrillar collagen has been found to be identical to BMP-1 (Li *et al.*, 1996).

Our knowledge on specific inhibitors for E-24.18 is rather limited. Actinonin is a potent (35 nM gave 50% inhibition in an assay with insulin B-chain), though non-specific inhibitor (Choudry and Kenny, 1991). Different amino acid hydroxamates have been found to inhibit the enzyme, with an aromatic hydroxamate being the most effective (Wolz *et al.*, 1991). In addition, peptide hydroxamates binding to both the S subsites in a non-competitive fashion and the S′ subsites in a competitive manner have been identified as effective inhibitors (Wolz, 1994). Clearly, more detailed studies are needed to determine substrate and inhibitor specificities and the recent report on the structure of astacin with a transition-state analogue inhibitor points in the direction such studies are likely to go (Grams *et al.*, 1996; see also Chapter 2).

10.4 Biosynthesis

10.4.1 Expression of E-24.18 in organ-cultured tissue

First insights into the biosynthesis and post-translational processing of E-24.18 were obtained from studies with the human enzyme (Sterchi *et al.*, 1988a). Intestinal explants were organ cultured, labelled with [^{35}S]-methionine and analysed by sodium dodecyl sulphate-polyacrylamide gel electrophoresis (SDS-PAGE) and fluorography after immunoprecipitation with monoclonal antibodies (Hauri *et al.*, 1985). Using this approach, it was shown that the enzyme dimerized in the rough endoplasmic reticulum (RER), and was secreted into the organ culture medium. More recently, cloning of the two subunits allowed the investigation of their biosynthesis and the processing in heterologous expression systems. The cDNAs coding for the α- and β-subunits were used to transfect COS-1, Madin–Darby canine kidney (MDCK) and 293 cells (Corbeil *et al.*, 1993; Dumermuth *et al.*, 1993; Grünberg *et al.*, 1993; Johnson and Hersh, 1994; Milhiet *et al.*, 1994). As the synthesis and post-translational processing of human α- and β-subunits in transfected mammalian cells were essentially similar to those for the rodent enzyme, the discussion in Section 10.4.2 refers to data for the human enzyme. Differences to the rodent enzyme are discussed.

10.4.2 Expression in transfected mammalian cells

Synthesis and processing of human E-24.18 in transiently transfected COS-1 cells is shown in *Figure 10.3*. Two days after transfection, cells were pulse-labelled for 15 min with [^{35}S]-methionine/cysteine and chased for the times indicated at the top of *Figure 10.3a*. The α-subunit is synthesised as a polypeptide of approximately 100 kDa (*Figure 10.3a*, lane 1) in a high-mannose glycosylated form (PPHα_h), indicating that it does not exit the RER. A second species (PPHα_h*), also endo H-sensitive, appears later and represents a proteolytically cleaved form of the 100 kDa form. After 2 h a complex glycosylated form of PPHα_{sec} appears in the culture medium (*Figure 10.3a*, lane 8). Secretion of the α-subunit therefore requires proteolytic removal of the C-terminal hydrophobic anchor peptide. This transmembrane domain appears to contain a signal for retention of the membrane-bound α-subunit in the RER and, once removed, the protein is transported very rapidly to the cell surface and secreted. Cotranslationally, the α-subunit forms disulphide-bridged homodimers which *in vivo* may form non-covalently linked tetramers with either α- or β-subunits (Sterchi *et al.*, 1988a).

Analysis of the proteins expressed after transfection of cells with cDNA for the β-subunit of human E-24.18 reveals two protein species of 95 and 104 kDa (*Figure 10.3b*). The

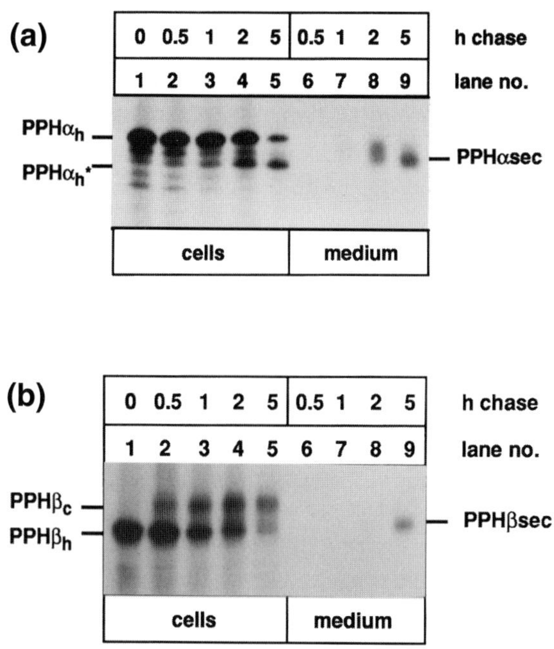

Figure 10.3. Pulse chase kinetics of E-24.18 α-subunit (a) and β-subunit (b) transfected COS-1 cells. COS-1 cells transfected with the cDNA of subunit α (a) or subunit β (b) were metabolically labelled for 15 min and chased for the indicated times. The cells and the media were harvested and the proteins were immunoprecipitated with subunit specific polyclonal antibodies. The immunoprecipitates were subjected to sodium dodecyl sulphate-polyacrylamide gel electrophoresis (7.5%) under reducing conditions and the proteins were detected by fluorography.

95 kDa band (PPHβ_h) is endo H-sensitive and represents an immature form located in the RER (*Figure10.3b,* lanes 1–4) whereas the 104 kDa form is endo H-resistant (PPHβ_c) and represents a mature form (*Figure 10.3b*, lanes 2–5). Small amounts of the latter species are secreted into the culture medium (*Figure 10.3b*, lane 9). Secretion of the β-subunit of human E-24.18 has also been observed in transfected polarized MDCK cells and is in contrast with the β-subunit of rodent E-24.18, which is not secreted (Marchand *et al.*, 1995). E-24.18 β-subunits also form disulphide-bridged homodimers when expressed alone in mammalian cells (J.A. Eldering *et al.*, unpublished observations).

10.4.3 Subunit assembly

To examine how the two subunits of E-24.18 assemble, COS-1 cells were co-transfected with subunit-specific cDNAs. *Figure 10.4* shows distinctly different protein patterns in immunoprecipitates from a single specific antibody. Subunit α-specific and β-specific antibodies clearly immunoprecipitate the respective subunit from separately transfected COS-1 cells (*Figure 10.4*, lanes 1 and 2). However, in cells co-transfected with both subunits, β-subunit-specific mAb immunoprecipitates both subunits (*Figure 10.4*, lane 3), demonstrating that the α-subunit is associated with the β-subunit.

Johnson and Hersh (1994) demonstrated that secretion of the α-subunit into the culture medium was prevented by co-expression of the α- and β-subunits of rat E-24.18 in 293 cells. Treatment of cells with dithiothreitol released the α-, but not the β-subunit into the supernatant, indicating that the α-subunit was covalently linked by disulphide

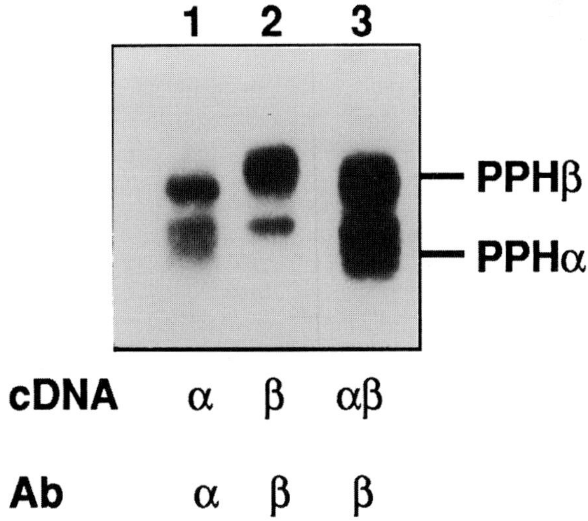

Figure 10.4. Coexpression of E-24.18 α- and β-subunits in COS-1 cells. COS-1 cells were either transfected with α-subunit (lane 1), β-subunit (lane 2) or both subunits (lane 3) and pulse labelled for 5 h. The cells were harvested and the expressed subunits were immunoprecipitated with α-specific antibody (lane 1) and β-specific antibody (lanes 2 and 3). The proteins were analysed under reducing conditions on 7.5% SDS-PAGE and subsequent fluorography.

bridges to the β-subunit. Therefore the β-subunit is necessary for retention of the α-subunit at the cell membrane. The same observation was made by Marchand *et al.* (1994). These investigators could further show that the released α-subunit lacks the EGF-like, the transmembrane and the cytosolic C-terminal domains. These findings are consistent with the hypothesis that the transmembrane domain of the α-subunit has to be removed for transport and secretion of the α-subunit.

10.4.4 Proteolytic processing of the α-subunit

The difference in C-terminal processing of the two subunits may be explained by the occurrence of the inserted domain (I domain) of 56 amino acids that is present only in the α-subunit. It is located immediately adjacent to the EGF-like domain and contains several potential processing sites including dibasic residues as well as a sequence for furin (RQKR for human and mice, RPKR for rat). Building the I domain sequence into the β-subunit just in front of the EGF domain leads to the secretion of this subunit. Deletion of the I domain on the other hand leads to the retention of this mutated α-subunit in the RER (Marchand *et al.*, 1995). These results indicate that the I domain is essential for proteolytic processing and that the transmembrane and/or the C-terminal domain may contain an ER retention motif (Marchand *et al.*, 1995). Results from experiments to define the furin cleavage site are contradictory. In the mouse α-subunit, mutation of this furin cleavage site did not prevent secretion of the subunit, suggesting that more than one possible cleavage site exists or that a specially exposed region is used by the proteolytic enzyme (Marchand *et al.*, 1995). In contrast, mutations of the RPKR furin site in the rat α-subunit prevented secretion of this protein (Milhiet *et al.*, 1995).

10.4.5 Subunit turnover

Biosynthesis and degradation of E-24.18 has been investigated in ICR (meprin A) and C3H/He (meprin B) mice. The rate of incorporation and loss of radiolabel ([^{35}S]-methionine/cysteine) was studied in organ cultured kidney slices and the half-lives of the α- and β-subunits were determined. The rate of synthesis of the α-subunit was found to be two-fold greater than that of the β-subunit. The half-lives of the subunits were comparable, but shorter than the half-lives of the average protein in kidney cells (Hall *et al.*, 1993).

10.4.6 Activation of E-24.18

In *in vitro* expression systems both subunits are expressed as inactive zymogens. Their activation may be achieved by limited treatment with trypsin (Corbeil *et al.*, 1993; Grünberg *et al.*, 1993; Johnson and Hersh, 1994; Milhiet *et al.*, 1994). Trypsin cleaves off the propeptide which, in analogy to astacin, makes the N-terminal Asn available for the formation of a salt bridge with a glutamic acid residue near the active site. *In vivo*, both subunits of intestinal E-24.18 are activated by luminal trypsin. Activation of the kidney enzyme (α-subunit) on the other hand requires a so far unidentified protease. The β-subunit of rodent E-24.18 is present in a latent form, and it is not clear under what circumstances, or by what means, it is activated.

10.5 Distribution in cells and tissues

A wide range of different mammalian cells express many metalloproteases. In contrast, E-24.18 seemed to be restricted mainly to kidney and intestine, the organs from which the enzyme initially has been purified (Barnes *et al.,* 1989; Craig *et al.*, 1987; Sterchi *et al.*, 1988a). With the discovery recently of other sites of expression, such as salivary glands of the rat and mouse (Barnes *et al.,* 1989; Craig *et al.*, 1991) and craniofacial structures in the rat embryo (Spencer-Dene *et al.*, 1994) this restricted view may have to be modified. One major rule may be drawn from the expression pattern so far identified in different species: E-24.18 appears to be confined to epithelial cells and, more specifically, to the apical domain of polarized cells.

10.5.1 Distribution in different species: the E-24.18 gene is found from bird to man

By Southern blot analysis the E-24.18 α-subunit gene has been detected in human, monkey, rat, mouse, dog, cow, rabbit and chicken, but not in yeast (Jiang *et al.*, 1993). The gene has been conserved among a wide range of animals, extending into non-mammalian species.

10.5.2 Expression in kidney: high and low E-24.18 activity

Kidney E-24.18 has been examined mainly in rodents (mouse and rat). It makes up 6% of the microvillus membrane protein in the rat kidney (Kenny and Ingram, 1987) and 5–20% of renal brush border protein in the mouse (Craig *et al.*, 1987). The enzyme has been immunolocalized to the brush border of proximal tubuli in a subpopulation of nephrons in the deeper renal cortex, the so-called juxtamedullary region (Barnes *et al.*, 1989; Craig *et al.*, 1987; Yamaguchi *et al.*, 1993), which differ functionally and structurally from those residing in the outer renal cortex.

Some inbred laboratory mouse strains (C3H, CBA) secrete only low E-24.18 activity into the urine. These strains are deficient in the E-24.18 α-subunit but still express the β-subunit (meprin B) in the kidney and intestine (Butler and Bond, 1988; Craig *et al.*, 1987; Gorbea *et al.*, 1993). Interestingly, in contrast to mouse (Bankus and Bond, 1996; Jiang *et al.*, 1993) the α-subunit was not detected in human kidney by reverse transcriptase-polymerase chain reaction (RT-PCR) (Sterchi *et al.*, 1994) (see *Table 10.2*). Conversely, in one study, E-24.18 protein and enzymic activity was found in human kidney (Yamaguchi *et al.*, 1994b).

10.5.3 Expression in intestine: proteolytic attack from the intestinal brush border

After the kidney, the gut is the second major site of expression for E-24.18. The enzyme has been detected at the luminal surface of epithelial cells from duodenum to rectum in mouse and rat using immunocytochemical techniques (Barnes *et al.*, 1989). Initially, Northern blot analysis of mouse small intestinal RNA failed to detect the E-24.18 α-subunit (Jiang *et al.*, 1993) but confirmed the expression of the β-subunit (Gorbea *et al.*, 1993). Recently, the intestinal expression pattern of E-24.18 was investigated in mouse strains with high and low expression of the α-subunit (Bankus and Bond, 1996). In ICR mice with high E-24.18 α-activity, this subunit was detected in the

Table 10.2. Expression of endopeptidase-24.18 mRNA in human, mouse and rat: mRNA was detected by Northern blot analysis (Dumermuth *et al.*, 1993; Jiang *et al.*, 1993; Sterchi *et al.*, 1994)

	Human		Mouse		Rat	
Tissue	α	β	α	β	α	β
Intestine	+	+	+/–[a]	+	+	+
Kidney	–[b]	–[b]	+/–[a]	+	+	+
Pancreas	–	–	n.d.	n.d.	n.d.	n.d.
Liver	–	–	–	–	–	–
Placenta	–	–	n.d.	n.d.	n.d.	n.d.
Testis	n.d.	n.d.	–	–	–	–
Spleen	n.d.	n.d.	–	–	–	–
Brain	–	–	–	–	–	–
Lung	–	–	–	–	–	–
Heart	–	–	–	–	–	–
Skeletal muscle	–	–	–	–	–	–

[a] Mouse strain-dependent.
[b] E-24.18 protein has been found in human kidney (Yamaguchi *et al.*, 1994b).
+, detected; – not detectable; n.d., not determined.

gut in addition to the β-subunit by Western blot analysis, using subunit-specific antibodies. Northern blot analysis from different gut regions revealed highest expression of both α- and β-subunits in the ileum with lower but still significant expression in proximal small intestine and large intestine. In mice with low E-24.18 α-activity (C3H strain), mRNA levels of the α-subunit in the kidney and the gut were generally very low, but the β-subunit was expressed essentially as in ICR mice (see *Table 10.2*).

Both endopeptidase α- and β-subunits have been immunoprecipitated from human small intestine (Sterchi *et al.*, 1988a,b). Immunostaining of human small intestinal tissue sections localized E-24.18 to the apical side of mature epithelial cells (*Figure 10.5*). A proximal to distal gradient with highest enzyme activity in the ileum has been observed in human intestine (Sterchi *et al.*, 1988b) and is consistent with recent findings from mouse intestine (Bankus and Bond, 1996). The α-subunit was detected by RT-PCR in human small intestine, human colon and colon tumour tissue (Sterchi *et al.*, 1994) (see *Table 10.2*).

10.5.4 Embryonic expression in rat: E-24.18 as an early factor in craniofacial development

The expression pattern of E-24.18 has been studied in the rat embryo along with endopeptidase-24.11 (Spencer-Dene *et al.*, 1994). Immunoreactivity for E-24.18 was always localized at the apical/luminal domain of epithelial cells. At embryonic day 12 (E12), E-24.18 was detected in neuroepithelial cells, more specifically the choroid plexus and ependymal lining of the developing brain. At E14 and E16, epithelial cells of the stria vascularis in the inner ear and sometimes the epithelia of oral and nasal cavity stained for E-24.18. A weak signal was also observed in the developing eye lens and the pigmented layer of the retina. Notably, immunoreactivity was also observed in embryo bladder, ureter and gut. The temporally and spatially tightly regulated expression pattern

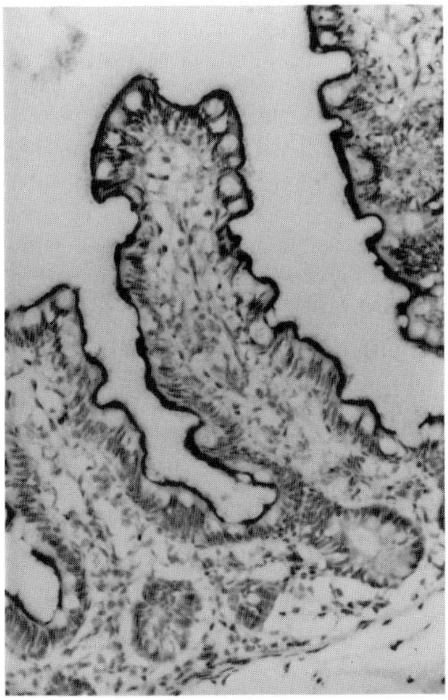

Figure 10.5. Immunostaining of the E-24.18 β-subunit in human small intestine. Cryostat sections were obtained from surgical specimens. The E-24.18 β-subunit was detected by immunohistochemistry using α-subunit-specific polyclonal rabbit antiserum, a biotinylated second antibody, an avidin–biotin–peroxidase complex and DAB (diaminobenzidine) as substrate. Counterstain was with Giemsa.

during embryogenesis is consistent with a possible involvement of this enzyme in the control of tissue differentiation.

10.5.5 Expression in carcinoma cells: variant β′-subunit mRNA produced through alternative splicing

A clone of a differentiating human colonic carcinoma cell line, HT29-18C1, and two undifferentiated mouse teratocarcinoma cell lines (phenotypically similar to embryonic ectodermal cells), F9 and Null-SSC-1, transcribe a variant β′-subunit message of bigger size (2.7 kb compared with 2.5 kb for the normal β mRNA) (Dietrich *et al.*, 1996). The β′ message is up-regulated by the differentiating agent, retinoic acid in Null-SSC-1 cells. The deduced amino acid sequence varies significantly in the N-terminal portion, including the signal peptide and part of the propeptide. By *in vitro* translation, it was demonstrated that the β′-subunit is translocated in microsomes and glycosylated. The different N terminus may confer to the β′-subunit a different targeting and/or activation mode. It was suggested that differential promoter usage and alternative splicing is responsible for generating the β′ mRNA in tumour cell lines. Gene expression may be controlled by such mechanisms, depending on the differentiation/proliferative status of the cells.

10.5.6 Other tissues expressing E-24.18: are there still sites to be discovered?

Mouse and rat E-24.18 have also been detected immunohistochemically in salivary glands and in the thyroid gland of mouse and rat (Barnes *et al.*, 1989; Craig *et al.*, 1991). In salivary glands, the enzyme could have a function similar to urinary E-24.18, namely the control of the release of pheromones. The observation of sex-specific differences in expression, with male mice generally showing stronger immunoreactivity, is consistent with this hypothesis.

10.6 Physiology and pathology

To this day no clear experimental evidence for a physiological substrate for E-24.18 is available and statements about the function of the enzyme must remain mostly speculative. In rodents, certain urinary proteins, like some MHC-molecules and so-called MUPS (major urinary proteins) (Finlayson *et al.*, 1965; Shaw *et al.*, 1983), were proposed to be cleaved by E-24.18 secreted into the urine, thereby releasing pheromones which control social behaviour, mating selection and recognition of 'self' and 'non-self' (Flannery *et al.*, 1990). E-24.18 may contribute to the catabolism of filtered or secreted biologically active peptides in the urine. Bradykinin, angiotensin I, II and III, substance P, luliberin and insulin B-chain (Stephenson and Kenny, 1988; Sterchi *et al.*, 1988b), neuropeptide Y (Price *et al.*, 1991), transforming growth factor-α (Choudry and Kenny, 1991), parathyroid hormone (PTH) and parathyroid hormone-related protein (PTHrP) (Yamaguchi *et al.*, 1994a, 1995) are all *in vitro* substrates. It has been suggested that E-24.18 is the crucial renal brush border protease degrading PTH and PTHrP, both involved in the regulation of calcium metabolism.

By analogy, intestinal E-24.18 could inactivate or activate luminal hormones and/or growth factors and regulate epithelial cell proliferation and differentiation (Dumermuth *et al.*, 1991). This may be particularly important during fetal and post-natal life. As the ileum is very rich in lymphatic tissue, ileal E-24.18 may also have an immunological function such as antigen processing and presentation (Bankus and Bond, 1996; Sterchi *et al.*, 1988b).

In a screen for gelatin-degrading proteases in renal cortex membrane preparations, E-24.18 was also isolated and shown to degrade laminin, collagen type IV and fibronectin *in vitro* (Kaushal *et al.*, 1994). Modification of extracellular matrix compounds by such a membrane-bound protease from epithelial cells could play a role in renal cell injury and regeneration.

Mouse strains with low and high E-24.18 activity and rats have been assessed for their susceptibility to toxic or ischaemic experimental renal disease (Reckelhoff *et al.*, 1988; Trachtman *et al.*, 1993, 1995). The results were contradictory and without a clear correlation of mortality or renal injury to E-24.18 activity.

The MEP1A locus is linked to MHC loci in human and mouse, which in turn are linked to numerous diseases. A role of E-24.18 in the pathogenesis of these diseases has been discussed (Jiang *et al.*, 1995).

10.7 Genetics

The existence of meprin-deficient mouse strains (these are related inbred strains, CBA/J, CBA/CaJ, C3H/HeJ and C3H/HeN) that have 2% or less specific meprin activities in kidney when compared with normal activities in C57BR/cdJ, C57BL/6J, BALB/cJ, A/J, DBA/1J, CD/1 Swiss and ICR strains (Beynon and Bond, 1983; Bond *et al.*, 1983), made it possible to investigate the inheritance of the protease. Crossbreeding experiments have shown that meprin deficiency is inherited as an autosomal recessive trait (Beynon and Bond, 1985; Bond *et al.*, 1984).

The fact that all the meprin-deficient mice were of the H-2k haplotype (Beynon and Bond, 1985; Bond *et al.*, 1984), led to the suggestion that the gene that controls meprin activity (*Mep-1*) is linked to the major histocompatibility complex (H-2 complex) and is therefore also located on chromosome 17. Using recombinant and congenic strains of mice in linkage studies the *Mep-1* gene was subsequently localized telomeric to H-2D and close to the *Pgk-2* gene on chromosome 17 (Bond *et al.*, 1984). Mice with high or low meprin activity were designated as follows: mice with a high level of activity are of the genotype $Mep\text{-}1^a/Mep\text{-}1^a$, mice with low activity $Mep\text{-}1^b/Mep\text{-}1^b$. Mice with the genotype $Mep\text{-}1^a/Mep\text{-}1^b$ have the high meprin activity phenotype (Reckelhoff *et al.*, 1985).

The structural genes for the α- and β-subunits have been mapped in the mouse and the human genome. It was possible by interspecific backcross analysis to map the structural gene for mouse meprin β (*Mep-1b*) to the proximal region of chromosome 18, linked to *Ncad*, *Palb*, *Apc* and *Fgfa* (Gorbea *et al.*, 1993). The structural gene for the α-subunit (*Mep-1a*) was mapped to chromosome 17 near the H-2 complex. The fact that the gene controlling meprin activity in mouse kidney is also localized on chromosome 17 to the right of the D end of H2 leads to the possibility that the regulation of expression of the α-subunit is dependent on the structural gene itself or a regulatory element close to it (Jiang *et al.*, 1993). Using radiation and somatic cell hybrids the structural gene for the human β-subunit *MEP1B* was mapped to region 18q12.2–q12.3 proximal to the *TTR/PALB* gene on chromosome 18 and the gene for the α-subunit *MEP1A* was mapped at position 6p11–p12.2 between the centromere and *GSTA2* of chromosome 6p (Bond *et al.*, 1995). Subsequent fine mapping localized *MEP1A* on human chromosome 6p21.2–p21.1 telomeric to *GSTA2* and closely linked to the HLA loci encoding proteins of the major histocompatibility complex (Jiang *et al.*, 1995). Thus the structural genes for the α- and β-subunits are unlinked in the human and the mouse genome and both genes for the α-subunits are linked to the MHC in mouse and human.

Acknowledgements

The work carried out in our laboratory was supported by grant no. 32–40571.94 from the Swiss National Science Foundation. We thank Eric Dumermuth, Joyce A. Eldering, Jürgen Grünberg, Ursula Luginbühl and Elizabeth Mandac, who during the last years have all contributed to the work on the human E-24.18.

References

Appella E, Weber IT, Blasi F. (1988) Structure and function of epidermal growth factor-like regions in proteins. *FEBS Lett.* **231**: 1–4.

Bankus JM, Bond JS. (1996) Expression and distribution of meprin protease subunits in mouse intestine. *Arch. Biochim. Biophys.* **331**: 87–94.

Barnes K, Ingram J, Kenny AJ. (1989) Proteins of the kidney microvillar membrane. Structural and immunochemical properties of rat endopeptidase-2 and its immunohistochemical localization in tissues of rat and mouse. *Biochem. J.* **264**: 335–346.

Beckman G, Bork P. (1993) An adhesive domain detected in functionally diverse receptors. *Trends Biochem. Sci.* **18**: 40–41.

Beynon RJ, Bond JS. (1983) Deficiency of a kidney metalloproteinase in inbred mouse strains. *Science* **219**: 1351–1353.

Beynon RJ, Bond JS. (1985) Expression of the Mep-1 gene regulating meprin, a kidney brush border proteinase. *Prog. Clin. Biol. Res.* **180**: 185–194.

Beynon RJ, Shannon JD, Bond JS. (1981) Purification and characterization of a metallo-endoproteinase from mouse kidney. *Biochem. J.* **199**: 591–598.

Bode W, Gomis Rüth FX, Huber R, Zwilling R, Stöcker W. (1992) Structure of astacin and implications for activation of astacins and zinc-ligation of collagenases. *Nature* **358**: 164–167.

Bode W, Gomis Rüth FX, Stöcker W. (1993) Astacins, serralysins, snake venom and matrix metalloproteinases exhibit identical zinc-binding environments (HEXXHXXGXXH and Met-Turn) and topologies and should be grouped into a common family, the metzincins. *FEBS Lett.* **331**: 134–140.

Bond JS, Shannon JD, Beynon RJ. (1983) Certain mouse strains are deficient in a kidney brush-border metallo-endopeptidase activity. *Biochem. J.* **209**: 251–255.

Bond JS, Beynon RJ, Reckelhoff JF, David CS. (1984) Mep-1 gene controlling a kidney metallo-endopeptidase is linked to the major histocompatibility complex in mice. *Proc. Natl Acad. Sci. USA* **81**: 5542–5545.

Bond JS, Rojas K, Overhauser J, Zoghbi HY, Jiang W. (1995) The structural genes, MEP1A and MEP1B, for the alpha and beta subunits of the metallo-endopeptidase meprin map to human chromosome 6p and 18q, respectively. *Genomics* **25**: 300–303.

Butler PE, Bond JS. (1988) A latent proteinase in mouse kidney membranes. Characterization and relationship to meprin. *J. Biol. Chem.* **263**: 13419–13426.

Butler PE, McKay MJ, Bond JS. (1987) Characterization of meprin, a membrane-bound metallo-endopeptidase from mouse kidney. *Biochem. J.* **241**: 229–235.

Campell ID, Bork P. (1993) Epidermal growth factor-like modules. *Curr. Opin. Struct. Biol.* **3:** 385–392.

Choudry Y, Kenny AJ. (1991) Hydrolysis of transforming growth factor-α by cell-surface peptidases *in vitro. Biochem. J.* **280:** 57–60.

Chevallier S, Ahn J, Boileau G, Crine P. (1996) Identification of the cysteine residues implicated in the formation of alpha2 and alpha/beta dimers of rat meprin. *Biochem. J.* **317**: 731–738.

Clarke B, Ofosu F, Sridhara S, Bona R, Rickles F, Blajchman M. (1992) The first epidermal growth factor domain of human coagulation factor VII is essential for binding with tissue factor. *FEBS Lett.* **298**: 206–210.

Corbeil D, Gaudoux F, Wainwright S, Ingram J, Kenny AJ, Crine P. (1992) Molecular cloning of the alpha-subunit of rat endopeptidase-24.18 (endopeptidase-2) and co-localization with endopeptidase-24.11 in rat kidney by *in situ* hybridization. *Febs Lett.* **309**: 203–208

Corbeil D, Milhiet PE, Simon V, Ingram J, Kenny AJ, Boileau G, Crine P. (1993) Rat endopeptidase-24.18 alpha subunit is secreted into the culture medium as a zymogen when expressed by COS-1 cells. *FEBS Lett.* **335**: 361–366.

Craig SS, Reckelhoff JF, Bond JS. (1987) Distribution of meprin in kidneys from mice with high- and low-meprin activity. *Am. J. Physiol.* **253**: C535–C540.

Craig SS, Mader C, Bond JS. (1991) Immunohistochemical localization of the metalloproteinase meprin in salivary glands of male and female mice. *J. Histochem. Cytochem.* **39**: 123–129.

Davis C. (1990) The many faces of epidermal growth factor repeats. *New Biol.* **2:** 410–419.

Dietrich J, Jiang W, Bond J. (1996) A novel meprin beta′ mRNA in mouse embryonal and human colon carcinoma cells. *J. Biol. Chem.* **271**: 2271–2278.

Dumermuth E, Sterchi EE, Jiang WP, Wolz RL, Bond JS, Flannery AV, Beynon RJ. (1991) The astacin family of metalloendopeptidases. *J. Biol. Chem.* **266**: 21381–21385

Dumermuth E, Eldering JA, Grünberg J, Jiang W, Sterchi EE. (1993) Cloning of the PABA peptide hydrolase alpha subunit (PPH alpha) from human small intestine and its expression in COS-1 cells. *FEBS Lett.* **335**: 367–375.

Elaroussi MA, DeLuca HF. (1994) A new member to the astacin family of metallo-endopeptidases: a novel 1,25-dihydroxyvitamin D-3-stimulated mRNA from chorioallantoic membrane of quail. *Biochim. Biophys. Acta* **1217**: 1–8.

Finelli AL, Bossie CA, Xie T, Padgett RW. (1994) Mutational analysis of the *Drosophila* tolloid gene, a human BMP-1 homolog. *Development* **120**: 861–870

Finlayson JS, Asofsky R, Potter M, Runner CC. (1965) Major urinary protein complex of normal mice: origin. *Science* **149**: 981–982.

Flannery AV, Dalzell GN, Beynon RJ. (1990) Proteolytic activity in mouse urine: relationship to the kidney metallo-endopeptidase, meprin. *Biochim. Biophys. Acta* **141**: 64–70.

Fukagawa M, Suzuki N, Hogan BL, Jones CM. (1994) Embryonic expression of mouse bone morphogenetic protein-1 (BMP-1), which is related to the *Drosophila* dorsoventral gene tolloid and encodes a putative astacin metalloendopeptidase. *Dev. Biol* **163**: 175–183

Gomis-Rüth FX, Stöcker W, Huber R, Zwilling R, Bode W. (1993) Refined 1.8 Å X-ray crystal structure of astacin, a zinc- endopeptidase from the crayfish *Astacus astacus* L. Structure determination, refinement, molecular structure and comparison with thermolysin. *J. Mol. Biol.* **229**: 945–968.

Gorbea CM, Flannery AV, Bond JS. (1991) Homo- and heterotetrameric forms of the membrane-bound metallo-endopeptidases meprin A and B. *Arch. Biochem. Biophys.* **290**: 549–553.

Gorbea CM, Marchand P, Jiang W, Copeland NG, Gilbert DJ, Jenkins NA, Bond JS. (1993) Cloning, expression, and chromosomal localization of the mouse meprin beta subunit. *J. Biol. Chem.* **268**: 21035–21043.

Grams F, Dive V, Yitakis A, Yiallouros I, Vassiliou S, Zwilling R, Bode W, Stöcker W. (1996) Structure of astacin with a transition-state analogue inhibitor. *Nature Struct. Biol.* **3**: 671–675.

Grünberg J, Dumermuth E, Eldering JA, Sterchi EE. (1993) Expression of the alpha subunit of PABA peptide hydrolase (EC 3.4.24.18) in MDCK cells. Synthesis and secretion of an enzymically inactive homodimer. *FEBS Lett.* **335**: 376–379.

Hall JL, Sterchi EE, Bond JS. (1993) Biosynthesis and degradation of meprins, kidney brush border proteinases. *Arch. Biochem. Biophys.* **307**: 73–77.

Hauri HP, Sterchi EE, Bienz D, Fransen JA, Marxer A. (1985) Expression and intracellular transport of microvillus membrane hydrolases in human intestinal epithelial cells. *J. Cell Biol.* **101**: 838–851.

Hwang SP, Partin JS, Lennarz WJ. (1994) Characterization of a homolog of human bone morphogeneticprotein 1 in the embryo of the sea urchin, *Strongylocentrotus purpuratus. Development* **120**: 559–568

Jiang W, Gorbea CM, Flannery AV, Beynon RJ, Grant GA, Bond JS. (1992) The alpha subunit of meprin A. Molecular cloning and sequencing, differential expression in inbred mouse strains, and evidence for divergent evolution of the alpha and beta subunit. *J. Biol. Chem.* **267**: 9185–9193.

Jiang W, Sadler PM, Jenkins NA, Gilbert DJ, Copeland NG, Bond JS. (1993) Tissue-specific expression and chromosomal localization of the alpha subunit of mouse meprin A. *J. Biol. Chem.* **268**: 10380–10385.

Jiang W, Dewald G, Brundage E, Mucher G, Schildhaens HU, Zerres K, Bond JS. (1995) Fine mapping of MEP1A, the gene encoding the alpha subunit of the metallo-endopeptidase meprin, to human chromosome 6P21. *Biochem. Biophys. Res. Commun.* **216**: 630–635.

Johnson GD, Hersh LB. (1992) Cloning a rat meprin cDNA reveals the enzyme is a heterodimer. *J. Biol. Chem.* **267**: 13505–13512.

Johnson GD, Hersh LB. (1993) Cloning of rat meprin cDNA reveals the enzyme is a heterodimer. *J. Biol. Chem.* **268**: 17647.

Johnson GD, Hersh LB. (1994) Expression of meprin subunit precursors. Membrane anchoring through the beta subunit and mechanism of zymogen activation. *J. Biol. Chem.* **269**: 7682–7688.

Kaushal GP, Walker PD, Shah SV. (1994) An old enzyme with a new function: purification and characterization of a distinct matrix-degrading metalloproteinase in rat kidney cortex and its identification as meprin. *J. Cell Biol.* **126**: 1319–1327.

Kenny AJ, Ingram J. (1987) Proteins of the kidney microvillar membrane. Purification and properties of the phosphoramidon-insensitive endopeptidase ('endopeptidase-2') from rat kidney. *Biochem. J.* **245**: 515–524.

Kenny AJ, Fulcher IS, Ridgwell K, Ingram J. (1981) Microvillar membrane neutral endopeptidases. *Acta Biol. Med. Ger.* **40**: 1465–1471.

Kitamoto Y, Yuan X, Wu Q, McCourt DW, Sadler JE. (1994) Enterokinase, the initiator of intestinal digestion, is a mosaic protease composed of a distinctive assortment of domains. *Proc. Natl Acad. Sci. USA* **91**: 7588–7592.

Kounnas MZ, Wolz RL, Gorbea CM, Bond JS. (1991) Meprin-A and -B. Cell surface endopeptidases of the mouse kidney. *J. Biol. Chem.* **266**: 17350–17357.

Lepage T, Ghiglione C, Gache C. (1992) Spatial and temporal expression pattern during sea urchin embryogenesis of a gene coding for a protease homologous to the human protein BMP-1 and to the product of the *Drosophila* dorsal–ventral patterning gene tolloid. *Development* **114**: 147–163

Li SW, Sieron AL, Fertala A, Hojima Y, Arnold WV, Prockop DJ. (1996) The C-proteinase that processes procollagens to fibrillar collagens is identical to the protein previously identified as bone morphogenic protein-1. *Proc. Natl Acad. Sci. USA* **93**: 5127–5130.

Maeno M, Xue Y, Wood TI, Ong RC, Kung HF. (1993) Cloning and expression of cDNA encoding *Xenopus laevis* bone morphogenetic protein-1 during early embryonic development. *Gene* **134**: 257–261

Marchand P, Tang J, Bond JS. (1994) Membrane association and oligomeric organization of the alpha and beta subunits of mouse meprin A. *J. Biol. Chem.* **269**: 15388–15393.

Marchand P, Tang J, Johnson G, Bond J. (1995) COOH-terminal proteolytic processing of secreted and membrane forms of the alpha subunit of the metalloprotease Meprin A. Requirement of the I domain for processing in the endoplasmic reticulum. *J. Biol. Chem.* **270**: 5449–5456.

Marchand P, Volkmann M, Bond J. (1996) Cysteine mutations in the MAM domain result in monomeric meprin and alter stability and activity of the protease. *J. Biol. Chem.* **271:** 24236–24241.

Milhiet PE, Corbeil D, Simon V, Kenny AJ, Crine P, Boileau G. (1994) Expression of rat endopeptidase-24.18 in COS-1 cells: membrane topology and activity. *Biochem. J.* **300**: 37–43.

Milhiet PE, Chevallier S, Corbeil D, Seidah N, Crine P, Boileau G. (1995) Proteolytic processing of the alpha-subunit of rat endopeptidase-24.18 by furin. *Biochem. J.* **309**: 683–688.

Nguyen T, Jamal J, Shimell MJ, Arora K, O'Connor MB. (1994) Characterization of tolloid-related-1: a BMP-1-like product that is required during larval and pupal stages of *Drosophila* development. *Dev. Biol.* **166**: 569–586

Price JS, Kenny AJ, Huskisson NS, Brown MJ. (1991) Neuropeptide Y (NPY) metabolism by endopeptidase-2 hinders characterization of NPY receptors in rat kidney. *Br. J. Pharmacol.* **104**: 321–326.

Reckelhoff JF, Bond JS, Beynon RJ, Savarirayan S, David CS. (1985) Proximity of the Mep-1 gene to H-2D on chromosome 17 in mice. *Immunogenetics* **22**: 617–623.

Reckelhoff JF, Craig SS, Beynon RJ, Bond JS. (1988) Meprin phenotype and cyclosporin A toxicity in mice. *Adv. Exp. Med. Biol.* **240**: 293–304.

Reynolds SD, Angerer LM, Palis J, Nasir A, Angerer RC. (1992) Early mRNAs, spatially restricted along the animal–vegetal axis of sea urchin embryos, include one encoding a protein related to tolloid and BMP-1. *Development* **114**: 769–786

Sato SM, Sargent TD. (1990) Molecular approach to dorsoanterior development in *Xenopus laevis. Dev. Biol.* **137**:135–141

Shaw PH, Held WA, Hastie ND. (1983) The gene family for major urinary proteins: expression in several secretory tissues of the mouse. *Cell* **32**: 755–761.

Shimell MJ, Ferguson EL, Childs SR, O'Connor MB. (1991) The *Drosophila* dorso-ventral patterning gene tolloid is related to human bone morphogenetic protein 1. *Cell* **67**: 469–481

Spencer-Dene B, Thorogood P, Nair S, Kenny AJ, Harris M, Henderson B. (1994) Distribution of, and putative role for, the cell-surface neutral metallo-endopeptidases during mammalian craniofacial development. *Development* **120**: 3213–3226.

Stephenson S, Kenny AJ. (1988) The metabolism of neuropeptides. Hydrolysis of peptides by the phosphoramidon-insensitive rat kidney enzyme, 'endopeptidase-2', and by rat kidney microvillar membranes. *Biochem. J.* **255**: 45–51.

Sterchi EE, Green JR, Lentze MJ. (1982) Non-pancreatic hydrolysis of *N*-benzoyl-L-tyrosyl-*p*-aminobenzoic acid (PABA-peptide) in the human small intestine. *Clin. Sci.* **62**: 557–560.

Sterchi EE, Green JR, Lentze MJ. (1983) Nonpancreatic hydrolysis of *N*-benzoyl-L-tyrosyl-*p*-aminobenzoic acid (PABA peptide) in the rat small intestine. *J. Pediatr. Gastroenterol. Nutr.* **2**: 539–457.

Sterchi EE, Naim HY, Lentze MJ. (1988a) Biosynthesis of *N*-benzoyl-L-tyrosyl-*p*-aminobenzoic acid hydrolase: disulfide-linked dimers are formed at the site of synthesis in the rough endoplasmic reticulum. *Arch. Biochem. Biophys.* **265**: 119–127.

Sterchi EE, Naim HY, Lentze MJ, Hauri HP, Fransen JA. (1988b) *N*-benzoyl-L-tyrosyl-*p*-aminobenzoic acid hydrolase: a metalloendopeptidase of the human intestinal microvillus membrane which degrades biologically active peptides. *Arch. Biochem. Biophys.* **265**: 105–118.

Sterchi E, Dumermuth E, Eldering J, Grünberg J. (1994) *Paba-peptide Hydrolase (PPH): Structure and Expression of a Metalloendopeptidase from Human Small Intestinal Epithelial Cells*, pp. 141–151. Thieme Medical Publishers, New York.

Stöcker W, Bode W. (1995) Structural features of a superfamily of zinc-endopeptidases: the metzincins. *Curr. Opin. Struct. Biol.* **5**: 383–390.

Stöcker W, Zwilling R. (1995) Astacin. *Methods Enzymol.* **248**: 305–325.

Stöcker W, Gomis-Rüth FX, Bode W, Zwilling R. (1993) Implications of the three-dimensional structure of astacin for the structure and function of the astacin family of zinc-endopeptidases. *Eur. J. Biochem.* **214**: 215–231.

Takahara K, Lyons GE, Greenspan DS. (1994) Bone morphogenetic protein-1 and a mammalian tolloid homologue (mTld) are encoded by alternatively spliced transcripts which are differentially expressed in some tissues. *J. Biol. Chem.* **269**: 32572–32578

Tarentino Al, Quinones G, Grimwood BG, Hauer CR, Plummer Th Jr. (1995) Molecular cloning and sequence analysis of flavastacin: an O–glycosylated prokaryotic zinc metalloendopeptidase. *Arch. Biochem. Biophys.* **319**: 281–285

Trachtman H, Greenwald R, Moak S, Tang J, Bond JS. (1993) Meprin activity in rats with experimental renal disease. *Life Sci.* **53**: 1339–1344.

Trachtman H, Valderrama E, Dietrich J, Bond J. (1995) The role of meprin A in the pathogenesis of acute renal failure. *Biochem. Biophys. Res. Comm.* **208**: 498–505.

Wolz RL. (1994) A kinetic comparison of the homologous proteases astacin and meprin A. *Arch. Biochem. Biophys.* **310**: 144–151.

Wolz RL, Bond JS. (1990) Phe5(4-nitro)-bradykinin: a chromogenic substrate for assay and kinetics of the metallo-endopeptidase meprin. *Analyt. Biochem.* **191**: 314–320.

Wolz RL, Harris RB, Bond JS. (1991) Mapping the active site of meprin-A with peptide substrates and inhibitors. *Biochemistry* **30**: 8488–8493.

Wozney JM, Rosen V, Celeste AJ, Mitsock LM, Whitters MJ, Kriz RW, Hewick RM, Wang EA. (1988) Novel regulators of bone formation: molecular clones and activities. *Science* **242**: 1528–1534

Yamaguchi T, Kido H, Kitazawa R, Kitazawa S, Fukase M, Katunuma N. (1993) A membrane-bound metallo-endopeptidase from rat kidney: its immunological characterization. *J. Biochem. (Tokyo)* **113**: 299–303.

Yamaguchi T, Fukase M, Kido H, Sugimoto T, Katunuma N, Chihara K. (1994a) Meprin is predominantly involved in parathyroid hormone degradation by the microvillar membranes of rat kidney. *Life Sci.* **54**: 381–386.

Yamaguchi T, Fukase M, Sugimoto T, Kido H, Chihara K. (1994b) Purification of meprin from human kidney and its role in parathyroid hormone degradation. *Biol. Chem. Hoppe-Seyler* **375**: 821–824.

Yamaguchi T, Fukase M, Sugimoto T, Chihara K. (1995) Hydrolysis of a carboxy-terminal fragment of parathyroid hormone-related protein by rat kidney: evidence for a crucial role of meprin. *Horm. Metab. Res.* **27**: 131–136.

Yan L, Pollock GH, Nagase H, Sarras MPJ. (1995) A 25.7 × 10^3 M_r *Hydra* metalloproteinase (HMP1), a member of the astacin family, localizes to the extracellular matrix of *Hydra vulgaris* in a head-specific manner and has a developmental function. *Development* **121**: 1591–1602.

Yasumasu S, Yamada K, Akasaka K, Mitsunaga K, Iuchi I, Shimada H, Yamagami K. (1992) Isolation of cDNAs for LCE and HCE, two constituent proteases of the hatching enzyme of *Oryzias latipes*, and concurrent expression of their mRNAs during development. *Dev. Biol.* **153**: 250–258

Zhong D, Smith K, Birktoft J, Bajaj S. (1994) First epidermal growth factor-like domain of human blood coagulation factor IX is required for its activation by factor VIIa/tissue factor but not by factor XIa. *Proc. Natl Acad. Sci. USA* **91**: 3574–3578.

Chapter 11

Aminopeptidase N

Ove Norén, Hans Sjöström and Jørgen Olsen

11.1 Introduction

Aminopeptidase N (EC 3.4.11.2, AP-N) is a zinc-dependent membrane-bound peptidase. The enzyme was first isolated by Pfleiderer and Celliers (1963) from pig kidney. It is known under several different names (see *Table 11.1*) and is probably the most extensively studied aminopeptidase. By comparison of the primary structures of human AP-N (Olsen *et al.*, 1988) with CD13 (Look *et al.*, 1989), as deduced from the cDNA sequences, it was evident that these two molecules were identical. Likewise, the cloning of the cDNA encoding p161, a murine membrane protein which is expressed on mast cells and on some macrophages, showed that p161 is identical to AP-N.

AP-N has been purified using traditional chromatographic methods from a pig kidney microsomal fraction as a hydrophilic, proteinase-released form (Wacker *et al.,* 1971). A little later, it was shown to be a major constituent of the brush border membranes of kidney proximal tubule cells (George and Kenny, 1973) and of enterocytes (Louvard *et al.*, 1973). Microvillar membrane vesicles, isolated from these tissues using a divalent ion (Mg^{2+} or Ca^{2+}) to precipitate and remove the intracellular and basolateral membranes, were then used to purify the amphiphilic, detergent-solubilized form of AP-N (Sjöström *et al.,* 1978), taking advantage of immunoadsorbent columns eluted by a hypotonic buffer (Danielsen *et al.*, 1982). This immuno-isolation technique made it possible to prepare quickly milligram amounts of intact amphiphilic AP-N of high purity.

AP-N preferentially releases neutral amino acids from oligopeptides and thus has a

Table 11.1. Cloned and sequenced AP-N cDNA from different species and tissues

Species	Tissue/cell type	Number of deduced amino acid residues	References
Human	Caco-2 cells	967	Olsen *et al.* (1988)
	Myeloid cells		Look *et al.* (1989)
Rat	Kidney	966/965	Malfroy *et al.* (1989)
			Watt and Yip (1989)
Rabbit	Kidney/intestine	965	Yang *et al.* (1993)
Pig	Small intestine	963	Delmas *et al.* (1994)
Cat	Fcwf-cells	967	Tresnan *et al.* (1996)
Mouse	CFTL-cells	966	Chen *et al.* (1996)

Cell-Surface Peptidases in Health and Disease, edited by A.J. Kenny and C.M. Boustead.

specificity similar to the cytosolic leucine aminopeptidase (EC 3.4.11.1; Hanson and Frohne, 1976). AP-N is present in almost all tissues (Rehfeld *et al.,* 1967). High levels are found in kidney proximal tubule cells and in enterocytes. In the small intestine, AP-N has a role in the final phase of protein digestion, releasing amino acids from oligopeptides generated by the action of peptidases secreted from the stomach and pancreas. In the kidney, the function of AP-N is supposed to be the hydrolysis of small peptides present in the glomerular filtrate.

There are several reviews on AP-N dating from before the introduction of recombinant DNA technology (Danielsen *et al.,* 1984; Kenny and Maroux, 1982; McDonald and Barrett, 1986; Norén *et al.*, 1986; Semenza, 1986). They summarize knowledge of substrate specificity and structure (obtained from enzymic and 'classical' protein studies, as well as results obtained from studies on biosynthesis). Reviews emphasizing aminopeptidases as a functional group of enzymes have also been published (Sanderink *et al.*, 1988; Taylor, 1993).

In this review we summarize early studies on the structure, function and biosynthesis of AP-N and try to integrate the earlier information with the new data obtained by DNA recombinant technology. We also review recent knowledge on the gene organization and regulation of expression of AP-N.

11.2 General structure and topology

AP-N exists as a homodimeric 'lollipop-like' structure in the membrane, with the major part of its mass on the ecto-side (*Figure 11.1*). This model has emerged from a

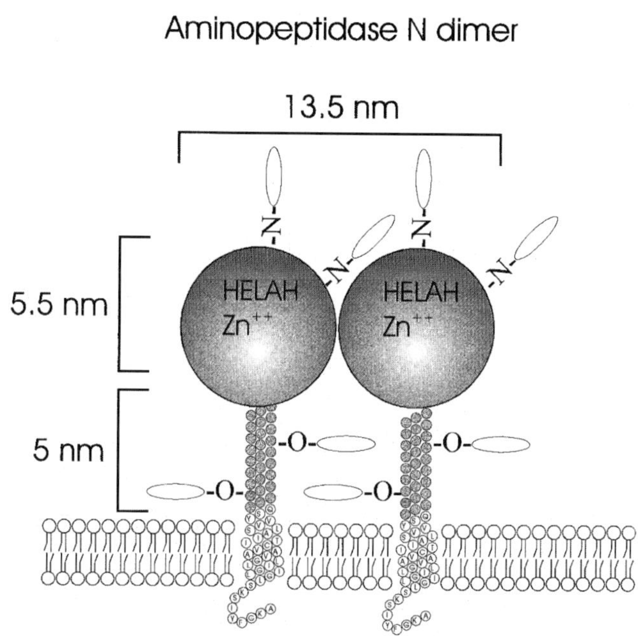

Figure 11.1. Topology of aminopeptidase N in the membrane.

combination of biochemical and ultrastructural data. Thus, it has been demonstrated by hydrodynamic studies of the hydrophilic form under native conditions, that the molecular weight of AP-N is about 280 kDa, whereas under denaturing conditions, as studied by sodium dodecyl sulphate polyacrylamide gel electrophoresis (SDS-PAGE), it is about 140 kDa (Maroux *et al.*, 1973; Sjöström *et al.*, 1983; Wacker *et al.*, 1971). Crosslinking experiments (Svensson *et al.*, 1982) demonstrated that the predominant form in the membrane and in solution is a dimer. Electron microscopy of purified amphiphilic AP-N reconstituted into liposomes (Hussain *et al.*, 1981) confirmed its dimeric structure in the membrane. The latter study also revealed that the bulk of the protein is separated from the membrane by a space of about 5 nm, attributable to the stalk connecting the ectodomain with the transmembrane and cytosolic domains of AP-N. AP-N is a transmembrane protein, as demonstrated by chemical labelling (Booth and Kenny, 1980). The stalk is sensitive to proteinases, such as trypsin and papain. Treatment of microvillar vesicles by either of these enzymes releases enzymically active AP-N from the membrane. This hydrophilic (proteolytic) form is about 4 kDa smaller than the detergent-solubilized form (Ferraci and Maroux, 1980), a difference corresponding to the size of the transmembrane and cytosolic domains and compatible with the view that AP-N spans the membrane only once. The anchor of AP-N is strongly hydrophobic: it can be labelled by a hydrophobic reagent (iodo-naphthylazide; Norén and Sjöström, 1980) and it binds detergent (Sjöström *et al.*, 1978). AP-N has its N terminus on the cytosolic side of the plasma membrane and is thus a class II integral membrane protein. This has been shown by the finding that the N-terminal amino acid residues of the detergent-released and proteolytically released forms differ (Louvard and Maroux, 1976; Sjöström and Norén, 1982). The deduced amino acid sequences from all the cloned cDNAs (listed in *Table 11.1*) corroborate this model.

11.3 Biosynthesis, intracellular transport and sorting

AP-N is synthesised on the rough endoplasmic reticulum (RER), in common with exocytosed proteins (Danielsen *et al.*, 1984). Docking, a necessary step in this process, is mediated by the N-terminal hydrophobic stretch (Desnuelle, 1979) which forms the transmembrane domain. It thus has a dual function, acting both as an uncleaved signal sequence and also an anchor to the membrane (Kenny and Booth, 1978). AP-N is co-translationally N-glycosylated by the addition of high-mannose oligosaccharides, a step that renders AP-N sensitive to endoglycosidase H (Danielsen, 1982; Ferraci *et al.*, 1985; Look *et al.*, 1985). During translation, AP-N segregates into the cisternae of the RER, where it is inaccessible to proteinases (Danielsen and Cowell, 1985). Dimerization of AP-N occurs while the enzyme is in its high-mannose glycosylated state (indicating that this process occurs in the RER) and is rather slow, hence rate-limiting for the transport of the protein from the RER to the Golgi (Danielsen, 1990). The high-mannose, transient form of AP-N is enzymically active (Sjöström *et al.*, 1985), indicating that native folding is completed and the zinc ion is associated with the molecule very early in the biosynthetic process. The conclusion that correct folding occurs very early during the process of biosynthesis and that it is dependent on correct N-glycosylation has been demonstrated by use of an antibody reacting exclusively with the native form of AP-N and by perturbing glycosylation with fructose (Danielsen, 1992). AP-N is found in the Golgi apparatus (Hansen *et al.*, 1987), the transport from RER to the *cis*-Golgi apparatus probably occuring via ERGIC (endoplasmic reticulum Golgi intermediate compartment; Hauri and Schweizer, 1992). In the Golgi apparatus, the N-linked oligosaccharides are sequentially modified by trimming and addition of terminal hexoses such as *N*-acetyl-glucosamine, galactose and sialic acid, thus making the enzyme resistant to endoglycosidase H (Danielsen, 1982). Here, too, AP-N becomes O-glycosylated (Norén *et al.,* 1997) and tyrosine-sulphated (Danielsen, 1987).

AP-N is distributed to the different domains of the plasma membrane (the basolateral and apical membranes) probably via the trans-Golgi network (TGN). The proportions transported directly to the two domains depend on cell type. Thus in Madin–Darby canine kidney cells, recombinantly expressing pig AP-N, about 80% is directed to the apical membrane (Vogel *et al.,* 1997), whereas in Caco-2 cells the corresponding figure is 20% (L.K. Vogel, personal communication 1995). AP-N molecules directed to the basolateral membrane are endocytosed and transported to the apical membrane probably via endosomes (Vogel *et al.,* 1995; see also Chapter 4). Endocytosis, and probably transcytosis, also occur from the apical to the basolateral membrane (Vogel *et al.*, 1995). Experiments performed by expressing membrane and secretory forms of AP-N in cell lines have given some clues for the existence of sorting signals. Thus, information for apical distribution has been located to the ectodomain (Vogel *et al.,*1992a,b) and the information for basolateral endocytosis and subsequent apical transport is located in the membrane-spanning domain or the stalk (Vogel *et al.*, 1995). The mechanism of sorting is unclear. AP-N is present in glycolipid rafts in intestinal epithelial cells (Danielsen, 1995) and the association to the rafts occurs before AP-N reaches the apical membrane. Interestingly, Kundu *et al.* (1996) demonstrated that the information for apical targetting of influenza virus neuraminidase (another type II membrane protein) resides in the membrane-spanning domain. A kinetic model has been proposed stating that it is the balance in the traffic between the different subcellular compartments that determines the steady-state distribution of AP-N, rather than the fractions of material distributed by TGN (see Chapter 4).

11.4 Detailed structure

11.4.1 Deduced amino acid sequence and functional domains

The cDNA-deduced amino acid sequences of AP-N have not only confirmed earlier models of the protein, but have also added new information. The N-terminal cytoplasmic tail, eight residues in length, contains two basic amino acids and one tyrosine and lacks any of the known signals for intracellular transport or sorting. This domain is followed by a hydrophobic stretch, 22 residues in length, sufficient to span the membrane once. This membrane-spanning part is assumed to exist in a helical form, as has been suggested for another class II microvillar membrane protein, sucrase–isomaltase (Spiess *et al.,* 1982). However, this is not revealed by Chou–Fasman analysis. This predominantly hydrophobic sequence is well conserved between the mammalian AP-Ns; it ends with an acidic residue, probably important for the correct membrane insertion during translation (High, 1995). The stalk, positioned on the ecto-side of the membrane and containing about 40 amino acid residues (rich in Ser, Thr and Pro) is also less conserved. The frequent occurrence of proline residues strongly argues against a helix, instead favouring a more extended structure. The stretch reveals a high probability for surface exposition, consistent with it forming the stalk. The notion that this stretch forms a separate domain is supported by the fact that it is possible to effectively express AP-N in cell lines, containing either the stalk plus the catalytic domain, or the catalytic domain alone, but not intermediary forms (Vogel *et al.,* 1992b).

The catalytic domain has two different regions which can be defined by the very distinct cleavage of the pig AP-N by trypsin into an N-terminal fragment (the B subunit; Sjöström *et al.*, 1978; Wacker *et al.,* 1976) and a C-terminal fragment (the C-subunit; Sjöström *et al.,* 1978; Wacker *et al.*, 1976). The subunits can be dissociated at non-denaturing concentrations of SDS. This distinct cleavage has not been described for AP-N isolated from other species, explained by the occurrence of Arg575–Pro576, the imido bond being resistant to trypsin (numbering refers to human AP-N; Delmas *et al.,* 1994). Residues 216–228 (numbering refers to human AP-N) form a domain which is completely conserved in human AP-N and aminopeptidases A (AP-A), indicating an important stretch, as yet with no known function.

Apart from the active site motif no other known functional motifs have been recognized by computer analysis. A long stretch of α-helix seems to be present in the region 800–865 which encompasses a highly conserved region between residues 852–862. Likewise a high probability for α-helix formation is noticed in the stretch 940–985 (all numbers refer to human AP-N). The binding site for the pig-specific coronavirus TGEV is located to residues 717–813 in the C subunit of pig AP-N (Delmas *et al.,* 1994). The human-specific coronavirus 229E, however, binds to quite another domain (residues 260–353) in the human AP-N (Kolb *et al.,* 1996). The sequences of both of these virus-binding stretches indicate a high probability for surface exposure (for more details of AP-N as a coronavirus receptor see Chapter 25).

11.4.2 Glycosylation sites, sulphation sites, S–S bridges

The stalk of AP-N is rich in Ser, Thr and Pro and contains at least two consensus-sequences for O-glycosylation (Elhammar *et al.*, 1993). At least one has recently been shown to be functional (Norén *et al.,* 1997). The catalytic domain of AP-N from all species so far analysed contains several sites for N-glycosylation, some of which are used (Danielsen,1982; Look *et al.,* 1985). By alignment of the different sequences, it is evident that most of the Asn-X-Thr/Ser N-glycosylation consensus sequences are differently located. Only three of the N-glycosylation sites of AP-N are conserved; one is also conserved between AP-N and AP-A. N-glycosylation is obligatory for the effective expression of AP-N, since addition of tunicamycin to intestinal explants abolished biosynthesis (Danielsen and Cowell, 1984). The biosynthesis was partially restored when a proteinase inhibitor (leupeptin) was added to the system, indicating that a proteolytic system effectively degrades non-glycosylated nascent peptide chains.

The catalytic domain of human AP-N also contains three predicted sites for tyrosine-sulphation (Tyr176, Tyr634 and Tyr692) and at least one has been shown to be functional (Danielsen, 1987). Disulphide bonds probably are of minor importance for the stabilization of the AP-N structure since only two cysteines (human AP-N residues Cys223 and Cys761, located on the B and C subunits) out of seven are found to be completely conserved between AP-N and AP-A. However, these two conserved cysteines do not form an S–S bridge, as subunits B and C of pig AP-N can be dissociated from each other without reduction (Sjöström and Norén, 1982). If any S–S bridges are present in the AP-N molecule, Cys113 and Cys223 (numbering of human AP-N) may be linked. It may be that Cys761, Cys768, Cys798 and Cys834 of subunit C of pig aminopeptidase (and corresponding stretches of rat, human and rabbit AP-N) are paired. However, no data concerning the presence of S–S bridges and free -SH groups on aminopeptidases are available in the literature.

11.4.3 The zinc-binding motif

The B subunit contains the zinc-binding motif HEXXH X_{18}E (start: His388, human AP-N) which defines the catalytic site of AP-N. This has been confirmed by site-directed mutagenesis changing one or two of the His residues into Ala or Leu (Delmas *et al.*, 1994). This motif is characteristic for the superfamily of metallopeptidases named gluzincins (Hooper, 1994) which encompasses the aminopeptidase family, including AP-A (see Chapter 12) and AP-N, leukotriene A_4-hydrolase (Funk *et al.*, 1987), thyrotrophin releasing hormone (TRH)-degrading enzyme (see Chapter 15), puromycin-sensitive aminopeptidase (Constam *et al.* 1995) and placental leucine aminopeptidase (vp165; oxytocinase; Keller *et al.,* 1995). The gluzincin superfamily also encompasses the thermolysin family, endopeptidase-24.11 family and the angiotensin-converting enzyme family (Hooper, 1994), differing only by the distance of the third zinc ligand from the HEXXH motif, indicating that all these enzymes have a common catalytic mechanism functioning as described for thermolysin (Matthews, 1988; Mock and Aksamawati, 1994), the only enzyme in this family so far crystallized.

11.5 Substrate specificity and inhibitors

11.5.1 Specificity

AP-N has a broad specificity and is most effective in the pH range 7–9. Thus it releases unblocked N-terminal L-amino acids, with the exception of proline, from oligopeptides. Aminoacylarylamide and nitro-anilide derivatives are often used as specific substrates for AP-N, as the hydrolysis of these compounds can easily be measured photometrically (Pfleiderer, 1970) and have the additional advantage that they are not hydrolysed by the cytosolic leucine aminopeptidase (Kohno *et al.,* 1986). However, other membrane-bound aminopeptidases, such as AP-A (see Chapter 12), oxytocinase, vp165 (Keller *et al.*, 1995), puromycin-sensitive aminopeptidase (Constam *et al.,* 1995) and TRH-degrading enzyme (Schauder *et al.,* 1994), are able to hydrolyse substrates of this type. AP-N is most effective on substrates with bulky hydrophobic and basic N-terminal residues and is less effective for acidic residues. K_m values for substrates with hydrophobic side chains are in the 0.1–0.3 mM range. AP-N also hydrolyses dipeptides (Gaertner *et al.,* 1981; Kania *et al.,* 1977; Kim *et al.,* 1976; Maroux *et al.,* 1973). Surprisingly, isopeptides, formed by the ε-amino group of lysine and the 4- or 5-carboxyl groups of aspartic and glutamic acids, that can constitute covalent crosslinks in proteins, are effectively hydrolysed by AP-N (Gaertner *et al.,* 1981). In a detailed analysis of substrate specificity, Kania *et al.* (1977) delineated the topography of the active site. They found that the active site has four substrate binding points, S_1, S_1',S_2', and S_3' [using the convention of Schechter and Berger (1967); see *Figure 1.2*]; k_{cat} is most effectively increased and K_m decreased by the binding of a substrate side chain to S_1, but substrate binding to S_1', S_2', and, to a lesser extent, S_3', positively influences the enzymic activity. Peptides longer than four residues do not additionally increase the enzymic activity.

Among natural peptides, it has been demonstrated that AP-N can hydrolyse Met-Lys- and Lys-bradykinin as well as [Met^5]- and [Leu^5]-enkephalins (Bausback and Ward, 1988; Giros *et al.,* 1986; Hersch, 1985; Sidorowicz *et al.,* 1981). In an extensive study using immunopurified human placental AP-N, Mitzutani *et al.,* (1993) analysed the ability of AP-N to hydrolyse a large series of biologically active peptides. The results confirmed the findings of other investigators who had reported AP-N hydrolysis of Met-Lys- and Lys-bradykinin and the enkephalins. In addition, it was reported that AP-N has the ability to hydrolyse thympentin, splenopentin, tuftsin, somatostatin and neurokinin A. However, other peptides such as the B chain of insulin, substance P, bradykinin and casein were unaffected, even after long periods of incubation (Kim *et al.,* 1976; Mitzutani *et al.,* 1993; Xu *et al.,* 1995).

11.5.2 Inhibitors

Chelating reagents such as EDTA and 1,10-phenanthroline inhibit AP-N, as is usual for most zinc metallopeptidases. Probestin is very potent (IC_{50} values in the range of 10 nM–50 nM), but also inhibits AP-A and aminopeptidase W (AP-W). Amastatin is relatively potent (IC_{50} values reported in the range of 150 nM–3 μM), but it also inhibits AP-A and AP-W. Bestatin is an order of magnitude less potent than amastatin. Actinonin is comparable with amstatin in potency and also fails to inhibit the two other aminopeptidases (Gee and Kenny, 1987; Matsas *et al.,* 1985; Tieku and Hooper, 1992), however it is not specific, endopeptidase-24.18, for example, being strongly inhibited (Choudry and Kenny, 1991). AP-N is inhibited by a general metallopeptidase inhibitor, JMV390–1 (K_i value, 1 μM; Kitabgi *et al.,* 1992). Recently, other inhibitors, more selective for AP-N and with K_i values in the range 0.01–0.05 μM have been designed (Schalk *et al.*, 1994) and it has been demonstrated that naturally occurring peptides, like substance P (K_i 0.5 μM), bradykinin (K_i 9 μM), as well as the morphine agonist morphiceptin (Xu *et al.*, 1995), can act as inhibitors. Some antibodies directed to human, pig and mouse AP-N are inhibitory (Delmas *et al.,* 1992; Hansen *et al.,* 1993; Look *et al.,* 1989; Matsas *et al.,* 1985).

11.6 Structure of the gene

Both the human and porcine AP-N genes have been cloned (Look *et al.,* 1986; Olsen *et al.*, 1989; Shapiro *et al.*, 1991). The coding part of the human gene is estimated to span approximately 35 kb and is composed of 20 exons (Lerche *et al.*, 1996; Look *et al.*, 1986). The human AP-N gene has been assigned to the long arm of chromosome 15 (Kruse *et al.*, 1988; Look *et al.*, 1986). The porcine gene is of similar size and has been localized to the long arm of chromosome 7 (Poulsen *et al.*, 1991). The sizes of the exons vary (78–666 bp) with the majority of the exons being in the range of 120–150 bp. Five regions of increased homology between AP-Ns from different species and AP-A are identified, indicating that these represent functional domains. These are located to exon 2, exons 5–6, exons 13–14, exons 17–18 and exon 20. In the human and pig genes, exon 1 encodes the cytosolic domain, the transmembrane domain and the stalk, indicating the existence of a functional genetic cassette. The zinc-binding motif HEXXH and the glutamic acid, 18 residues downstream, that is the third zinc ligand, are encoded on separate exons (exons 5 and 6, respectively). Thus, the organization of the gene encoding the active site is very similar for AP-N and AP-A (see Section 12.4).

Analysis of the transcriptional initiation sites used in different tissues has led to the identification of two different promoters, which control the AP-N gene (Olsen *et al.*, 1991; Shapiro *et al.*, 1991). The two promoters are arranged in tandem and are separated by approximately 8 kb (in the case of the human gene). The promoter which is closest to the coding part of the gene drives transcription in epithelial cell-types, such as enterocytes, kidney proximal tubule cells and hepatocytes (Olsen *et al.*, 1991; Shapiro *et al.*, 1991), and accordingly referred to as the *epithelial* promoter. The upstream promoter is active mainly in myeloid cell-types such as monocytes and myeloid leukaemic cells (Shapiro *et al.*, 1991) and this promoter is referred to as the *myeloid* promoter. Transcripts originating from the myeloid and the epithelial promoters encode the same protein. They differ only in the 5′ untranslated region. Transcriptional initiation from the epithelial promoter results from one major transcriptional initiation site, located 45 bp upstream of the initiator methionine codon (Olsen *et al.*, 1989; Shapiro *et al.*, 1991). Initiation from the myeloid promoter is heterogeneous with multiple initiation sites distributed within a region of 70 bp. The region immediately downstream of the myeloid promoter encodes a short non-translated exon, which is separated from the next exon by an intron measuring 8 kb. Thus mature mRNA originating from the human myeloid promoter is assembled from at least 21 exons.

11.7 Regulation of the gene

11.7.1 The myeloid promoter

The sequence of the myeloid promoter is CpG rich and lacks a TATA box (Shapiro *et al.*, 1991). Deletional analyses of the promoter have demonstrated that the region between positions –411 and –291 confers the specificity of the promoter in myeloid cells (Shapiro, 1995). This region contains one potential binding site for the Myb transcription factor and, using nuclear extracts from cells transfected with a Myb expression vector, it was demonstrated that Myb can bind to this site in electrophoretic mobility shift assays (Shapiro, 1995). In addition, the –411 to –291 region contains two potential binding sites for the Ets transcription factor and bacterially produced Ets was shown to bind to the region, although the precise binding site was not mapped (Shapiro, 1995). Co-transfection of Myb and Ets expression vectors with the myeloid promoter/reporter construct resulted in a synergistic activation, demonstrating the functional significance of the Myb and Ets binding sites (Shapiro, 1995).

The Myb transcription factor family contains three family members A-Myb, B-Myb and C-Myb, the latter being the most intensively investigated. The Myb proteins contain three functionally important domains: a DNA-binding domain, a transactivation domain and a negative regulation domain (Graf *et al.*, 1992; Luscher and Eisenman, 1990). The Ets transcription factor family is large and consists of members with a common DNA-binding motif – the Ets domain. The Myb proteins are mainly expressed in haematopoietic cells (Luscher and Eisenman, 1990), whereas the Ets proteins are found in a wide variety of cell types, although the individual members may show restricted tissue distribution (Wasylyk *et al.*, 1993). Both Myb and Ets are involved in the regulation of proliferation and differentiation of haematopoietic cells, moreover, Ets and Myb are found as a fusion protein in the avian erythroblastosis virus E26, which transforms immature haematopoietic cells into multipotent cells with the capacity to differentiate into both myeloid and erythroid cells (Graf *et al.*, 1992). Thus the cooperativity in the response of the myeloid promoter to the co-expression of Myb and Ets might explain the expression of AP-N seen in myeloid cell-types. The expression of AP-N in haematopoietic cells is not limited to myeloid cells, as expression in T cells has also been reported (Lendeckel *et al.*, 1994; Wex *et al.*, 1995). In these cells it has been demonstrated that the AP-N gene is induced by mitogenic activation.

11.7.2 The epithelial promoter

Structural analysis of the epithelial promoter has revealed four distinct *cis* elements (Olsen *et al.*, 1991). The UF element (positions –112 to –90), the HNF1 element (positions –85 to –58), the Sp1 element (positions –53 to –30) and the TATA box (positions –29 to –24). The UF element most likely interacts with the COUP transcription factor (Olsen *et al.*, 1995), which is a member of the steroid hormone receptor family (Wang *et al.*, 1989). Although the UF binding activity is found in many tissues (Olsen *et al.*, 1991, 1995), no function of the element could be demonstrated in transient transfection assays (Olsen *et al.*, 1991, 1994). In contrast, the deletional analysis of the epithelial promoter in intestinal (Caco-2) or hepatic (HepG2) cells showed that the HNF1 and Sp1 sites, in combination with the TATA box, are necessary and sufficient for full promoter activity. Moreover, single mutations of either of the three sites resulted in severely reduced transcription from the promoter (J. Olsen, unpublished results).

The TATA box, found in many RNA polymerase II-transcribed genes, interacts with the TATA box-binding protein, which forms a part of the general RNA polymerase II transcription machinery (for review see Latchman, 1995). The HNF1 element interacts with homo- and heterodimers of HNF1α and HNF1β. These two HNF1 family members are expressed in a variety of epithelial cell-types such as enterocytes, hepatocytes and kidney proximal tubular cells (for review, see Tronche and Yaniv, 1992). The HNF1 proteins form dimers in solution via an N-terminally located dimerization domain and they bind to DNA via an extra large homeodomain. HNF1 proteins are important for the transcription of a large number of genes, which are expressed in the liver, such as albumin, α1-antitrypsin and γ-fibrinogen genes (Tronche and Yaniv, 1992). In the small intestine, the HNF1 proteins are also involved in the regulation of microvillar glycosidase genes, such as sucrase–isomaltase (Wu *et al.*, 1994) and lactase–phlorizin hydrolase (T. Troelsen and N. Spodsberg, personal communication).

Sp1 is a ubiquitously expressed transcription factor, which binds to DNA via a region containing three zinc fingers (Kadonaga *et al.*, 1988) and it is important for the expression of a multitude of genes showing both restricted, as well as wider tissue distribution (Kadonaga *et al.*, 1986). The requirement of both Sp1 and HNF1 for epithelial promoter activity, in combination with the tissue distribution of these transcription factors, imply that the HNF1 element restricts the activity of the epithelial promoter to a limited number of epithelial cell-types expressing HNF1 proteins. Indirect confirmation of this was achieved from studies in the rabbit endometrium. In this tissue the expression of aminopeptidase N is up-regulated following human chorionic gonadotropin-induced pseudopregnancy (Classen-Linke *et al.*, 1987). The induction of AP-N was shown to be the result of increased level of AP-N mRNA. In parallel with this, a dramatic increase in the amounts of HNF1 was found in endometrial nuclei (Olsen *et al.*, 1995).

11.7.3 The AP-N enhancer

Recently a *bona fide* enhancer has been localized 2.7 kb upstream from the porcine epithelial promoter and Southern blot analysis suggests that the enhancer is conserved in the human gene as well (Olsen *et al.*, 1997). The enhancer was found to be active in liver, intestinal and myeloid cell types, but not in cervical or mammary carcinoma cells, suggesting that the enhancer from its position between the two AP-N promoters is involved in the cell-type specific transcriptional regulation of both promoters.

11.8 Pattern of expression

The enterocytes and the epithelial cells of the kidney proximal tubule express AP-N at very high levels (Kenny and Maroux, 1982). In the intestine, the level of AP-N increases from duodenum to ileum (Freeman, 1995; Skovbjerg, 1981). Along the crypt–villus axis, AP-N is best expressed at the base of the villi (Freeman, 1995; Hansen *et al.*, 1994; Norén *et al.*, 1989). By the use of monoclonal CD13 antibodies, AP-N has been shown to be present on a large variety of cells. Thus, AP-N was found on granulocyte-monocyte progenitors and on cells of the granulocytic and monocytic lineages at all stages of differentiation (Griffin *et al.*, 1981). It is also present on leukaemic blasts in acute myeloid leukaemias and by a smaller subset of acute lymphoid leukaemias. Bone marrow and endometrial stromal cells, osteoclasts, fibroblasts and hepatocytes also express AP-N (Imai *et al.*, 1992; Look *et al.*, 1989). In the nervous system AP-N is found on endothelial and synaptic membranes as well as on the surface of astrocytes and pericytes (Barnes *et al.*, 1994; Kunz *et al.*, 1994). AP-N is also present in human serum, being a proteolytic form released from the surface of cells expressing AP-N (Watanabe *et al.*, 1995).

11.9 Physiology and pathophysiology

AP-N is an ectoenzyme on the apical membrane of the enterocytes, where it constitutes about 5% of the microvillar membrane proteins. The localization and the abundance of the enzyme strongly imply that AP-N is functioning in the final phase of protein digestion by generating free amino acids and, to a lesser extent, di- and tripeptides which are then transported into the enterocyte. Likewise, it is believed that the function of AP-N in the proximal tubule cells is to hydrolyse peptides in the glomerular filtrate. The function of AP-N in the central nervous system (see Section 19.3.1) and in other cellular locations is less clear. It has been suggested that AP-N is a part of a series of peptidases that are able to modulate the effects of biologically active peptides such as enkephalins and tuftsin. AP-N might also participate in the elimination of the chemotactic peptide fMet–Leu–Phe during neutrophil-mediated inflammatory response (Painter *et al.,* 1988).

A very interesting role has been proposed for AP-N, as a component of the proteolytic machinery responsible for the processing of antigens to be presented to T cells (Mouritsen *et al.,* 1992). This is based on the finding that AP-N is present on the surface of MHC class II-bearing antigen-presenting cells (Hansen *et al.,* 1993) and that major histocompatibility complex (MHC) class II antigens can protect the core of a model peptide from degradation by AP-N. The proposition is further supported experimentally by the finding that an inhibitory monoclonal antibody to AP-N suppresses the T-cell stimulation via the presentation of a model peptide by a matching antigen-presenting cell line. In addition, metal-chelating agents, but not serine, cysteine or aspartate proteinase inhibitors, were found to inhibit the T-cell stimulation dramatically (Larsen *et al.,* 1996).

It has also been suggested that AP-N has a role in the tissue invasion by malignant cells. This is based on the observation that AP-N is present in patches on the surface of colorectal tumour cells and on malignant melanoma cells (Menrad *et al.,* 1993; Real *et al.,* 1992). A role in tissue invasion is further supported by the finding that AP-N is closely associated with the extracellular matrix and that an immunopurified AP-N preparation was able to partially degrade Matrigel® (Menrad *et al.,* 1993; Saiki *et al.,* 1993). Furthermore, the invasion of malignant melanoma, renal carcinoma and fibrosarcoma cell lines through an extracellular matrix-coated membrane was inhibited by monoclonal and polyclonal AP-N antibodies known also to inhibit AP-N. Similar results have been obtained by aminopeptidase inhibitors, not affecting the matrix metallo-proteinases (Fujii *et al.*, 1996; Menrad *et al.*, 1993; Saiki *et al.*, 1993). Penetration of tumour cells through basement membranes involves attachment of tumour cells and action by degradative enzymes localized on the surface of the tumour cells followed by migration of the cells into the target tissue. As AP-N-specific monoclonal antibodies had no effect on the attatchment of the tumour cells to the extracellular matrix and no effect on tumour cell motility, it was concluded that AP-N has a role in a degradative cascade of extracellular matrices (Saiki *et al.*, 1993) also involving the degradation of collagen IV. This conclusion is endorsed by the finding that overexpression of AP-N in a transfected human melanoma cell line increased its invasive capacity and type IV collagenolytic activity and also increased its ability to colonize lungs (Fujii *et al.*, 1995).

AP-N is present on tumour-infiltrating lymphocytes of some malignant tumours of lung and kidney (Riemann *et al.,* 1994). The functional importance of this finding is not clear, but AP-N on these cells might also be a part of a cascade facilitating tumour invasion. Alternatively, AP-N may degrade some biologically active peptides involved in the maintenance of tumour growth. AP-N is present on acute myeloid leukaemia cells (Griffin and Schlossman, 1984). It has been elegantly demonstrated that AP-N expression on leukaemic cells can be transcriptionally suppressed by culturing on bone marrow stroma cells (Saito *et al.*, 1995). As AP-N seems to have importance for tumour invasion, a defect in the suppressive signal-transduction pathway might have significant importance for the spreading capacity of a tumour. Thus, several studies support the notion that AP-N, in cooperation with other peptidases, has an important role in tumour invasion and is involved in the spreading and colonization of metastases.

AP-N seems to be an important molecule in early events of cytomegalovirus infection (Söderberg *et al.,* 1996a,b; see also Chapter 25). During infection, AP-N becomes attached to the cytomegalovirus particles and then MHC-restricted peptides, originating from AP-N, are believed to be presented to AP-N-non-tolerant resting B-cells. This causes circulating auto-antibodies found in a majority of cytomegalovirus-infected patients and might be a mechanism for viral induction of autoimmune manifestations connected to cytomegalovirus infections (Söderberg *et al.*, 1996a). It has been suggested that this autoimmunity may contribute to specific tissue damage during severe forms of chronic graft vs. host disease seen in bone marrow-transplanted patients (Söderberg *et al.,* 1996b).

References

Barnes K, Kenny AJ, Turner AJ. (1994) Localization of aminopeptidase N and dipeptidyl peptidase IV in pig striatum and in neuronal and glial cell cultures. *Eur. J. Neurosci.* **6:** 531–537.

Bausback HH, Ward PE. (1988) Kallidin and bradykinin metabolism by isolated cerebral microvessels. *Biochem. Pharmacol.* **37:** 2973–2978.

Booth A, Kenny AJ. (1980) Proteins of the kidney microvillar membrane. Asymmetric labelling of the membrane by lactoperoxidase-catalysed radioiodination and by photolysis of 3,5-di[^{125}I] iodo-4-azidobenzenesulphonate. *Biochem. J.* **187:** 31–44.

Chen H, Kinzer CA, Paul WE. (1996) P161, a murine membrane protein expressed on mast cells and some macrophages, is mouse cd13/aminopeptidase N. *J. Immunol.* **157:** 2593–2600.

Choudry Y, Kenny AJ. (1991) Hydrolysis of transforming growth factor-[alpha] by cell-surface peptidases *in vitro. Biochem. J.* **280:** 57–60

Classen-Linke I, Denker H-W, Winterhager E. (1987) Apical plasma membrane-bound enzymes of rabbit uterine epithelium. Pattern changes during the periimplantation phase. *Histochemistry* **87:** 517–529.

Constam DB, Tobler AR, Rensing-Ehl A, Kemler I, Hersh LB, Fontana A. (1995) Puromycin-sensitive aminopeptidase. Sequence analysis, expression and functional characterization. *J. Biol. Chem.* **270:** 26931–26939.

Danielsen EM. (1982) Biosynthesis of intestinal microvillar proteins. Pulse-chase labelling studies on aminopeptidase N and sucrase-isomaltase. *Biochem. J.* **204:** 639–645.

Danielsen EM. (1987) Tyrosine sulfation: a post-translational modification of microvillar enzymes in the small intestinal enterocyte. *EMBO J.* **6:** 2891–2896.

Danielsen EM. (1990) Biosynthesis of intestinal microvillar proteins. Dimerization of aminopeptidase N and lactase-phlorizin hydrolase. *Biochemistry* **29**: 305–308

Danielsen EM. (1992) Folding of intestinal brush border enzymes. Evidence that high-mannose glycosylation is an essential early event. *Biochemistry* **31:** 2266–2272.

Danielsen EM. (1995) Involvement of detergent-insoluble complexes in the intracellular transport of intestinal brush border enzymes. *Biochemistry* **34**: 1596–1605

Danielsen EM, Cowell GM. (1984) Biosynthesis of intestinal microvillar proteins. Further characterization of the intracellular processing and transport. *FEBS Lett.* **166:** 28–32.

Danielsen EM, Cowell GM. (1985) Biosynthesis of intestinal microvillar proteins. Evidence for an intracellular sorting taking place in, or shortly after, exit from the Golgi complex *Eur. J. Biochem.* **152:** 493–499.

Danielsen EM, Sjöström H, Norén O. (1982) Hypotonic elution, a new desorption principle in immunoadsorbent chromatography. *J. Immunol. Meth.* **52**: 223–232.

Danielsen EM, Cowell GM, Norén O, Sjöström H. (1984) Biosynthesis of microvillar proteins. *Biochem. J.* **221**: 1–14.

Delmas B, Gelfi J, L'Haridon G, Vogel LK, Sjöström H, Norén O, Laude H. (1992) Aminopeptidase N is a major receptor for the enteropathogenic coronavirus TGEV. *Nature* **357:** 417–420

Delmas B, Gelfi J, Kut E, Sjöström H, Norén O, Laude H. (1994) Determinants essential for the transmissible gastroenteritis virus-receptor interaction reside within a domain of aminopeptidase-N that is distinct from the enzymic site. *J. Virol.* **68:** 5216–5224.

Desnuelle P. (1979) Intestinal and renal aminopeptidase: a model of a transmembrane protein. *Eur. J. Biochem.* **101**: 1–11.

Elhammar ÅP, Poorman RA, Brown E, Maggiora LL, Hoogerheide JG, Kezdy FJ. (1993) The specificity of UDP-GalNAc: polypeptide *N*-acetylgalactosaminyltransferase as inferred from a database of *in vivo* substrates and from the *in vitro* glycosylation of proteins and peptides. *J. Biol. Chem.* **268:** 10029–10038.

Feracci H, Maroux S. (1980) Rabbit intestinal aminopeptidase N. Purification and molecular properties. *Biochim. Biophys. Acta* **599:** 448–463.

Feracci H, Rigal A, Maroux S. (1985) Biosynthesis and intracellular pool of aminopeptidase N in rabbit enterocytes. *J. Memb. Biol.* **83:** 139–146.

Freeman TC. (1995) Parallel patterns of cell-specific gene expression during enterocyte differentiation and maturation in the small intestine of the rabbit. *Differentiation* **59:** 179–192.

Fujii H, Nakajima M, Saiki I, Yoneda J, Azuma I, Tsuruo T. (1995) Human melanoma invasion and metastasis enhancement by high expression of aminopeptidase N/CD13. *Clin. Exp. Metastasis* **13:** 337–344.

Fujii H, Nakajima M, Aoyagi T, Tsuruo T. (1996) Inhibition of tumor cell invasion and matrix degradation by aminopeptidase inhibitors. *Biol. Pharmacol. Bull.* **19:** 6–10.

Funk CD, Radmark O, Fu JY, Matsumoto T, Jornvall H. Shimuzu T, Samuelson B. (1987) Molecular cloning and amino acid sequence of leukotriene A4 hydrolase. *Proc. Natl Acad. Sci. USA* **84:** 6677–6681.

Gaertner H, Puigserver A, Maroux S. (1981) Hydrolysis of the isopeptide bond of ϵ-*N*-L-methionyl-L-lysine by intestinal aminopeptidase N. *FEBS Lett.* **133:** 135–138.

Gee NS, Kenny AJ. (1987) Proteins of the kidney microvillar membrane. Enzymic and molecular properties of aminopeptidase W. *Biochem. J.* **246:** 97–102.

George SG, Kenny AJ. (1973) Studies on the enzymology of purified preparations of brush border from rabbit kidney. *Biochem. J.* **134:** 43–57.

Giros B, Gros C, Solhonne B, Schwartz JC. (1986) Characterization of aminopeptidases responsible for inactivating endogenous [Met5] enkephalin in brainslices using peptidase inhibitors and anti aminopeptidase M inhibitors. *Mol. Pharmacol.* **29:** 281–287.

Graf T, McNagny K, Brady G, Frampton J. (1992) Chicken 'erythroid' cells transformed by the Gag-Myb-Ets-encoding E26 leukemia virus are multipotent. *Cell* **70:** 201–213.

Griffin JD, Ritz J, Nadler LM, Schlossman SF. (1981) Expression of myeloid differentiation antigens on normal and malignant myeloid cells. *J. Clin. Invest.* **68:** 932–941.

Griffin JD, Mayer RJ, Weinstein HJ, Rosenthal DS, Beveridge RP, Schlossman SF. (1983) Surface analysis of acute myeloblastic leukaemia: identification of differentation-associated phenotypes. *Blood* **62:** 557–563.

Hansen GH, Sjöström H, Norén O, Dabelsteen E. (1987) Immunomicroscopic localization of aminopeptidase N in the pig enterocyte. Implications for the route of intracellular transport. *Eur. J. Cell Biol.* **43:** 252–259.

Hansen AS, Norén O, Sjöström H, Werdelin O. (1993) Mouse aminopeptidase N is a marker for antigen presenting cells and appears to be co-expressed with major histocompatibility complex class II molecules. *J. Immunol.* **23:** 2358–2364.

Hansen GH, Niels-Christiansen LL, Poulsen MD, Norén O, Sjöström H. (1994) Distribution of three microvillar enzymes along small intestinal crypt–villus axis. *J. Submicros. Cytol. Pathol.* **26:** 453–460.

Hanson JH, Frohne M. (1976) Crystalline leucine aminopeptidase from lens. *Meth. Enzymol.* **45**: 504–510.

Hauri HP, Schweizer A. (1992) The endoplasmic reticulum–Golgi intermediate compartment. *Curr. Opin. Cell Biol.* **4:** 600–608.

Hersh LB. (1985) Characterization of membrane-bound aminopeptidases from rat brain: identification of the enkephalin-degrading aminopeptidase. *J. Neurochem.* **44:** 1427–1435.

High S. (1995) Protein translocation at the membrane of the endoplasmic reticulum. *Prog. Biophys. Mol. Biol.* **63:** 233–250.

Hooper NM. (1994) Families of zinc metalloproteinases. *FEBS Lett.* **354:** 1–6.

Hussain MM, Tranum-Jensen J, Norén O, Sjöström H, Christiansen K. (1981) Reconstitution of purified amphiphilic pig intestinal microvillus aminopeptidase. Mode of membrane insertion and morphology. *Biochem. J.* **199**: 179–186.

Imai K, Kanzaki H, Fujiwara H, Kariya M, Okamoto N, Takakura K, Maeda M, Mori T. (1992) Expression of aminopeptidase N and neutral endopeptidase on the endometrial stromal cells in endometriosis and adenomyosis. *Hum. Repro.* **7:** 1326–1328.

Kadonaga JT, Katherine A, Tjian R. (1986) Promoter-specific activation of RNA polymerase II transcription by Sp1. *Trends Biochem. Sci.* **11:** 20–23.

Kadonaga JT, Courey AJ, Ladika J, Tjian R. (1988) Distinct regions of Sp1 modulate DNA binding and transcriptional activation. *Science* **242:** 1566–1570.

Kania RK, Santiago NA, Gray GM. (1977) Intestinal surface amino-oligopeptidases. II Substrate kinetics and and topography of the active site. *J. Biol. Chem.* **252:** 4929–4934.

Keller SR, Scott HM, Mastick CC, Aebersold R, Lienhard GE. (1995) Cloning and characterization of a novel insulin-regulated aminopeptidase from Glut-4 vesicles. *J. Biol. Chem.* **270:** 23612–23618.

Kenny AJ, Booth AG. (1978) Microvilli: their ultrastructure, enzymology and molecular organization. *Essays Biochem.* **14:** 1–44.

Kenny AJ, Maroux S. (1982) Topology of microvillar membrane hydrolases of kidney and intestine. *Physiol. Rev.* **62:** 91–128.

Kim YS, Brophy EJ, Nicholsen A. (1976) Rat intestinal brush border membrane peptidases. II. Enzymic properties, immunochemistry, and interaction with lectins of two different forms of the enzyme. *J. Biol. Chem.* **251:** 3206–3212.

Kitabgi P, Dubuc I, Nouel D, Costentin J, Cuber JC, Fulerand H, Doulut S, Rodriguez M, Martinez J. (1992) Effects of thiophan, bestatin and a novel metallopeptidase inhibitor, JMV390–1, on the recovery of neurotensin and neuromedin N released from mouse hypothalamus. *Neurosci. Lett.* **142:** 200–204.

Kohno H, Kanda S, Kanno T. (1986) Immunoaffinity purification and characterization of leucine aminopeptidase from human liver. *J. Biol. Chem.* **261:** 10744–10748.

Kolb AF, Maile J, Heister A, Siddell SG. (1996) Characterization of functional domains in the human coronavirus HCV 229E receptor. *J. Gen. Virol.* **77:** 2515–2521.

Kruse TA, Bolund, L, Grzeschik KH, Ropers HH, Olsen J, Sjöström H, Norén O. (1988) Assignment of the human aminopeptidase N (peptidase E) gene to chromosome 15q13-qter. *FEBS Lett.* **239:** 305–308.

Kundu A, Avalos RT, Sanderson CM, Nayak DP. (1996) Transmembrane domain of influenza virus neuraminidase, a type II protein, possessing an apical sorting signal in polarized MDCK cells. *J. Virol.* **70:** 6508–6515

Kunz J, Krause D, Kremer M, Dermietzel R. (1994) The 140-kDa protein of blood-brain barrier-associated pericytes is identical to aminopeptidase N. *J. Neurochem.* **62:** 2375–2386.

Larsen SL, Pedersen LØ, Buus S, Stryhn A. (1996) T cell responses affected by aminopeptidase N (CD13)-mediated trimming of major histocompatibility complex class II-bound peptides. *J. Exp. Med.* **184:** 183-189.

Latchman DS. (1995) Transcription factors and constitutive transcription. In: *Eukaryotic Transcription Factors* (ed. D.S. Latchman), pp. 45–70. Academic Press, London.

Lendeckel U. Wex T, Kahne T, Frank K, Reinhold D, Ansorge S. (1994) Expression of the aminopeptidase N (CD13) gene in the human T cell lines HuT78 and H9. *Cell. Immunol.* **153:** 214-226.

Lerche, C, Vogel LK, Shapiro, LH, Norén O, Sjöström H. (1996) Human aminopeptidase N is encoded by 20 exons. *Mamm. Genome* **7:** 712-713.

Look AT, Peiper SC, Rebentisch MB, Ashmun RA, Roussel MF, Rettenmier CW, Sherr CJ. (1985) Transfer and expression of the gene encoding a human myeloid membrane antigen (gp150). *J. Clin. Invest.* **75:** 569- 579.

Look AT, Peiper SC, Rebentisch MB, Ashmun RA, Roussel MF, Lemons RS, Le Beau MM, Rubin CM, Sherr CJ. (1986) Molecular cloning, expression, and chromosomal localization of the gene encoding a human myeloid membrane antigen (gp150). *J. Clin. Invest.* **78:** 914–921.

Look AT, Ashmun RA, Shapiro LH, Peiper SC. (1989) Human myeloid plasma glycoprotein CD13 (gp150) is identical to aminopeptidase N. *J. Clin. Invest.* **83:** 1299–1307.

Louvard D, Maroux S. (1976) On the hydrophobic part of aminopeptidase and maltases which bind the enzyme to the intestinal brush border membrane. *Biochim. Biophys. Acta* **419:** 189–195.

Louvard D, Maroux S. Baratti J, Desnuelle P, Mutaftschiev S. (1973) On the preparation and some properties of closed membrane vesicles from hog duodenal and jejunal brush border. *Biochim. Biophys. Acta* **291:** 747–763.

Luscher B, Eisenman RN. (1990) New light on Myc and Myb. Part II. Myb. *Genes Devel.* **4:** 2235–2241.

McDonald JK, Barrett AJ. (1986) Microsomal alanyl aminopeptidase. In: *Mammalian Proteinases* Vol. 2, pp. 59–71. Academic Press, New York.

Malfroy B, Kado-Fong H, Gros C, Giros B, Schwartz J-C, Helmiss R. (1989) Molecular cloning and amino acid sequence of rat kidney aminopeptidase M: a member of a super family of zinc-metallohydrolases. *Biochem. Biophys. Res. Commun.* **161:** 236–241.

Maroux S, Louvard D, Baratti J. (1973) The aminopeptidase from hog intestinal brush border. *Biochem. Biophys. Acta* **321:** 282–295.

Matsas R, Stephenson SL, Hryszko J, Kenny AJ, Turner AJ. (1985) The metabolism of neuropeptides. Phase separation of synaptic membrane preparations with Triton X-114 reveals the presence of aminopeptidase N. *Biochem. J.* **231:** 445–449.

Matthews BW. (1988) Structural basis of the action of thermolysin and related Zn-peptidases. *Acc. Chem. Res.* **21:** 333–340.

Menrad A, Speicher D, Wacker J, Herlyn M. (1993) Biochemical and functional characterization of aminopeptidase N expressed by human melanoma cells. *Cancer Res.* **53:** 1450–1455.

Mizutani S, Goto K, Nomura S, Ino K, Goto S, Kikkawa F, Kurauchi O, Goldstein G, Tomoda Y. (1993) Possible action of human placental aminopeptidase N in feto-placental unit. *Res. Commun. Chem. Path. Pharmacol.* **82:** 65–80.

Mock W, Aksamawati M. (1994) Binding to thermolysin of phenolate-containing inhibitors necessitates a revised mechanism of catalysis. *Biochem J.* **302:** 57–68.

Mouritsen S, Meldal M, Werdelin O, Hansen AS, Buus S. (1992) MHC molecules protect T cell epitopes against proteolytic destruction. *J. Immunol.* **149:** 1987–1993.

Norén O, Sjöström H. (1980) The insertion of pig microvillus aminopeptidase into the membrane as probed by [^{125}I]iodonaphthylazide. *Eur. J. Biochem.* **104:** 25–31.

Norén O, Sjöström H, Danielsen EM Cowell GM, Skovbjerg H. (1986) The enzymes of the enterocyte plasma membrane. In: *Molecular and Cellular Basis of Digestion* (eds P. Desnuelle, H. Sjöström and O. Norén) pp. 335–365. Elsevier, Amsterdam.

Norén O, Dabelsteen E, Hoyer PE, Olsen J, Sjöström H, Hansen GH. (1989) Onset of transcription of the aminopeptidase N (leukemia antigen CD13) gene at the crypt–villus axis transition zone during rabbit enterocyte differentiation. *FEBS Lett.* **259**: 107–112.

Norén K, Hansen GH, Clausen H, Norén O, Sjöström H, Vogel LK. (1997) Defectively N-glycosylated and non-O-glycosylated aminopeptidase N (CD13) is normally expressed at the cell surface and has full enzymic activity. *Exp. Cell Res.* **231:** 112–118.

Olsen J, Sjöström H, Norén O. (1989) Cloning of the pig aminopeptidase N gene. Identification of possible regulatory elements and the exon distribution in relation to the membrane-spanning region. *FEBS Lett.* **251:** 275–281.

Olsen, J, Laustsen L, Kärnström U, Sjöström H, Norén O. (1991) Tissue-specific interactions between nuclear proteins and the aminopeptidase N promoter. *J. Biol. Chem.* **266:** 18089–18096.

Olsen J, Cowell GM, Kønigshøfer E *et al.* (1988) Complete amino acid sequence of human intestinal aminopeptidase N as deduced from cloned cDNA. *FEBS Lett.* **238**: 307–314.

Olsen J, Laustsen, L, Troelsen J. (1994) HNF1a activates the aminopeptidase N promoter in intestinal (Caco-2) cells. *FEBS Lett.* **342:** 325–328.

Olsen J, Classen-Linke I, Sjöström H, Norén O. (1995) Pseudopregnancy induces the expression of hepatocyte nuclear factor-1 beta and its target gene aminopeptidase N in rabbit endometrium via the epithelial promoter. *Biochem. J.* **312:** 31–37.

Olsen J, Kokholm K, Troelsen JT, Lausten L. (1997) An enhancer with cell-type dependent activity is located between the myeloid and the epithelial aminopeptidase N promoters. *Biochem. J.* **232:** 899–908.

Painter RG, Dukes R, Sullivan J, Carter R, Erdös EG, Johnson AR. (1988) Function of neutral endopeptidase on the cell membrane of human neutrophils. *J. Biol. Chem.* **263:** 9456–9461.

Pfleiderer G. (1970) Particle-bound aminopeptidase from pig kidney. *Meth. Enzymol.* **19:** 514–521.

Pfleiderer G, Celliers PG. (1963) Isolierung einer aminopeptidase aus nierenpartikeln. *Biochem. Z.* **339:** 186-189.

Poulsen PH, Thomsen PD, Olsen J. (1991) Assignment of the porcine aminopeptidase N (PEPN) gene to chromosome 7cen–q21. *Cytogen. Cell Genet.* **57:** 44–46.

Real FX, Xu M, Vila MR, de Bolos C. (1992) Intestinal brush border associated enzymes: co-ordinated expression in colorectal cancer. *Int. J. Cancer* **51:** 173–181.

Rehfeld N, Peters JE, Giesecke H, Beier L, Haschen RJ. (1967) Studies on aminoacid-arylamidases. I. Distribution and isoenzymes of the aminoacid-arylamidase in the human organism. *Acta Biol. Med. Germ..* **19:** 809–818.

Riemann D, Gohring B, Langner J. (1994) Expression of aminopeptidase N/CD13 in tumour-infiltrating lymphocytes from human renal cell carcinoma. *Immunol. Lett.* **42:** 19–23.

Saiki I, Fujii H, Yoneda J, Abe F, Nakajima M, Tsuruo T, Azuma I. (1993) Role of aminopeptidase N (CD13) in tumor-cell invasion and extracellular matrix degradation. *Int. J. Cancer* **54:** 137–143.

Saito M, Kumagai M, Okazaki T, Nakazawa S, Shapiro LH, Look AT, Campana D. (1995) Stromal cell-mediated transcriptional regulation of the CD13/aminopeptidase N gene in leukemic cells. *Leukemia* **9:** 1508-1516.

Sanderink G-J, Artur Y, Siest G. (1988) Human aminopeptidases: a review of the literaure. *J. Clin. Chem. Clin. Biochem.* **26:** 795–807.

Schalk C, d'Orchimont H, Jauch C. (1994) 3-Amino-2-tetralone derivatives: novel and selective inhibitors of aminopeptidase M (EC 3.4.11.2). *Arch. Biochem. Biophys.* **311:** 42–46.

Schauder B, Schomburg L, Köhrle J. Bauer K. (1994) Cloning of a cDNA encoding an ectoenzyme that degrades thyrotropin-releasing hormone. *Proc. Natl Acad. Sci. USA* **91:** 9534–9538.

Schechter I, Berger A. (1967) On the size of the active site in proteases. I. Papain. *Biochem. Biophys. Res. Commun.* **27:** 157–162.

Semenza G. (1986) Anchoring and biosynthesis of stalked brush border membrane proteins: glycosidases and peptidases of enterocytes and renal tubuli. *Annu. Rev. Cell Biol.* **2:** 255–313.

Shapiro, LH. (1995) Myb and Ets proteins cooperate to transactivate an early myeloid gene. *J. Biol. Chem.* **270:** 8763–8771.

Shapiro LH, Ashmun RA, Roberts WM, Look AT. (1991) Separate promoters control transcription of the human aminopeptidase N gene in myeloid and intestinal epithelial cells. *J. Biol. Chem.* **266:** 11999–12007.

Sidorowicz W, Zownir O, Behal FJ. (1981) Action of human pancreas alanine aminopeptidase on biologically active peptides: kinin converting activity. *Clin. Chim. Acta* **111:** 69–79

Sjöström H, Norén O. (1982) Changes of the quaternary structure of microvillus aminopeptidase in the membrane. *Eur. J. Biochem.* **122:** 245–250.

Sjöström H, Norén O, Jeppesen L, Staun M, Svensson B, Christiansen L. (1978) Purification of different amphiphilic forms of a microvillus aminopeptidase from pig small intestine using immunoadsorbent chromatography. *Eur J. Biochem.* **88:** 503–511.

Sjöström H, Norén O, Danielsen EM, Skovbjerg H. (1983) Structure of microvillar enzymes in different phases of their life cycles. In: *Brush Border Membranes,* vol. 95 (eds R. Porter and G. Collins), pp. 50–69. Pitman, London.

Sjöström H, Norén O, Danielsen EM. (1985) Enzymic activity of "high-mannose" glycosylated forms of intestinal microvillar hydrolases. *J. Ped. Gastroenterol. Nutr.* **4:** 980–983.

Skovbjerg H. (1981) Immunoelectrophoretic studies on human small intestinal brush border proteins. The longitudinal distribution of peptidases and disaccharidases. *Clin. Chim. Acta* **112:** 205–212.

Söderberg C, Sumitran-Karuppan S, Ljungman P, Möller E. (1996a) CD13-specific autoimmunity in cytomegalovirus-infected immunocompromised patients. *Transplantation* **61:** 594–600.

Söderberg C, Larsson S, Rozell BL, Sumitran-Karuppan S, Ljungman P, Möller E. (1996b) Cytomegalovirus-induced CD13-specific autoimmunity – a possible cause of chronic graft-versus-host disease. *Transplantation* **61:** 600–609.

Spiess M, Brunner J, Semenza G. (1982) Hydrophobic labeling, isolation, and partial characterization of the NH_2-terminal membranous segment of sucrase-isomaltase complex. *J. Biol. Chem.* **257:** 2370–2377.

Svensson B, Sjöström H, Norén O. (1982) Structural studies on microvillus aminopeptidase from pig small intestine. *Eur. J. Biochem.* **126:** 481–488.

Taylor A. (1993) Aminopeptidases: structure and function. *FASEB J.* **7:** 290–298.

Tieku S, Hooper NM. (1992) Inhibition of aminopeptidases N, A and W. A re-evaluation of the action of bestatin and inhibitors of angiotensin converting enzyme. *Biochem. Pharmacol.* **44:** 1725–1730.

Tresnan DB, Levis R, Holmes KV. (1996) Feline aminopeptidase N serves as a receptor for feline, canine, porcine and human coronaviruses in serogroup I. *J. Virol.* **70:** 8669–8674.

Tronche F, Yaniv M. (1992) HNF1, a homeoprotein member of the hepatic transcription regulatory network. *BioEssays* **14:** 579–587.

Vogel LK, Spiess M, Sjöström H, Norén O. (1992a) Evidence for an apical sorting signal on the ectodomain of human aminopeptidase N. *J. Biol. Chem.* **267:** 2794–2797.

Vogel LK, Norén O, Sjöström H. (1992b) The apical sorting signal on human aminopeptidase N is not located in the stalk but in the catalytic head group. *FEBS Lett.* **308:** 14–17.

Vogel LK, Norén O, Sjöström H. (1995) Transcytosis of aminopeptidase N in Caco-2 cells is mediated by a non-cytoplasmic signal. *J.Biol.Chem.* **270:** 22933–22938.

Wacker H, Lehky P, Fischer EM, Stein EA. (1971) Physical and chemical characterization of pig kidney particulate aminopeptidase. *Helv. Chim. Acta* **54:** 473–485.

Wacker H, Lehky P, Vanderhaeghe F, Stein EA. (1976) On the subunit structure of particulate aminopeptidase from pig kidney. *Biochim. Biophys. Acta* **429:** 546–554.

Wang LH, Tsai SY, Cook RG, Beattie WG, Tsai MJ, O'Malley BW. (1989) COUP transcription factor is a member of the steroid receptor superfamily. *Nature* **340:** 163–166.

Wasylyk B, Hahn SL, Giovane A. (1993) The Ets family of transcription factors. [Erratum in *Eur. J. Biochem.* (1993) **215:** 907.] *Eur. J. Biochem.* **211:** 7–18.

Watanabe Y, Iwaki-Egawa S, Mizukoshi H, Fujimoto Y. (1995) Identification of an alanine aminopeptidase in human maternal serum as a membrane-bound aminopeptidase N. *Biol. Chem. Hoppe-Seyle* **376:** 397–400.

Watt VM, Yip CC. (1989) Amino acid sequence deduced from a rat kidney cDNA suggests it encodes the Zn-peptidase aminopeptidase N. *J. Biol. Chem.* **264:** 5480–5487.

Wex T, Lendeckel U, Wex H, Frank K, Ansorge S. (1995) Quantification of aminopeptidase N mRNA in T cells by competitive PCR. *FEBS Lett.* **374:** 341–344.

Wu GD, Chen, L, Forslund K, Traber PG. (1994) Hepatocyte nuclear factor-1 alpha (HNF-1 alpha) and HNF-1 beta regulate transcription via two elements in an intestine-specific promoter. *J. Biol. Chem.* **269:** 17080–17085.

Xu Y, Wellner D, and Scheinberg D. (1995) Substance P and bradykinin are natural inhibitors of CD13/aminopeptidase N. *Biochem. Biophys. Res. Comm.* **208:** 664–674.

Yang XF, Milhiet PE, Gaudoux F, Crine P, Boileau G. (1993) Complete sequence of rabbit kidney aminopeptidase N and mRNA localization in rabbit kidney by *in situ* hybridization. *Biochem. Cell Biol.* **7:** 278–287.

Chapter 12

Structure and function of aminopeptidase A

Jiyang Wang and Max D. Cooper

12.1 Introduction

The use of monoclonal antibodies over the past few years has allowed the identification and characterization of many cell-surface glycoproteins that are expressed during defined stages in lymphoid and myeloid development. Biochemical and molecular analysis of these glycoproteins has revealed some of them to be membrane-associated enzymes. One such enzyme, the murine B-lymphocyte differentiation antigen BP-1/6C3, has been identified as aminopeptidase A (AP-A or glutamyl aminopeptidase, EC 3.4.11.7), that catalyses the release of acidic amino acid residues from the amino terminus of protein or peptide substrates. Although the BP-1/6C3 antigen is expressed by cells in many different tissues, such as the kidney and intestine, among haematopoietic cell lineages, it is preferentially expressed by progenitor and precursor B cells. Recent studies suggest that BP-1/6C3 antigen may serve not only as an early B-lineage marker, but may also be involved in the differentiation of these cells. In this chapter, we summarize information on the tissue distribution, biochemical and molecular characteristics and gene organization of AP-A/BP-1/6C3.

12.2 Tissue distribution of AP-A

The BP-1 antigen was identified initially as a cell-surface glycoprotein expressed by murine pre-B cells (Cooper *et al.*, 1986). A leukaemic pre-B-cell antigen identified by the 6C3 antibody (Pillemer *et al.*, 1984) was found to be the same molecule (Wu *et al.*, 1989). Immunohistochemical staining with the BP-1 antibody confirmed the wide tissue distribution indicated previously by histo-enzymic assay of AP-A activity (*Table 12.1;* Li *et al.*, 1993b, and references therein). AP-A is highly expressed on the intestinal brush border of the small intestine, renal glomeruli, proximal renal tubules, and vascular endothelium in many organs. Other tissues expressing the BP-1/AP-A antigen include stromal cells in the thymus cortex, bile canaliculi in the liver, gall bladder epithelium, pulmonary alveolar cells, interlobular ducts in the pancreas, the ovarian theca interna, the basement membrane of the epididymis and the splanchnopleure region of the placenta. AP-A enzymic activity has been detected in most of these

Cell-Surface Peptidases in Health and Disease, edited by A.J. Kenny and C.M. Boustead.

Table 12.1. Correlation of BP-1 antigen distribution with reported AP-A activity

Tissues	BP-1 antibody staining	AP-A activity by histochemical activity
Small intestine	Enterocyte brush border	Same
Kidney	Glomeruli	Same
	Proximal tubule brush border	
	Juxtaglomerular apparatus	
Thymus	Cortical stromal cells	Same
Lung	Alveoli wall	Same
Liver	Bile canaliculi	Same
Gall bladder	Epithelium	ND
Pancreas	Interlobular ducts	Same
	Interlobular artery	
Salivary gland	Not seen	Intercalated and striated duct epithelial cells
Ovary	Theca intema	Same
Placenta	Splanchnopleure	Not found
	Spongiotrophoblast	
Epididymis	Basement membrane	ND
Vesicular gland	ND	Epithelial cells
Brain	Leptomeninges	Same
	ND	Choroid plexus
Eye	Choroidea	Same
	Retina	
Aorta	Media	Muscle cells of media
Multiple organs	Capillary endothelium media	Capillary endothelium and/or muscle cells

locations, in keeping with identification of the BP-1/6C3 antigen as glutamyl aminopeptidase.

Among haematopoietic cells, AP-A expression is lineage- and differentiation stage-specific (*Figure 12.1*; Cooper *et al.*, 1986). It is expressed on early B-lineage cells including both progenitor (pro-B) and precursor (pre-B) cells. The BP-1 antigen, in combination with B220, CD43 and HSA antigens, serves as a useful marker that subdivides early B-cell differentiation stages. Pro-B cells can be subdivided into fractions A, B and C, based on their differential expression of B220, CD43, HSA and BP-1 antigens (Hardy *et al.*, 1991). *In vitro* analysis using the isolated subpopulations of bone marrow cells suggests that pro-B-cell differentiation proceeds from fraction A (B220+CD43+HSA-BP-1–) to fraction B (B220+CD43+HSA+BP-1–) to fraction C (B220+CD43+HSA+BP-1+) and then on to the pre-B-cell stage (fraction D, B220+CD43-HSA+BP-1+; Hardy *et al.*, 1991). The expression of AP-A/BP-1 is dramatically down-regulated in IgM+ mature B cells and remains extinguished in terminally differentiated plasma cells (Cooper *et al.*, 1986). BP-1 expression is thus confined to the pro-B, pre-B, and immature B-cell stages in the B-cell pathway. BP-1/6C3 is also expressed by bone marrow and thymic stromal cells (Li *et al.*, 1993b; Whitlock *et al.*, 1987). This expression pattern suggests that BP-1/6C3/AP-A may function during early B- and T-cell differentiation.

Growth and differentiation of pro-B and pre-B cells are regulated by interactions with stromal cells and their cytokine products (Kincade, 1994). The stromal cell

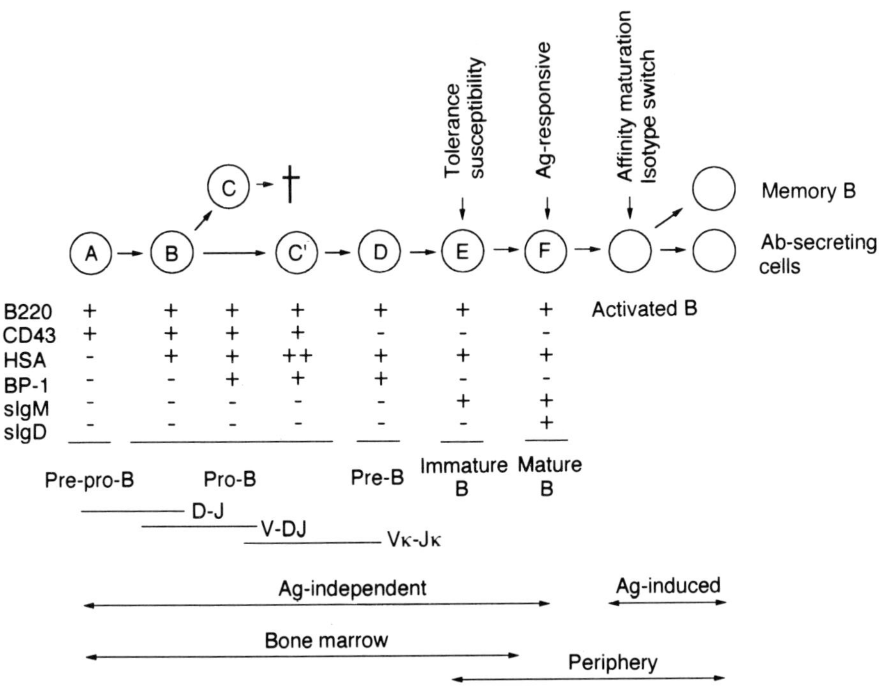

Figure 12.1. Regulated expression of AP-A/BP-1 during B-cell development.

cytokine interleukin-7 (IL-7) (Namen *et al.*, 1988) is required for both B- and T-lymphocyte development in mice. Interestingly, IL-7 treatment induces both proliferation of early B-lineage cells and increased expression of AP-A, suggesting that this ectoenzyme might be involved in regulating the proliferation of pre-B cells (Sherwood and Weissman, 1990; Welch *et al.*, 1990). Virus-transformed pre-B cells also express high levels of the BP-1/6C3 antigen (Cooper *et al.*, 1986; Pillemer *et al.*, 1984), suggesting a link between elevated AP-A expression and cell growth. However, while IL-7 induces both pre-B-cell growth and increased AP-A expression, the type I interferons α and β (IFNα/β) up-regulate BP-1/AP-A expression while inhibiting IL-7-induced proliferation (Wang *et al.*, 1996). This indicates that up-regulated BP-1 expression can be disassociated from pre-B-cell growth.

12.3 Biochemical and molecular characterization

AP-A is a homodimeric, phosphorylated integral membrane protein composed of two disulphide-linked 140 kDa subunits. When AP-A molecules immunoprecipitated by the BP-1 antibody are treated with endoglycosidase F to remove N-linked oligosaccharides, the size of the BP-1/AP-A monomer is reduced to 115 kDa (Cooper *et al.*, 1986). The deduced amino acid sequence of BP-1/AP-A cDNAs indicates that it is a type II integral membrane protein composed of 945 amino acids (Wu *et al.*, 1990; *Figure 12.2*).

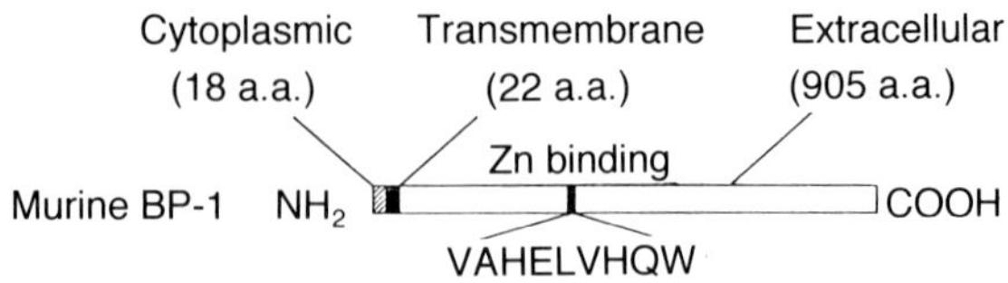

Figure 12.2. Schematic illustration of the predicted murine AP-A/BP-1/6C3 protein structure. a.a., amino acid residues.

The predicted size of 108 kDa correlates well with the actual protein size after removal of N-linked oligosaccharides. BP-1 molecules immunoprecipitated from murine pre-B cells exhibit AP-A activity that is inhibitable by AP-A inhibitors (Wu *et al.*, 1991) and recombinant mouse and human BP-1 exhibit AP-A activity (Li *et al.*, 1993a; Wang and Cooper, 1993).

AP-A, AP-N and LTA4 hydrolase share a highly conserved zinc-coordination motif, His-Glu-Xaa-Xaa-His-(18 amino acid residues)-Glu. This zinc-binding motif is shared by many zinc metalloproteases other than the aminopeptidases. These include thermolysin (a neutral protease), elastase, endopeptidase-24.11 and others (*Figure 12.3*; Vallee and Auld, 1989, 1990). The zinc-binding motif has even been found in *Avena sativa* phytochromes 3 and 4, putative zinc-binding proteins, and in tetanus toxin and melanotransferrin (Jongeneel *et al.*, 1989). The zinc-binding motif may not be related *per se* to substrate specificity, but rather may reflect an evolutionary relationship among these proteins. For example, leucyl aminopeptidase, unlike AP-A and AP-N, uses different zinc coordination ligands. Examination of the thermolysin crystal structure revealed that the two histidine residues in the zinc-binding motif (His142 and His146), one

	142 146	166
Thermolysin *(B. thermoproteolyticus)*	VVAHELTHAVT	GAINEAISD
Thermolysin *(B. stearothermophilus)*	VVGHELTHAVT	GAINEAMSD
Neutral protease *(B. subtilis)*	VTAHEMTHGVT	GALNESFSD
Elastase *(P. aeruginosa)*	VAAHEVSHGFT	GGMNEAFSD
Neutral endopeptidase (human, rat and rabbit)	VIGHEITHGFD	NTLGENIAD
Microsomal alanyl AP (AP-N) (human and rat)	VIAHELAHQWF	LWLNEGFAS
Microsomal alanyl AP (AP-N) *(E. coli)*	VIGHEYFHNWT	LSLKEGLTV
Glutamyl AP (AP-A) (human and mouse)	VVAHELVHQWF	LWLNEGFAS
LTA4 hydrolase (human and mouse)	VIAHEISHSWT	FWLNEGHTV

Figure 12.3. Comparison of zinc ligands in the zinc-binding motifs of various metallopeptidases. The numbers represent the position of the three zinc ligands (shown in bold) in thermolysin (His142, His146 and Glu166).

glutamic acid residue 20 amino acids downstream (Glu166), and a water molecule together bind a Zn^{2+} to form a catalytic zinc motif. The glutamic acid residue between the two histidines appears to be the active catalyst (Vallee and Auld, 1990). Mutation of one of the histidine residues abolishes both enzymic activity and zinc binding, while mutation of Glu143 abolishes enzymic activity without affecting zinc binding. AP-A is highly homologous with thermolysin only in the zinc-binding motif. Mutation of the second histidine residue to phenylalanine completely abolishes AP-A activity, thus confirming the importance of this motif in aminopeptidases (*Figure 12.4a* and *b*). This mutation does not perceptibly alter antigenic integrity since three different AP-A antibodies recognize the mutant as well as wild-type molecules (*Figure 12.4c*; Wang and Cooper, 1993).

A remarkable characteristic of AP-A is its activation by Ca^{2+} and inhibition by Zn^{2+} (Danielsen *et al.*, 1980; Tobe *et al.*, 1980; Yamada *et al.*, 1988). However, the mutant AP-A described above does not exhibit enzyme activity even in the presence of Ca^{2+}, suggesting that the zinc-binding motif is critical and Ca^{2+} may serve to enhance the activity. Transitional metal ions, including Zn^{2+}, clearly inhibit AP-A activity, however. When cells

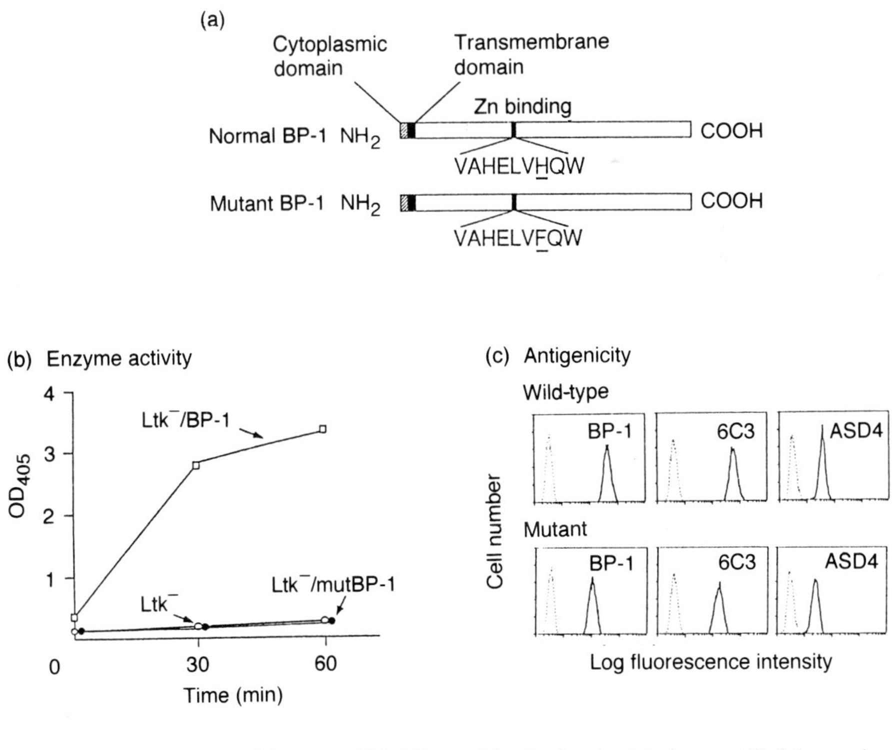

Figure 12.4. Point mutation of the second histidine residue in the zinc-binding motif of the murine AP-A abrogates enzyme activity. (a) The second His in this motif was mutated to Phe. (b) Cell-surface AP-A activity is lost for the mutant form but not for the wild-type BP-1/AP-A molecules expressed in murine L cells. (c) Both the mutant and wild-type BP-1/AP-A are recognized on the transfected L cells by three different BP-1/AP-A antibodies, one of which (ASD4) efficiently inhibits AP-A enzymic activity.

expressing recombinant or natural AP-A are treated with EDTA or 1,10-phenanthroline to remove metal ions, Zn^{2+} ions in low concentration enhance AP-A activity in the presence or absence of Ca^{2+}. In contrast, higher concentrations of Zn^{2+} inhibit AP-A activity (Wang and Cooper, 1993). These observations suggest that while Zn^{2+} is responsible for AP-A activity, excess Zn^{2+} is inhibitory. The acquisition of environmental zinc may account for the inhibitory effect of Zn^{2+} seen with native AP-A. Excess Zn^{2+} may competitively bind to the putative Ca^{2+} regulatory motif, thereby inhibiting enzymic activity. Like many other aminopeptidases, the enzymic activity of AP-A is inhibited by chelating agents such as 1,10-phenanthroline or EDTA, but relatively specific inhibitors are also known for AP-A, including amastatin (Tieku and Hooper, 1992).

12.4 Gene organization and regulation of expression

Analysis of genomic clones encoding the *AP-A/BP-1* gene (Wang *et al.*, 1996) indicates that the gene spans more than 100 kb and contains 20 exons (*Figure 12.5*). Except for the first and the last, the gene is composed of small exons, ranging from 56 bp to 171 bp in size, that are separated by introns ranging from less than 100 bp to approximately 10 kb. The zinc-binding motif HEXXH and the glutamic acid residue 19 amino acids downstream that also binds zinc are encoded separately in exons 5 and 6, respectively. Whether this is a common feature for other aminopeptidases is presently unknown. The *BP-1/AP-A* genomic structure resembles that of other peptidase genes, *CD10* (endopeptidase-24.11; D'Adamio *et al.*, 1989) and *CD26* (dipeptidyl peptidase IV; Bernard *et al.*, 1994), in that each contains many small exons separated by introns ranging from 100 bp to tens of kilobases.

Primer extension analysis indicates a common major transcriptional start site in pre-B cells, bone marrow stromal cells and kidney cells. The promoter region contains a TATA-like element and potential DNA-binding motifs for lymphocyte-specific transcription factors including Ikaros, BSAP, PU.1 and octamer-binding proteins, as well as DNA-binding motifs for several ubiquitous transcription factors. An interferon-responsive element in the promoter region appears to be functional, since type I interferons (IFNα/β) can up-regulate BP-1 expression in pre-B-cell lines. The presence of silencers or transcriptional suppressors is suggested by the extinction of BP-1 expression when pre-B cells were fused with plasma cells (Wu *et al.*, 1990). However, when a 2.1 kb 5′ flanking promoter fragment was fused to the luciferase reporter gene, we found transcriptional activity not only in pre-B cells, which normally express BP-1, but also in plasma cells, which do not express BP-1. Potential silencer regions outside of the 2.1 kb fragment may be required for the inhibition of *BP-1/AP-A* gene expression in mature B and plasma cells. The regulated expression of

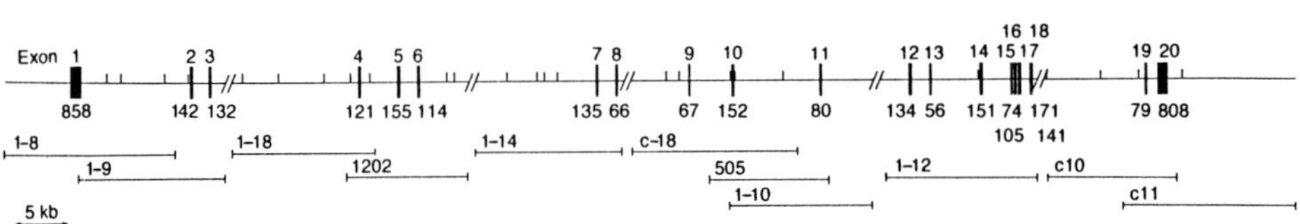

Figure 12.5. Gene organization of mouse *AP-A/BP-1*.

the *BP-1/AP-A* gene during B-cell development is likely to be governed by a combination of the available transcription factors, gene methylation status (Bird, 1992) and other regulatory mechanisms. Further studies are required to elucidate the mechanism that directs lineage- and stage-specific *BP-1/AP-A* expression among haematopoietic cells while allowing relatively ubiquitous expression in many other tissues.

The *BP-1/AP-A* gene is localized in the distal region of mouse chromosome 3 (*Figure 12.6*) in a region that lacks those mouse mutations associated with a phenotype indicating defective B-lymphocyte development or aberrant blood pressure regulation. The distal region of mouse chromosome 3 shares a region of homology with human chromosome 4q (*Figure 12.6*). Recent studies (L. Li and M.D. Cooper, unpublished) have localized the *BP-1/AP-A* gene to human chromosome 4q25 near the complement factor 1 gene.

12.5 Biological function

A well-known physiological substrate of AP-A is angiotensin II (AII), an important regulator of blood pressure. Hence, AP-A has also been called angiotensinase A (Nagatsu *et al.*, 1970; Sakura *et al.*, 1983). The biosynthetic pathway of angiotensin is

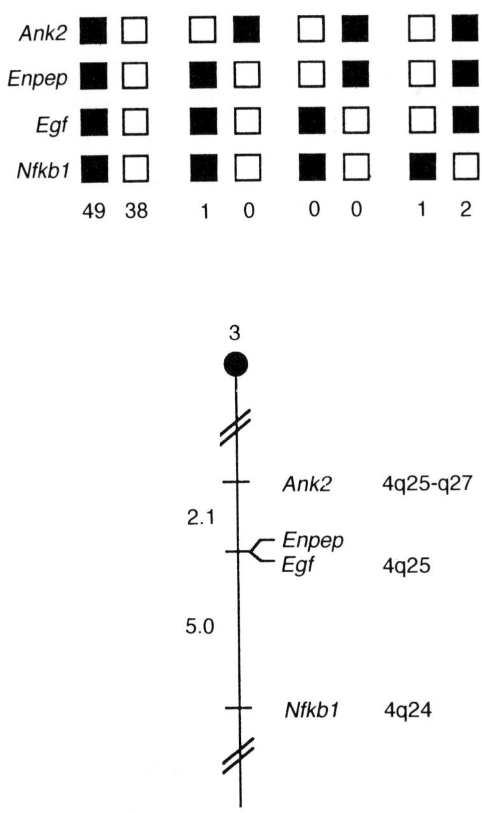

Figure 12.6. Chromosomal localization of mouse *AP-A/BP-1* gene.

outlined in *Figure 12.7.* AP-A converts AII to angiotensin III (AIII), one of the steps leading to the inactivation of AII. AIII may have a different function than AII in certain tissues. For example, AIII appears to be an active neuropeptide in the brain (Healy and Wilk, 1993; Song *et al.*, 1993).

A role for BP-1/AP-A has also been suggested in early B-cell development. The expression of BP-1/AP-A is restricted to fractions C, C′ and D during early B-cell differentiation and is up-regulated by treatment with IL-7, one of the most important growth factors for murine B-cell progenitors. This expression pattern suggests that BP-1 may function during these early stages of B-cell development. Alternatively, BP-1/AP-A might serve as an adhesion molecule for pre-B cells in the bone marrow. Both IL-7 and the IL-7 receptor contain an N-terminal glutamic acid (Goodwin *et al.*, 1990; Namen *et al.*, 1988) and therefore could potentially serve as substrates for BP-1/AP-A. Cleavage of the N-terminal glutamic acid residue on either of these molecules, or the interaction of BP-1 with them, might alter the IL-7/IL-7 receptor binding affinity. An alteration in pre-B-cell proliferation would be the expected consequence of these events. A BP-1 antibody has been shown to inhibit IL-7 driven proliferation of B-cell precursors and it has been suggested that AP-A might cleave a natural peptide inhibitor (Welch, 1995). AP-A, like other peptidases, could also function to hydrolyse regulatory peptides that act on pre-B or stromal cells. In support of this idea, bone marrow stromal cells have been shown to express the angiotensin type I receptor (Lin *et al.*, 1994). The effect of AII on stromal cells is still unknown, but the AP-A expressed by stromal cells and pre-B cells would be expected to convert AII to AIII, resulting in a change of AII activity or leading to its degradation (*Figure 12.7*).

In order to examine the *in vivo* role of AP-A in immune system development and in other organ systems, we have recently generated BP-1-deficient mice through homologous recombination in embryonic stem cells (Q. Lin, J. Wang and M.D. Cooper, unpublished). The BP-1$^{-/-}$ mice develop and breed normally, and do not express detectable AP-A protein or enzyme activity. These mice developed normal numbers of B and T cells in the bone marrow, spleen, lymph node and thymus. Phenotypic analysis of their bone marrow lymphocytes reveals a normal pattern of early B-lineage differentiation. *In vitro* experiments indicate normal B lymphopoiesis in Whitlock-Witte cultures of AP-A-deficient bone marrow cells, and the proliferative response to IL-7 is not impaired in bone marrow cells from AP-A knock-out mice. Although detailed analysis is required to clarify the role of AP-A during the early B-cell differentiation pathway, it is possible that the function of AP-A in BP-1$^{-/-}$ mice may have been compensated for by

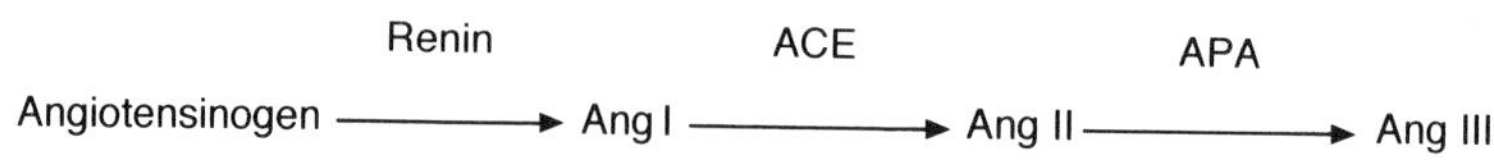

Figure 12.7. Schematic outline of the components of the renin–angiotensin system. Renin hydrolyses angiotensinogen to angiotensin I (Ang I), which is then converted to angiotensin II (Ang II) by angiotensin-converting enzyme (ACE) (see Chapter 7). AP-A cleaves the NH_2-terminal aspartic acid of Ang II to generate Ang III thus diminishing its pressor activity and facilitating further degradation by endopeptidases.

other peptidases. For example, B-cell precursors express CD10 or endopeptidase-24.11 in addition to BP-1/AP-A. It will be interesting to cross BP-1-deficient mice with mice deficient in other peptidases to examine the combined effects.

12.6 Conclusions and future perspectives

Many of the cell-surface proteins expressed by lymphoid and myeloid cells have proved to represent ectoenzymes formerly shown to be expressed by cells of the intestinal epithelium, renal tubules, liver and various other tissues. Molecular cloning of the genes encoding these ectoenzymes promises to greatly facilitate the exploration of structure–function relationships for these ectoenzymes. Monoclonal antibodies including ones that inhibit enzymic activity, will also facilitate the study of their physiological roles. The development of specific enzyme inhibitors may prove invaluable in pharmacotherapy as well as determining their biological roles. Manipulation of the gene(s) by gene deletion and construction of transgenic mouse models will also accelerate the elucidation of *in vivo* functions of these ectoenzymes.

Acknowledgements

We thank Peter Burrows, Li Li and Qun Lin for critical reading of the manuscript and helpful discussions, and Ann Brookshire for help in preparing the manuscript. J. Hidies in our laboratory was supported in part by USPHS grants AI39816. MDC is an investigator at the Howard Hughes Medical Institute.

References

Bernard AM, Mattei MG, Pierres M, Marguet D. (1994) Structure of the mouse dipeptidyl peptidase IV (CD26) gene. *Biochemistry* **33:** 15204–15214.

Bird A. (1992) The essentials of DNA methylation. *Cell* **70:** 5–8.

Cooper MD, Mulvaney D, Coutinho A, Cazenave PA. (1986) A novel cell-surface molecule on early B-lineage cells. *Nature* **321:** 616–618.

D'Adamio L, Shipp MA, Masteller EL, Reinherz EL. (1989) Organization of the gene encoding common acute lymphoblastic leukemia antigen (neutral endopeptidase 24.11): multiple miniexons and separate 5′ untranslated regions. *Proc. Natl Acad. Sci. USA* **86:** 7103–7107.

Danielsen EM, Norén O, Sjöström H, Ingram J, Kenny AJ. (1980) Aspartate aminopeptidase: purification by immunoadsorbent chromatography and properties of the detergent- and proteinase-solubilized forms. *Biochem. J.* **189:** 591–603

Goodwin RG, Friend D, Ziegler SF *et al.* (1990) Cloning of the human and murine interleukin-7 receptors: demonstration of a soluble form and homology to a new receptor superfamily. *Cell* **60:** 941–951.

Hardy RR, Carmack CE, Shinton SA, Kemp JD, Hayakawa K. (1991) Resolution and characterization of pro-B and pre-pro-B cell stages in normal mouse bone marrow. *J. Exp. Med.* **173:** 1213–1225.

Healy DP, Wilk S. (1993) Localization of immunoreactive glutamyl aminopeptidase in rat brain. II. Distribution and correlation with angiotensin II. *Brain Res.* **606:** 295–303.

Jongeneel CV, Bouvier J, Bairoch A. (1989) A unique signature identifies a family of zinc-dependent metallopeptidases. *FEBS Lett.* **242:** 211–214.

Kincade PW. (1994) B lymphopoiesis: global factors, local control. *Proc. Natl Acad. Sci. USA* **91:** 2888–2889.

Li L, Wang J, Cooper MD. (1993a) cDNA cloning and expression of human glutamyl aminopeptidase (aminopeptidase A). *Genomics* **17:** 657–664.

Li L, Wu Q, Wang J, Bucy RP, Cooper MD. (1993b) Widespread tissue distribution of aminopeptidase A, an evolutionary conserved ectoenzyme recognized by the BP-1 antibody. *Tissue Antigens* **42:** 488–496.

Lin Q, Wang J, Cooper MD. (1994) Angiotensin receptor (AT-1) expression in murine stromal cell lines. *FASEB J.* **8:** 1444.

Nagatsu I, Nagatsu T, Yamamoto T, Glenner GG, Mehl JW. (1970) Purification of aminopeptidase A in human serum and degradation of angiotensin II by the purified enzyme. *Biochim. Biophys. Acta* **198:** 255–270.

Namen AE, Lupton S, Hjerrild K, *et al.* (1988) Stimulation of B-cell progenitors by cloned murine interleukin-7. *Nature* **333:** 571–573.

Pillemer E, Whitlock C, Weissman IL. (1984) Transformation-associated proteins in murine B-lymphomas that are distinct from Abelson virus gene products. *Proc. Natl Acad. Sci. USA* **81:** 4434–4438.

Sakura H, Kobayashi H, Mizutani S, Sakura N, Hashimoto T, Kawashima, Y. (1983) Kinetic properties of placental aminopeptidase A: N-terminal degradation of angiotensin II. *Biochem. Int.* **6:** 609–615.

Sherwood PJ, Weissman IL. (1990) The growth factor IL-7 induces expression of a transformation-associated antigen in normal pre-B cells. *Int. Immunol.* **2:** 399–406.

Song L, Wilk E, Wilk S, Healy DP. (1993) Localization of immunoreactive glutamyl aminopeptidase in rat brain. I. Association with cerebral microvessels. *Brain Res.* **606:** 286–294.

Tieku S, Hooper NM. (1992) Inhibition of aminopeptidases N, A and W. A re-evaluation of the actions of bestatin and inhibitors of angiotensin converting enzyme. *Biochem. Pharmacol.* **44:** 1725–1730.

Tobe H, Kojima F, Aoyagi T, Umezawa H. (1980) Purification by affinity chromatography using amastatin and properties of aminopeptidase A from pig kidney. *Biochim. Biophys. Acta* **613:** 459–468.

Vallee BL, Auld DS. (1989) Short and long spacer sequences and other structural features of zinc binding sites in zinc enzymes. *FEBS Lett.* **257:** 138–140.

Vallee BL, Auld DS. (1990) Zinc coordination, function, and structure of zinc enzymes and other proteins. *Biochemistry* **29:** 5647–5659.

Wang J, Cooper MD. (1993) Histidine residue in the zinc-binding motif of aminopeptidase A is critical for enzymic activity. *Proc. Natl Acad. Sci. USA* **90:** 1222–1226.

Wang J, Walker H, Lin Q, Jenkins N, Copeland NG, Watanabe T, Burrows PD, Cooper MD. (1996) The mouse BP-1 gene: structure, chromosomal localization, and regulation of expression by type I interferons and interleukin-7. *Genomics* **33:** 167–176.

Welch PA, Burrows PD, Namen AE, Gillis S, Cooper MD. (1990) Bone marrow stromal cells and interleukin-7 induce coordinate expression of the BP-1/6C3 antigen and pre-B cell growth. *Int. Immunol.* **2:** 697–705.

Welch PA. (1995) Regulation of B cell precursor proliferation by aminopeptidase A. *Int. Immunol.* **7:** 737–746.

Whitlock CA, Tidmarsh GF, Muller-Sieburg C, Weissman IL. (1987) Bone marrow stromal cell lines with lymphopoietic activity express high levels of pre-B neoplasia-associated molecule. *Cell* **48:** 1009–1021.

Wu Q, Tidmarsh GF, Welch PA, Pierce JH, Weissman IL, Cooper MD. (1989) The early B lineage antigen BP-1 and the transformation-associated antigen 6C3 are on the same molecule. *J. Immunol.* **143:** 3303–3308.

Wu Q, Lahti JM, Air GM, Burrows PD. Cooper MD. (1990) Molecular cloning of the murine BP-1/6C3 antigen: a member of the zinc-dependent metallopeptidase family. *Proc. Natl Acad. Sci. USA* **87:** 993–997.

Wu Q, Li L, Cooper MD, Pierres M, Gorvel JP. (1991) Aminopeptidase A activity of the murine B-lymphocyte differentiation antigen BP-1/6C3. *Proc. Natl Acad. Sci. USA,* **88:** 676–680.

Yamada R, Mizutani S, Kurauchi O, Okano K, Imaizumi H, Narita O, Tomoda Y. (1988) Purification and characterization of human placental aminopeptidase A. *Enzyme* **40:** 223–230.

Chapter 13

Dipeptidyl peptidase IV (DPP-IV/CD26): biochemistry and control of cell-surface expression

Germain Trugnan, Tounsia Ait-Slimane, Frédéric David, Laurent Baricault, Tsouria Berbar, Christelle Lenoir and Catherine Sapin

13.1 Background

1996 was the 30th anniversary of the discovery of dipeptidyl peptidase IV (DPP-IV) by Hopsu-Havu and Glenner (1966). First described as a dipeptide naphthylamidase, from its ability to hydrolyse glycyl-prolyl-β-naphthylamide, the enzyme was later referred to as dipeptidyl aminopeptidase IV (DAP-IV) and finally by its E.C. name, dipeptidyl peptidase IV (EC 3.4.14.5.). Immunologists have described a cluster differentiation antigen, CD26, that was shown to be identical to DPP-IV at the protein (Hegen *et al.*, 1990) and DNA levels (Darmoul *et al.*, 1992; Marguet *et al.*, 1992). DPP-IV is a plasma membrane peptidase, hence an 'ectopeptidase'. In contrast, the other dipeptidyl peptidases so far identified are not ectoenzymes, since they are found either in lysosomes (DPP-I and DPP-II) or in the cytoplasm (DPP-III). Because of its cell-surface localization, its wide tissue distribution and its specificity for proline (a turn-making amino acid) as the P_1 residue, DPP-IV has been postulated to have important functions and has been extensively studied during the last three decades, a period in which more than 1000 papers on this enzyme have been published. A recent review on DPP-IV/CD26 is that by Yaron and Naider (1993). In this chapter we will summarize the present knowledge, with particular attention to the mechanisms that may control expression of DPP-IV at the cell surface.

13.2 Molecular structure

13.2.1 Biochemical data

DPP-IV belongs to the serine peptidase family (Kenny *et al.*, 1976). Electrophoretic migration of native (detergent-solubilized) DPP-IV purified from kidney indicated an apparent M_r of 130 000 (Macnair and Kenny, 1979). Several studies have shown that DPP-IV varies from 100 kDa to 150 kDa, depending on species, tissue, the developmental status and the pathophysiological state of the animal. Comparison of the properties of detergent-solubilized DPP-IV with that released by proteinase treatment

Cell-Surface Peptidases in Health and Disease, edited by A.J. Kenny and C.M. Boustead.

(autolysis) showed that DPP-IV is a transmembrane protein with a very short N-terminal tail. Chemical analysis indicated that DPP-IV is a glycosylated protein and contains several cysteine residues. Electron microscopy showed the protein to be dimeric and the model then proposed (Macnair and Kenny, 1979) was confirmed 10 years later by molecular cloning.

13.2.2 Cloning of DPP-IV

The first cloning data were published by Hong and Doyle (1987), who sequenced a cDNA from a rat library that encoded gp110, identified at that time as a 'bile canaliculus domain specific membrane glycoprotein'. Two years later Ogata *et al.* (1989) isolated and characterized a full length cDNA clone for rat DPP-IV that encompassed 3.5 kb, including the whole coding region for a 767 residue protein. *In vitro* translation of this cDNA gave an 87 kDa protein. A first partial human sequence was published 1 year later (Darmoul *et al.*, 1990). Mouse thymocyte activation molecule (THAM) cDNA was then cloned and subsequently shown to be identical to mouse DPP-IV (Marguet *et al.*, 1992). The encoded mouse protein contained 760 residues. At the same time we published the first complete human DPP-IV cDNA sequence (Darmoul *et al.*, 1992) that corresponded to a 766 residue protein with a calculated M_r of 88 277. Several other groups (Misumi *et al.*, 1992; Tanaka *et al.*, 1992) have published data on human DPP-IV cDNA. The human enzyme exhibits 84.4% identity with that of mouse and rat (Darmoul *et al.*, 1992; Marguet *et al.*, 1992; Ogata *et al.*, 1989).

13.2.3 Molecular organization

Primary structure. The biochemical and molecular data support a model for DPP-IV as shown in *Figure 13.1*. The main characteristics of this model are:

(i) a very short (six residues) cytoplasmic tail that contains basic residues and a threonine that may be a phosphorylation site;
(ii) a 22 residue hydrophobic stretch that most likely corresponds to the transmembrane domain;
(iii) a stalk region proposed by Fleischer (1994) that allows the ectodomain to be flexible in the extracellular space;
(iv) a highly glycosylated domain that contains most of the potential N-glycosylation sites (seven out of nine for the human protein and six out of eight for rat and mouse proteins);
(v) a cysteine-rich domain that contains seven of the 11 cysteine residues in all species;
(vi) a 200 residue C-terminal stretch of highly conserved sequence among the different species and other proteins of the DPP-IV subfamily that includes the catalytic domain;
(vii) the catalytic domain itself containing the active site serine (position 630 for the human, 624 for the mouse and 631 for the rat);
(viii) each DPP-IV monomer associates with a second one to form a dimer. However, tetrameric molecules have also been observed (D. Marguet, personal communication and G. Trugnan, unpublished results).

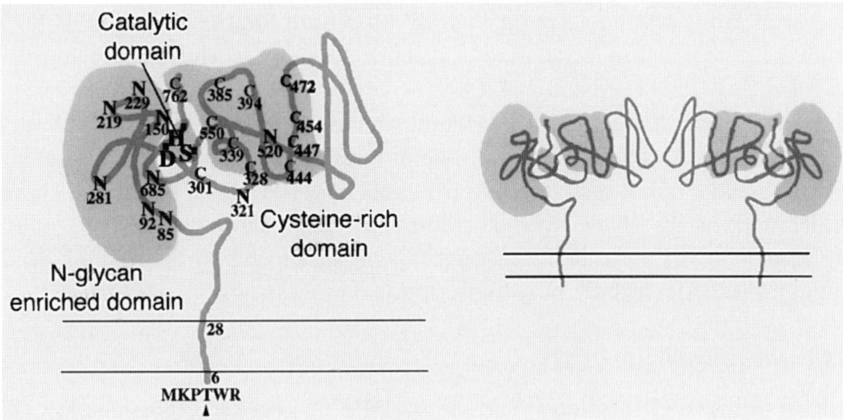

Figure 13.1. Putative models for monomer and dimer organization of DPP-IV.

Glycosylation. Interestingly, the eight or nine asparagine residues forming the putative N-glycosylation sites are found in regions with a high solvent accessibility coefficient, hence they are probably at the surface of the molecule. This correlates well with the fact that all the glycosylation sites seem to be occupied, as judged by endo H- and endo F-induced molecular weight shifts (see e.g. Darmoul *et al.*, 1992; Hauri *et al.*, 1985).

An uncleavable signal peptide is the transmembrane anchor domain. The identity of the N-terminal sequence of the DPP-IV protein with that predicted from the cDNA sequence suggests that the signal peptide is not cleaved and serves as a membrane anchor (Marguet *et al.*, 1992). Site-directed and deletion mutagenesis has shown that rough endoplasmic reticulum (RER) targeting information is contained within the 34 N-terminal residues and that the six residues of the cytoplasmic domain are needed to anchor the protein in the RER membrane (Hong and Doyle, 1990).

13.3 Genetics

13.3.1 Chromosome location

The DPP-IV gene has been mapped to the 2q23 region of human chromosome 2 by various methods, including a somatic cell hybrid panel approach (Darmoul *et al.*, 1990; Morrison *et al.*, 1993). *In situ* hybridization has shown that the mouse DPP-IV gene maps to the mouse 2C2–2D region (Marguet *et al.*, 1992).

13.3.2 Gene structure

Using cDNA probes, human (Abbott *et al.*, 1994) and mouse (Bernard *et al.*, 1994) DPP-IV genes have been isolated and characterized. Human and mouse DPP-IV genes have the same complex intron/exon distribution, with 26 exons (Marguet *et al.*, 1995). Other ectoenzymes also share this unusually large number of exons. The length of introns varies from 100 bp to more than 20 kb with a total gene length of 90 kb as determined for the mouse gene. Exons in both species are clustered in four groups (exons 1 and 2, 3–10, 11–23 and 24–26).

Interestingly, two distinct polyadenylation signals are present in mouse and rat DPP-IV mRNA, leading to the expression of two different transcripts: in the mouse, a major one of 3.3 kb and a minor one of 4.9 kb (Marguet *et al.*, 1995); in the rat 3.2 kb and 4.8 kb (Hildebrandt *et al.*, 1991). No such studies have been carried out in the human. The relative abundance of these two transcripts seems to depend on the tissue considered, thus suggesting tissue-specific post-transcriptional processing.

13.3.3 Gene regulation

Analysis of the 5′ UTR of human and mouse DPP-IV genes indicates that the DPP-IV promoter belongs to the TATAA- and CCAAT-boxes-deficient G+C-rich group (Abbott *et al.*, 1994; Bernard *et al.*, 1994). Several consensus sequences for transcriptional factors have been identified, such as multiple copies of Sp1 and AP-2, and also AP-1 GATA, LF-A1 and XREbf. No tissue-specific or differentiation-specific (except for a putative 'butyrate responsive element'; Böhm *et al.*, 1995) sequences have so far been identified. This correlates well with the ubiquitous distribution of DPP-IV and with the nature of the mechanisms that control DPP-IV expression as a function of cell differentiation.

Transfection of constructs containing portions of the human DPP-IV promoter and luciferase as a reporter have been carried out to delineate the functional properties of this promoter. It was shown that the region contained between –150 and –109, relative to the start of translation, is important for high level promoter activity (Böhm *et al.*, 1995).

13.3.4 Regulation of DPP-IV gene expression

The level of DPP-IV expression depends on the cell type, state of differentiation, developmental level and the activation status. In addition, DPP-IV expression may vary as a function of pathophysiological states. The mechanisms that underlie changes of DPP-IV expression vary according to the model and the biological process considered. Some of the mechanisms involved in the control of DPP-IV gene expression have been recently elucidated (Trugnan *et al.*, 1995). DPP-IV expression is high in normal, differentiated intestinal cells (Darmoul *et al.*, 1991) as well as in differentiated colon cancer cells (Darmoul *et al.*, 1992), whereas normal or undifferentiated cancer cells display low DPP-IV levels. The turn-over of DPP-IV is slow, with a half-life of more than 96 h, thus preventing rapid changes in the steady-state level. The DPP-IV mRNA steady-state level parallels protein expression, indicating that mRNA translation does not seem to be involved in the control of DPP-IV expression (Darmoul *et al.*, 1991, 1992; Hildebrandt *et al.*, 1991; Hong *et al.*, 1989; Reutter *et al.*, 1995; Trugnan *et al.*, 1995). Undifferentiated cells efficiently degrade DPP-IV mRNA, with a half-life of 12–15 h, whereas differentiated cells do not (Trugnan *et al.*, 1995). 'Run on' experiments demonstrate that gene transcription does not change as a function of cell differentiation. DPP-IV expression during differentiation is, therefore, controlled at the post-transcriptional level. DPP-IV gene expression may also be manipulated by changing the nutritional status for a 50% gelatin-containing diet (=proline-rich diet) instead of a 4% casein diet. Such a switch induces an increase of DPP-IV mRNA followed by an increased enzyme activity (Suzuki *et al.*, 1993).

13.3.5 A genetic defect

Fischer 344 (F344) rats from Charles River in Kanagawa, Japan, exhibit a deficiency (0.1% of normal activity) of DPP-IV, while other ectoenzymes appear to be unaffected (Watanabe *et al.*, 1987). F344 rats from USA are normal (Tiruppathi *et al.*, 1990), while F344 rats from German sources are like those from Japan (Thompson *et al.*, 1991). The deficient rats do not present any obvious abnormality, but this may be explained by the high level of redundancy in the peptidase family. It appears that another mechanism of control of gene expression is involved in deficient rats since they have normal steady-state mRNA levels (Erickson *et al.*, 1992). The defect is due to a single base mutation that causes a Gly to Arg change in the active site (at position 633), a mutation that leads to degradation within the endoplasmic reticulum (ER; Tsuji *et al.*, 1992).

13.3.6 DPP-IV-related proteins

DPP-IV cDNA and gene cloning show that two novel classes of proteins, namely fibroblast-activating protein-α (FAP-α; Scanlan *et al.*, 1994) and dipeptidyl peptidase X (DPP-X; Wada *et al.*, 1992), display high sequence homology with DPP-IV. These two groups of proteins, lacking enzymic activity and of unknown function, resemble DPP-IV in molecular structure, having probably evolved from a common ancestor, the more recent evolutionary link being between DPP-IV and FAP-α, since they map to the same chromosomal region (2q23), whereas DPP-X was assigned to chromosome 7q.

13.4 Active site/catalytic mechanism

13.4.1 Active site and catalytic triad

Cloning of the cDNA has identified the active-site serine (Darmoul *et al.*, 1992; Marguet *et al.*, 1992; Misumi *et al.*, 1992) and its role has been confirmed by site-directed mutagenesis (Fujiwara *et al.*, 1992; Ogata *et al.*, 1992; Tsuji *et al.*, 1992). It has been proposed that DPP-IV might represent a novel family of non-classical serine hydrolases in which serine, aspartic acid and histidine of the catalytic triad are arranged in reverse order compared with classical serine proteases, such as chymotrypsin or subtilisin, where the catalytic triad is His, Asp, Ser (Marguet *et al.*, 1992). Using a site-directed mutagenesis approach, the catalytic triad of DPP-IV has been unambiguously identified as Ser624, Asp702 and His734. The active serine is part of a consensus Gly–X–Ser–X–Gly (common to esterases and lipases; Brenner, 1988) and a 200 residue stretch near the C terminus shares a high degree of homology with acyl-amino acid hydrolase and prolyl-oligopeptidase (David *et al.*, 1993).

13.4.2 Catalytic mechanism

In common with other classical serine peptidases, DPP-IV cleaves the peptide bond of its substrates through nucleophilic attack by the serine that has been activated by aspartic acid and histidine residues from the catalytic triad (for review, see Demuth and Heins, 1995). An acyl–enzyme intermediate is then formed and deacylation of the enzyme is the rate-limiting step.

13.5 Specificity (synthetic and natural substrates)

13.5.1 Peptide bond specificity

DPP-IV is one of a group of peptidases able to split an X–Pro bond, a bond which is generally resistant to cleavage. DPP-IV preferentially cleaves dipeptides from the N terminus of oligopeptides of at least three residues, with Pro as the P_1 residue. Pro may be replaced by Ala and to a lesser extent by Hyp, Ser, Gly, Val and Leu. A variety of different residues is tolerated at P_2 (for review see Demuth and Heins, 1995). Other requirements are an unsubstituted N terminus and a *trans* conformation of the scissile bond.

13.5.2 Assays using synthetic substrates

The general structure of these substrates consists of a dipeptide, X–Pro, linked to a chromogenic amine: 2-naphthylamide (Hino *et al.*, 1975; Hopsu-Havu and Glenner, 1966), or 4-nitroanilides (Nagatsu *et al.*, 1976). Others substrates such as 4-phenylazoanilides, methylcoumarylamides or 4-methoxy-2-naphthylamides have also been used. All the assays depend on colorimetric or fluorometric detection.

13.5.3 Natural substrates in vivo *and* in vitro

Insects. Promellitin is transformed by repeated attack by DPP-IV into mellitin, a major component of bee venom. The 10 steps, from the N terminus are: Ala–Pro-/-Glu–Pro-/-Glu–Pro-/-Ala–Pro-/-Glu–Pro-/-Glu–Ala-/-Glu–Ala-/-Asp–Ala-/-Glu–Ala-/-Asp–Pro- (reviewed by Yaron and Naider, 1993). Other insect anti-bacterial peptides (cecropins A and B) have been recently identified as DPP-IV substrates (Boman *et al.*, 1989).

Mammals. DPP-IV can attack several biologically active peptides, including substance P (Arg–Pro–Lys–Pro–Gln–Gln–Phe–Phe–Gly–Leu–Metamide) and β-casomorphin, a morphomimetic peptide. DPP-IV may also be involved in the cleavage of fibrin α-chain, pro-colipase and trypsinogen, however the rates of hydrolysis decrease with increasing chain length (Hoffmann *et al.*, 1993). DPP-IV is also involved in the metabolism of the VIP-glucagon peptide family, and inactivates plasma growth releasing factor (GRF) *in vivo* (Frohman *et al.*, 1989). Among the other peptides of this family, DPP-IV cleaves those with Ala as the P_1 residue [gastric inhibitory peptide (GIP) or glucagon-like peptide-1 (GLP-1) but not those with a Ser in this position [VIP, glucagon, secretin, pituitary adenylyl cyclase activating peptide (PACAP)] (Mentlein *et al.*, 1993b)]. The neuropeptide Y/peptide YY (NPY/PYY) family of hormonal peptides is attacked by DPP-IV (Mentlein *et al.*, 1993a). Although the cleaved peptides cannot bind Y1-type receptors, they retain the ability to bind and activate Y2- and Y3-type receptors, thus suggesting that DPP-IV may modulate Y1-mediated functions.

A number of polypeptides have X–Pro as the N terminus (e.g. some members of the cytokine family and several growth factors). However, these polypeptides are not cleaved by DPP-IV (Hoffmann *et al.*, 1993). In contrast, several peptides produced by endoproteolysis of these cytokines or growth factors become DPP-IV substrates when their lengths are below 24 residues (Ansorge *et al.*, 1995).

13.5.4 Is there a DPP-IV endopeptidase activity?

It has been suggested that DPP-IV may interact with internal proline residues of components of the extracellular matrix (e.g. collagen and fibronectin; Hanski *et al.*, 1985; Piazza *et al.*, 1989). The entry of human immunodeficiency virus (HIV) may require an attack by DPP-IV on the proline-rich V3 loop of HIV gp120 (Callebaut *et al.*, 1993), actions that imply an endopeptidase attack. However, an attack on purified gp120 has not yet been demonstrated. Reutter *et al.* (1995) observed that DPP-IV may cleave collagen I *in vitro*, but they were cautious in interpreting this result, since it was not clear whether the DPP-IV preparation used was free of prolyl endopeptidase activity, which may contaminate even highly purified samples (Macnair and Kenny, 1979).

13.6 Inhibitors

In addition to group-specific serine hydrolase inhibitors, such as di-isopropylfluorophosphate (DFP) and phenylmethylsulphonyl fluoride (Kenny *et al.*, 1976), a number of specific chemical DPP-IV inhibitors have been described (see reviews by Yaron and Naider, 1993, and Demuth and Heins, 1995). They may be classified into four groups: (i) oligopeptides (e.g. diprotin A, Ile–Pro–Ile, and B, Val–Pro–Leu) containing an N-terminal X–Pro sequence, which are competitive and reversible; (ii) irreversible, 'suicide' inhibitors, the *N*-peptidyl-*O*-acyl hydroxylamines, containing an X–Pro–peptidyl sequence; (iii) pyrrolidide and thiazolidide aminoacyl derivatives mimicking the structure of a proline residue, are very potent and specific, with K_i values in the micromolar to nanomolar range; (iv) inhibitors containing an aminoacyl-boroPro group are effective in the picomolar range (Flentke *et al.*, 1991).

13.7 Tissue and cellular distribution of DPP-IV

13.7.1 Tissue distribution

Initially described as a brush border peptidase (see for example Hopsu-Havu and Glenner, 1966; Kenny and Maroux, 1982; Nordström *et al.*, 1968), DPP-IV is a ubiquitous enzyme present on almost all cell types (Fleischer, 1994; Reutter *et al.*, 1995). The localization has been studied in detail in the rat (Hartel *et al.*, 1988): DPP-IV is strongly expressed on epithelial cells (kidney, intestine, liver, bile duct, pancreas, prostate, uterus, lung, thyroid, placenta, endometrium, etc.), on almost all endothelial cells (e.g. lung, kidney, heart, skeletal muscle, spleen), in some connective tissues (skin, oesphagus, uterus and penis), in few neuronal preparations, and in almost all the white blood cell lines (Ansorge *et al.*, 1995; Lojda, 1977), especially those of the lymphocyte/thymocyte lineage. However, some of these studies were performed using assays which did not take advantage of the new highly potent inhibitors, or using antibodies that had not been well characterized.

13.7.2 Cellular distribution

DPP-IV expression is mostly associated with the cell-surface, although it has also been possible to purify it from rat liver lysosomes where DPP-IV may represent as much as 3% of lysosomal membrane proteins (Kyouden *et al.*, 1992). In epithelial cells, DPP-IV is mainly present at the apical pole, although it has also been demonstrated at the basolateral surface (Baricault *et al.*, 1995; Darmoul *et al.*, 1991, 1992; Le Bivic *et al.*, 1990; Low *et al.*, 1991a,b; Matter *et al.*, 1990a; Weisz *et al.*, 1992).

13.7.3 The question of the soluble form

DPP-IV has been detected in the serum as well as in other biological fluids, such as saliva, urine and uterine secretion. About 40% of the lysosomal DPP-IV was found to be soluble (Kyouden *et al.*, 1992). Biochemical and molecular analysis suggested that these soluble molecules may be derived from the membrane-bound form through proteolytic digestion, since microsequencing indicated that the soluble protein starts at position 35 of the membrane form, at a papain cleavage site. Another form of soluble DPP-IV may exist that does not derive from the membrane-bound protein (Duke-Cohan *et al.*, 1995). However, no cloning data have been produced and it is not possible to judge whether this novel form of DPP-IV is the product of a new gene, or of one of the mRNAs previously observed on Northern blotting (Hildebrandt *et al.*, 1991; Hong *et al.*, 1989).

13.8 Physiological and pathological roles

13.8.1 Functions depending on enzymic activity

Putative roles in the hydrolysis of biological peptides and other polypeptides have been discussed in Section 13.5.3. The ability of DPP-IV to generate free dipeptides at the surface of epithelia in the kidney and intestine has raised the possibility that it may have a role in transport, in respect of which the DPP-IV-deficient rats (F344) have provided a useful experimental model (Tiruppathi *et al.*, 1990). These deficient animals were unable to perform the terminal digestion of proline-rich peptides such as gliadin (Tiruppathi *et al.*, 1993) which may account for the increased excretion of Gly–Pro–Hyp-containing peptides in the urine of these rats (Watanabe *et al.*, 1993).

13.8.2 Interactions of DPP-IV with proteins independent of enzymic activity

The interaction of DPP-IV with collagen may depend on the cysteine-rich region (Loster *et al.*, 1995), although it has also been suggested that an endopeptidase attack may be involved (Reutter *et al.*, 1995). Other interactions are those between DPP-IV and adenosine deaminase (Kameoka *et al.*, 1993); pathophysiological implications of this binding have been explored on T cells. It is suggested that adenosine deaminase has an immunomodulatory role by changing local adenosine concentrations (Blanco *et al.*, 1996; Dong *et al.*, 1996) and it would be of interest to explore this phenomenon in other cellular systems where adenosine deaminase (Torimoto *et al.*, 1991) binds to DPP-IV. Whether binding of CD45 (protein tyrosine phosphatase) to DPP-IV will also explain some of its biological functions remains to be explored in non-immunological systems. The binding to tat protein from HIV may explain why infection causes an antigen-specific inhibition of T-lymphocyte proliferation (Gutheil *et al.*, 1994). Binding to, and subsequent cleavage of, gp120 has been suggested as a mechanism for HIV entry into T cells (Callebaut *et al.*, 1993) but awaits further confirmation (Alizon and Dragic, 1994). Several authors have raised the possibility that CD26 may be involved in acquired immune deficiency syndrome (AIDS) pathogeny (Dalgleish, 1995; Oravecz *et al.*, 1995; Morimoto *et al.*, 1994).

13.8.3 DPP-IV as a pathological marker

DPP-IV concentration has been shown to vary in several pathological states such as cancer (Morrison *et al.*, 1993; Zweibaum *et al.*, 1984), HIV (Oravecz *et al.*, 1995), psoriasis (Raynaud *et al.*, 1992) and cirrhosis (Matsumoto *et al.*, 1992). However, the significance of these changes and the mechanisms remain unclear.

13.9 Control of cell-surface expression

Since DPP-IV is a protein with a long half-life, the only way to adapt the amount of active enzyme rapidly is to modify its concentration at the cell surface. Several processes contribute to the control of DPP-IV cell-surface expression.

13.9.1 Intracellular traffic

In common with other membrane glycoproteins, DPP-IV is synthesised in the RER, and co- and post-translationally N-glycosylated as the protein moves from the RER to the Golgi cisternae and to the surface (Baricault *et al.*, 1995; Hauri *et al.*, 1985). Exit from the ER does not require dimerization of DPP-IV, a feature that is specific and relatively rare (Jascur *et al.*, 1991). We recently found that within the highly conserved 200-amino-acid C-terminal stretch, Asp599 in the mouse protein is essential for correct protein folding and normal exit from the ER (David *et al.*, 1996). Site-directed mutagenesis of this Asp for an Ala residue induces an ER retention that has been vizualized using confocal microscopy (*Figure 13.2a*) and this retention is associated with the rapid degradation of the mutant protein (David *et al.*, 1996). An additional β-strand was generated by the mutation (*Figure 13.2b*), suggesting a pivotal role for this region in the control of DPP-IV stability.

13.9.2 Sorting pathway

In polarized cells, DPP-IV is mostly present at the apical pole (Darmoul *et al.*, 1991; Kenny and Maroux, 1982), indicating that the protein must be specifically sorted from the trans-Golgi network. This process strongly depends on the cell type, since DPP-IV is

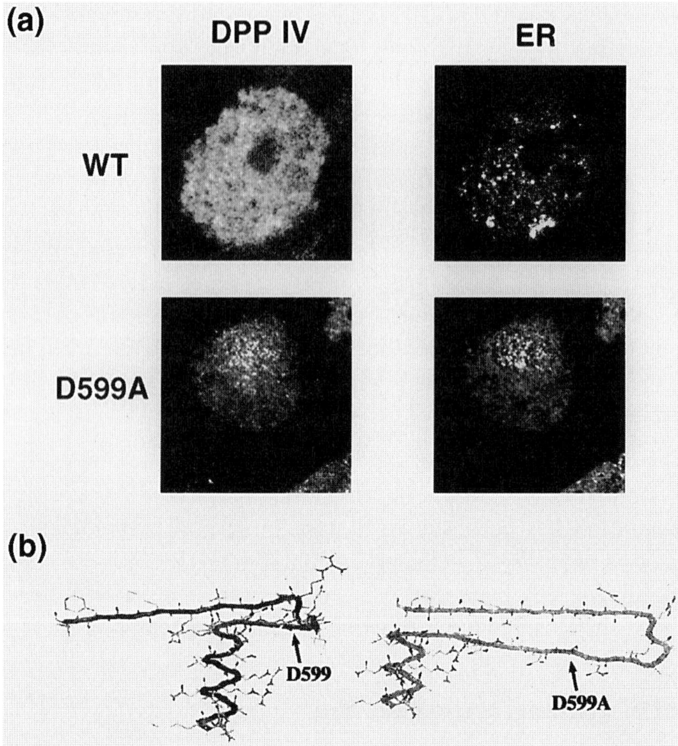

Figure 13.2. (a) Confocal image showing DPP-IV and ER staining in L cells transpected with WT or DA599 mutated mouse DPP-IV. (b) D to A mutation at residue 599 dramatically changes the 3-D organization of the molecule by introducing an additional β leaflet.

sorted directly in kidney-derived Madin–Darby canine kidney cells (Casanova *et al.*, 1991; Low *et al.*, 1991b), whereas it is routed to the basolateral surface of hepatocytes and then re-addressed to the apical pole (Bartles *et al.*, 1987), and it may follow both routes to a variable extent in intestinal Caco-2 cells (Baricault *et al.*, 1995; Le Bivic *et al.*, 1990; Matter *et al.*, 1990a) and in kidney LLCPK cells (Low *et al.*, 1991a). The use of soluble and chimeric DPP-IV proteins has suggested that the protein might contain a dominant apical signal in its extracellular domain and also a competing basolateral signal in the transmembrane and/or cytoplasmic tail (Weisz *et al.*, 1992). However, other work has shown that this transmembrane and/or cytoplasmic tail may induce Golgi retardation of the chimeric proteins (Low *et al.*, 1994). The sorting signal has not yet been identified (see also Chapter 4).

13.9.3 Recycling

One feature of DPP-IV traffic is its ability to enter into an endocytosis–exocytosis cycle (Baricault *et al.*, 1995; Kreisel *et al.*, 1988; Le Bivic *et al.*, 1990; Matter *et al.*, 1990b). During recycling, DPP-IV glycosylation changes (Kreisel *et al.*, 1988), indicating that

DPP-IV re-enters the Golgi apparatus (Volz *et al.*, 1995). In polarized cells, DPP-IV endocytosis takes place from both poles of the cell (Baricault *et al.*, 1995).

13.9.4 Control of traffic and sorting

As a first step towards understanding a possible 'hormonal regulation' of sorting, experiments have been done using forskolin, a protein kinase A inhibitor and phorbol myristate acetate, a protein kinase C stimulator. Forskolin and phorbol myristate acetate rapidly blocks exocytosis of DPP-IV and other surface proteins through an autophagic-like process (Baricault *et al.*, 1993, 1995; Sapin *et al.*, 1997). Whether this process represents a regulation pathway remains to be explored.

13.10 Future prospects

Future research on DPP-IV may clarify a number of points:

(i) the molecular mechanisms involved in the control of DPP-IV expression, in particular why some tissues express two mRNAs and by what means cell- and tissue-specific regulation is achieved.
(ii) Which residues are involved in the recognition of peptide bond specificity?
(iii) What is the mechanism of dimerization?
(iv) Which domains bind collagen and other proteins? Is there a role for cysteines or N-glycans?
(v) Is there a second soluble form of DPP-IV and what is its molecular structure?

Three-dimensional structural analysis of the molecule may help to answer these questions. *In vivo* studies are also needed on novel DPP-IV-deficient animals, such as transgenic mice.

References

Abbott C, Baker E, Sutherland G, McCaughan G. (1994) Genomic organization, exact localization and tissue expression of the human CD26 (dipeptidyl peptidase-IV) gene. *Immunogenetics* **40**: 331–338.

Alizon M, Dragic T. (1994) CD26 antigen and HIV fusion. *Science* **264**: 1161–1162.

Ansorge S, Buhling F, Hoffmann T, Kahne T, Neubert K, Reinhold D. (1995) DPP-IV/CD26 on human lymphocytes: functional role in cell growth and cytokine regulation. In: *Dipeptidyl Peptidase-IV (CD26) in Metabolism and the Immune Response* (ed. B. Fleischer), pp. 163–184. RG Landes, Austin, TX.

Baricault L, Garcia M, Cibert C, Sapin S, Geraud G, Codogno C, Trugnan G. (1993) Forskolin blocks the apical expression of dipeptidyl peptidase-IV in Caco-2 cells and induces its retention in Lamp-1-containing vesicles. *Exp. Cell Res.* **209:** 277–287.

Baricault L, Fransen J, Garcia M, Sapin C, Codogno P, Ginsel L, Trugnan G. (1995) Rapid sequestration of DPP-IV/CD26 and other cell-surface proteins in an autophagic-like compartment in Caco-2 cells treated with forskolin. *J. Cell Sci.* **108**: 2109–2121.

Bartles J, Feracci H, Stieger B, Hubbard A. (1987) Biogenesis of the rat hepatocyte plasma membrane *in vivo*: comparison of the pathways taken by apical and basolateral proteins using subcellular fractionation. *J. Cell Biol.* **105:** 1241–1251.

Bernard A, Mattei M, Pierres M, Marguet D. (1994) Structure of the mouse dipeptidyl peptidase-IV (CD26) gene. *Biochemistry* **33:** 15204–15214.

Blanco J, Marie I, Callebaut C, Jacotot E, Krust B, Hovanessian A. (1996) Specific binding of adenosine deaminase but not HIV-1 transactivator protein Tat to human CD26. *Exp. Cell. Res.* **225:** 102–111.

Böhm S, Gum J, Erickson R, Hicks J, Kim Y. (1995) Human dipeptidyl peptidase-IV gene promoter: tissue specific regulation from a TATA-less GC-rich sequence characteristic of a housekeeping gene promoter. *Biochem. J.* 311: 835–843.

Brenner S. (1988) The molecular evolution of genes and proteins: a tale of two serines. *Nature* **334:** 528–530.

Callebaut C, Krust B, Jacotot E, Hovanessian AG. (1993) T-cell activation antigen, CD26, as a cofactor for entry of HIV in CD4+ cells. *Science* **262:** 2045–2050.

Casanova J, Misumi Y, Ikehara Y, Hubbard A, Mostov K. (1991) Direct apical sorting of rat liver dipeptidyl peptidase-IV expressed in MDCK cells. *J. Biol. Chem.* **266**: 24428–24432.

Dalgleish A. (1995) HIV and CD26. *Nature Medicine* **1**: 881–882.

Darmoul D, Lacasa M, Chantret I, Swallow D, Trugnan G. (1990) Isolation of a cDNA probe for the human intestinal dipeptidyl peptidase-IV and assignment of the gene locus DPP4 to chromosome 2. *Ann. Hum. Genet.* **54:** 191–197.

Darmoul D, Rouyer-Fessard C, Blais A *et al.* (1991) Dipeptidyl peptidase-IV expression in rat jejunal crypt–villus axis is controled at the mRNA level. *Am. J. Physiol.* **261**: G763–G769.

Darmoul D, Lacasa M, Baricault L, Marguet D, Sapin C, Trotot P, Barbat A, Trugnan G. (1992) Dipeptidyl peptidase-IV (CD 26) gene expression in enterocyte-like colon cancer cell lines HT-29 and Caco-2. Cloning of the complete human coding sequence and changes of dipeptidyl peptidase-IV mRNA levels during cell differentiation. *J. Biol. Chem.* **267**: 4824–4833.

David F, Bernard A, Pierres M, Marguet D. (1993) Identification of serine 624, aspartic acid 702 and His 734 as the catalytic triad residues of mouse dipeptidyl peptidase-IV (CD26). A member of a novel family of nonclassical serine hydrolases. *J. Biol. Chem.* **268**: 17247–17252.

David F, Baricault L, Sapin C, Gallet X, Marguet D, Thomas-Soumarmon A, Trugnan G. (1996) Reduced cell-surface expression of a mutated dipeptidyl peptidase-IV (DPP-IV/CD26) correlates with the generation of a beta strand in its C-terminal domain. *Biochem. Biophys. Res. Comm.* **222**: 833–838.

Demuth H, Heins J. (1995) Catalytic mechanism of dipeptidyl peptidase-IV. In: *Dipeptidyl Peptidase-IV (CD26) in Metabolism and the Immune Response* (ed. B. Fleischer), pp. 1–35. RG Landes, Austin, TX.

Dong R, Kameoka J, Hegen M, Tanaka T, Xu Y, Schlossman S, Morimoto C. (1996) Characterisation of adenosine deaminase binding to human CD26 on T cells and its biologic role in immune response. *J. Immunol.* **156**: 1349–1355.

Duke-Cohan J, Morimoto C, Rocker J, Schlossman S. (1995) A novel form of dipeptidyl peptidase-IV found in human serum. Isolation, characterization, and comparison with T-lymphocyte membrane dipeptidyl peptidase-IV (CD26). *J. Biol. Chem.* **270**: 14107–14114.

Erickson R, Suzuki Y, Sedlmayer A, Kim Y. (1992) Biosynthesis and degradation of altered immature forms of intestinal dipeptidyl peptidase-IV in a rat strain lacking the enzyme. *J. Biol. Chem.* **267**: 21623–21629.

Fleischer B. (1994) CD26: a surface protease involved in T cell activation. *Immunol. Today* **15**: 180–184.

Flentke G, Munoz E, Huber B, Plaut A, Kettner C, Bachovchin W. (1991) Inhibition of dipeptidyl peptidase-IV (DP-IV) by Xaa–boroPro dipeptides and use of these inhibitors to examine the role of DP-IV in T cell function. *Proc. Natl Acad. Sci. USA* **88**: 1556–1559.

Frohman L, Downs T, Heimer E, Felix A. (1989) Dipeptidyl peptidase-IV and trypsin-like enzymatic degradation of human growth hormone-releasing hormone in plasma. *J. Clin. Invest.* **83**: 1533–1540.

Fujiwara T, Tsuji E, Misumi Y, Takami N, Ikehara Y. (1992) Selective cell-surface expression of dipeptidyl peptidase-IV with mutations at the active site sequence. *Biochem. Biophys. Res. Commun.* **185**: 776–784.

Gutheil W, Subramanyam M, Flentke G, Sanford D, Munoz E, Huber B, Bachovchin W. (1994) Human immunodeficiency virus 1 Tat binds to dipeptidyl aminopeptidase-IV (CD26): a possible mechanism for Tat's immunosuppressive activity. *Proc. Natl Acad. Sci. USA* **91**: 6594–6598.

Hanski C, Huhle T, Reutter W. (1985) Involvement of plasma membrane dipeptidyl peptidase-IV in fibronectin-mediated adhesion of cells to collagen. *Biol. Chem. Hoppe-Seyler* **366**: 1169–1176.

Hartel S, Gossrau R, Hanski C, Reutter W. (1988) Dipeptidyl peptidase-IV (DPP-IV) in rat organs. Comparison of immunochemistry and enzyme activity. *Histochemistry* **89**: 151–161.

Hauri HP, Sterchi EE, Bienz D, Fransen JAM, Marxer A. (1985) Expression of intracellular transport of microvillus membrane hydrolases in human intestinal epithelial cells. *J. Cell Biol.* **101**: 838–851.

Hegen M, Niedobitek G, Klein C, Stein H, Fleisher B. (1990) The T cell triggering molecule, Tp 103, is associated with dipeptidyl aminopeptidase-IV activity. *J. Immunol.* **144**: 2908–2914.

Hildebrandt M, Reutter W, Gitlin J. (1991) Tissue specific regulation of dipeptidyl peptidase-IV expression during development. *Biochem. J.* **277**: 331–334.

Hino M, Nagatsu T, Kakumu S, Okuyama S, Yoshii Y, Nagatsu I. (1975) Glycylprolyl β-naphthyl-amidase activity in human serum. *Clin. Chim. Acta* **62**: 5–11.

Hoffmann T, Faust J, Neubert K, Ansorge S. (1993) Dipeptidyl peptidase-IV (CD26) and aminopeptidase N (CD13) catalyzed the hydrolysis of cytokines and peptides with N-terminal cytokines sequences. *FEBS Lett.* **336**: 61–64.

Hong W, Doyle D. (1987) cDNA cloning for a bile canaliculus domain-specific membrane glycoprotein of rat hepatocytes. *Proc. Natl Acad. Sci. USA* **84**: 7962–7966.

Hong W, Doyle D. (1990) Molecular dissection of the NH_2-terminal signal/anchor sequence of rat dipeptidyl peptidase-IV. *J. Cell Biol.* **111**: 323–328.

Hong W, Petell J, Swank D, Sanford J, Hixson D, Doyle D. (1989) Expression of dipeptidyl peptidase-IV in rat tissue is mainly regulated at the mRNA levels. *Exp. Cell Res.* **182**: 256–266.

Hopsu-Havu V, Glenner G. (1966) A new dipeptide naphthylamidase hydrolysing glycyl-prolyl-beta-naphthylamide. *Histochimie* **7**: 197–201.

Jascur T, Matter K, Hauri H. (1991) Oligomerization and intracellular transport: dimerization of intestinal dipeptidyl peptidase-IV occurs in the Golgi apparatus. *Biochemistry* **30**: 1908–1915.

Kameoka J, Tanaka T, Nojima Y, Schlossman SF, Morimoto C. (1993) Direct association of adenosine deaminase with a T-cell activation antigen, CD26. *Science* **261**: 466–469.

Kenny AJ, Maroux S. (1982) Topology of microvillar membrane hydrolases of kidney and intestine. *Physiol. Rev.* **62**: 91–128.

Kenny J, Booth AG, George SG, Ingram J, Wood EJ, Young AR. (1976) Dipeptidyl peptidase-IV, a kidney brush border serine peptidase. *Biochem. J.* **157**: 169–182.

Kreisel W, Hanski C, Tran Thi T, Katz N, Decker K, Reutter W, Gerok W. (1988) Remodeling of rat hepatocyte plasma membrane glycoprotein. De- and reglycosylation of dipeptidyl peptidase-IV. *J. Biol. Chem.* **263**: 11736–11742.

Kyouden T, Himeno M, Ishikawa T, Oshumi Y, Kato K. (1992) Purification and characterization of dipeptidyl peptidase-IV in rat liver lysosomal membranes. *J. Biochem. (Tokyo)* **111**: 770–777.

Le Bivic A, Quaroni A, Nichols B, Rodriguez-Boulan E. (1990) Biogenetic pathways of plasma membrane proteins in Caco-2, a human intestinal epithelial cell line. *J. Cell Biol.* **111**: 1351–1361.

Lojda Z. (1977) Studies on glycylproline naphthylamidase. I. Lymphocytes. *Histochemistry* **54**: 299–309.

Loster K, Zeilinger K, Schuppan D, Reutter W. (1995) The cysteine-rich region of dipeptidyl peptidase-IV (CD26) is the collagen binding site. *Biochem. Biophys. Res. Commun.* **217**: 341–348.

Low SH, Wong SH, Tang BL, Hong W. (1991a) Involvement of both vectorial and transcytotic pathways in the preferential apical cell-surface localization of rat dipeptidyl peptidase-IV in transfected LLC-PK1 cells. *J. Biol. Chem.* **266**: 19710–19716.

Low SH, Wong SH, Tang BL, Subramanian VN, Hong W. (1991b) Apical cell-surface expression of rat dipeptidyl peptidase-IV in transfected Madin Darby canine kidney cells. *J. Biol. Chem.* **266**: 13391–13396.

Low SH, Tang B, Wong S, Hong W. (1994) Golgi retardation in MDCK and CHO cells of a transmembrane chimera of two surface proteins. *J. Biol. Chem.* **269**: 1985–1994.

Macnair D, Kenny A. (1979) Proteins of the kidney microvillar membrane. The amphiphatic form of dipeptidyl peptidase-IV. *Biochem. J.* **179**: 379–395.

Marguet D, Bernard AM, Vivier I, Darmoul D, Naquet P, Pierres M. (1992) cDNA cloning for mouse thymocyte-activating molecule – a multifunctional ecto-dipeptidyl peptidase-IV (CD26) included in a subgroup of serine proteases. *J. Biol. Chem.* **267**: 2200–2208.

Marguet D, Bernard A, David F, Lazaro-Trueba I, Pierres M. (1995) Structural organization of the DP-IV gene and its relationship with DPX and FAP-α transcripts. In: *Dipeptidyl Peptidase-IV (CD26) in Metabolism and the Immune Response* (ed. B. Fleischer), pp. 37–53. RG Landes, Austin, TX.

Matsumoto Y, Bishop G, McCaughan G. (1992) Altered zonal expression of CD26 antigen (dipeptidyl peptidase-IV) in human cirrhotic liver. *Hepatology* **15**: 1048–1053.

Matter K, Brauchbar M, Bucher K, Hauri HP. (1990a) Sorting of endogenous plasma membrane proteins occurs from two sites in cultured human intestinal epithelial cells (Caco-2). *Cell* **60**: 429–437.

Matter K, Stieger B, Klumperman J, Ginsel L, Hauri HP. (1990b) Endocytosis, recycling, and lysosomal delivery of brush border hydrolases in cultured human intestinal epithelial cells (Caco-2). *J. Biol. Chem.* **265**: 3503–3512.

Mentlein R, Dahms P, Grandt D, Krüger R. (1993a) Proteolytic processing of neuropeptide Y and peptide YY by dipeptidyl peptidase-IV. *Regul. Pep.* **49**: 133–144.

Mentlein R, Gallwitz B, Schmidt W. (1993b) Dipeptidyl peptidase-IV hydrolyses gastric inhibitory peptide (GIP), glucagon-like peptide-1(7–36) amide, peptide histidine methionine (PHM) and is responsible for their degradation in human serum. *Eur. J. Biochem.* **214**: 829–835.

Misumi Y, Hayashi Y, Arakawa F, Ikehara Y. (1992) Molecular cloning and sequence analysis of human dipeptidyl peptidase-IV, a serine protease on the cell-surface. *Biochim. Biophys. Acta* **1131**: 333–336.

Morimoto C, Lord CI, Zhang C, Duke-Cohan JS, Letvin NL, Schlossman SF. (1994) Role of CD 26/dipeptidyl peptidase-IV in human immunodeficiency virus type 1 infection and apoptosis. *Proc. Natl Acad. Sci. USA* **91**: 9960–9964.

Morrison M, Vijayasaradhi S, Engelstein D, Albino A, Houghton A. (1993) A marker for neoplastic progression of human melanocytes is a cell-surface ectopeptidase. *J. Exp. Med.* **177**: 1135–1143.

Nagatsu T, Hino M, Fuyamada H, Hayakawa T, Sakakibara S, Nakagawa Y, Takemoto T. (1976) New chromogenic substrates for X-prolyl dipeptidyl aminopeptidase. *Analyt. Biochem.* **74**: 466–476.

Nordström C, Dahlqvist A, Josephsson L. (1968) Quantitative determination of enzymes in different parts of the villi and crypts of rat small intestine. Comparison of alkaline phosphatase, disaccharidases and dipeptidases. *J. Histochem. Cytochem.* **15**: 713–720.

Ogata S, Misumi Y, Ikehara Y. (1989) Primary structure of rat liver DPP-IV deduced from its cDNA and identification of the NH_2-terminal signal sequence as the membrane anchoring domain. *J. Biol. Chem.* **264**: 3596–3601.

Ogata S, Misumi Y, Ikehara Y. (1992) Identification of the active site residues in dipeptidyl peptidase-IV by affinity labeling, and site directed mutagenesis. *Biochemistry* **31**: 2582–2587.

Oravecz T, Roderiquez G, Koffi J, Wang J, Ditto M, Chequer Bou-Habib D, Lusso P, Norcross M. (1995) CD26 expression correlates with entry, replication and cytopathicity of monocytotropic HIV-1 strains in a T-cell line. *Nature Medicine* **1**: 919–926.

Piazza GA, Callanan HM, Mowery J, Hixson DC. (1989) Evidence for a role of dipeptidyl peptidase-IV in fibronectin mediated interactions of hepatocytes with extracellular matrix. *Biochem. J* **262**: 327–334.

Raynaud F, Bauvois B, Gerbaud P, Evain-Brion D. (1992) Characterization of specific proteases associated with the surface of human skin fibroblast and their modulation in pathology. *J. Cell. Physiol.* **151**: 378–385.

Reutter W, Baum O, Löster K, Fan H, Bork J, Bernt K, Hanski C, Tauber R. (1995) Functional aspects of the three cellular domains of dipeptidyl peptidase-IV: characterisation of glycosylation events, of the collagen binding site and of endopeptidase activity. In: *Dipeptidyl Peptidase-IV*

(CD26) in Metabolism and the Immune Response (ed. B. Fleischer), pp. 55–78. RG Landes, Austin, TX.

Sapin C, Baricault I, Trugnan G. (1997) PKC-dependent long term effect of PMA on protein cell surface expression in Caco 2 cells. *Exp. Cell Res.* **231:** 308–318.

Scanlan M, Raj B, Calvo B, Garin-Chesa P, Sanz-Moncasi M, Healey J, Old L, Rettig W. (1994) Molecular cloning of fibroblast activation protein alpha, a member of the serine protease family selectively expressed in stromal fibroblasts of epithelial cancers. *Proc. Natl Acad. Sci. USA* **91**: 5657–5661.

Suzuki Y, Erickson, RH, Sedlmayer A, Chang SK, Ikehara Y, Kim, YS. (1993) Dietary regulation of rat intestinal angiotensin converting enzyme and dipeptidyl peptidase-IV. *Am. J. Physiol.* **264**: G1153–G1159.

Tanaka T, Camerini D, Seed B, Torimoto Y, Dang N, Kameoka J, Dahlberg H, Schlossman S, Morimoto C. (1992) Cloning and functional expression of T-cell activation antigen CD26. *J. Immunol.* **149**: 481–486.

Thompson N, Hixson D, Callanan H, Panzica M, Flanagan DF RA, Hong, W, Hartel-Shenk S, Doyle D. (1991) A Fischer rat substrain deficient in dipetidyl peptidase-IV activity makes normal steady state RNA levels and an altered protein. Use as a liver cell-transplantation model. *Biochem. J.* **273**: 497–502.

Tiruppathi C, Miyamoto Y, Ganapathy V, Leibach F. (1993) Genetic evidence for a role of DPP-IV in intestinal hydrolysis and assimilation of prolyl peptides. *Am. J. Physiol.* **265**: G81–G89.

Tiruppathi C, Miyamoto Y, Ganapathy V, Roesel R, Whitford G, Leibach F. (1990) Hydrolysis and transport of proline-containing peptides in renal brush border membrane vesicles from dipeptidyl peptidase-IV-positive and dipeptidyl peptidase-IV-negative rat strains. *J. Biol. Chem.* **265**: 1476–1483.

Torimoto Y, Dang N, Vivier E, Tanaka T, Schlossman S, Morimoto C. (1991) Coassociation of CD26 (dipeptidyl peptidase-IV with CD45 on the surface of human lymphocytes. *J. Immunol.* **147**: 2514–2517.

Trugnan G, Baricault L, David F, Garcia M, Sapin C. (1995) Control of dipeptidyl peptidase-IV/CD26 cell-surface expression in intestinal cells. In: *Dipeptidyl Peptidase-IV (CD26) in Metabolism and the Immune Response* (ed. B. Fleischer), pp. 79–98. RG Landes, Austin, TX.

Tsuji E, Misumi Y, Fujiwara T, Takami N, Ogata S, Ikehara Y. (1992) A active site mutation (Gly^{633}→Arg) of dipeptidyl peptidase-IV causes its retention and rapid degradation in the endoplasmic reticulum. *Biochemistry* **31**: 11921–11927.

Volz B, Orberger G, Porwoll S, Hauri H, Tauber R. (1995) Selective reentry and recycling of cell-surface glycoproteins to the biosynthetic pathway in human hepatocarcinoma HepG2 cells. *J. Cell Biol.* **130**: 537–551.

Wada K, Yokotani N, Hunter C, Doi K, Wenthold R, Shimazaki S. (1992) Differential expression of two different mRNA encoding members of a dipeptidyl aminopeptidase family. *Proc. Natl Acad. Sci. USA* **89**: 197–201.

Watanabe Y, Kojima T, Fujimoto Y. (1987) Deficiency of membrane-bound dipeptidyl aminopeptidase-IV in a certain rat strain. *Experientia* **43:** 400–401.

Watanabe Y, Kojima-Komatsu T, Iwaki-Egawa S, Fujimoto Y. (1993) Increased excretion of proline-containing peptides in DPP-IV-deficient rats. *Res. Comm. Chem. Pathol. Pharmacol.* **81:** 323–330.

Weisz OA, Machamer CE, Hubbard AL. (1992) Rat liver dipeptidylpeptidase-IV contains competing apical and basolateral targeting information. *J. Biol. Chem.* **267**: 22282–22288.

Yaron A, Naider F. (1993) Proline dependent structural and biological properties of peptides and proteins. *Crit. Rev. Biochem. Mol. Biol.* **28:** 31–81.

Zweibaum A, Hauri HP, Sterchi E, Chantret I, Haffen K, Bamat J, Sordat B. (1984) Immunohistological evidence, obtained with monoclonal antibodies, of small intestinal brush border hydrolases in human colon cancers and foetal colons. *Int. J. Cancer* **34**: 591–598.

Chapter 14

Three glycolipid-anchored peptidases: membrane dipeptidase, aminopeptidase P and carboxypeptidase M

Anthony J. Turner and Nigel M. Hooper

14.1 Introduction

The discovery of the glycolipid-anchored peptidases arose serendipitously from studies on the mode of anchorage of angiotensin-converting enzyme (ACE; EC 3.4.15.1) to the brush border membranes of pig kidney. However, ACE is also found as a soluble form circulating in plasma, probably generated by an endogenous membrane activity, which we have termed 'ACE secretase' (Hooper *et al.*, 1987a; Oppong and Hooper, 1993; S. Parvathy, S.Y. Oppong, E. Karran, D.R. Buckle, A.J. Turner and N.M. Hooper, unpublished observations) (for a fuller account of the anchors of membrane peptidases see Chapter 3). Trypsin treatment of membrane ACE can generate a hydrophilic form resembling plasma ACE. By comparing the N-terminal sequences of the amphipathic and hydrophilic forms of ACE we deduced that the protein was anchored through its C terminus rather than by an uncleaved N-terminal signal peptide (Hooper *et al.*, 1987a). At about this time, several glycolipid anchored proteins were being characterized in which the novel glycosyl phosphatidylinositol (GPI) anchor was shown to be attached to the C terminus of the mature protein. However, we were unable to detect any release of ACE from brush border membranes by treatment with phosphatidylinositol-specific phospholipase C (PI-PLC) and we concluded that ACE was probably a type I membrane protein, anchored by a C-terminal transmembrane polypeptide (Hooper *et al.*, 1987a), a view subsequently confirmed by cloning of the enzyme (Soubrier *et al.*, 1988). Attempts to release ACE from a membrane preparation by PI-PLC treatment did, however, reveal an activity that was able to degrade His–Leu, one of the products of the standard assay for ACE using Hip–His–Leu as substrate. This 'His–Leu degrading activity' was released in a time- and concentration-dependent fashion by bacterial PI-PLC, suggesting that it may be GPI-anchored. Subsequent characterization of this activity with selective substrates and inhibitors identified it as the brush border enzyme renal (or microsomal) dipeptidase (also referred to as dehydropeptidase I) (Hooper *et al.*, 1987b). The enzyme has been renamed 'membrane dipeptidase' (MDP) by the

Cell-Surface Peptidases in Health and Disease, edited by A.J. Kenny and C.M. Boustead.

Enzyme Commission and reclassified as EC 3.4.13.19. Subsequently MDP was confirmed as the first GPI-anchored peptidase. A survey of other brush border peptidases revealed that only aminopeptidase P (AP-P; X-Pro aminopeptidase; EC 3.4.11.9) was susceptible to release by PI-PLC from porcine and human renal membranes (Hooper and Turner, 1988a). Although we failed to detect release of carboxypeptidase M (CP-M; EC 3.4.17.12) from porcine kidney membranes, other studies, particularly with Madin–Darby canine kidney (MDCK) cells, have established that CP-M is GPI-anchored in some cell lines and tissues, a conclusion supported by the cloning and sequencing of the human cDNA (see Section 14.4). Thus, there are three identified mammalian GPI-anchored peptidases: MDP, AP-P and CP-M and this chapter summarizes their key features. There is no clear explanation why these peptidases are so anchored: most likely it provides a ready means for their selective release from the cell surface, a process that may be hormonally or developmentally regulated. *Table 14.1* provides a summary of some of the properties of MDP, AP-P and CP-M.

Table 14.1. Summary of properties of the GPI-anchored cell-surface peptidases MDP, AP-P and CP-M

Characteristics	MDP	AP-P	CP-M
Subunit M_r (kDa)	47	91	62
Oligomeric structure	Covalent dimer	Non-covalent (trimer?)	Monomer
Human chromosomal location	16q24	?	?
Specificity	↓ O–O	↓ O–Pro–	↓ –O–Arg/Lys
Inhibitors	Cilastatin	Apstatin	MGTA GEMSA
Activators	PI-PLC cleavage	Mn^{2+}	Co^{2+}
Substrates	Dipeptides, leukotriene D_4, imipenem, Gly-D-Phe	Bradykinin, substance P, neuropeptide Y, peptide YY, Gly–Pro–Hyp	Bradykinin, dynorphin A(1–13), anaphylatoxin C5a, epidermal growth factor
Localization (major sites)	Kidney, lung, small intestine, pancreas (zymogen granules)	Kidney, lung, intestine	Widely distributed in tissues, cells and body fluids

14.2 Membrane dipeptidase

14.2.1 Discovery of MDP

MDP was originally described as a dehydropeptidase activity in which the substrate glycyl-dehydrophenylalanine was hydrolysed to phenylpyruvate, glycine and ammonia (Bergmann and Schleich, 1932) and the enzyme was first purified to homogeneity from pig kidney by Robinson *et al.* (1953). The purified enzyme was capable of cleaving efficiently a wide range of dipeptides hence its classification as a dipeptidase. It can also hydrolyse dipeptides in which the C-terminal amino acid is in the D configuration. Appreciation of the physiological roles of MDP has been slow to emerge. MDP may play a role in a final step of the degradation of oligopeptides to amino acids, prior to their uptake into cells. However, a more specific function has been established in the metabolism of glutathione (McIntyre and Curthoys, 1982). The major site for glutathione metabolism is the renal brush border where γ-glutamyl transpeptidase (γGT) first hydrolyses the tripeptide glutathione to glutamate and the dipeptide cysteinyl-glycine. The latter is then further hydrolysed to its constituent amino acids by either MDP or aminopeptidase N (EC 3.4.11.2), each contributing equally to this process (Curthoys, 1987). A corollary to the metabolism of glutathione by MDP and γGT is its involvement in the metabolism of the peptidyl leukotriene D_4 (LTD_4) which is a potent broncho- and vasoconstrictor. LTD_4 is formed in two steps from LTA_4 by the addition of glutathione followed by γGT hydrolysis of the intermediate LTC_4. The action of MDP on LTD_4 produces an inactive product (LTE_4) and thus MDP serves to switch off the leukotriene signal at the cell surface. Although MDP and γGT co-localize in some tissues (e.g. kidney and small intestine), there are mismatches in their distribution. For example, MDP is expressed at high levels in lung where γGT is not found and the reverse is true in seminal vesicles and fetal liver (Habib *et al.*, 1996). This suggests additional roles for MDP independent of γGT action.

14.2.2 Metabolism of carbapenem antibiotics by MDP

The carbapenems represent a group of β-lactam antibiotics in which the *trans*-conformation of the β-lactam ring produces resistance to conventional bacterial β-lactamases. The prototype antibiotic in this class was imipenem which, although highly effective

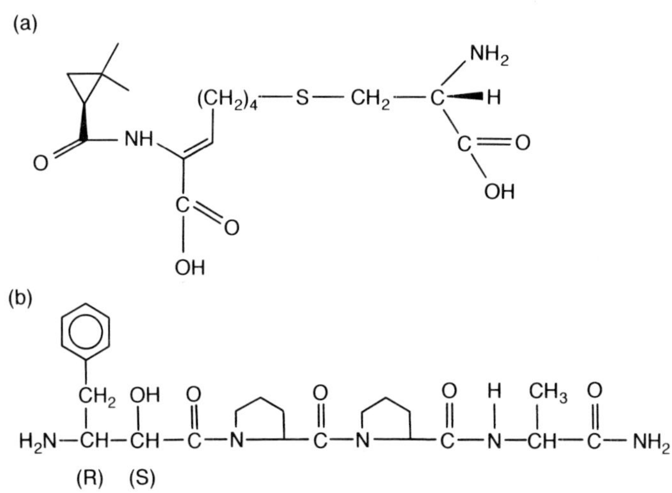

Figure 14.1. The structures of the MDP inhibitor, cilastatin (a) and of the AP-P inhibitor, apstatin (b).

against many bacterial infections, was ineffective against urinary tract infections due to rapid metabolism of the antibiotic by the kidney (Kropp *et al.*, 1982). Co-purification of the β-lactamase activity with glycyl-dehydrophenylalanine-hydrolysing activity from kidney established that the enzyme responsible for imipenem inactivation was MDP. This was the first and only example of a mammalian β-lactamase, although MDP acts in a stereospecific manner, failing to hydrolyse the *cis*-conformation of the β-lactam ring as seen in penicillin. The development of selective inhibitors of MDP, especially the compound cilastatin (*Figure 14.1a*), led to the formulation of the combination of imipenem with cilastatin which, to date, has proved an effective antibiotic mixture with a broad spectrum of anti-bacterial potency against Gram-positive and Gram-negative aerobic and anaerobic bacteria. A new threat to antibiotic therapy, however, are the bacterial metallo-β-lactamases some of which (e.g. the *cphA* gene product of *Aeromonas hydrophila*) can hydrolyse imipenem but are unable to cleave efficiently the *cis* conformation of the β-lactam ring of the classical penicillins and cephalosporins. This led us to examine possible similarities in structural features and catalytic mechanisms between MDP and the *A. hydrophila* β-lactamase (Keynan *et al.*, 1995). Although the bacterial enzyme was unable to hydrolyse dipeptides or dehydropeptides, it was sensitive to inhibition by cilastatin, though at higher concentrations than was MDP (Keynan *et al.*, 1995). Cilastatin could therefore provide a lead compound for the development of more potent and specific inhibitors of this subgroup of metallo-β-lactamases. No significant sequence similarity is seen between MDP and the *cphA* gene product but a recently described dipeptidase from *Acinetobacter calcoaceticus* (ACDP) shows 48.8% similarity (25.2% identity) in amino acid sequence with human MDP although it is not inhibited by cilastatin and does not hydrolyse dehydropeptides (Adachi and Tsujimoto, 1995) (see Section 14.2.4). The ability of the *A. calcoaceticus* enzyme to hydrolyse β-lactam compounds was not examined.

14.2.3 Affinity purification and molecular characterization of MDP

The recognition of MDP as a GPI-anchored peptidase (Hooper *et al.*, 1987b) allowed us to exploit the ability of bacterial PI-PLC to release selectively and efficiently a hydrophilic form of the enzyme as an aid to purification. Previous solubilization of MDP had utilized treatment of brush border membranes with papain or 1-butanol (Kim and Campbell, 1983). A single-step affinity purification on immobilized cilastatin was then able to produce a homogeneous preparation of both porcine and human MDP protein after PI-PLC treatment (Hooper *et al.*, 1990b; Littlewood *et al.*, 1989). By using this two-step method, milligram quantities of the purified hydrophilic form of the enzyme could readily be produced from 200 g of pig kidney cortex. Release of MDP from microvillar membranes by PI-PLC also results in substantial activation of the enzyme (an effect that is not seen with other GPI-anchored enzymes) possibly the result of relaxation of conformational constraints on the active site which exist when the enzyme is anchored in the lipid bilayer (Brewis *et al.*, 1994). The amphipathic form of MDP, containing the intact GPI anchor, can be solubilized from brush border membranes by treatment with detergent. However, the choice of detergent is critical since those of low critical micellar concentration, such as Triton X-100, fail to solubilize

MDP and a detergent with a high critical micellar concentration, such as octyl glucoside, is required (Hooper and Turner, 1988b). This Triton-insolubility of MDP originally led to the suggestion that it was tightly associated with the cytoskeleton (Kim and Campbell, 1983) but it appears to be a general property of GPI-anchored proteins and is a reflection of their association in Triton-insoluble domains within the plasma membrane, which may represent caveolae (Parkin *et al.*, 1996). Purification of the detergent-solubilized form of MDP can also be effected by affinity chromatography on cilastatin–Sepharose, although in much lower yield since the amphipathic form of the enzyme binds much less efficiently to the immobilized inhibitor (Hooper and Turner, 1989a). Purified MDP has been used as starting material to obtain the complete structure of the GPI anchor of the porcine enzyme and the glycan core structure has been compared with that of the human enzyme (Brewis *et al.*, 1995; see Chapter 3).

Both the porcine and human enzymes exist as disulphide-linked homodimers, of subunit M_r 47 000 and 59 000 respectively, due to extensive differences in the level of glycosylation (Hooper *et al.*, 1990b). N-terminal sequence information on porcine MDP was used to design two degenerate oligonucleotides in order to amplify an internal fragment of 80 bp by polymerase chain reaction (PCR). This amplified fragment was used to screen a pig kidney cortex cDNA library and isolate a full length clone with an open reading frame of 1227 nucleotides encoding 409 amino acids of which the first 16 constituted the signal peptide (Rached *et al.*, 1990). When MDP mRNA was subsequently expressed in *Xenopus* oocytes, a fully functional GPI-anchored MDP activity was detected at the cell surface. Simultaneously, Adachi *et al.* (1990) succeeded in cloning the human MDP cDNA and the predicted protein differed in length by only two amino acids, confirming that the observed difference in size of the mature human and porcine MDPs was due to differences in glycosylation. The porcine enzyme possessed two putative N-linked glycosylation sites compared with four in the human sequence. Subsequently MDP has been cloned from a number of other species (rabbit, rat, mouse and sheep) (Adachi *et al.*, 1992; An *et al.*, 1994; Igarashi and Karniski, 1991; Satoh *et al.*, 1993a) and the predicted protein sequences show high homology (e.g. 83% identity between human and porcine sequences). Northern blot analysis of the mRNA from various rat tissues originally revealed 1.6 and 2.2 kb transcripts in lung and kidney but only the 1.6 kb transcript in small intestine. MDP mRNA expression was also detected, although at lower levels, in stomach, pancreas and testis but not in brain, heart, liver or spleen (Adachi *et al.*, 1992). A more detailed analysis of the expression of MDP mRNA in the mouse has detected four distinct RNA species which are generated by transcription from two alternative promoters in the 5′-non-coding region of the gene, and by the use of two different poly(A) addition sites (Habib *et al.*, 1996) and which allow tissue-specific expression (*Figure 14.2*). Promoter I is used predominantly in the small intestine and kidney, and promoter II in lung and kidney. RNase protection assays indicate that additional MDP mRNAs must exist but it is unclear whether these derive from the use of other promoters or from alternative splicing (Habib *et al.*, 1996). The sequence of a genomic clone of human MDP has been reported (Satoh *et al.*, 1993b) which was shown to contain nine introns and 10 exons. The human MDP gene has been mapped to chromosome 16q24 (Nakagawa *et al.*, 1992). The mouse MDP gene is single copy and similar in structure to the human gene

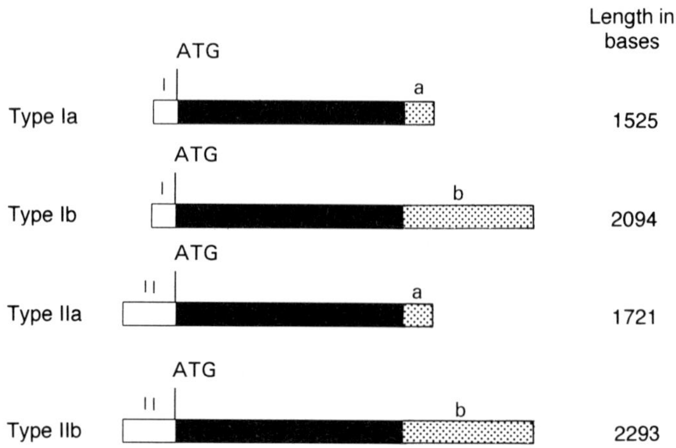

Figure 14.2. Schematic representation of MDP mRNAs synthesised by mouse kidney. Types Ia, Ib, IIa and IIb represent the various combinations of MDP mRNAs expressed in kidney. Open boxes indicate the 5′-non-coding regions of the mRNAs (I and II). Solid boxes show the coding region with ATG as indicated. Shaded boxes show the 3′-untranslated regions (a and b). Adapted from Habib *et al.* (1996).

but consists of 11 exons (Habib *et al.*, 1996). In addition to an N-terminal cleavable signal peptide, the predicted sequence of all the cloned MDPs revealed a stretch of hydrophobic residues at the C terminus consistent with a GPI anchor addition signal (see Chapter 3). Subsequent sequencing of the C-terminal peptide of porcine MDP has identified Ser368 as the amino acid to which the anchor is attached (Brewis *et al.*, 1995).

Recently, we have identified MDP as one of the major integral proteins of the pancreatic zymogen granule membrane, where, alongside glycoprotein-2 (GP-2), it is GPI-anchored (Hooper *et al.*, 1997). The role of MDP in this site is not clear, although its GPI anchor appears to be susceptible to an endogenous phospholipase A activity that promotes the release of the enzyme from the membrane. A soluble form of MDP has been identified in pancreatic secretions which may result from this releasing activity (N.M. Hooper and D. LeBel, unpublished observations). MDP has also been identified in another unexpected site, on the surface of 3T3-L1 adipocytes (S. Movahedi and N.M. Hooper, unpublished observations). Upon differentiation of these cells from fibroblasts to adipocytes MDP activity is increased over 20-fold. Again, the enzyme is GPI-anchored as shown by its release by bacterial PI-PLC and subsequent recognition by anti-cross-reacting determinant antibodies. Interestingly, MDP is also released from these cells in response to insulin treatment, through the action of a phospholipase C.

14.2.4 Site-directed mutagenesis of MDP

A surprising finding from the cloning and sequencing of MDP was that it lacked any of the classical signatures of zinc-dependent metallopeptidases [e.g. the HEXXH motif of the thermolysin family or the HXXEH of the pitrilysin family (Hooper,

1994)]. However, comparison of the sequences of MDP with other zinc enzymes revealed a sequence (DHLDH; residues 269–273) similar to a sequence (DHTH) which contains the active site zinc ligands in the D-alanyl–D-alanine-cleaving carboxypeptidase of *Streptomyces albus* (Joris *et al.*, 1983). A further zinc ligand for the bacterial enzyme has been identified as another histidine 40 residues distant on the N-terminal side of the DHXH motif. In MDP a conserved histidine (His219) 50 residues N-terminal to the putative motif could fulfil this role. As a first step to identifying the catalytic residues in MDP, we mutated D269, H270, D272 and H273 individually and expressed each of the mutant enzymes in COS-1 cells (Keynan *et al.*, 1994). All these mutants were fully active, thereby excluding a role for these residues in catalysis and negating any similarity between MDP and the *Streptomyces* carboxypeptidase. However, His219 in MDP was found to be a critical residue, since mutation to Arg, Lys or Leu resulted in complete abolition of enzyme activity, although the mutant was expressed at normal levels at the cell-surface. Subsequently, we have mutated all the remaining five histidines in porcine MDP that are conserved among different species and have identified three further critical histidines (20, 152 and 198) (S. Keynan, N.M. Hooper and A.J. Turner, unpublished observations). Mutants H20L and H198K were still able to bind to cilastatin–Sepharose, suggesting that these two histidines are involved in catalysis. The His152L mutant was both inactive and failed to bind inhibitor, suggesting that this residue is important for substrate/inhibitor binding. A sequence comparison of the recently cloned ACDP (Adachi and Tsujimoto, 1995) with MDP reveals that His20, His198 and His219 of MDP are conserved in the bacterial enzyme and lie in regions of high similarity (*Figure 14.3*). His152 of MDP, which appears to be involved in cilastatin binding, is not conserved in ACDP consistent with that enzyme being insensitive to the inhibitor. Unlike many zinc peptidases, therefore, neither MDP nor ACDP contains a pair of very closely spaced histidine residues acting

20

pMDP	1	DQFRDLAVRIMQDT**PV**I**DGHND**LPWQ	26
hMDP	1	DFFRDEAERIMRDS**PV**I**DGHND**LPWQ	26
ACDP	1	MKPSHI**PV**F**DGHND**ALTR	18

198 219

pMDP	191	GVM**ID**LA**H**VSVATMRAALK**L**SQA**P**VIFS**HS**S**A**YSLC	226
hMDP	191	GVL**ID**LA**H**VSVATMKATLQ**L**SRA**P**VIFS**HS**S**A**YSVC	226
ACDP	198	KMV**ID**VS**H**MNEQAFWDTVD**L**LQQ**P**IVAT**HS**N**A**HALC	233

Figure 14.3. Partial alignment of the amino acid sequences of porcine and human MDP and *A. calcoaceticus* dipeptidase (ACDP). Regions of the sequence of porcine MDP (pMDP, amino acids 1–26 and 191–226; Rached *et al.*, 1990) are aligned with those of human MDP (hMDP; Adachi *et al.*, 1990) and the dipeptidase from *A. calcoaceticus* (ACDP; Adachi and Tsujimoto, 1995). Residues conserved between all species are in bold. The conserved histidine residues in MDP are indicated above the sequences by their position number. The alignment was generated by using the Wisconsin GCG program, BESTFIT.

as zinc ligands. An additional residue in MDP, Glu125, has also been shown to be critical for enzyme activity (Adachi *et al.*, 1993) and is conserved in ACDP.

Site-directed mutagenesis has also been used to identify the cysteine residue(s) involved in disulphide-linked dimerization in MDP (Keynan *et al.*, 1996). All six fully conserved cysteines in porcine MDP were mutated either to glycine or alanine. Only the Cys361 mutant migrated on SDS gels as a monomer (47 kDa) under non-reducing conditions. Thus, the subunits form a symmetrical dimer with Cys361 alone being involved in interchain disulphide linkage. This places the disulphide link close to the site of GPI anchor addition (Ser368) and to the membrane surface. Titration of porcine MDP with the thiol reagent 2-nitro-5-thiosulphabenzoate indicated that there are two additional intra-chain disulphide bonds in the protein.

14.2.5 MDP and Wilms' tumour

Wilms' tumour (WT) is an embryonic malignancy of the kidney which can appear both in inherited and non-hereditary forms. In patients in whom the WT protein is missing, there is also a loss of a region from chromosome 16, including a specific loss of MDP mRNA. This observation has implicated a portion of the chromosomal region which contains the MDP gene in tumour initiation or progression by acting as a suppressor gene (Austry *et al.*, 1993). In order to establish whether MDP itself is directly involved in this process, we have examined the expression of MDP in a number of transformed and non-transformed cell lines (S. Keynan, A. Asipu, N.M. Hooper, A.J. Turner and G.E. Blair, unpublished observations). Although MDP enzyme activity and mRNA were detected in primary baby rat kidney (BRK) cells maintained in culture for up to 4 weeks, negligible MDP activity or mRNA could be detected in a number of stable transformed cell lines, derived either by transformation with the DNA tumour viruses SV40 or adenovirus, or human tumour cell lines. In addition, MDP expression in several clonal BRK cell lines transformed by the temperature-sensitive SV-40-tsA mutant, expressing a thermolabile mutant T antigen, was investigated. MDP activity was not detectable in tsSV-BRK cell lines grown at the permissive temperature (33°C). However, after 5–14 days of incubation at the non-permissive temperature (39.5°C), MDP enzyme activity and protein were detected. Thus, it would appear that MDP may be a transformation-sensitive protein whose expression is actively repressed by DNA viral oncogenes such as SV40 T antigen.

14.3 Aminopeptidase P (AP-P)

14.3.1 Discovery and isolation of AP-P

AP-P was first identified in pig kidney as an enzyme capable of releasing the N-terminal amino acid from peptides with a prolyl residue in the penultimate position (Dehm and Nordwig, 1970). Peptides of the X–Pro–Y sequence were hydrolysed most readily and a synthetic substrate (glycyl-prolyl-hydroxyproline; Gly–Pro–Hyp) was routinely used to monitor enzyme activity. AP-P has been located in the lungs and brush border membranes of kidney and intestine in a number of species (Hooper and Turner, 1988a; Lasch *et al.*,

1986; Orawski and Simmons, 1995; Orawski *et al.*, 1987; Ryan *et al.*, 1994) and soluble forms of AP-P have been identified in serum (Holtzman *et al.*, 1987; Ryan *et al.*, 1992) and brain (Harbeck and Mentlein, 1991). The serum form of AP-P is probably released from the membrane by phospholipase or proteinase action. A cytosolic form, while exhibiting similar specificity and other characteristics to membrane AP-P (Harbeck and Mentlein, 1991), is most likely a separate gene product; it has been much less studied and will not be discussed further here. Studies on bovine lung (Orawski *et al.*, 1987) and rat intestine (Lasch *et al.*, 1986) showed that AP-P was an integral membrane protein and work on the pig and human kidney enzymes revealed AP-P to be GPI anchored. Bacterial PI-PLCs were shown to release a hydrophilic form of AP-P from a membrane preparation (Hooper and Turner, 1988a) as also did a preparation of serum phospholipase D (Hooper and Turner, 1989b). Subsequently, AP-P has been solubilized by bacterial PI-PLC from a number of other species and tissues (Koelsch *et al.*, 1994; Orawski and Simmons, 1995; Ryan *et al.*, 1994; Simmons and Orawski, 1992) indicating that GPI anchorage of AP-P is widespread. The selective solubilization of AP-P from membrane preparations by bacterial PI-PLC was utilized as an important initial step in purification, which was first achieved for the porcine kidney enzyme (Hooper *et al.*, 1990a). No inhibitor affinity chromatographic step has yet been developed for the isolation of AP-P and purification depended on a series of conventional chromatographic procedures. More recently guinea pig AP-P has been purified by immunoaffinity chromatography using a monoclonal antibody developed to the enzyme (Ryan *et al.*, 1994). On sodium dodecyl sulphate-polyacrylamide gel electrophoresis (SDS-PAGE), pig kidney AP-P migrates as a polypeptide of M_r 90 000–95 000 under reducing or non-reducing conditions (Hooper *et al.*, 1990a). Similar values have been obtained for the bovine lung (Simmons and Orawski, 1992) and rat lung enzymes (Orawski and Simmons, 1995). On gel filtration, pig kidney AP-P exhibited an M_r of 280 000 (Hooper *et al.*, 1990a). This would suggest that AP-P exists as a trimer in solution unless the protein behaves anomalously on gel filtration. The rat brain cytosolic AP-P has been reported to exist as a dimer or trimer in solution depending on ionic strength (Harbeck and Mentlein, 1991). Up to 25% of the weight of AP-P in both pig kidney and rat lung is due to the presence of N-linked sugars as revealed by enzymic deglycosylation (Hooper *et al.*, 1990a; Orawski and Simmons, 1995). The presence of a GPI anchor on PI-PLC-cleaved AP-P has been confirmed immunologically (Hooper *et al.*, 1990a), but the structure of the glycolipid anchor has not yet been determined.

14.3.2 Characterization of AP-P

The observation that chelating agents, such as ethylene diamine tetraacetic acid (EDTA) and 1,10-phenanthroline, were able to inhibit AP-P activity, suggested that AP-P was a metalloenzyme (Hooper *et al.*, 1990a; Lasch *et al.*, 1988; Orawski *et al.*, 1987) and purified pig kidney AP-P was subsequently shown to contain 1 mol of zinc per mol of protein (Hooper *et al.*, 1992). However, AP-P is unusual among zinc peptidases in containing an additional metal requirement. Dehm and Nordwig (1970) observed that Mn^{2+} ions were required for optimal activity and assays of the enzyme have routinely included Mn^{2+} in the buffer, with Gly–Pro–Hyp as substrate. More detailed studies on the pig kidney enzyme showed that Mn^{2+} and Co^{2+}, and to a lesser extent Ca^{2+} and Mg^{2+}, stimulated enzyme activity whereas certain other divalent cations (e.g. Cu^{2+}, Ni^{2+} and Zn^{2+}) were inhibitory (Hooper *et al.*, 1990a). Unlike some other aminopeptidases, AP-P is not inhibited by actinonin, amastatin, bestatin and puromycin although certain ACE inhibitors (in micromolar concentrations), such as enalaprilat and ramiprilat, can inhibit AP-P (Hooper *et al.*, 1992). A specific inhibitor of AP-P, apstatin (N[(2S, 3R)-3-amino-2-hydroxy-4-phenylbutanoyl]-L-prolyl-L-prolyl-L-alaninamide) (*Figure 14.1b*) has recently been developed and exhibits a K_i value of 2.6 μM against the rat lung enzyme (Prechel *et al.*, 1995).

The first study of the substrate specificity of AP-P used a partially purified preparation of the enzyme from pig kidney (Dehm and Nordwig, 1970) and assays were performed in the presence of 2 mM Mn^{2+}. Peptides of the type X–Pro–Y were cleaved, with hydrolysis of Gly–Pro–Y peptides showing the fastest reaction rates of those tested. Subsequent studies on the bovine lung enzyme (Orawski *et al.*, 1987, 1989; Simmons and Orawski, 1992) confirmed the requirement for proline in the penultimate position of any peptide substrate and hydrolysis could also be detected, although at a much slower rate, when Pro was replaced by Ala. In these studies, Arg–Pro–Pro was seen to be cleaved most rapidly by the enzyme whereas Gly–Pro–Hyp was a relatively poor substrate. The reason for the discrepancy with the original study by Dehm and Nordwig (1970) was that Mn^{2+} was not included in the assay buffer in studies on the bovine lung enzyme. As we and others have subsequently shown (Lloyd *et al.*, 1996; Orawski and Simmons, 1995) the presence or absence of Mn^{2+} is critical to the apparent substrate specificity of AP-P (see below).

The initial observation that AP-P could hydrolyse peptides of the type X–Pro–Y, with greatest reaction rates towards peptides of the sequence Gly–Pro–Y– led to the suggestion that the enzyme may have a role in the degradation of collagen (Dehm and Nordwig, 1970) which contains repeating Gly–Pro–Y sequences. A report of the association of AP-P deficiency with an inherited metabolic disease characterized by massive urinary excretion of iminooligopeptides such as Gly–Pro–Hyp–Gly (Blau *et al.*, 1988) provides some support for this hypothesis. However, most attention has focused on the involvement of AP-P in the metabolism of the vasodilator bradykinin (Arg–Pro–Pro–Gly–Phe–Ser–Pro–Phe–Arg). Bradykinin is an excellent substrate for AP-P *in vitro*, hydrolysis occurring at the Arg1–Pro2 bond, which results in loss of biological activity. Prechel *et al.* (1995) have employed the AP-P inhibitor apstatin to assess the relative contributions of AP-P and ACE to bradykinin degradation. [^{3}H]-bradykinin was perfused through the isolated rat lung in the presence or absence of apstatin and/or an ACE inhibitor. Quantitative analysis of bradykinin and metabolites in the perfusate by high performance liquid chromatography (HPLC) revealed that the bradykinin-degrading activity in the rat pulmonary vascular bed can be fully accounted for by a combination of AP-P (30%) and ACE (70%). Several other regulatory peptides possess a penultimate prolyl residue at their N termini suggesting they may be susceptible to inactivation by AP-P. These include substance P, interleukin-2, as well as members of the pancreatic polypeptide family such as neuropeptide Y and peptide YY. N-terminal processing of peptide YY and neuropeptide Y generates receptor-selective forms of the peptides and AP-P has been shown to contribute to this processing

(Medeiros and Turner, 1994, 1996). In view of the diverse substrates available to AP-P, its physiological function may be tissue-specific, contributing to vasoactive peptide metabolism in the lung, digestion of proline-containing peptides and proteins in the intestine and possibly cytokine metabolism in the immune system. The availability of a selective inhibitor of AP-P should now allow these various roles to be examined in more detail.

The role of Mn^{2+} in AP-P activity has recently been shown to be substrate-specific (Lloyd *et al.*, 1996; Orawski and Simmons, 1995). Thus, the hydrolysis of certain substrates (Gly–Pro–Hyp; substance P) is substantially activated by Mn^{2+}, with optimum activity being detected at 4–10 mM of the cation in agreement with the original report of Dehm and Nordwig (1970). However, the hydrolysis of bradykinin and Arg–Pro–Pro shows no activation at any concentration of Mn^{2+}, which is inhibitory above 10^{-5} M (Lloyd *et al.*, 1996). There appears to be no simple dependence of the effect on the nature of the amino acid in the P_1 position in the substrate since both substance P and bradykinin possess an N-terminal arginine yet the hydrolysis of one is activated and the other is inhibited. Length of peptide substrate also does not appear to be a critical factor in determining Mn^{2+} sensitivity (Lloyd *et al.*, 1996). The inhibition of AP-P by certain ACE inhibitors is also a substrate- and Mn^{2+}-dependent phenomenon. The hydrolysis of bradykinin by AP-P is only inhibited by ACE inhibitors when Mn^{2+} is present in the assay buffer. Substrates for AP-P can therefore be divided into two groups based on their responses to certain inhibitors and cation activators, leading to the possibility that AP-P may possess two active sites. Consistent with this hypothesis is a marked non-linearity in the kinetics of hydrolysis of bradykinin by purified AP-P and different sensitivities of bradykinin and Gly–Pro–Hyp hydrolysis to EDTA and the thiol reagent *p*-chloromercuriphenylsulphonate (Lloyd *et al.*, 1996).

14.3.3 AP-P is a member of the proline peptidase family

Bazan and colleagues (Bazan *et al.*, 1994) have identified structural and limited sequence similarities between the structure of *Escherichia coli* methionine aminopeptidase and the C-terminal domain of creatinase from *Pseudomonas putida*. In particular, both display the so-called 'pitta-bread' fold that is quite distinct from the fold present in, for example, leucine aminopeptidase. Further sequence comparisons (Bazan *et al.*, 1994) have suggested that this pitta-bread fold may be present in several other hydrolytic enzymes. These include agropine synthase, prolidase (proline dipeptidase) and *E. coli* AP-P and all these proteins have been assigned to the 'proline peptidase' family although not all have peptidase activity. A comparison of the sequences of prolidase, *E. coli* AP-P, first cloned by Yoshimoto *et al.* (1989), and limited partial amino acid sequences of guinea pig AP-P (Denslow *et al.*, 1994), suggest that mammalian AP-P is also a member of this family. By using a combination of Edman degradation and mass spectrometry, Vergas Romero *et al.* (1995) were recently able to obtain the full amino acid sequence of AP-P purified to homogeneity from pig kidney. This sequence, of course, lacked the N- or C-terminal signal sequences required for membrane targeting and GPI anchor addition. Sequence comparisons further emphasized the similarity between pig kidney AP-P and other members of the proline peptidase family, although the similarities were restricted to the C-terminal half of AP-P, suggesting that the active site lies in this region. Knowledge of the amino acid sequence of purified AP-P revealed that it did not contain any recognized zinc signature and therefore must represent a novel zinc peptidase. Within the proline peptidase family, *E. coli* methionine aminopeptidase is a dual-metal ion enzyme containing a pair of closely linked metal ions (reportedly Co^{2+} for optimal activity; Roderick and Matthews, 1993). The eukaryotic methionyl aminopeptidases are also cobalt-dependent, but only distantly related to the prokaryotic enzyme (Arfin *et al.*, 1995). Prolidase is also proposed to require the action of two metal ions in the active site (Mock and Liu, 1995). The assignment of mammalian AP-P to the proline peptidase family is therefore consistent with the Mn^{2+} dependence of the enzyme in addition to the presence of 1 mol zinc per mol protein. The nature of any additional metal ion in AP-P is unknown at present but pig kidney AP-P does not contain Co^{2+} (Lloyd *et al.*, 1996).

14.3.4 Cloning and expression of AP-P

AP-P has now been cloned from a porcine kidney cortex cDNA library following the use of PCR to identify sub-libraries enriched in AP-P clones. A full length clone, containing an insert of 3563 bp was isolated and sequenced completely (Hyde *et al.*, 1996). The full sequence of the clone comprised 345 nucleotides of 5′-untranslated region, an open reading frame of 2019 nucleotides, and 1199 nucleotides of 3′-untranslated region, including 24 nucleotides of the poly(A) tail (*Figure 14.4*). The full length cDNA sequence encodes a protein of 673 amino acids with a calculated M_r of 75 755, compared with an estimated M_r of 91 000 for the purified protein on SDS gels. Thus approximately 17% of the mass is due to glycosylation and the sequence contains six putative N-linked glycosylation sites. A hydropathy plot of the protein sequence reveals potential transmembrane-spanning hydrophobic regions at both N and C termini. At the N terminus is a typical cleavable signal sequence but it is not possible to identify the N terminus of the mature protein unequivocally. Two alternative residues are predicted (His22 or Lys24) with almost equal weight by the method of Von Heijne (1986). Since the purified protein is blocked at the N terminus it has not been possible to confirm the true initial residue. The N terminus of the sequence of AP-P reported by Vergas Romero *et al.* (1995) was in error due to partial proteolysis as the authors acknowledged. Another error in that sequence was also detected from the cloned cDNA: Cys530 was reported as a glutamine residue in Vergas Romero *et al.* (1995). As expected for a GPI-

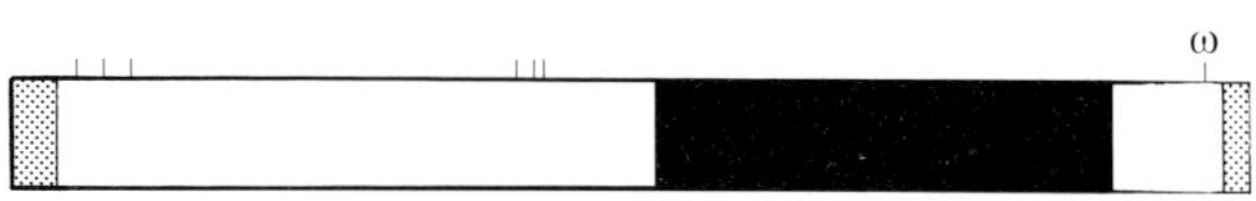

Figure 14.4. Schematic representation of the primary structure of AP-P. The model shows the complete amino acid sequence (deduced from the cDNA sequence). The hydrophobic domains at the N and C termini are shown in grey. The vertical lines indicate the positions of the six putative N-linked glycosylation sites. ω is the site of the attachment of the GPI anchor. The black box indicates the region of the sequence with the highest homology to other members of the proline peptidase family.

anchored protein, there is a stretch of predominantly hydrophobic residues at the C terminus which serves as the signal for attachment of the GPI anchor. By using the method of Udenfriend and Kodukula (1995), the ω residue to which the GPI moiety is attached is predicted to be Ala649. This places Arg and Ala in the ω+1 and ω+2 positions consistent with the established criteria. The protein sequence of mature AP-P determined by Vergas Romero *et al.* (1995) terminates at Ala649 further supporting the location of the ω residue. Overall, the sequence of pig kidney AP-P (Hyde *et al.*, 1996) has sequence similarities with the AP-Ps from *E. coli* (44.8%) (Butler *et al.*, 1993) and *S. lividans* (50.3%) (Yoshimoto *et al.*, 1989), especially in the C-terminal half of the protein. There are also marked similarities in this region with other members of the proline peptidase family, as noted above.

The cloned pig kidney AP-P cDNA has been successfully expressed in COS-1 cells (Hyde *et al.*, 1996) producing an activity that could hydrolyse both Gly–Pro–Hyp and bradykinin at the predicted bonds. The hydrolysis of Gly–Pro–Hyp was stimulated by Mn^{2+} and inhibited by the ACE inhibitor enalaprilat. Substrate hydrolysis was shown to be inhibited also by EDTA and apstatin. The protein was expressed at the surface of COS-1 cells and could be released in a hydrophilic form by treatment with PI-PLC (Hyde *et al.*, 1996). On a Western blot a single polypeptide (M_r 91 000) was detected. Thus a fully functional AP-P was expressed showing all the expected kinetic and immunological properties of the previously purified enzyme.

As with the protein sequence reported by Vergas Romero *et al.* (1995), no likely zinc-binding or catalytic residues are predicted in AP-P and site-directed mutagenesis will be required to identify critical residues in the protein. In particular, chemical modification of the purified pig kidney enzyme with diethylpyrocarbonate has indicated the presence in the protein of at least two critical histidine residues whose modification is prevented by protection with substrate or inhibitor (Lim and Turner, 1996). Comparison of the C-terminal half of pig kidney AP-P (residues 350–623) with other members of the proline peptidase family identifies seven histidines of which only four are conserved (residues 403, 493, 497 and 506) and which therefore represent primary targets for mutagenesis. These studies are now in progress. Region 491–507 of pig kidney AP-P is compared with other members of the proline peptidase family in *Figure 14.5.*

pkAPP	YG**HG**T**GH**GI**G**NFLEV**HE**
ecAPP	FM**HG**LS**H**WL**G**L--D**VHD**
huPRO	MP**HG**L**GH**FL**G**I--D**VHD**
ecPRO	MP**HG**I**GH**PL**G**L--Q**VHD**
ppCRE	YG**H**SF**G**------TLSHY

Figure 14.5. Comparison of a partial sequence of pig kidney AP-P (residues 491–507) with other members of the proline peptidase family. pkAPP, pig kidney AP-P; ecAPP, *E. coli* AP-P; huPRO, human prolidase; ecPRO, *E. coli* prolidase; ppCRE, *Pseudomonas putida* creatinase. Identical residues conserved between species are indicated in bold. A full comparison of the C-terminal sequences of these proteins is provided in Vergas Romero *et al.* (1995). Reproduced from Hyde *et al.* (1996) with permission from The Biochemical Society.

Overall the structural studies on AP-P indicate that it is a member of a novel zinc metallopeptidase family. Much remains to be learnt about mammalian AP-P: the catalytic mechanism and the explanation of the anomalous behaviour of certain substrates to cations and inhibitors; its relation to the soluble forms of AP-P including the cytosolic form; its functional roles other than in bradykinin metabolism, and a more precise description of its tissue and cellular localization.

14.4 Carboxypeptidase M (CP-M)

14.4.1 Discovery of CP-M

Skidgel *et al.* (1984) reported the presence of a CP-N-like activity in membrane fractions from various human and bovine tissues, and in cultured cells, including placenta, kidney, lung and pulmonary arterial endothelial cells. This membrane-bound basic carboxypeptidase activity was subsequently purified from human placental microvilli following solubilization with 3-[(3-cholamidopropyl)-dimethylammonio]-1-propanesulphonate (CHAPS) (Skidgel *et al.*, 1989), and shown to be structurally, catalytically and immunologically distinct from other known carboxypeptidases and it was named carboxypeptidase M, to indicate that it was membrane-bound. Human CP-M is a single chain protein with an M_r of 62 000 as deduced by SDS-PAGE, and an M_r of 73 000 upon gel filtration (Skidgel *et al.*, 1989). Like all cell-surface peptidases, CP-M is a glycoprotein, as shown by a reduction in mass to an M_r of 47 600 after chemical deglycosylation. Subsequent cDNA cloning and sequencing of CP-M revealed the presence of six potential N-linked glycosylation sites (see Section 14.4.2) consistent with its glycoprotein nature. CP-M is present in a number of tissues, the highest levels being found in lung and placenta, with lower amounts in kidney, intestine, brain, peripheral nerves and blood vessels.

14.4.2 Cloning of CP-M

CP-M was cloned from a human placental cDNA library following N-terminal sequencing of the protein and the design of complementary oligonucleotide probes (Tan *et al.*, 1989). Sequencing of the cDNA revealed an open reading frame of 1317 base pairs that coded for a protein of 439 amino acids. The N-terminal protein sequence matched the deduced amino acid sequence starting from residue 14. Although the initiating Met was not encoded by the cDNA, hydropathy plot analysis revealed an N-terminal hydrophobic region that is absent from the mature protein. In addition, a second hydrophobic region was identified at the C terminus which could act as a signal for GPI anchor attachment. That CP-M was indeed GPI anchored came from metabolic labelling of MDCK cells with [^{3}H]-ethanolamine followed by immunoprecipitation of radiolabelled CP-M, and its release from MDCK cell membranes and human placental microvillar membranes by bacterial PI-PLC (Deddish *et al.*, 1990). To date, the site of GPI anchor attachment in CP-M has not been reported. However, by using the method of Udenfriend and Kodukula (1995), a possible ω site at Ser406 with Ala in the ω+1 and ω+2 positions would be the most favourable.

In addition to the membrane-bound form, CP-M is also present as a soluble form in various body fluids such as urine, seminal fluid and amniotic fluid (Skidgel *et al.*, 1984, 1988). The levels of the soluble form of the enzyme are seen to increase in bronchoalveolar lavage in human lung disease (Dragovic *et al.*, 1995). The mechanism of release of CP-M has not been determined but could, because of the presence of the GPI anchor, be due to the action of a phospholipase. However, unlike MDP and AP-P, CP-M is readily released from the membrane by trypsin (Deddish *et al.*, 1990) indicating that it has a protease-susceptible stalk region. Thus, the action of an endogenous protease in generating the soluble form cannot be ruled out.

14.4.3 The carboxypeptidase family

Structural and mechanistic studies on, in particular, pancreatic CP-A have provided much insight into the catalytic mechanism of zinc metallopeptidases, and clearly revealed those residues involved in binding the zinc atom. The carboxypeptidase family has a unique zinc-binding motif consisting of two histidines and a glutamic acid residue. The first His is separated from the Glu by only two residues, while the second His is some 100 or so residues downstream (*Figure 14.6*).

Molecular cloning revealed that CP-M had higher sequence similarity to plasma CP-N and carboxypeptidase H (enkephalin convertase) (41% sequence identity), than to pancreatic CP-A and -B (15% sequence identity) (Tan *et al.*, 1989). Sequence analysis of all the mammalian carboxypeptidases that have been cloned (Hooper, 1994) shows that the enzymes fall into two groups: those that have high sequence identity with the pancreatic carboxypeptidases and those that do not. The first group comprises pancreatic CP-A

(a)	1 4		108–135
CPM	VANM**H**GD**E**TVGRE	....	L**H**GGAL
CPN	VGNM**H**GN**E**ALGRE	....	L**H**GGAV
CPH	IGNM**H**GN**E**AVGRE	....	L**H**GGDL
AEBP1	TAGI**H**GN**E**VLGRE	....	L**H**GGER
dCPa	VGNM**H**GD**E**PLARP	....	L**H**GGSV
dCPb	IGNM**H**GN**E**VVGRE	....	L**H**GGSL
dCPc	VAGI**H**GNAPVGTE	....	LDGGSL

(b)			
CPA1	DTGI**H**SR**E**WVTQA	....	I**H**SYSQ
CPA2	DAGI**H**AR**E**WVTQA	....	L**H**SYSQ
mcCPA	DCGI**H**AR**E**WISPA	....	F**H**SYSQ
CPB	DCGF**H**AR**E**WISPA	....	I**H**SYSQ
CPU	DTGI**H**AR**E**WISPA	....	M**H**SYSQ

Figure 14.6. Zinc-binding residues identified in the carboxypeptidases. Sequences are aligned starting from the HXXE motif. (a) Carboxypeptidases M (CPM), N (CPN) and H (CPH), the adipocyte transcriptional repressor (AEBP1) and the three domains of the recently identified duck liver carboxypeptidase (dCPa, dCPb, dCPc); (b) carboxypeptidases A1 (CPA), A2 (CPA2), B (CPB), U (CPU) and mast cell carboxypeptidase A (mcCPA). Modified from Hooper (1994).

and -B, mast cell carboxypeptidase and carboxypeptidase U, while the second group consists of CP-H and -N, and CP-M. A recent addition to this second group was revealed with the cloning and sequencing of a cDNA that encodes a M_r 180 000 membrane-bound glycoprotein from duck liver that binds the duck hepatitis B virus (Kuroki *et al.*, 1995). This large protein contains three carboxypeptidase-like domains, of which the first two contain all the putative active site residues identified in CP-H, -M and -N, whereas the third domain lacks two of the zinc ligands (*Figure 14.6*). This membrane-bound carboxypeptidase, which would appear to be enzymically active (Kuroki *et al.*, 1995), is probably anchored through a hydrophobic sequence of amino acids near the C terminus. A more recent addition to the carboxypeptidase family is a eukaryotic transcriptional repressor involved in adipocyte differentiation that shows similarity (40% identity) to CP-H, -M and -N (He *et al.*, 1995), retaining all the zinc ligands and displaying enzyme activity (*Figure 14.6*).

14.4.4 Characterization of CP-M activity

Like the other basic carboxypeptidases, CP-M is activated by Co^{2+} ions and is inhibited by the arginine analogues, D,L-2-mercaptomethyl-3-guanidinoethylthiopropanoic acid (MGTA) and guanidinoethylmercaptosuccinic acid (GEMSA) (Deddish *et al.*, 1990; Skidgel *et al.*, 1989). However, no selective inhibitor of CP-M has yet been reported. Both the Co^{2+} activation and inhibitor binding of CP-M are enhanced at low pH (Deddish *et al.*, 1989). CP-M readily cleaves the basic residues Arg or Lys from the C terminus of both synthetic and natural peptides (Deddish *et al.*, 1990; Skidgel *et al.*, 1989), with C-terminal Arg being hydrolysed faster than Lys. Natural peptides that have been shown to be hydrolysed by CP-M, at least *in vitro*, include bradykinin, anaphylatoxin C5a, dynorphin A (1–13), Met^5-Lys^6-enkephalin, Met^5-Arg^6-enkephalin and Leu^5-Arg^6-enkephalin. In the case of bradykinin and the histamine-releasing anaphylatoxin C5a, removal of the C-terminal Arg by CP-M would lead to an alteration of the biological properties of the peptide, although it has yet to be established whether CP-M is involved in this process *in vivo*. Another possible role for CP-M is in the metabolism of growth factors, such as epidermal growth factor, that contain C-terminal basic residues. Although epidermal growth factor is cleaved by a carboxypeptidase in liver and fibroblasts either at the cell surface or in early endosomes, and CP-M has been shown to readily remove the C-terminal Arg from epidermal growth factor *in vitro* (McGwire and Skidgel, 1995), the role of this conversion *in vivo* is not known. Recently, CP-M has been shown to be identical to the differentiation regulated antigen MAX.1 on the surface of macrophages and lymphocytes (de Saint-Vis *et al.*, 1995; Rehli *et al.*, 1995), although as yet the functional role of the enzyme during monocytic differentiation and activation has not been elucidated.

Acknowledgements

We thank the Wellcome Trust and the British Heart Foundation for support of this work.

References

Adachi H, Tsujimoto M. (1995) Cloning and expression of dipeptidase from *Acinetobacter cacoaceticus* ATCC 23055. *J. Biochem.* **118**: 555–561.

Adachi H, Tawaragi Y, Inuzuka C, Kubota I, Tsujimoto M, Nishihara T, Nakazato H. (1990) Primary structure of human microsomal dipeptidase deduced from molecular cloning. *J. Biol. Chem.* **265**: 3992–3995.

Adachi H, Ishida N, Tsujimoto M. (1992) Primary structure of rat renal dipeptidase and expression of its mRNA in rat tissues and COS-1 cells. *Biochim. Biophys. Acta* **1132**: 311–314.

Adachi H, Katayama T, Nakazato H, Tsujimoto M. (1993) Importance of Glu-125 in the catalytic activity of human renal dipeptidase. *Biochim. Biophys. Acta* **1163**: 42–48.

An S, Schmidt FJ, Campbell BJ. (1994) Molecular cloning of sheep lung dipeptidase: a glycosyl phosphatidylinositol-anchored ectoenzyme that converts leukotriene D_4 to leukotriene E_4. *Biochim. Biophys. Acta* **1226**: 337–340.

Arfin SM, Kendall RL, Hall L, Weaver LH, Stewart AE, Matthews BW, Bradshaw RA. (1995) Eukaryotic methionyl aminopeptidases: two classes of cobalt-dependent enzymes. *Proc. Natl Acad. Sci. USA* **92**: 7714–7718.

Austry E, Cohen-Salmon M, Antignac C, Beroud C, Henry I, Van Cong N, Brugieres L, Junien C, Jeanpierre C. (1993) Isolation of kidney complementary DNAs down-expressed in Wilms' tumour by a subtractive hybridization approach. *Cancer Res.* **53**: 2888–2894.

Bazan JF, Weaver LH, Roderick SL, Huber R, Matthews BW. (1994) Sequence and structure comparison suggest that methionine aminopeptidase, prolidase, aminopeptidase P, and creatinase share a common fold. *Proc. Natl Acad. Sci. USA* **91**: 2473–2477.

Bergmann M, Schleich H. (1932) Uber die enzymatische Spaltung dehydrierter Peptide. Auffindung einer Dehydropeptidase. *Z. Physiol. Chem.* **205**: 65–82.

Blau N, Niederwieser A, Shmerling DH. (1988) Peptiduria presumably caused by aminopeptidase P deficiency. A new inborn error of metabolism. *J. Inher. Metab. Dis.* **11**: 240–242.

Brewis IA, Turner AJ, Hooper NM. (1994) Activation of the glycosyl-phosphatidylinositol-anchored membrane dipeptidase upon release from porcine kidney membranes by phospholipase C. *Biochem. J.* **303**: 633–638.

Brewis IA, Ferguson MAJ, Mehlert A, Turner AJ, Hooper NM. (1995) Structures of the glycosyl-phosphatidylinositol anchors of porcine and human membrane dipeptidase. Interspecies comparison of the glycan core structures and further structural studies on the porcine anchor. *J. Biol. Chem.* **270**: 22946–22956.

Butler MJ, Bergeron A, Soostmeyer G, Zimny T, Malek LT. (1993) Cloning and characterization of an aminopeptidase P-encoding gene from *Streptomyces lividans*. *Gene* **123**: 115–119.

Curthoys NP. (1987) Extracellular catabolism of glutathione. In: *Mammalian Ectoenzymes* (eds A.J. Kenny and A.J. Turner), pp. 249–264, Elsevier, Amsterdam.

de Saint-Vis B, Cupillard L, Padrau-Garcia D, *et al.* (1995) Distribution of carboxypeptidase M on lymphoid and myeloid cells parallels the other zinc-dependent proteases CD10 and CD13. *Blood* **86**: 1098–1105.

Deddish PA, Skidgel RA, Erdös EG. (1989) Enhanced Co^{2+} activation and inhibitor binding of carboxypeptidase M at low pH. Similarity to carboxypeptidase H (enkephalin convertase). *Biochem. J.* **261**: 289–291.

Deddish PA, Skidgel RA, Kriho VB, Li X-Y, Becker RP, Erdös EG. (1990) Carboxypeptidase M in Madin-Darby canine kidney cells. Evidence that carboxypeptidase M has a phosphatidylinositol glycan anchor. *J. Biol. Chem.* **265**: 15083–15089.

Dehm P, Nordwig A. (1970) The cleavage of prolyl peptides by kidney peptidases. Partial purification of a 'X-prolyl-aminopeptidase' from swine kidney microsomes. *Eur. J. Biochem.* **17**: 364–371.

Denslow ND, Ryan JW, Nguyen HP. (1994) Guinea pig membrane-bound aminopeptidase P is a member of the proline peptidase family. *Biochem. Biophys. Res. Commun.* **205**: 1790–1795.

Dragovic T, Schraufnagel DE, Becker RP, Sekosan M, Votta-Velis EG, Erdös EG. (1995) Carboxypeptidase M activity is increased in bronchoalveolar lavage in human lung disease. *Am. J. Respir. Crit. Care Med.* **152**: 760–764.

Habib GM, Barrios R, Shi Z-Z, Lieberman MW. (1996) Four distinct membrane-bound dipeptidase RNAs are differentially expressed and show discordant regulation with γ-glutamyl transpeptidase. *J. Biol. Chem.* **271**: 16273–16280.

Harbeck H-T, Mentlein R. (1991) Aminopeptidase P from rat brain. Purification and action on bioactive peptides. *Eur. J. Biochem.* **198**: 451–458.

He G-P, Muise A, Li AW, Ro H-S. (1995) A eukaryotic transciptional repressor with carboxypeptidase activity. *Nature* **378**: 92–96.

Holtzman EJ, Pillay G, Rosenthal T, Yaron A. (1987) Aminopeptidase P activity in rat organs and human serum. *Analyt. Biochem.* **162**: 476–484.

Hooper NM. (1994) Families of zinc metalloproteases. *FEBS Lett.* **354**: 1–6.

Hooper NM, Turner AJ. (1988a) Ectoenzymes of the kidney microvillar membrane. Aminopeptidase P is anchored by a glycosyl-phosphatidylinositol moiety. *FEBS Lett.* **229**: 340–344.

Hooper NM, Turner AJ. (1988b) Ectoenzymes of the kidney microvillar membrane. Differential solubilization by detergents can predict a glycosyl-phosphatidylinositol membrane anchor. *Biochem. J.* **250**: 865–869.

Hooper NM, Turner AJ. (1989a) Ectoenzymes of the kidney microvillar membrane. Isolation and characterization of the amphipathic form of renal dipeptidase and hydrolysis of its glycosyl-phosphatidylinositol anchor by an activity in plasma. *Biochem. J.* **261**: 811–818.

Hooper NM, Turner AJ. (1989b) Hydrolysis of the glycosyl-phosphatidylinositol anchors of renal microvillar peptidases by a plasma phospholipase D. *Biochem. Soc. Trans.* **17**: 885–886.

Hooper NM, Keen J, Pappin DJC, Turner AJ. (1987a) Pig kidney angiotensin converting enzyme. Purification and characterization of amphipathic and hydrophilic forms of the enzyme establishes C-terminal anchorage to the plasma membrane. *Biochem. J.* **247**: 85–93.

Hooper NM, Low MG, Turner AJ. (1987b) Renal dipeptidase is one of the membrane proteins released by phosphatidylinositol-specific phospholipase C. *Biochem. J.* **244**: 465–469.

Hooper NM, Hryszko J, Turner AJ. (1990a) Purification and characterization of pig kidney aminopeptidase P. A glycosyl-phosphatidylinositol-anchored ectoenzyme. *Biochem. J.* **267**: 509–515.

Hooper NM, Keen JN, Turner AJ. (1990b) Characterization of the glycosyl-phosphatidylinositol-anchored human renal dipeptidase reveals that it is more extensively glycosylated than the pig enzyme. *Biochem. J.* **265**: 429–433.

Hooper NM, Hryszko J, Oppong SY, Turner AJ. (1992) Inhibition by converting enzyme inhibitors of pig kidney aminopeptidase P. *Hypertension* **19**: 281–285.

Hooper NM, Cook S, Laine J, LeBel D. (1997) Identification of membrane dipeptidase as a major protein of the pancreatic zymogen granule membrane, and evidence for cleavage of its GPI anchor by phospholipase A. *Biochem. J.*, in press.

Hyde R, Hooper NM, Turner AJ. (1996) Molecular cloning and expression in Cos-1 cells of porcine kidney aminopeptidase P. *Biochem. J.* **319**: 197–200.

Igarashi P, Karniski LP. (1991) Cloning of cDNAs encoding a rabbit renal brush border membrane protein immunologically related to band 3. Sequence similarity with microsomal dipeptidase. *Biochem. J.* **280**: 71–78.

Joris B, Van Beeumen J, Casagrande F, Gerday C, Frere JM, Ghuysen JM. (1983) The complete amino acid sequence of the Zn^{2+}-containing D-alanyl-D-alanine-cleaving carboxypeptidase of *Streptomyces albus* G. *Eur. J. Biochem.* **130**: 53–69.

Keynan S, Hooper NM, Turner AJ. (1994) Directed mutagenesis of pig renal membrane dipeptidase. His^{219} is critical but the DHXXH motif is not essential for zinc binding or catalytic activity. *FEBS Lett.* **349**: 50–54.

Keynan S, Hooper NM, Felici A, Amicosante G, Turner AJ. (1995) The renal membrane dipeptidase (dehydropeptidase I) inhibitor, cilastatin, inhibits the bacterial metallo-β-lactamase enzyme CphA. *Antimicrob. Agents Chemother.* **39**: 1629–1631.

Keynan S, Habgood NT, Hooper NM, Turner AJ. (1996) Site-directed mutagenesis of conserved cysteine residues in porcine membrane dipeptidase. Cys 361 alone is involved in disulphide-linked dimerization. *Biochemistry* **35**: 12511–12517.

Kim HS, Campbell BJ. (1983) Association of renal dipeptidase with the triton-insoluble fraction of kidney microvilli. *J. Membr. Biol.* **75**: 115–122.

Koelsch R, Gottwald S, Lasch J. (1994) Release of GPI-anchored membrane aminopeptidase P by enzymes and detergents has some peculiarities. *Biochim. Biophys. Acta* **1190**: 170–172.

Kropp H, Sundelof JG, Hajdu R, Kahan FM. (1982) Metabolism of thienamycin and related carbapenum antibiotics by the renal dipeptidase, dehydropeptidase-I. *Antimicr. Agents Chemother.* **22**: 62–70.

Kuroki K, Eng F, Ishikawa T, Turck C, Harada F, Ganem D. (1995) gp180, a host cell glycoprotein that binds duck hepatitis B virus particles, is encoded by a member of the carboxypeptidase gene family. *J. Biol. Chem.* **270**: 15022–15028.

Lasch J, Koelsch R, Ladhoff A-M, Hartrodt B. (1986) Is the proline-specific aminopeptidase P of the intestinal brush border an integral membrane enzyme? *Biomed. Biochim. Acta* **45**: 833–843.

Lasch J, Koelsch R, Steinmetzer J, Neumann V, Demuth H-V. (1988) Enzymic properties of intestinal aminopeptidase P: a new continuous assay. *FEBS Lett.* **227**: 171–174.

Lim J, Turner AJ. (1996) Chemical modification of porcine kidney aminopeptidase P indicates the involvement of two critical histidine residues. *FEBS Lett.* **381**: 188–190.

Littlewood GM, Hooper NM, Turner AJ. (1989) Ectoenzymes of the kidney microvillar membrane. Affinity purification, characterization and localization of the phospholipase C-solubilized form of renal dipeptidase. *Biochem. J.* **257**: 361–367.

Lloyd GS, Hryszko J, Hooper NM, Turner AJ. (1996) Inhibition and metal ion activation of pig kidney aminopeptidase P. Dependence on nature of substrate. *Biochem. Pharmacol.* **52**: 229–236.

McGwire GB, Skidgel RA. (1995) Extracellular conversion of epidermal growth factor (EGF) to [des-Arg^{53}]EGF by carboxypeptidase M. *J. Biol. Chem.* **270**: 17154–17158.

McIntyre T, Curthoys NP. (1982) Renal catobolism of glutathione. Characterisation of a particulate rat renal dipeptidase that catalyses the hydrolysis of cyteinylglycine. *J. Biol. Chem.* **257**: 11915–11921.

Medeiros MS, Turner AJ. (1994) Processing and metabolism of peptide YY: pivotal roles for dipeptidyl peptidase IV, aminopeptidase P and endopeptidase-24.11. *Endocrinology* **134**: 2088–2094.

Medeiros MS, Turner AJ. (1996) Metabolism and functions of neuropeptide Y. *Neurochem. Res.* **21**: 1131–1138.

Mock WL, Liu Y. (1995) Hydrolysis of picolinylprolines by prolidase. A general mechanism for the dual-metal ion containing aminopeptidases. *J. Biol. Chem.* **270**: 18437–18446.

Nakagawa H, Inazawa J, Inoue K, Misawa S, Kashima K, Adachi H, Nakazato H, Abe T. (1992) Assignment of the human renal dipeptidase gene (DPEP1) to band q24 of chromosome 16. *Cytogen. Cell. Genet.* **59**: 258–260.

Oppong SY, Hooper NM. (1993) Characterization of a secretase activity which releases angiotensin-converting enzyme from the membrane. *Biochem. J.* **292**: 597–603.

Orawski AT, Simmons WH. (1995) Purification and properties of membrane-bound aminopeptidase P from rat lung. *Biochemistry* **34**: 11227–11236.

Orawski AT, Susz JP, Simmons WH. (1987) Aminopeptidase P from bovine lung: solubilization, properties and potential role in bradykinin degradation. *Mol. Cell. Biochem.* **75**: 123–132.

Orawski AT, Susz JP, Simmons WH. (1989) Metabolism of bradykinin by multiple co-existing membrane-bound peptidases in lung: techniques for investigating the role of each peptidase using specific inhibitors. In: *Kinins V, Part B* (eds K. Abe, H. Moriya and S. Fujii), pp. 355–364.

Parkin ET, Turner AJ, Hooper NM. (1996) Isolation and characterization of two distinct low-density, Triton-insoluble complexes from porcine lung membranes. *Biochem. J.* **319**: 887–896.

Prechel MM, Orawski AT, Maggiora LL, Simmons WH. (1995) Effect of a new aminopeptidase P inhibitor, apstatin, on bradykinin degradation in the rat lung. *J. Pharmacol. Exp. Ther.* **275**: 1136–1142.

Rached E, Hooper NM, James P, Semenza G, Turner AJ, Mantei N. (1990) cDNA cloning and expression in *Xenopus laevis* oocytes of pig renal dipeptidase, a glycosyl-phosphatidylinositol-anchored ectoenzyme. *Biochem. J.* **271**: 755–760.

Rehli M, Krause SW, Kreutz M, Andreesen R. (1995) Carboxypeptidase M is identical to the MAX.1 antigen and its expression is associated with monocyte to macrophage differentiation. *J. Biol. Chem.* **250**: 15644–15649.

Robinson DS, Birnbaum SM, Greenstein JP. (1953) Properties of an aminopeptidase from kidney cellular particulates. *J. Biol. Chem.* **202**: 1–26.

Roderick SL, Matthews BW. (1993) Structure of the cobalt-dependent methionine aminopeptidase from *E. coli*: a new type of proteolytic enzyme. *Biochemistry* **32**: 3907–3912.

Ryan JW, Valido F, Berryer P, Chung AYK, Ripka JE. (1992) Purification and characterisation of guinea pig serum aminoacylproline hydrolase (aminopeptidase P). *Biochim. Biophys. Acta* **1119**: 140–147.

Ryan JW, Denslow ND, Greenwald JA, Rogoff MA. (1994) Immunoaffinity purifications of aminopeptidase P from guinea pig lungs, kidney and serum. *Biochem. Biophys. Res. Commun.* **205**: 1796–1802.

Satoh S, Keida Y, Konta Y, Maeda M, Matsumoto Y, Niwa M, Kohsaka M. (1993a) Purification and molecular cloning of mouse renal dipeptidase. *Biochim. Biophys. Acta* **1163**: 234–242.

Satoh S, Kusunoki C, Konta Y, Niwa M, Kohsaka M. (1993b) Cloning and structural analysis of genomic DNA for human renal dipeptidase. *Biochim. Biophys. Acta* **1172**: 181–183.

Simmons WH, Orawski AT. (1992) Membrane bound aminopeptidase P from bovine lung. Its purification, properties and degradation of bradykinin. *J. Biol. Chem.* **267**: 4897–4903.

Skidgel RA, Johnson AR, Erdos EG. (1984) Hydrolysis of opioid hexapeptides by carboxypeptidase N. Presence of carboxypeptidase in cell membranes. *Biochem. Pharmacol.* **33**: 3471–3478.

Skidgel RA, Deddish PA, Davis, RA. (1988) Isolation and characterisation of a basic carboxypeptidase from human seminal plasma. *Arch. Biochem. Biophys.* **267**: 660–667.

Skidgel RA, Davis RM, Tan F. (1989) Human carboxypeptidase M. Purification and characterisation of a membrane-bound carboxypeptidase that cleaves peptide hormones. *J. Biol. Chem.* **264**: 2236–2241.

Soubrier F, Alhenc-Gelas F, Hubert C, Allegrini J, John M, Tregear G, Corvol P. (1988) Two putative active centers in human angiotensin I-converting enzyme revealed by molecular cloning. *Proc. Natl Acad. Sci. USA* **85**: 9386–9390.

Tan F, Chan S-J, Steiner DF, Schilling JW, Skidgel RA. (1989) Molecular cloning and sequencing of the cDNA for human membrane bound carboxypeptidase M. Comparison with carboxypeptidases A, B, H and N. *J. Biol. Chem.* **264**: 13165–13170.

Udenfriend S, Kodukula K. (1995) Prediction of ω site in nascent precursor of glycosylphosphatidylinositol protein. *Methods Enzymol.* **250**: 571–582.

Vergas Romero C, Neudorfer I, Mann K, Schafer W. (1995) Purification and amino acid sequence of aminopeptidase P from pig kidney. *Eur. J. Biochem.* **229**: 262–269.

Von Heijne G. (1986) A new method for predicting signal sequence cleavage sites. *Nucleic Acid Res.* **14**: 4683–4690.

Yoshimoto T, Tone H, Honda T, Osatomi K, Kobayashi R, Tsura D. (1989) Sequencing and high expression of aminopeptidase P gene from *E. coli* HB101. *J. Biochem.* **105**: 412–416.

Chapter 15

Inactivation of thyrotrophin-releasing hormone (TRH) by a TRH-specific ectoenzyme

Karl Bauer, Heike Heuer, Frank Iffländer, Anne Peters, Stephanie Schmitmeier, Lutz Schomburg, Stefan Turwitt and Matthias Wilkens

15.1 Introduction

The isolation and characterization of the tripeptideamide, pyroGlu–His–Pro–NH_2, in 1969 ushered in the modern area of neuroendocrinology since it was the first hypothalamic hypophysiotrophic peptide hormone to be structurally elucidated (for review see Guillemin, 1978; Schally, 1978). This peptide was designated thyrotrophin-releasing hormone (TRH) since it was identified by virtue of its capacity to stimulate the secretion of thyrotrophin (TSH) from mammalian anterior pituitary. Later, it was found that this peptide was equally effective in stimulating both synthesis and secretion of prolactin from pituitary cells (Tashjian *et al.*, 1971). Moreover, TRH also elicits the 'paradoxical' release of growth hormone under pathological conditions but it does not stimulate growth hormone secretion in normal subjects (for review see Harvey, 1990). In addition to these neuroendocrine functions, TRH elicits a variety of behavioural changes and neuropharmacological as well as electrophysiological effects in the central nervous system (CNS), suggesting that it may be involved in various forms of cellular communication and may act as a putative neuromodulator or even as a neurotransmitter (Griffiths, 1985; Horita *et al.*, 1986). This interpretation is supported by numerous biochemical studies and by the observation that TRH and TRH receptors are widely distributed in extrahypothalamic brain areas throughout the CNS as well as in the gastrointestinal tract (for reviews see Jackson *et al.*, 1990; Sharif, 1989).

Since substances of high biological activity are potentially extremely toxic, we should expect there to be very efficient degradation and/or elimination systems for TRH, ensuring its rapid inactivation after release. Such mechanisms would be particularly important in terminating the action of TRH as a putative neurotransmitter. Although TRH was found to be resistant to various proteolytic enzymes, rapid inactivation by blood and tissue enzymes was demonstrated by Burgus and Guillemin (1970) even before the structure of TRH had been elucidated. The catabolism of TRH has been extensively studied by various investigators (for reviews see O'Cuinn *et al.*, 1990; O'Leary and O'Connor, 1995).

Cell-Surface Peptidases in Health and Disease, edited by A.J. Kenny and C.M. Boustead.

The degradation of TRH, and thus its inactivation, can be initiated by several enzymes (*Figure 15.1*). There are two soluble enzymes, namely a pyroglutamate aminopeptidase (EC 3.4.19.3, pyroglutamyl-peptidase I) capable of cleaving the pyroglutamyl residue from various pyroglutamyl-containing peptides and a TRH-deamidating enzyme which has been identified as a proline endopeptidase (EC 3.4.21.26), an enzyme with a broad substrate specificity that cleaves a variety of peptides at the carboxyl side of proline residues. However, as cytosolic enzymes, neither of these peptidases can be important for the inactivation of a peptidergic transmitter substance after its release.

Initial studies demonstrated that in serum the catabolism of TRH to its constituent amino acids is initiated by enzymic cleavage of the pyroGlu–His bond. This cleavage is catalysed by an enzyme which exhibits chemical characteristics distinctly different from the cytosolic pyroglutamate aminopeptidase (Bauer and Nowak, 1979; Taylor and Dixon, 1978). Of special biological interest was the observation that the serum TRH-degrading enzyme exhibits an extraordinarily high degree of substrate specificity (Bauer *et al.*, 1981).

Using homogenates from guinea-pig brain, Browne *et al.* (1981) first reported that the hydrolysis of TRH at the pyroGlu–His bond is also catalysed by a particulate enzyme distinctly different from the cytosolic pyroglutamate aminopeptidase. The existence of a particulate enzyme was later confirmed by the same group (O'Connor and O'Cuinn, 1984) and by other laboratories (Elmore *et al.*, 1990; Garat *et al.*, 1985; Horsthemke *et al.*, 1984; O'Leary and O'Connor, 1995; Wilk and Wilk, 1989). Subsequent studies (see below) demonstrated that the particulate TRH-degrading enzyme (EC 3.4.19.6, also known as pyroglutamyl-peptidase II and thyroliberinase) exhibits the same chemical characteristics as the serum enzyme and identical properties with respect to the substrate specificity. Based on these data it has been suggested that both enzymes are derived from the same gene (O'Connor and O'Cuinn, 1985). This notion is further supported by protein sequence data and by the findings that the serum TRH-degrading enzyme is recognized by monoclonal and polyclonal antibodies that were generated by immunizing the animals with the purified membrane-bound TRH-degrading enzyme (K. Bauer, unpublished results).

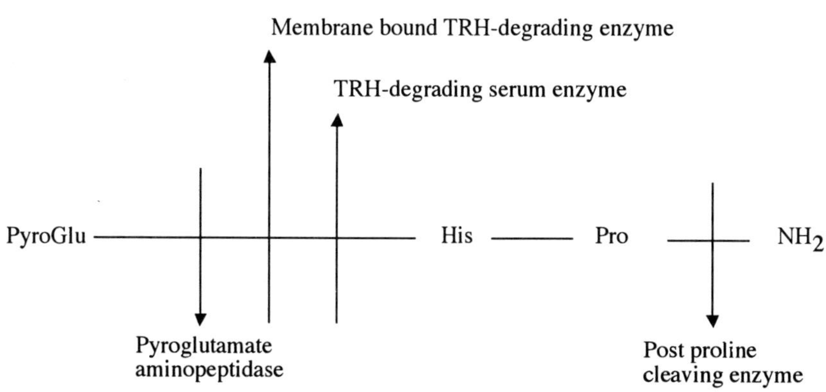

Figure 15.1. Points of attack by various peptidases capable of hydrolysing TRH.

15.2 Molecular structure

15.2.1 Identification of the membrane-bound TRH-degrading enzyme as an ectopeptidase

For the enzymic inactivation of a peptidergic signal, the minimal condition requires that the enzyme be localized on the plasma membrane with its active site orientated towards the extracellular space. Subcellular fractionation studies first demonstrated that the particulate TRH-degrading enzyme is located in synaptosomal (Elmore *et al.*, 1990; Garat *et al.*, 1985; O'Connor and O'Cuinn, 1984; O'Leary and O'Connor, 1995) and adenohypophyseal (Horsthemke *et al.*, 1984) plasma membranes. Studies with intact synaptosomes and various cell culture systems subsequently revealed that the enzyme is indeed a true ectoenzyme which is located on the surface of neuronal and pituitary cells, but not on glial cells (Bauer, 1987b; Bauer *et al.*, 1990; Charli *et al.*, 1988; Cruz *et al.*, 1991).

15.2.2 Purification

The TRH-degrading enzyme is not released by washing the membranes under various conditions, even by treatment at alkaline pH. Since it is also not anchored by a phosphatidylinositol-glycan residue, these data strongly indicated that the enzyme represents an integral membrane protein. The enzyme was effectively solubilized by some detergents but, unfortunately, the enzymic activity could not be stabilized and decreased very rapidly. Release of the enzymically active peptidase was finally achieved when the membranes were treated with proteolytic enzymes (Bauer, 1994; O'Connor and O'Cuinn, 1985; Wilk and Wilk, 1989). This enzyme preparation retained its enzymic activity for long periods of time and thus the peptidase could be purified by conventional chromatographic methods (O'Connor and O'Cuinn, 1985; Wilk and Wilk, 1989). After a 200 000-fold purification, the enzyme could finally be isolated in electrophoretically homogeneous form (Bauer, 1994).

For the proteolytically truncated enzyme a molecular mass of 230 kDa could be estimated by gel filtration chromatography. By SDS-PAGE under reducing and also under non-reducing conditions, a single band with an apparent molecular mass of 116 kDa could be detected, indicating that the enzyme may consist of two identical subunits which are not covalently linked. In this respect the TRH-degrading ectoenzyme resembles other cell-surface peptidases (e.g. aminopeptidase N and endopeptidase-24.11), which also form tightly assembled homodimers that are held together solely by non-covalent interaction.

As expected for a cell-surface enzyme, the TRH-degrading ectoenzyme was identified as a glycoprotein by lectin-binding analysis and by the reduction of the molecular mass to 97 kDa upon treatment of the denatured protein with endoglycosidases. In its native form, however, the enzyme was only partially deglycosylated and retained full enzymic activity.

15.2.3 Cloning

After chemical and proteolytic fragmentation of the purified enzyme, several peptides could be isolated and sequenced. This information provided the basis for screening cDNA libraries from rat brain and pituitary. Finally, a cDNA clone could be isolated that encoded the entire enzyme and, upon transfection in mammalian cells, a functionally active enzyme could be expressed, displaying the characteristics of the TRH-degrading enzyme (Schauder *et al.,* 1994).

According to the cDNA-deduced amino acid sequence *(Figure 15.2)*, the protein consists of 1025 amino acid residues and contains a short, presumably cytosolic region at its amino terminus with a potential phosphorylation site that could be important for the inactivation of the enzyme. This region is followed by a stretch of 24 hydrophobic residues with the characteristics of the transmembrane-spanning region of an integral membrane protein. The large, most likely extracellular domain contains 12 potential N-glycosylation sites, a consensus sequence for tyrosine sulphation and, most importantly, the consensus sequence of the zinc-dependent metallopeptidase family, namely the HEXXH motif, with a second glutamic acid separated by 18 amino acids as well as tyrosine and arginine residues at the expected distance relative to the HEXXH sequence. Within the zinc-binding motif, the enzyme contains a cysteine residue (HEICH), a special feature not generally found among the members of the zinc-containing metallopeptidases. Whether this cysteine residue is of functional significance with respect to the unusually high degree of substrate specificity of the enzyme remains to be investigated by site-directed mutagenesis.

Recently, we also succeeded in cloning the cDNA of the human enzyme. The deduced amino acid sequence shows a 96% overall identity when compared with the rat enzyme. Most of the amino acid substitutions are conservative and 15 out of the 40 exchanges are concentrated in the presumed stalk region. Most remarkably, a domain of 225 amino acids including the HEXXH consensus sequence is completely conserved. These data indicate that almost the entire extracellular domain is likely to be of functional importance.

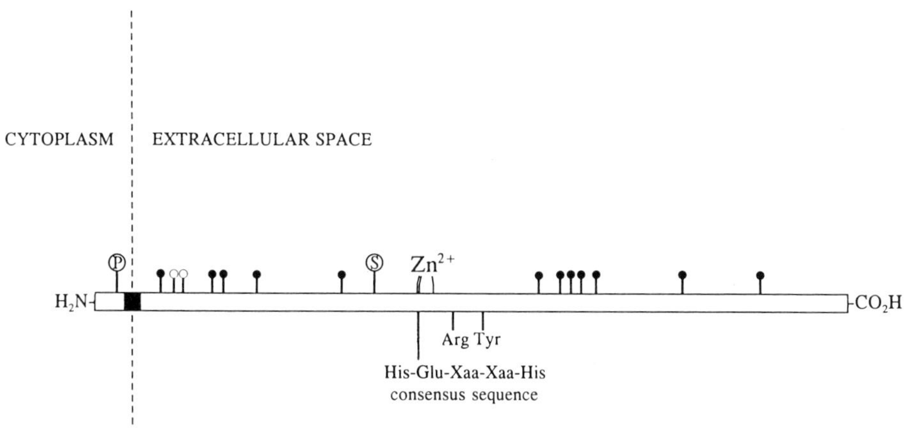

Figure 15.2. Model of the primary structure of the membrane-bound TRH-degrading enzyme. The membrane spanning domain is shown in black; P and S indicate putative phosphorylation and sulphation sites; the 'lollipops' indicate potential sites of O- and N-glycosylation respectively.

Interestingly, the computer search revealed that the amino acid sequence of the TRH-degrading ectoenzyme shares significant homology with the sequences of aminopeptidase N (34%) and aminopeptidase A (32%). Moreover, within the extracellular domain there are sequences that display a high degree of homology and there are two stretches each with seven amino acids that are completely conserved, indicating that these enzymes share common ancestry and belong to the same family.

15.2.4 Active site – catalytic mechanism

Previous studies by various investigators clearly demonstrated that the TRH-degrading enzymes are insensitive towards serine- and cysteine-specific reagents or the peptide inhibitors of aspartic endopeptidases. With chelating agents a time-dependent inhibition could finally be demonstrated (Czekay and Bauer, 1993). The slow inhibition by these agents indicated that the enzyme contains the catalytically active metal ion in a tightly complexed form. The enzymic activity could be restored by addition of Zn^{2+} and, as observed for other metallopeptidases, by Co^{2+}.

Studies with protein-modifying reagents are in full agreement with the classification of the enzyme as a metallopeptidase. Using a wide panel of these reagents and the competitive inhibitor pyroGlu–His–Trp as an active site protective agent, O'Connor and O'Cuinn (1987) convincingly demonstrated that tyrosine, histidine and arginine residues are constituents of the active site of the enzyme. Chemical modification studies with carbodiimides subsequently revealed the presence of an essential acidic amino acid residue which is protected by luteinizing hormone-releasing hormone (LHRH), a competitive inhibitor of the enzyme (Czekay and Bauer, 1993), indicating that the active site contains a glutamic acid residue. As shown for other metallopeptidases, a reactive glutamic acid residue most likely serves as a proton acceptor for the hydrolysis of the substrate.

15.3 Substrate specificity

In general, peptidases do not exhibit a high degree of substrate specificity but recognize only certain structural features (e.g. an amino or carboxyl terminus, certain side chains, etc.) of the peptides that are accepted as substrates. A limited number of 'general' peptidases could thus provide a highly efficient degradation system for an unlimited number of biologically active peptides, provided the peptidases are ectoenzymes located at specific target sites. The TRH-inactivating enzyme seems to be one of the rare exceptions to this rule. Like the serum enzyme (Bauer *et al.*, 1981), the TRH-degrading ectoenzyme exhibits a high degree of substrate specificity (Elmore *et al.*, 1990; O'Connor and O'Cuinn, 1984; O'Leary and O'Connor, 1995; Wilk and Wilk, 1989). Studies with TRH analogues suggested that the enzyme recognizes the entire entity of the TRH molecule. It is therefore not surprising that this enzyme, unlike the pyroglutamate aminopeptidase, does not degrade other naturally occurring pyroGlu-containing peptides (e.g. neurotensin, gastrin, etc.) nor even the TRH-related peptide pyroGlu–Glu–Pro–NH_2, recently isolated from rabbit prostate and rat anterior pituitary glands (for review see Cockle, 1995). The decapeptideamide LHRH (a hypothalamic neuropeptide hormone that stimulates the secretion of the pituitary hormones, luteinizing hormone and follicle-stimulating hormone) with its amino-terminal sequence pyroGlu–His is also not degraded. However, although not hydrolysed, this decapeptideamide is obviously recognized by the enzyme since it inhibits the degradation of TRH quite effectively. Interestingly, this peptide exhibits an even higher affinity for the enzyme than TRH itself ($K_m \approx 40\ \mu$M for TRH; $K_i \approx 20\ \mu$M for LHRH) (Elmore *et al.*, 1990).

15.4 Inhibitors

Enzyme-specific inhibitors that are suitable for pharmacological studies would be valuable tools for defining the biological functions of the TRH-inactivating enzyme. Given the strict substrate specificity of the enzyme, distinct endocrine, neuropharmacological and behavioural effects should become apparent if it is involved in the mechanisms anticipated. In contrast to the pharmacological effects of inhibitors for neuropeptide-unspecific peptidases, these effects should be directly related to the interference of the TRH signalling system. However, the strict substrate specificity renders the synthesis of specific enzyme inhibitors extremely difficult, since only minor modifications of the basic structure will be tolerated. It is therefore not surprising that so far only a few inhibitors have been synthesised which affect the enzymic activity *in vitro* (Charli *et al.,* 1989; Czekay and Bauer, 1993), albeit with insufficient efficiency to be suitable for *in vivo* pharmacological studies. α-*N*-(1-carboxy-2-phenylethyl)-1-benzylhistidinyl-β-naphthylamide has been described as a substrate-related inhibitor ($K_i \approx 8\ \mu$M) (Charli *et al.,* 1989), but it seems likely that the inhibition is not due to a specific site-directed interaction of the compound with the enzyme, but probably due to unspecific hydrophobic interactions of the enzyme with the aromatic structures of the compound.

Based on the identification of the enzyme as a Zn^{2+}-metallopeptidase and the observation that the enzyme is weakly inhibited by the TRH cleavage products (pyroGlu and His–Pro-NH_2), various compounds have been synthesised. While hydroxamic acid derivatives of peptides or amino acids are known to inhibit various metallopeptidases very effectively, L-pyroglutamyl hydroxamic acid was found to act as a relatively weak competitive inhibitor with a K_i value of 160 μM. Likewise, the thiol derivative (S)-5-mercaptomethyl-2-pyrolidinon exhibits a K_i of only 100 μM in sharp contrast to the strong inhibition of aminopeptidase N which is inhibited by homologous thiol derivatives of amino acids in the nanomolar range (Fournié-Zaluski *et al.*, 1992). 5-Phosphono-2-pyrolidinon and phosphonamide derivatives of His–Pro-NH_2 were also not effective and even the highly reactive chloromethylketone derivative of pyroglutamic acid acted only as a competitive (K_i 110 μM) and not as an irreversible inhibitor, as expected. With *N*-bromoacetyl-His–Pro-NH_2, a time-dependent inactivation of the enzyme was also not observed, but only weak competitive inhibition. Other derivatives of pyroGlu or His–Pro-NH_2 bearing various functional groups that were expected to interact with the catalytic metal ion (Powers and Harper, 1986; Rich, 1990) also proved to be rather ineffective.

15.5 Tissue and cellular distribution

The peptidase activities in brain are generally rather low in comparison to other tissues. The opposite is true for the TRH-degrading ectoenzyme. The highest activities of this enzyme are found in the brain and significant activities are present throughout the central nervous system. Within the brain, highest activities are found in the olfactory bulb, cerebral cortex and hippocampus. Considerably lower, but significant activities are also found in spinal cord, cerebellum and posterior pituitary (Bauer, 1987b; Friedman and Wilk, 1986; Vargas *et al.,* 1987). In agreement with its neuronal localization and its association with synaptosomal membranes, the enzyme is detectable very early during fetal life at the developmental stages when TRH-containing neurons appear (Bauer and Schulz, 1992; Vargas *et al.,* 1992).

In the anterior pituitary basal enzyme activities are very low. Cell fractionation studies revealed a close correlation between enzyme activity and prolactin content, indicating that the enzyme is preferentially, if not exclusively, associated with lactotrophic cells (Bauer *et al.,* 1990).

In the peripheral organs significant activities are detectable in retina (Wilk *et al.,* 1988), lung and liver, while other tissues are almost devoid of this peptidase activity. This is even true for the kidneys and the intestines, generally the richest sources of membrane peptidases. In the liver, significant activities of the membrane-bound enzyme have been detected in some species (e.g. rat, pig, chicken) but not in others (e.g. rabbit and cow) (S. Schmitmeier *et al.,* unpublished; Vargas *et al.,* 1992). Interestingly, in the latter species the serum does not contain significant enzyme activity, providing further evidence that the serum TRH-degrading enzyme is of liver origin. This interpretation is supported by studies on the ontogenesis of these enzymes. In the rat, the serum TRH-degrading enzyme (Neary *et al.,* 1976) and the particulate TRH-degrading liver enzyme develop very late (Bauer and Schulz, 1992; Scharfman and Aratan-Spire, 1991). Almost absent at birth, the activities of these enzymes steadily increase after 2–3 weeks and reach maximal levels very late in adult life. Whether the serum enzyme is generated by alternative splicing of the mRNA and secreted as a soluble enzyme or whether it is released from the cell surface by shedding remains to be investigated.

15.6 Physiological functions

The high degree of substrate specificity and the special features of the enzyme strongly suggest that it may serve specialized functions. In the brain, the high enzymic activity and the specific localization on neuronal cells suggest that the enzyme could be important for the termination, and thus the transmission, of TRH signals. The analogy to acetylcholinesterase is evident, but further studies will be necessary to define precisely the biological functions within the TRH-signalling pathway.

For the adenohypophyseal TRH-degrading ectoenzyme, the present data strongly suggest that it might serve regulatory functions since the enzyme itself is regulated by peripheral hormones. *In vivo* studies first demonstrated that the activity of the adenohypophyseal enzyme is stringently regulated by thyroid hormones (Bauer, 1987; Ponce *et al.,* 1988;

Schomburg and Bauer, 1995; Suen and Wilk, 1989). When rats were rendered hypothyroid by treatment with goitrogenic agents, the activity and the mRNA levels of the enzyme rapidly declined within a few days. Conversely, when the animals were made hyperthyroid by injecting triiodothyronine, the enzyme activity rapidly increased, preceded by a drastic increase in the mRNA levels, and returned to basal levels within physiologically meaningful time periods. Interestingly, an inverse relationship was observed with regard to the regulation of the TRH receptor by thyroid hormones indicating that both elements cooperate in regulating the responsiveness of adenohypophyseal target cells.

The observation that the transcript levels and the enzymic activities of the adenohypophyseal enzyme of male rats are considerably higher (by a factor of three) than those of female rats already indicated that the enzyme is also regulated by oestradiol. Subsequent *in vivo* studies substantiated this interpretation. When male rats received a single injection of oestradiol, the mRNA levels and the enzymic activity rapidly and transiently decreased. Conversely, after ovariectomy the activity rapidly increased and decreased again after treatment with oestradiol. Interestingly, these effects were specific for oestradiol and were not observed with other steroid hormones (Bauer, 1988; L. Schomburg and K. Bauer, unpublished).

It is worth noting that these hormonal effects were tissue-specific and were not noticed in brain (Bauer, 1987a; Emerson and Wu, 1987; Ponce *et al.,* 1988) nor in any other tissues that contain detectable activities of this enzyme. This was true even when the animals received multiple injections of the hormones at pharmacological levels. In serum, however, significant changes in the activity of the TRH-degrading enzyme were noticed when rats were treated for prolonged periods of time with thyroid hormones and the goitrogenic agent propylthiouracil, respectively (Bauer, 1976; Dupont *et al.,* 1976; Emerson and Wu, 1987; White *et al.,* 1976). Prolonged treatment produced a five-fold difference between hypothyroid and hyperthyroid rats. In the human some differences in serum activity were noticed depending on the sex, the thyroid status and in certain cases of chronic illness (Iversen, 1995). However, due to the considerable inter-individual variations, these differences were not significant. The observed changes in the plasma clearance rate seemed to reflect alterations in the pharmacokinetics of TRH (Iversen, 1995). In a recent report, it has been suggested that in human serum the enzyme activity correlates positively with body mass (Friedman *et al.,* 1995) but not with the thyroid status. Since the blood of some species does not contain significant activities of the TRH-degrading serum enzyme, the biological function of this enzyme still remains obscure.

References

Bauer K. (1976) Regulation of degradation of thyrotropin-releasing hormone by thyroid hormones. *Nature* **259:** 591–593.

Bauer K. (1987a) Adenohypophysial degradation of thyrotropin-releasing hormone regulated by thyroid hormones. *Nature* **330:** 375–377.

Bauer K. (1987b) Degradation and biological inactivation of thyrotropin-releasing hormone and other neuropeptides. In: *Integrative Neuroendocrinology: Molecular, Cellular and Clinical Aspects* (eds S.M. McCann and R.I. Weiner), pp. 102–114. Karger, Basel.

Bauer K. (1988) Degradation and biological inactivation of thyrotropin-releasing hormone (TRH): regulation of the membrane-bound TRH-degrading enzyme from rat anterior pituitary by estrogens and thyroid hormones. *Biochimie* **70:** 69–74.

Bauer K. (1994) Purification and characterization of the thyrotropin-releasing hormone-degrading ectoenzyme. *Eur. J. Biochem.* **224:** 387–396.

Bauer K, Nowak P. (1979) Characterization of a thyroliberin-degrading serum enzyme catalyzing the hydrolysis of thyroliberin at the pyroglutamyl-histidine bond. *Eur. J. Biochem.* **99:** 239–246.

Bauer K, Schulz M. (1992) Inactivation of the thyrotropin releasing hormone. *Acta Med. Austr.* **19(S1):** 79–82.

Bauer K, Nowak P, Kleinkauf H. (1981) Specificity of a serum peptidase-hydrolyzing thyroliberin at the pyroglutamyl-histidine bond. *Eur. J. Biochem.* **118:** 173–176.

Bauer K, Carmeliet P, Schulz M, Baes M, Denef C. (1990) Regulation and cellular localization of the membrane-bound thyrotropin-releasing hormone-degrading enzyme in primary cultures of neuronal, glial and adenohypophyseal cells. *Endocrinology* **127:** 1224–1233.

Browne P, Phelan J, O'Cuinn G. (1981) TRH pyroglutamate aminopeptidase activity in two different fractions of guinea-pig brain. *Ir. J. Med. Sci.* **150:** 348–349.

Burgus R, Guillemin R. (1970) Chemistry of thyrotropin-releasing factor. In: *Hypophysiotropic Hormones of the Hypothalamus* (ed. Meites J.), pp. 227–241, Williams and Wilkins, Baltimore, MD.

Charli JL, Cruz C, Vargas MA, Joseph-Bravo P. (1988) The narrow specificity pyroglutamate aminopeptidase degrading TRH in rat brain is an ectoenzyme. *Neurochem. Int.* **13:** 237–242.

Charli JL, Mendez M, Vargas MA, Cisneros M, Assai M, Joseph-Bravo P, Wilk S. (1989) Pyroglutamyl peptidase II inhibition specifically increases recovery of TRH released from rat brain slices. *Neuropeptides* **14:** 191–196.

Cockle SM. (1995) Fertilization-promoting peptide: a novel peptide, structurally similar to TRH, with potent physiological activity. *J. Endocrinol.* **146:** 3–8.

Cruz C, Charli JL, Vargas MA, Joseph-Bravo P. (1991) Neuronal localization of pyroglutamate aminopeptidase II in primary cultures of fetal mouse brain. *J. Neurochem.* **56:** 1594–1601.

Czekay G, Bauer K. (1993) Identification of the thyrotropin-releasing-hormone-degrading ectoenzyme as a metallopeptidase. *Biochem. J.* **290:** 921–926.

Dupont A, Labrie F, Levasseur L, Dussault JH, Schally AV. (1976) Effect of thyroxine on the inactivation of thyrotropin-releasing hormone by rat and human plasma. *Clin. Endocrinol.* **5:** 323–328.

Elmore MA, Griffiths EC, O'Connor B, O'Cuinn G. (1990) Further characterization of the substrate specificity of a TRH-hydrolyzing pyroglutamate aminopeptidase from guinea-pig brain. *Neuropeptides* **15:** 31–36.

Emerson CH, Wu CF. (1987) Thyroid status influences rat serum but not brain TRH pyroglutamyl aminopeptidase activities. *Endocrinology* **120:** 1215–1217.

Fournié-Zaluski MC, Coric P, Turcaud S, Bruetschy L, Lucas E, Noble F, Roques BP. (1992) Potent and systemically active aminopeptidase N inhibitors designed from active-site investigation. *J. Med. Chem.* **35:** 1259–1266.

Friedman TC, Wilk S. (1986) Delineation of a particulate thyrotropin-releasing hormone-degrading enzyme in rat brain by the use of specific inhibitors. *J. Neurochem.* **46:** 1231–1239.

Friedman TC, Yanovski JA, Jayasvasti V, Yanovski SZ, Koenig RJ, Wilk S. (1995) Pyroglutamyl peptidase-II (thyroliberinase) activity in human serum – influence of weight and thyroid status. *J. Clin. Endoc. Metab.* **80(4):** 1086–1089.

Garat B, Miranda J, Charli JL, Joseph-Bravo P. (1985) Presence of membrane bound pyroglutamyl aminopeptidase degrading thyroliberin-releasing hormone in rat brain. *Neuropeptides* **6:** 27–40.

Griffiths EC. (1985) Thyrotropin-releasing hormone: endocrine and central effects. *Psychoneuroendocrinology* **10:** 225–235.

Guillemin R. (1978) Peptides in the brain: the new endocrinology of the neuron. *Science* **202:** 390–402.

Harvey S. (1990) Thyrotropin-releasing hormone: a growth hormone-releasing factor. *J. Endocrinol.* **125:** 345–358.

Horita A, Carino MA, Lai H. (1986) Pharmacology of thyrotropin-releasing hormone. *Ann. Rev. Pharmacol. Toxicol.* **26:** 311–332.

Horsthemke B, Leblanc P, Kordon C, Wattiaux-De Coninck S, Wattiaux R, Bauer K. (1984) Subcellular distribution of particle bound neutral peptidases capable of hydrolyzing gonadoliberin, thyroliberin, enkephalin and substance P. *Eur. J. Biochem.* **139:** 315–320.

Iversen E. (1995) Thyrotropin-releasing hormone: occurrence and role outside the central nervous system. *Dan. Med. Bull.* **42:** 257–268.

Jackson IMD, Lechan RM, Lee SL. (1990) TRH prohormone: biosynthesis, anatomic distribution and processing. *Front. Neuroendocrinol.* **11:** 267–312.

Neary JR, Kieffer JD, Federico P, Mover H, Maloof F, Soodak M. (1976) Thyrotropin-releasing hormone: development of inactivation system during maturation of the rat. *Science* **193:** 403–405.

O'Connor B, O'Cuinn G. (1984) Localization of a narrow specificity thyroliberin-hydrolyzing pyroglutamate aminopeptidase in synaptosomal membranes of guinea-pig brain. *Eur. J. Biochem.* **144:** 271–278.

O'Connor B, O'Cuinn G. (1985) Purification of and kinetic studies on a narrow specificity synaptosomal membrane pyroglutamate aminopeptidase from guinea-pig brain. *Eur. J. Biochem.* **150:** 47–52.

O'Connor B, O'Cuinn G. (1987) Active site studies on a narrow specificity thyroliberin-hydrolyzing pyroglutamate aminopeptidase purified from synaptosomal membrane of guinea-pig brain. *J. Neurochem.* **48:** 676–680.

O'Cuinn G, O'Connor B, Elmore M. (1990) Degradation of thyrotropin-releasing hormone and luteinizing hormone-releasing hormone by enzymes of brain tissue. *J. Neurochem.* **54:** 1–13.

O'Leary R, O'Connor B. (1995) Thyrotropin-releasing hormone. *J. Neurochem.* **65:** 953–963.

Ponce G, Charli JL, Pasten JA, Aceves C, Joseph-Bravo P. (1988) Tissue-specific regulation of pyroglutamte aminopeptidase II activity by thyroid hormones. *Neuroendocrinology* **48:** 211–213.

Powers JC, Harper JW. (1986) Inhibitors of metalloproteases. In: *Proteinase Inhibitors* (eds Barret A.J. and Salvesen G.), pp. 219–298, Elsevier, Amsterdam.

Rich DH. (1990) Peptidase Inhibitors. In: *Comprehensive Medical Chemistry. The Rational Design, Mechanistic Study and Therapeutic Application of Chemical Compounds,* Vol. 2 (eds C. Hansch, P.G. Sammes and J.B. Taylor), pp. 391–441. Pergamon Press, Oxford.

Schally AV. (1978) Aspects of hypothalamic regulation of the pituitary gland. *Science* **202:** 18–28.

Scharfman R, Aratan-Spire S. (1991) Ontogeny of two topologically distinct TRH-degrading pyroglutamate aminopeptidase activities in the rat liver. *Regul. Pept.* **32:** 75–83.

Schauder B, Schomburg L, Köhrle J, Bauer K. (1994) Cloning of a cDNA encoding an ectoenzyme that degrades thyrotropin-releasing hormone. *Proc. Natl Acad. Sci. USA* **91:** 9534–9538.

Schomburg L, Bauer K. (1995) Thyroid hormones rapidly and stringently regulate the messenger RNA levels of the thyrotropin-releasing hormone (TRH) receptor and the TRH-degrading ectoenzyme. *Endocrinology* **136:** 3480–3485.

Sharif NA. (1989) Quantitative autoradiography of TRH receptors in discrete brain regions of different mammalian species. *Ann. NY Acad. Sci.* **553:** 147–175.

Suen CS, Wilk S. (1989) Regulation of thyrotropin-releasing hormone-degrading enzymes in rat brain and pituitary by L-3,5,3′-triiodothyronine. *J. Neurochem.* **52:** 884–888.

Tashjian AH, Barowsky NJ, Jensen DK. (1971) Thyrotropin releasing hormone: direct evidence for stimulation of prolactin production by pituitary cells in culture. *Biochem. Biophys. Res. Commun.* **43:** 516–523.

Taylor WL, Dixon DE. (1978) Characterization of a pyroglutamate aminopeptidase from rat serum that degrades thyrotropin-releasing hormone. *J. Biol. Chem.* **253:** 6934–6940.

Vargas M, Mendiz M, Cisneros M, Joseph-Bravo P, Charli JL. (1987) Regional distribution of the membrane-bound pyroglutamate aminopeptidase-degrading thyrotropin-releasing hormone in rat brain. *Neurosci. Lett.* **79:** 311–314.

Vargas MA, Herrera J, Uribe RM, Charli JL, Josepha-Bravo P. (1992) Ontogenesis of pyroglutamyl peptidase II activity in rat brain, adenohypophysis and pancreas. *Dev. Brain Res.* **66:** 251–256.

White N, Jeffcoate SL, Griffiths EC, Hooper KC. (1976) Effect of thyroid status on the thyrotropin-releasing hormone-degrading activity of rat serum. *J. Endocrinol.* **71:** 13–19.

Wilk S, Wilk EK. (1989) Pyroglutamyl peptidase II, a thyrotropin-releasing hormone-degrading enzyme: purification and specificity studies of the rabbit brain enzyme. *Neurochem. Int.* **15:** 81–90.

Wilk S, Suen CS, Wilk EK. (1988) Occurrence of pyroglutamyl peptidase II, a specific TRH degrading enzyme in rabbit retinal membranes and in human retinoblastoma cells. *Neuropeptides* **12:** 43–47.

Chapter 16

Aminopeptidase W

A. John Kenny

16.1 A monoclonal antibody revealed a novel aminopeptidase

16.1.1 GK5C1 antibody

A programme to generate monoclonal antibodies to antigens in pig kidney brush border membranes was started in my laboratory with the expectation that useful clones might be isolated which would secrete IgGs to some of the cell-surface peptidases we were then studying, but also with a more distant hope that other clones might lead us to novel proteins in the membrane. The plan yielded an early success in providing a valuable antibody to endopeptidase-24.11 (E-24.11; Gee *et al.*, 1983), a product which was crucial in many aspects of the work on that enzyme. Further antibodies to E-24.11 and to aminopeptidase N (AP-N) followed (Gee and Kenny, 1985a), but one hybridoma, 5C1 fulfilled our distant hope and ultimately gave us a new ectopeptidase (Gee and Kenny, 1985b).

The antibody designated GK5C1 and shown to be an IgG_1, was partially purified from ascites fluid by diethylaminoethyl (DEAE)-cellulose chromatography, the main contaminant being transferrin. An immuno-affinity column was prepared by coupling to cyanogen bromide (CNBr)-Sepharose, which was effective in purifying the antigen from detergent-solubilized pig kidney micrivillus membranes. The conventional elution step involved treatment at pH 10.6 (a step which was routinely successful in purifying other peptidases from immuno-affinity columns), but which was later found to inactivate the aminopeptidase and had to be replaced by an elution with a hypo-osmotic buffer at pH 8.0.

16.1.2 GK5C1 antigen

An immunoradiometric assay (Gee *et al.*, 1985) was developed to monitor the purification of the antigen from pig kidney cortex. The antigen was 1000-fold purified from the starting material and was shown to migrate in reducing conditions as a 130 kDa polypeptide on sodium dodecyl sulphate (SDS) gel electrophoresis. The single band stained positively for carbohydrate by the periodic acid–Schiff reagent and further evidence that GK5C1 was indeed a glycoprotein was later obtained by incubation with *N*-glycanase, a treatment that diminished the apparent subunit M_r on SDS gels from 130 to

Cell-Surface Peptidases in Health and Disease, edited by A.J. Kenny and C.M. Boustead.

90 kDa (Gee and Kenny, 1987). Incubation of the microvillus membrane preparation with papain generated a smaller polypeptide (125 kDa), revealed by immunoblotting. This effect was consistent with the proteolytic clipping of a stalked surface membrane protein to release a slightly smaller hydrophilic form. The antigen (in its detergent-solubilized form) was reconstituted into liposomes and viewed by negative staining in the electron microscope. The protein had a globular appearance attached to the liposome by a stalk about 3 nm in length; dimers were frequently observed. The oligomeric structure was further investigated by chromatography on Ultrogel AcA-22, from which it eluted as a 520 kDa protein, somewhat larger than that expected for a dimer (a dimer associated with a single detergent micelle might give a value of about 360 kDa). This discrepancy has not been explained, but it is possible that two detergent micelles were bound to the amphipathic protein.

The 5C1 antigen was thus a glycoprotein comparable in size, oligomeric structure and electron microscopic appearance to other brush border ectoenzymes. It seemed highly probable that it was one of the hydrolases in microvilli and a long search began for evidence of its enzymic activity. Experiment after experiment ended in failure: no hydrolysis was detected of insulin B chain, nor of a variety of neuropeptides, nor of substrates for carboxypeptidases, esterases, nucleotidases, disaccharidases and so on. Weak hydrolysis was noted with fluorogenic substrates for AP-N and dipeptidyl peptidase IV (DPP-IV), but these activities were too feeble (about 5 nmol min^{-1} mg^{-1}) for the antigen to be identified with either peptidase. It seemed more likely that the preparation was contaminated with trace amounts of these two abundant and resilient enzymes. Success was finally achieved when the antigen was found to hydrolyse L-leucyl-L-tryptophan, the release of tryptophan being monitored by high performance liquid chromatography (HPLC).

16.1.3 Evidence for a novel aminopeptidase

Hydrolysis of 0.5 mM Leu–Trp by the purified antigen, over a broad pH range from 7 to 9, was repeatedly demonstrated, but the specific activity was not impressive, being only about 840 nmol min^{-1} mg^{-1}. At this stage it became important to demonstrate that the peptidase activity was associated with the 130 kDa protein and not the result of contaminating enzymes (e.g. AP-N or membrane dipeptidase). This was shown by rechromatographing the purified antigen by two different systems, Untrogel AcA-22 and DEAE-cellulose: in each case the activity peak coincided with the presence in the same fractions of the 130 kDa polypeptide on SDS gels, well removed from zones that might contain the 160 kDa AP-N or the 50 kDa membrane dipeptidase. When the stability of the activity was studied as a function of pH, at 4°C, it was discovered that all activity was lost in 1 h at pH 5 and pH 11. The purification from the immuno-affinity column was then redesigned to avoid an elution at pH 10.6. The specific activity of these later preparations eluted at pH 8.0 was some 20-fold higher than the earlier batches. The activity was clearly that of an aminopeptidase and not that of a dipeptidase since extended peptides were also hydrolysed, including Leu–Trp–Leu and Leu–Trp–Met–Arg, while blocked substrates such as Z–Leu–Trp resisted cleavage. Aminopeptidase activity diminished as the chain was lengthened, in marked contrast to the rates observed with AP-N, which preferentially hydrolyses longer oligopeptide substrates and has weaker activity for dipeptides.

16.2 Enzymology of aminopeptidase W

16.2.1 Substrate specificity and kinetics

The aminopeptidase hydrolysed a wide range of dipeptides (Gee and Kenny, 1987). Glu–Trp was the preferred substrate, but various other P_1 residues, including those with basic side chains as well as Gly and Pro, were tolerated (see *Table 16.1*). Trp was favoured as the P_1' residue, with Phe and Tyr giving somewhat lower rates, while Leu and Gly were relatively unfavourable residues. Detailed kinetics were calculated for five peptides with Trp as P_1', see *Table 16.2*. The two best substrates as judged by the specificity constant, k_{cat}/K_m, are Glu–Trp and Leu–Trp; Gly–Trp had a higher k_{cat} value, but the K_m was three-fold higher, yielding a lower specificity constant. Kinetic constants have been reported for aspartame (an artificial sweetener, Asp–Phe–methylester) and peptides derived from it (Asp–Phe and Asp–Phe–amide) (Hooper *et al.*, 1994). The K_m values for all three were an order of magnitude higher than that for Glu–Trp and the k_{cat} value for Asp–Phe was similar to that of Gly–Trp. The special role of the Trp (W) residue made the name aminopeptidase W (AP-W) highly appropriate, as a reminder of this preference for tryptophan. The E.C. name, X-Trp aminopeptidase, seems clumsy and is out of line with the naming of the other aminopeptidases in cell membranes (see Section 1.2.2). Analysis by atomic absorption spectrophotometry was consistent with the presence of one atom of zinc/subunit. Activity was highest in Tris–HCl buffer at pH

Table 16.1. Rates of hydrolysis of dipeptides by AP-W[a]

	P_1' residue				
P_1 residue	Trp	Phe	Tyr	Leu	Gly
Glu	100	—	24	—	—
Asp	—	54	—	—	—
Leu	89	53	41	—	—
Tyr	—	32	—	18	—
Gly	60	—	15	—	—
Phe	50	59	18	9	5
Ala	48	9	7	—	—
Ile	41	15	8	—	—
Arg	34	—	—	—	—
Trp	26	14	5	2	4
Val	20	6	—	—	—
Pro	14	4	2	—	—
Mean for P_1' residue	100	50±12	25±4.4	13	12

[a] The rates are expressed relative to that of Glu–Trp (=100). The effect of varying the P_1' residue is shown in the last line as mean values, compared with Trp as the P_1' residue (=100).
Data from Gee and Kenny (1987) reproduced by permission of The Biochemical Society.

Table 16.2. Kinetic constants for the hydrolysis of some X-Trp dipeptides by AP-W[a]

Substrate	K_m (μM)	V_{max} (μmol min^{-1}mg^{-1})	k_{cat} (min^{-1})	k_{cat}/K_m (min^{-1}μM^{-1})
Glu–Trp	568	52	6773	11.9 (100)
Leu–Trp	488	43	5557	11.4 (96)
Gly–Trp	1757	59	7590	4.3 (36)
Ala–Trp	625	19	2312	3.7 (31)
Pro–Trp	1700	14	1775	1.04 (9)

[a] The values of k_{cat}/K_m are expressed relative to that of Leu–Trp (=100).
Data from Gee and Kenny (1987) reproduced by permission of The Biochemical Society.

8.0, with a distinct maximum at 100 mM; by contrast activity in phosphate buffer was less than 10% that in Tris–HCl at the same pH value.

16.2.2 A fluorimetric assay

The assay requiring reverse-phase HPLC provided precise data, but was time consuming and not necessarily suitable for detection of the enzyme in crude tissue extracts. A two-step fluorimetric assay was designed using the preferred dipeptide substrate, Glu–Trp (Jackson *et al.*, 1988). In the first step glutamate was released by incubation with AP-W, the mixture was then heated to 100°C and glutamate determined in the second step by addition of glutamate dehydrogenase, NAD^+ and ADP, the generation of NADH being monitored by fluorescence. AP-A and -N and membrane dipeptidase are other peptidases in tissue extracts that might contribute to the hydrolysis of Glu–Trp. Their contributions were abolished by the inclusion in the first step of 5 mM 1,10-phenanthroline and 50 μM cilastatin; the former strongly inhibits the two aminopeptidases and the latter the dipeptidase, while AP-W is essentially unaffected by these agents.

16.2.3 Inhibitors

The identification of zinc in a sample of the protein suggested that AP-W is a Zn-metallopeptidase. Unexpectedly, AP-W resisted inhibition by ethylenediamine tetraacetic acid (EDTA) and 1,10-phenanthroline at concentrations as high as 10 mM. It is, however, sensitive to inhibition by some thiols: dithiothreitol, captopril and 2-mercaptoethanol (IC_{50} concentrations 0.1–0.6 mM). Although no inhibition was observed with 0.1 mM actinonin, AP-W is very sensitive to amastatin and bestatin (IC_{50} values 2 μM and 6 μM). Strong inhibition was observed with metal ions, especially Zn^{2+} and Co^{2+} and these ionic effects were pH- and buffer ion-dependent, the inhibition by Zn^{2+} being fully reversible by EDTA (Gee and Kenny, 1987). The studies on the specific inhibitors were later extended to include comparisons with AP-A and -N (Tieku and Hooper, 1992). All three enzymes were equally sensitive to amastatin, AP-W was more sensitive to bestatin, but 100-fold less sensitive to probestin. AP-W was the only one of the three to show inhibition by any of the ACE inhibitors, of which rentiapril was the most effective, giving an IC_{50} of 1.6 μM, comparable to that for amastatin.

16.3 Locations of AP-W

Long before the enzymic activity of the 130 kDa antigen had been revealed, an immunoradiometric assay had enabled a detailed survey of the antigen in pig tissues to be made. The assay was based on that employed for the detection of E-24.11 (Gee *et al.*, 1985), but in contrast to the wide distribution observed for E-24.11, AP-W was detected only in the kidney and small intestine (Gee and Kenny, 1985b). The antigen was most abundant in the mucosal scrapings from the ileum, also rich in kidney cortex, with less in the jejunum and very little in the duodenum. None was detected in stomach, spleen, lymph nodes, salivary gland, pancreas, thyroid, adrenal, liver, lung, heart and aorta.

Immunohistochemistry revealed the antigen in the renal proximal tubules (none in the glomeruli) and on the surface of the intestinal villi and the lining of the crypts. AP-W has, therefore, a brush border location in kidney and intestine and it was not surprising that it should be detected in two human colonic carcinoma cell lines, Caco-2 and HT-29, the activity in the former increasing as the cells differentiated on achieving confluency (Howell *et al.*, 1992). Weak activity has also been reported in membranes prepared from human osteoblast-like cells in culture (Howell *et al.*, 1993).

The choroid plexus also has a brush border and AP-W has been detected in membrane preparations, but with a specific activity about 1% of that in renal microvilli (Bourne *et al.*, 1989). However, this is not the only location of AP-W in the nervous system. A low activity was also noted in a microsomal fraction prepared from pig peripheral nerves, at a level corresponding to 0.3% of that in renal microvilli. This low level of AP-W proved to have an important significance when immunohistochemistry, now using a rabbit polyclonal antibody (shown to be monospecific by immunoblotting), revealed the antigen to be present in both myelinated and unmyelinated nerve fibres (Barnes *et al.*, 1991). This location in neurites is the more extraordinary in that similar experimental approaches have failed to reveal AP-W in brain sections. Furthermore, in peripheral nerve, AP-W alone among all the other ectopeptidases studied had a neuronal location, the others being associated with Schwann cells (E-24.11) or microvessels and/or connective tissue elements (AP-N, angiotensin-converting enzyme, DPP-IV). When a peripheral nerve is cut, the distal portion undergoes Wallerian degeneration, which persists unless the cut ends of the nerve are accurately rejoined so as to permit regeneration. As expected, in the degenerated segments AP-W was no longer detectable, since no neuronal elements remained. It was, however, now revealed in the activated macrophages that entered the degenerated region (Kenny and Bourne, 1991).

16.4 Some reflections on the discovery of AP-W

Serendipity led to the discovery of this previously undetected peptidase and it is interesting to speculate whether its existence might have been suspected in the course of studies on the enzymes of renal microvilli. Its properties argue against this likelihood. First, it is unremarkable in being a glycoprotein with 130 kDa subunits: DPP-IV and carboxypeptidase P are similar in these respects and DPP-IV is much more abundant in microvilli than is AP-W: there is no reason why it might have been noticed as a band on SDS gels (Hedeager-Sorensen and Kenny, 1985; Kenny *et al.*, 1976). Furthermore it has no specific substrate, other peptidases can also attack the peptides that AP-W hydrolyses, often with greater efficiency. So AP-W activity is readily masked by peptidases that may be more abundant and possess broader specificities. One can only be grateful to the splenocyte that managed to fuse successfully with the Sp2/0 myeloma cell to generate the hybridoma which fortunately did not die before it could be cloned!

References

Barnes K, Bourne A, Cook PA, Turner AJ, Kenny AJ. (1991) Membrane peptidases in the peripheral nervous system of the pig: their localization by immunohistochemistry at light and electron microscopic levels. *Neuroscience* **44**: 245–261.

Bourne A, Barnes K, Taylor BA, Turner AJ, Kenny AJ. (1989) Membrane peptidases in the porcine choroid plexus and on other cell membranes in contact with the cerebral spinal fluid. *Biochem. J.* **259**: 69–80.

Gee NS, Kenny AJ. (1985a) Dissection of renal brush border membranes with monoclonal antibodies. *Biochem. Soc. Trans.* **13**: 101–102.

Gee NS, Kenny AJ. (1985b) Proteins of the kidney microvillar membrane. The 130 kDa protein in pig kidney, recognised by monoclonal antibody GK5C1, is an ectoenzyme with aminopeptidase activity. *Biochem. J.* **230**: 753–764.

Gee NS, Kenny AJ. (1987) Proteins of the kidney microvillar membrane. Enzymic and molecular properties of aminopeptidase W. *Biochem. J.* **246**: 97–102.

Gee NS, Matsas R, Kenny AJ. (1983) A monoclonal antibody to kidney endopeptidase-24.11. Its application in immunoadsorbent purification of the enzyme and immunofluorescent microscopy of kidney and intestine. *Biochem. J.* **214**: 377–386.

Gee NS, Bowes MA, Buck P, Kenny AJ. (1985) An immunoradiometric assay for endopeptidase-24.11 shows it to be a widely distributed enzyme in pig tissues. *Biochem. J.* **228**: 119–126.

Hedeager-Sorensen S, Kenny AJ. (1985) Proteins of the kidney microvillar membrane. Purification and properties of carboxypeptidase P from pig kidneys. *Biochem. J.* **229**: 251–257.

Hooper NJ, Hesp RJ, Tieku S. (1994) Metabolism of aspartame by human and pig intestinal microvillar peptidases. *Biochem. J.* **298**: 635–639.

Howell S, Kenny AJ, Turner AJ. (1992) A survey of membrane peptidases in two human colonic carcinoma cell lines, Caco-2 and HT29. *Biochem. J.* **284**: 595–601.

Howell S, Caswell AM, Kenny AJ, Turner AJ. (1993) Membrane peptidases on human osteoclast-like cells in culture: hydrolysis of calcitonin and hormonal regulation of endopeptidase-24.11. *Biochem. J.* **290**: 159–164.

Jackson MC, Choudry Y, Bourne A, Woodley JF, Kenny AJ. (1988) A fluorimetric assay for aminopeptidase W. *Biochem. J.* **253**: 299–302.

Kenny AJ, Bourne A. (1991) Cellular reorganization of membrane peptidases in Wallerian degeneration of pig peripheral nerve. *J. Neurocytol.* **20**: 875–885.

Kenny AJ, Booth AG, George SG, Ingram J, Kershaw D, Wood EJ, Young AR. (1976) Dipeptidyl peptidase IV, a kidney brush border serine peptidase. *Biochem. J.* **157**: 169–182.

Tieku S, Hooper NM. (1992) inhibition of aminopeptidases N, A and W. A reevaluation of the actions of bestatin and inhibitors of angiotensin converting enzyme. *Biochem. Pharmacol.* **44**: 1725–1730.

Chapter 17

Hepsin

Kotoku Kurachi

17.1 Background

Hepsin was originally found by screening a human liver cDNA library with an oligonucleotide probe designed against the amino acid sequence MFCAG (Leytus *et al.*, 1988; see Kurachi *et al.*, 1994, for a comprehensive review). This sequence is highly conserved among serine proteases, and is located approximately 15 amino acid residues away from the active site serine residue towards the amino terminus. One out of 31 cDNA clones strongly hybridizing with this probe (out of about 14 000 recombinant cDNA colonies in the library) was identified, and found to have a single long reading frame encoding a putative, novel serine protease of 417 amino acid residues. It was named hepsin. Its predicted amino acid sequence suggested a potential membrane-spanning serine protease. Being originally identified as a putative protein derived from cDNA clones, however, no information was available regarding its biochemical properties and biological function. This novel and unique structure made hepsin of particular interest to us, leading to its subsequent characterization and to proposals concerning its potential biological roles. More recently, hepsin has increasingly attracted a wider research interest.

17.2 Nucleotide and amino acid sequences: topology at the cell surface

The nucleotide sequence of the human hepsin cDNA and the deduced amino acid sequence are described elsewhere (Leytus *et al.*, 1988). The original hepsin cDNA isolated is about 1.8 kb in length, which is in good agreement with its mRNA size (1.8–1.85 kb) as estimated by RNA blot analysis (Tsuji *et al.*, 1991). The predicted amino acid sequence shows that hepsin is synthesised as a single polypeptide chain of 417 amino acid residues including the first Met residue (amino acid +1). No typical signal peptide is contained, and instead, an internal hydrophobic sequence stretch (27 amino acid residues in length) is located at residues 18–44. The sequence, Arg–Ile–Val–Gly–Gly, which is typical for the proteolytic activation site of many serine protease zymogens is contained at residues 162–166. If this sequence is cleaved at the Arg–Ile bond, analogous to the activation mechanism of zymogens of other serine proteases, hepsin may be converted to an active form, consisting of a light chain (the amino-terminal half) and a heavy chain (carboxyl-terminal half, including the catalytic site) linked by a disulphide bond. The putative light chain (162 residues in length) is composed of the amino-terminal sequence (17

Cell-Surface Peptidases in Health and Disease, edited by A.J. Kenny and C.M. Boustead.

residues) followed by the 27 residue long hydrophobic sequence and a link sequence of 118 residues. This hydrophobic sequence is likely to be the membrane-spanning anchor conferring on hepsin the topology of a type II cell-surface protein. In this configuration the short N-terminal sequence is intracellular, with the large C-terminal portion at the cell surface (Tsuji *et al.*, 1991; *Figure 17.1*).

The putative heavy chain (the catalytic subunit) is composed of 255 amino acid residues (residues 163–417). The heavy chain contains all the features of a typical, functional serine protease, including the amino acid residues (His203, Asp257 and Ser353) involved in the formation of the active site triad. The overall similarities of the heavy chain to chymotrypsin, trypsin and the catalytic subunit of plasmin are 39, 43 and 42%, respectively. Similarity to other well characterized proteases is also in the range of 30–40%. These results have suggested that hepsin is a novel serine protease, which is membrane-associated. The putative light chain of hepsin has no significant similarity to other known protein sequences, except for the hydrophobic sequence which is similar to other membrane-spanning sequences.

The sequence of hepsin is predicted to have 19 cysteine residues, 10 in the catalytic subunit and nine in the non-catalytic subunit. Based on the structural similarity, nine out of 10 in the catalytic subunit may account for intra-subunit disulphide bonds (paired Cys residues, 188/204, 291/359, 322/338 and 349/381) and an inter-subunit disulphide bond involving Cys277. This leaves Cys372 with no analogous partner among other serine proteases. This residue may be involved in linking hepsin to another ligand or protein, including a second hepsin molecule. The latter possibility, however, is not supported by protein blotting analyses, which show only the monomeric form of hepsin under non-reducing conditions. Nine cysteine residues predicted in the non-catalytic subunit may be involved in forming four intra-subunit disulphide bonds and one inter-subunit disulphide bond (Cys153 and Cys277 in the catalytic subunit) as predicted from the similarity to other serine proteases.

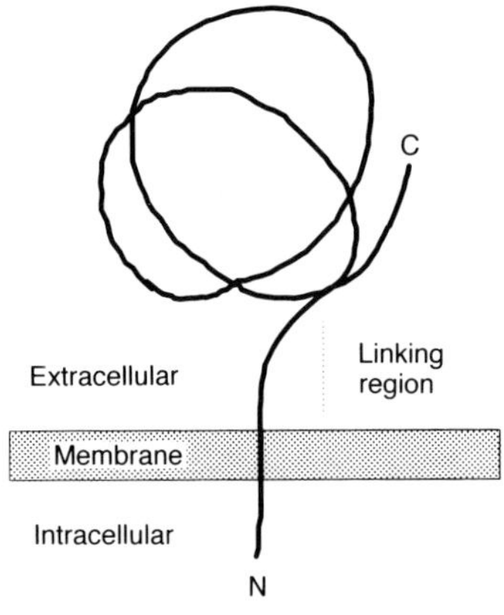

Figure 17.1. Schematic representation of hepsin as a type II membrane protein in the plasma membrane. The extracellular portion of hepsin contains the protease catalytic domain.

Hepsin contains at least one potential attachment site for an N-linked carbohydrate chain located at amino acid 112.

17.3 Genetics

The human hepsin gene is located at chromosome 19 at q13.1 (Tsuji *et al.*, 1991). Recently a rat hepsin cDNA has been cloned (Farley *et al.*, 1993). The cDNA is 1739 bp in size encoding rat hepsin of 416 amino acid residues, one amino acid residue fewer than human hepsin, due to the absence of the Glu3, which is present in human hepsin. Human and rat hepsin molecules share a very high overall similarity of 88.7% in their amino acid sequences.

No natural deficiency of hepsin in humans or animals has been reported to date, and hence, no information about the phenotype due to hepsin deficiency is currently available. In recent years, an increasing number of cell membrane-associated serine proteases have been identified (Kurachi *et al.*, 1994), which may help in delineating the genetics as well as the biological role of hepsin.

17.4 Active site and substrate specificity

Hepsin has all the necessary structures required for a functional serine protease including the critical amino acid residues, His203, Asp257 and Ser353, which are involved in the active site triad formation upon limited proteolysis at the sequence Arg–Ile–Val–Gly–Gly. These structural features suggested that hepsin is synthesised as a single polypeptide chain, and must be proteolytically cleaved to become active (Torres-Rosado *et al.*, 1993). During its purification, hepsin is partially activated by an unknown protease, whereas trypsin treatment *in vitro* significantly increases hepsin activity (Torres-Rosado *et al.*, 1993). Recently, Kazama *et al.* (1995) reported that the factor VII-activating activity of hepsin at the surface of BHK cells, transfected with hepsin expression vector, can be substantially increased by incubation of the cells with various concentrations of trypsin. However, the protease responsible for proteolytic activation of hepsin under the natural conditions has yet to be determined.

Synthetic substrates including *N*-benzoyl-Leu-Ser-Arg-pNA HCl, *N*-benzoyl-Ile-Glu-Phe-Ser-Arg-pNA HCl and *N*-benzoyl-Phe-Val-Arg-pNA HCl (pNA, para-nitroanilide) are cleaved by partially activated hepsin (Kurachi *et al.*, 1994; Torres-Rosado *et al.*, 1993; Tsuji *et al.*, 1991). Together with the structural features, this indicated that hepsin is a trypsin-type serine protease. A comprehensive study on its substrate specificity using arrays of synthetic substrates has yet to be made with highly purified hepsin. Recently, Kazama *et al.* (1995) have shown that when hepsin is overexpressed in BHK cells, the surface-associated hepsin of the cells can proteolytically convert factor VII to the activated form of factor VII. Under similar conditions, factor X, factor IX, prothrombin and protein C are not activated. These observations indicate that hepsin has a high natural substrate specificity suggesting a possibility that it may function as a processing enzyme for its natural substrate protein(s). Whether or not factor VII is among the natural substrates for hepsin, however, has yet to be determined.

17.5 Inhibitors

Several inhibitors are known for hepsin (Kazama *et al.*, 1995). 4-Aminidophenylmethanesulphonyl fluoride (1 mM concentration), aprotinin (8 μM), antithrombin III (3 μM), and antithrombin III (3 μM) with heparin (1 unit ml^{-1}) inhibit the ability of cell-surface hepsin to activate factor VII by 100%, 90%, 76% and 80% respectively. Soybean trypsin inhibitor (5 mM) and tissue factor pathway inhibitor (25 nM) without or with factor Xa (25 nM) do not show the inhibitory activity.

17.6 Tissue/cellular distribution and subcellular localization

High expression of hepsin was observed in various mammalian cells, including the hepatocyte-derived HepG2 cells and Alexander cells (PLC/PRF/5), breast cancer cells, and nerve cells (PC12) (Kurachi *et al.*, 1994; Torres-Rosado *et al.*, 1993). BHK cells express hepsin at low but significant levels, while hepsin was undetectable in endothelial cells (human umbilical cord or rat capillary). These data indicate that hepsin plays a crucial role(s) in most, if not all, different types of mammalian cells.

In HepG2 cells, hepsin is primarily present in plasma membranes as a molecular species of 51 kDa (major) and 28 kDa (minor), while in BHK cell membranes, only a 51 kDa species is present (Tsuji *et al.*, 1991). Hepsin is undetectable in cytosol as well as in culture medium. The *in vitro* translational products of hepsin mRNA, prepared from the cDNA with T7 promoter, gave a molecular size of about 45 kDa, which is in good agreement with the apparent molecular size of cellular hepsin (51 kDa), assuming that some carbohydrate chains are attached to cellular hepsin. Carbohydrate chains are absent in the *in vitro* translation product. The minor molecular species (28 kDa) observed for hepsin may be a specific degradation product of hepsin. A [^{3}H]-di-isopropylfluorophosphate (DFP)-labelling experiment showed that a hepsin fragment of 46 kDa which is derived from the intact 51 kDa by limited proteolysis contains the active site serine residue (Kurachi *et al.*, 1994). Whether or not the difference (5 kDa) in molecular weight between 51 kDa and 46 kDa can be accounted for by proteolytic release of a portion of the putative light chain, including the intracellular and membrane-spanning domains, is not clear. Such limited proteolysis would result in release of the extracellular portion (containing the catalytic subunit) with most of the linking sequence of the light chain (*Figure 17.1*). This fragment may still be proteolytically activated by an unidentified protease, thus making the fragment susceptible to the DFP labelling. Further study is required to establish the mechanism underlying hepsin activation as well as the protease responsible.

Hepsin is also present in other subcellular fractions, such as mitochondria or nuclei of HepG2 cells, indicating that, as described in the following sections, hepsin may be involved in multiple cellular functions (Tsuji *et al.*, 1991). The half-life of cellular hepsin is about 3.5 h, as estimated by pulse-labelling experiments using Alexander cells (Kurachi *et al.*, 1994).

Northern blot analysis of baboon tissues indicated that hepsin is ubiquitously expressed in most tissues, with particularly high levels in the liver, suggesting the possibility of a critical housekeeping function for hepsin (Tsuji *et al.*, 1991).

17.7 Biological roles

The precise biological role(s) of hepsin and underlying mechanisms have yet to be determined. Like other known cell-surface proteases (Delmas *et al.*, 1992; Kido *et al.*, 1991; Yeager *et al.*, 1992), it may be involved in membrane protein catabolism, processing and/or activation of other membrane-associated or extracellular proteins and peptides. It may even serve as a receptor for specific ligands. Its protease activity may or may not be essential for these potential biological functions, and hepsin may exert its biological activity by molecular properties other than its protease activity.

The first evidence to indicate the crucial biological role of hepsin in cell growth was obtained by using an affinity-purified anti-hepsin antibody (HAbP5), which was generated against the carboxyl-terminal region of hepsin (Torres-Rosado *et al.*, 1993). HAbP5 added to the culture medium suppressed the growth of hepatoma cells including HepG2 and Alexander (PLC/PRE/5) cells. Alexander cells were used in a series of experiments as a model cell line. On day 4, wells to which the antibody had been added had a reduced cell number (11% of that of control wells). This suppression of cell growth was associated with no noticeable changes in cell morphology or decrease in hepsin production, suggesting the importance of the presence of the accessible hepsin molecule on the cell surface for growth of these cells. The suppressive effect on cell growth was reversed upon removing the antibody from the culture medium. Other control antibodies (non-specific rabbit IgG or anti-angiogenin antibodies) did not show any such negative effects on cells, supporting the specificity of the suppressive effect of anti-hepsin antibody.

The importance of hepsin for cell growth was further demonstrated by a series of experiments employing hepsin-specific antisense oligonucleotides (Torres-Rosado *et al.*, 1993). The optimal concentration of antisense oligonucleotides for these cells is a range of 12.5–15 μM. Antisense oligonucleotides, ASpdoligo237 (phosphodiester regular oligonucleotide) and ASptoligo237 (phosphorothioate derivative), which are designed to the first ATG site area, showed gross inhibitory effects leading to a total arrest of Alexander cell growth after 3–4 days of treatment (*Figure 17.2*). The lag of the negative effects of antisense oligonucleotides is in good agreement with the estimated half-life of hepsin (approx. 3.5 h). Upon treatment with these oligonucleotides, gross changes in cell morphology (flattening and enlargement to give a pancake-like appearance) were induced. The arrest of cell growth and change in cell morphology were reversible by removing antisense oligonucleotides from the media. However, prolonged treatment of cells with the antisense oligonucleotides (longer than about 1 week) resulted in a gradual loss of cell viability. Various control oligonucleotides including sense strand oligonucleotides (SSpdoligo237 or SSptoligo237), oligonucleotides with randomized sequences of antisense oligonucleotides (RASpdoligo237 or RASptoligo237) or an unrelated antisense oligonucleotide which had been designed to the factor IX sequence (ASFIXoligo) showed no significant effects on cell growth or cell morphology, indicating the specific nature of the effects of antisense oligonucleotides. It was also noted that, upon antisense oligonucleotide treatment, some subcellular structures such as mitochondria and rough endoplasmic reticulum were substantially disturbed. These inhibitory effects of antisense oligonucleotides on cell growth are due to the specific inhibition of hepsin biosynthesis,

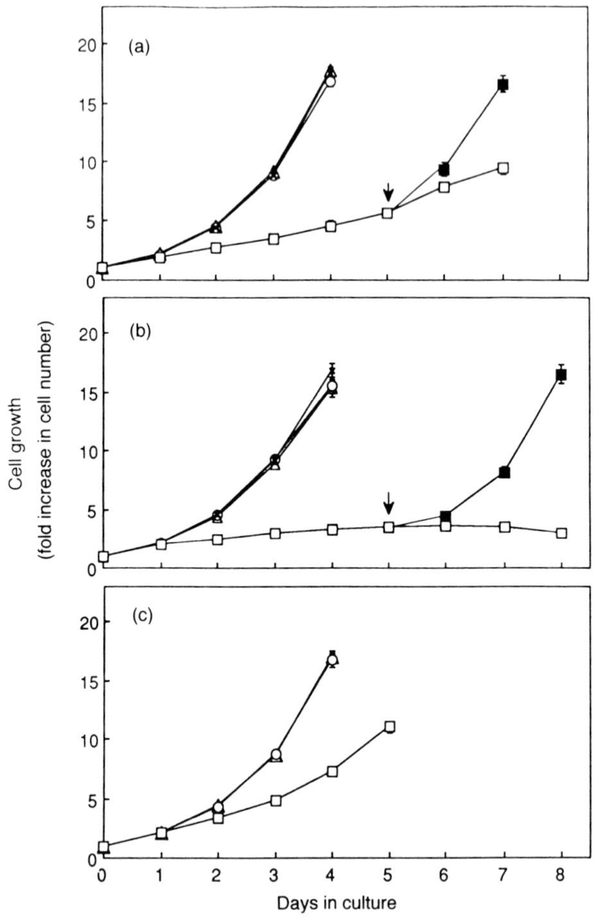

Figure 17.2. Effects of anti-hepsin antibodies and antisense oligonucleotides on the growth of Alexander cells. (a) Cells cultured in medium to which water (△), non-specific rabbit IgG (○), anti-angiogenin IgG (X), or anti-hepsin IgG (□) was added daily. Arrow indicates change from an anti-hepsin antibody-containing medium (□) to an anti-hepsin antibody-free medium (■). Bars indicate standard error ranges for days 1–5 and range of values for days 6 and 7. (b) Cells cultured in medium to which water (△), RAS-pt-oligo237 (○), SS-pt-oligo237 (X), or AS-pt-oligo237 (□) was added. Bars indicate standard error ranges. Arrow indicates a change from an oligo-containing (□) to an oligo-free (■) medium. (c) Cells cultured in medium to which water (△), RAS-pt-oligo952 (○), or AS-pt-oligo952 (□) was added. Mean values (*n*=5) were plotted. All other conditions are the same as in (b).

as shown by immunofluorescent staining of cells treated with antisense oligonucleotides. Cells treated with various control oligonucleotides did not show this effect. The hepsin level in Alexander cells upon exposure to antisense oligonucleotides was rapidly lowered to 29% after 1 day and to 8% after 2 days (Torres-Rosado *et al.*, 1993). The antisense oligonucleotide RASptoligo952, which is designed to a sequence much further downstream, but still in the coding region, also showed a specific, but weaker inhibitory effect on cell growth (about 60% inhibition). Antisense oligonucleotides designed to the general 3′-untranslated region away from the known functional region, showed no significant inhibitory effects. The weaker or non-inhibitory effects of antisense oligonucleotides designed to the downstream regions are in good agreement with the general inhibitory effects of this type of antisense oligonucleotide for biosynthesis of various gene products (Goodchild, 1989). These antisense oligonucleotides are considered to exert their inhibitory effects by passively interfering with mRNA binding to ribosomes.

These observations with anti-hepsin antibodies and hepsin-specific antisense oligonucleotides have demonstrated, for the first time, that hepsin plays essential roles in cell growth as well as in maintenance of the normal cell morphology. In relation to cell growth, it is interesting to find that hepsin gene expression is cell-cycle-dependent, as tested with synchronously growing Alexander cells (Kurachi *et al.*, 1994). High hepsin gene expression is induced in the middle of the S phase and its high level expression continues through the G_2 phase, followed by an abrupt decrease in the M phase, with only marginal expression through the G_1 phase. Since the half-life of hepsin is estimated to be about 3.5 h, this hepsin gene regulation may result in a maximal level of cellular hepsin during the middle of the G_2 phase until the M phase, which is then followed by a much reduced level throughout the G_1 phase and the most of the S phase.

As a protease, hepsin may be involved in the degradation of extracellular proteins, thus creating space for their growth which requires change in cell shape and migration. If hepsin plays such a role in cell growth, it may be produced at a significant level, particularly in actively growing and rearranging tissues. In the early stage of development (tested as early as day 10 of development), hepsin is expressed at a relatively low level throughout the embryo (Torres-Rosado *et al.*, 1993). By day 13, hepsin is sparsely expressed in the forming surface ectoderm and colonic epithelium, increasing significantly over the next several days, both in intensity and in the number of tissues stained, such as skin, liver, muscle and others. These observations have indicated that hepsin plays an important role in cell growth and tissue rearrangement, particularly at highly circumscribed developmental stages. The protease activity of hepsin is highly likely to be required in these processes to facilitate cell migration and process extension through the extracellular matrix and cell-filled milieu. Some cell-surface proteases have also been shown to function as receptors for viruses (see Chapter 25). Whether or not hepsin also assumes such a role has yet to be determined.

More recently, cell-surface hepsin has been shown to activate factor VII, which in turn appreciably activates factor X, resulting in thrombin formation in a mechanism significantly different from that involving the factor VIIa–tissue factor pathway (Kazama *et al.*, 1995). The hepsin-mediated factor VII activation may possibly have an important role in fibrin deposition *in situ* in certain cancer tissues. This is an interesting observation and warrants further studies to determine its significance. However, as descibed above, hepsin is expressed at a high level in the normal liver. Furthermore, hepsin is not only present at the cell surface, but is also present in subcellular organelles.

These observations strongly indicate that hepsin plays multiple critical roles in cell growth and maintenance of the normal cell morphology. However, the underlying mechanisms for these activities of hepsin have yet to be delineated.

References

Delmas B, Gelfi J, Haridon RL, Vogel LK, Sjostrom H, Noren O, Laude H. (1992) Aminopeptidase N is a major receptor for the enteropathogenic coronavirus TGEV. *Nature* **357**: 417–420.

Farley D, Reymond F, Nick H. (1993) Cloning and sequence analysis of rat hepsin, a cell surface serine proteinase. *Biochim. Biophys. Acta* **1173:** 350–352.

Goodchild J. (1989) In: *Oligodeoxynucleotides: Antisense Inhibitors of Gene Expression* (ed. J.S. Cohen), pp. 53–76. CRC Press, Boca Raton, FL.

Kazama Y, Hamamoto T, Foster DC, Kisiel W. (1995) Hepsin, a putative membrane-associated serine protease, activates human factor VII and initiates a pathway of blood coagulation on the cell surface leading to thrombin b formation. *J. Biol. Chem.* **270**: 66–72.

Kido H, Fukutomi A, Katunuma N. (1991) A novel membrane-bound serine esterase in human $T4^+$-lymphocytes is a binding protein of envelope glycoprotein gp120 of HIV-1. *Biomed. Biochim. Acta* **50:** 781–789.

Kurachi K, Torres-Rosado A, Tsuji A. (1994) Hepsin. *Meth. Enzymol.* **244:** 100–114.

Leytus SP, Loeb KR, Hagen FS, Kurachi K, Davie EW. (1988) A novel trypsin-like serine protease (hepsin) with a putative transmembrane domain expressed by human liver and hepatoma cells. *Biochemistry* **27:** 1067–1074.

Torres-Rosado AA, O'Shea KS, Tsuji A, Chou S-H, Kurachi K. (1993) Hepsin, a putative cell-surface protease, is required for mammalian cell growth. *Proc. Natl Acad. Sci. USA* **90:** 7181–7185.

Tsuji A, Torres-Rosado A, Arai T, Le Beau MM, Lemons RS, Chou S-H, Kurachi K. (1991) Hepsin, a cell membrane-associated protease. Characterization, tissue distribution and gene location. *J. Biol. Chem.* **266:** 16948–16953.

Yeager CL, Ashmun RA, Williams RK, Cardellichio CB, Shapiro LH, Look T, Holmes KV. (1992) Human aminopeptidase N is a receptor for human coronavirus 229E. *Nature* **357**: 420–422.

Chapter 18

Fertilin and the MDC family of related proteins

Len Hall

18.1 Background

In recent years, a novel family of mammalian membrane proteins has been identified which shows remarkable sequence homology to a number of snake venom haemorrhagic components possessing metalloproteinase and/or disintegrin activities. On the basis of this homology, these mammalian proteins have been alternatively referred to as either the MDC (metalloproteinase-like, disintegrin-like, cysteine-rich) family (Perry *et al.*, 1992, 1994, 1995a,b) or the ADAM (a disintegrin and metalloprotease) family (Wolfsberg *et al.*, 1995a,b).

The first full-length MDC cDNAs to be cloned and sequenced were those encoding rat and monkey (*Macaca fascicularis*) epididymal apical protein I (EAP I) (Perry *et al.*, 1992); an androgen-regulated 89 kDa transmembrane protein synthesised exclusively in the caput region of the epididymis (a highly convoluted tubule juxtaposed to the testis, involved in sperm maturation and storage) with a predominant localization on the apical surface of its principal epithelial cells. Soon after, full-length cDNAs encoding two other membrane proteins associated with the male reproductive tract, the α- and β-subunits of the fertilin complex, were shown to possess sequence homology to EAP I (Wolfsberg *et al.*, 1993).

Mammalian fertilization requires the species-specific recognition of morphologically distinct male and female gametes, followed by their fusion. Two levels of recognition appear to be involved in this process; initially sperm surface components interact with the zona pellucida, a glycoprotein matrix which surrounds the oocyte. This is then followed by the binding of further specific sperm membrane antigens to cognate receptors associated with the oolemma plasma membrane and gamete membrane fusion subsequently occurs. It is at this second level of sperm–egg interaction that the sperm surface fertilin complex (formerly known as PH30 antigen) has been proposed to play a major role, being implicated in both oolemma plasma membrane recognition and possibly also subsequent gamete membrane fusion (Blobel *et al.*, 1992).

The fertilin complex was first described in the guinea-pig (Blobel *et al.*, 1990; Primakoff *et al.*, 1987) where it was reported to exist as a heterodimeric glycoprotein composed of α- and β-subunits. Both subunits are initially synthesised in the testis as large precursors, but then undergo endoproteolytic processing to yield the mature products. In the

Cell-Surface Peptidases in Health and Disease, edited by A.J. Kenny and C.M. Boustead.

case of the fertilin α-subunit, this processing occurs exclusively in the testis, whereas fertilin β processing occurs partially in the testis and is completed in the epididymis after leaving the testis. The sequences of fertilins from a number of mammalian species are now available from the GENBANK and/or EMBL databases, including guinea-pig, monkey, mouse, rat and rabbit.

More recently, a large and rapidly growing number of additional MDC proteins have been described. Some appear to be specific to the male reproductive tract or are most abundantly expressed in this region, for example tMDC I (Barker *et al.*, 1994) or cyritestin (Heinlein *et al.*, 1994), tMDC II (Perry *et al.*, 1994), tMDC III (L. Hall and A.C.F. Perry, unpublished) and tMDC IV (Perry *et al.*, 1995b) suggesting a role in fertility; some are specific to other tissues, for example meltrin α (Yagami-Hiromasa *et al.*, 1995) in muscle which is implicated in myoblast fusion during myotube formation; whilst yet others exhibit a broad tissue distribution, for example MADM (Howard *et al.*, 1996), MDC9 (Weskamp *et al.*, 1996) and metargidin (Krätzschmar *et al.*, 1996), suggesting they may play a more general role.

18.2 Molecular structure of MDC proteins

All members of the mammalian MDC family of proteins (or their precursors) have a similar domain organization, as illustrated in *Figure 18.1*. In particular they all possess a zinc metalloproteinase-like domain and a disintegrin-like domain, similar to those associated with the haemorrhagic venoms of a wide range of viperid snakes (Hite *et al.*, 1994; Perry *et al.*, 1992). Within some snake venoms, for example that of habu (*Trimeresurus flavoviridis*), the disintegrin activity and the metalloproteinase activity reside within discrete polypeptide components (see *Figure 18.1*), where the former represents a small, cysteine-rich, integrin ligand (of about 70 amino acid residues) which binds to the platelet integrin $\alpha_{IIb}\beta_3$, thereby inhibiting fibrinogen binding and hence platelet aggregation, and the latter is a zinc proteinase (of approximately 200 amino acid residues) with basement membrane-hydrolysing activity. In other cases, for example the metalloproteinase jararhagin from the pit viper, *Bothrops jararaca* (Paine *et al.*, 1992), the disintegrin and metalloproteinase activities are present as domains within a single mosaic haemorrhagic protein (see *Figure 18.1*) composed of a prepro-sequence, a proteinase domain, a disintegrin domain and a cysteine-rich domain, similar to the arrangement found in the mammalian MDC proteins. In general, snake venom metalloproteinases (or reprolysins as they are now often called) can be divided into four classes based upon their structural organization (Rawlings and Barrett, 1995). Class I reprolysins contain a proteinase domain alone, class II contain a proteinase and a disintegrin domain, class III contain a proteinase, a disintegrin-like and a cysteine-like domain, and class IV additionally have a lectin-like domain at the C terminus. These multi-domain snake venom metalloproteinases, particularly the class III reprolysins, are generally more potent haemorrhagic agents, able to cause immediate bleeding, even at low dosage. In some cases it has been shown or suggested that the discrete disintegrin and metalloproteinase components found in some snake venoms arise by the post-translational proteolytic processing of these larger mosaic proteins (Hite *et al.*, 1994).

Unlike the snake venom mosaic proteins, almost all mammalian MDC proteins described to date are integral membrane proteins, possessing a single transmembrane

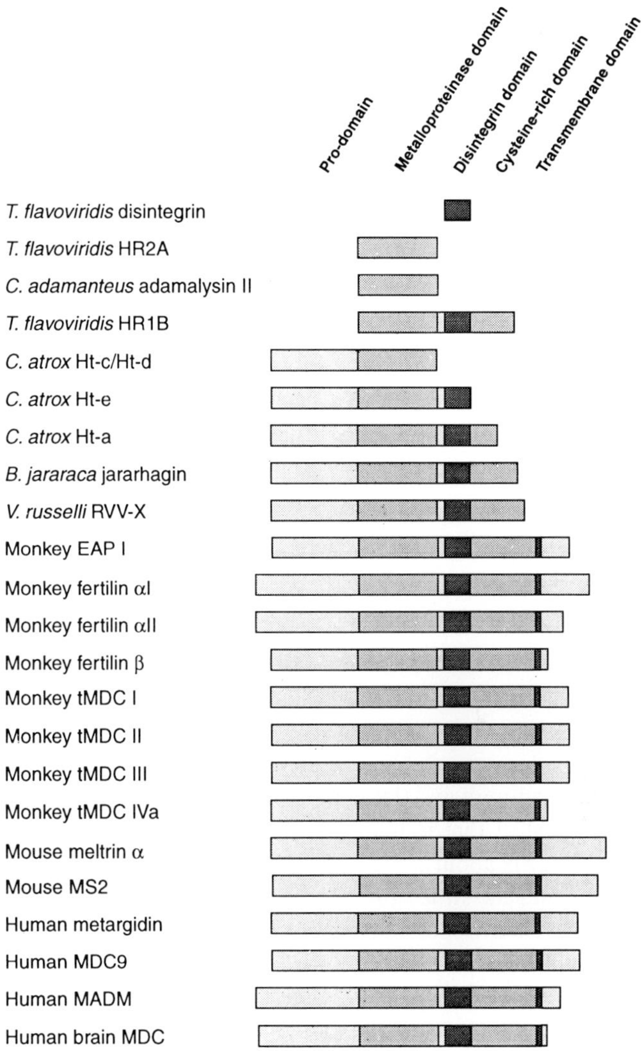

Figure 18.1. Diagrammatic representation of the mammalian MDC family of membrane proteins showing their homology to snake venom haemorrhagic proteins.

domain (of approximately 21 amino acid residues) immediately C-terminal to the cysteine-rich domain and this in turn is followed by a further hydrophilic domain of variable length depending on the particular MDC protein. In some cases (e.g. guinea-pig fertilin α) this C-terminal domain contains a number of repeated amino acid sequence motifs.

18.3 Genetics and tissue distribution

Until very recently, the most extensively studied members of the MDC family of membrane proteins were those which are expressed either exclusively or predominantly in the male reproductive tract (e.g. fertilin α, fertilin β, tMDC I or cyritestin), since these were the first to be identified and cloned and their abundance in these tissues has made them more amenable to study. However, it is now clear that a range of different MDCs are expressed in many mammalian tissues, some with a very broad tissue distribution (e.g. MADM, MDC9 and metargidin) and others with a more restricted pattern of expression. Indeed, reproductive tissues are by no means the only source of abundant, tissue-specific MDCs, meltrin α for example being predominantly expressed in neonatal muscle. Furthermore, the levels of some MDC transcripts in some tissues can differ markedly between different mammalian species (J. Jury, J. Frayne and L. Hall, unpublished). Very little is known about the regulation of specific MDCs, although epididymal apical protein I gene expression has been shown to be acutely androgen-dependent (Perry *et al.*, 1992).

Recent studies of the tissue distribution of some MDC transcripts, using the highly sensitive technique of reverse transcriptase-polymerase chain reaction (RT-PCR), have indicated that some MDCs which were originally believed to be tissue-specific (on the basis of Northern blot analysis, for example), are in fact expressed in a much broader range of tissues, albeit at a considerably lower level. Indeed fertilin α (but not fertilin β) transcripts have now been shown to be expressed at a low level in a wide range of non-reproductive tissues (including brain, kidney, liver, spleen, etc.) (Perry *et al.*, 1995b; Wolfsberg *et al.*, 1995a).

18.4 Active site

All members of the mammalian MDC family of proteins contain a metalloproteinase-like domain reminiscent of that found in the snake venom reprolysins. Whilst very little, if anything, is known about the function of the metalloproteinase-like domain of most MDC proteins and no structural information is yet available, sequence comparisons with the snake venom reprolysins might give us important clues as to which are likely to be proteolytically active and which are not.

Adamalysin II (EC 3.4.24.46) is a well-studied 24 kDa (203 amino acid residues) class I reprolysin from the eastern diamondback rattlesnake, *Crotalus adamanteus,* which is representative of both the simple metalloproteinases (class I reprolysins) and the proteinase domains of the larger mosaic haemorrhagic proteins (classes II–IV reprolysins) found in many snake venoms. Adamalysin II does not possess haemorrhagic activity, but is involved in the inactivation of serpins by limited proteolysis of their reactive-site loops; requiring both zinc and calcium for this activity (Kress, 1986). However, it is very similar to the low molecular weight zinc proteinases Ht-c and Ht-d (Hite *et al.*, 1994) found in the venom of *Crotalus atrox*, which degrade type IV collagen and some gelatins, and are regarded as haemorrhagic proteins.

Recent crystallographic studies of adamalysin II, to a resolution of 2 Å (Gomis-Rüth *et al.*, 1993), reveal an ellipsoid-shaped molecule with a shallow active site cleft separating a calcium-binding major domain (residues 6–150) made up of a five-stranded β-sheet and four α-helices, from an irregularly folded smaller subdomain (constituting the last 50 residues). The zinc-binding catalytic domain of adamalysin II comprises an extended MXHEXGHXLGXXHD motif, identical to that found in many other snake venom metalloproteinases and possessing the HEXXHXXGXXH consensus sequence common to the astacins (*Astacus astacus* astacin-like metalloproteinases; Titani *et al.*, 1987), serralysins (bacterial *Serratia*-like proteinases; Nakahama *et al.*, 1986) and matrixins (matrix metalloproteinases; Birkedal-Hansen *et al.*, 1993) as well as the more limited HEXXH motif of thermolysin (see *Figure 18.2*). The glutamate residue within the extended catalytic domain is implicated in the catalytic mechanism and the three histidine residues and a water molecule linked to the glutamate residue comprise the tetrahedrally co-ordinated zinc-binding ligand of adamalysin II (cf. the astacins and serralysins possess a fifth zinc co-ordinating position involving a tyrosine residue). Following the extended zinc-binding site, adamalysin II and astacin possess a similar 'methionine turn' which forms the hydrophobic basement of the active site.

The simple snake venom metalloproteinases and larger haemorrhagic mosaic proteins hydrolyse various matrix proteins. In general, long peptides which extend at least three residues beyond the scissile peptide bond are efficiently cleaved whilst short peptides are extremely poor substrates (Fox *et al.*, 1986). Cleavage occurs preferentially before medium-sized hydrophobic residues such as leucine (Fox *et al.*, 1986).

If we now turn to the metalloproteinase-like domains found in the mammalian MDC membrane proteins, it is found that some possess the zinc-binding HEXXHXXGXXH 'catalytic centre' consensus sequence common to the reprolysins, astacins, serralysins and matrixins, suggesting that they may be active metalloproteinases (see *Figure 18.2*). These include fertilin α, MS2 (macrophage-specific cysteine-rich transmembrane glycoprotein), metargidin, MDC9, meltrin α and MADM. Indeed all of these potential metalloproteinases contain some or all of the additional conserved residues associated with the extended MXHEXGHXLGXXHD consensus found in adamalysin II and many of the other related snake venom metalloproteinases. In contrast, the epididymal protein EAP I lacks the catalytic glutamate (E) residue suggesting that it is not likely to be catalytically active, even though all of the other residues associated with the extended 'catalytic centre' motif are conserved. However, it should still be borne in mind that the catalytic glutamate residue is replaced by a glutamine (Q) residue in EAP I, somewhat similar to the aspartate to asparagine substitution in the catalytic triad when comparing trypsin-like serine to papain-like cysteine proteinases. The remaining members of the MDC protein family depicted in *Figure 18.2* (i.e. tMDCI (cyritestin), tMDC II, tMDC III, tMDC IV, fertilin β and brain MDC) lack both the essential zinc-coordinating histidine residues and the catalytic glutamate residue, indicating that they could not function as zinc metalloproteinases.

18.5 Possible physiological roles

Whilst catalytic activity, if any, has yet to be established for many of the mammalian MDC proteins possessing the zinc-binding 'catalytic centre' motif, at least one member of this family, MADM, has been shown to be an active metalloproteinase. MADM was initially identified as a myelin basic protein-degrading activity, associated with brain myelin membranes. This myelin basic protein degradation was inhibited by

Trimeresurus flavoviridis HR2A	M T H E I G H N L G M E H D
Trimeresurus gramineus trigramin	M T H E M G H N L G M H H D
Trimeresurus flavoviridis HR1B	M T H E M G H N L G I P H D
Bothrops jararaca jararhagin	M A H E M G H N L G I H H D
Crotalus adamanteus adamalysin II	M A H E L G H N L G M E H D
Crotalus atrox Ht-c/Ht-d	M A H E L G H N L G M E H D
Crotalus atrox Ht-e	M T H E L G H N L G I H H D
Crotalus atrox Ht-a	M A H E L G H N L G M V H D
Agkistrodon contortrix fibrolase	M A H E L G H N L G M N H D
Vipera russelli RVV-X	M A H E L S H N L G M Y H D
Monkey fertilin α	M V H E L G H N L G I Q H D
Guinea-pig fertilin α	L V H E L G H N L G I R H D
Mouse MS2	M A H E L G H N L G M S H D
Human metargidin	I A H E L G H S L G L D H D
Human MDC9	V A H E L G H N L G M N H D
Mouse meltrin α	L A H E L G H N F G M N H D
Human MADM	F A H E V G H N F G S P H D
Monkey EAP I	M A H Q L G H N L G M Q H D
Rat EAP I	M A H Q L G H S L G M R H D
Monkey tMDC I	M A Q L L G I N L G L T Y D
Mouse cyritestin	L S Q L L G I N L G L A Y D
Monkey tMDC II	I V Q L L G L N L G L T Y D
Monkey tMDC III	I A Q L L G L N V G L T Y D
Human brain MDC	L A Q T L G Q N L G M M W N
Monkey fertilin β	L A Q L L S L S M G I P Y D
Guinea-pig fertilin β	L V Q L L S V S M G I A Y D
Monkey tMDC IV	T T Q T L M R S M G V E Y D
Astacus astacus astacin	I I H E L M H A I G F Y H E
Serratia sp. proteinase	F T H E I G H A L G L S H P
Human stromelysin	A A H E I G H S L G L F H S
B. stearothermophilus thermolysin	V A H E L T H A V T D Y T A

Figure 18.2. Comparison of the extended 'catalytic centre' motif of snake venom metalloproteinases with mammalian MDC proteins (or their precursors) and other types of zinc metalloproteinases. Conserved residues which form part of the extended MXHEXGHXLGXXHD reprolysin consensus sequence are shaded. All of these sequences may be found in the NBRF-PIR and/or SWISS-PROT protein databases.

phenanthroline and dithiothreitol (Chantry *et al.*, 1988), suggesting that MADM was a metalloproteinase. However, MADM was subsequently shown to be expressed not only in the brain, but also at similar levels in many other tissues (Chantry and Glynn, 1990), indicating that myelin is unlikely to be the natural substrate for this widely expressed metalloproteinase, although its true physiological substrate has yet to be established. Recently, cDNA clones encoding bovine, rat and human MADM have been isolated and sequenced (Howard *et al.*, 1996), and shown to belong to the mammalian MDC family. The pro-domain of MADM contains a single cysteine residue which may, by analogy with the matrixins and snake venom reprolysins, interact with the zinc ion at the metalloproteinase catalytic centre, thereby maintaining the proenzyme in a latent state. However, this cysteine residue is not part of the characteristic P(R/K)XCGV 'cysteine switch' motif associated with the reprolysins. Since the prodomain of MADM ends with a tetrabasic furin-like protease motif (RKKR), loss of the pro-domain may occur during transport through the Golgi complex, thereby presenting a functionally active, membrane-bound metalloproteinase at the cell surface. Such a pathway is different from the secretion of latent matrixin proenzymes, but is reminiscent of the processing of prostromelysin-3 (Pei and Weiss, 1995).

Recent isolation and sequence analysis of the *Drosophila kuzbanian* (*kuz*) gene (Rooke *et al.*, 1996), a gene essential for the partitioning of neural and non-neuronal cells during development of the central and peripheral nervous systems in *Drosophila*, has shown that *kuz* encodes an MDC protein, most closely related to bovine MADM (with a similar metalloprotease-like catalytic centre motif), raising the possibility that some mammalian MDCs may play a role in neurogenesis.

Sequence comparison of the monkey, guinea-pig, rabbit, mouse and rat fertilin α subunit precursors reveals that the metalloproteinase-like domain is the most highly conserved region of this protein, even though this domain is lost during fertilin α processing in the testis. For example, the metalloproteinase-like domains of the monkey and guinea-pig fertilin α precursors exhibit 83% nucleotide sequence identity, whereas the pro-domain, disintegrin-like domain and cysteine-rich plus C-terminal domains exhibit only 55%, 54% and 37% identity respectively (see *Figure 18.3*). In contrast, the metalloproteinase-like domains of the monkey and guinea-pig fertilin β-subunit precursors (which lack the 'catalytic centre' histidine and glutamate residues) show only 64% nucleotide sequence identity, a value identical to that for the disintegrin-like domain and not much greater than that of the N- and C-terminal domains (see *Figure 18.3*).

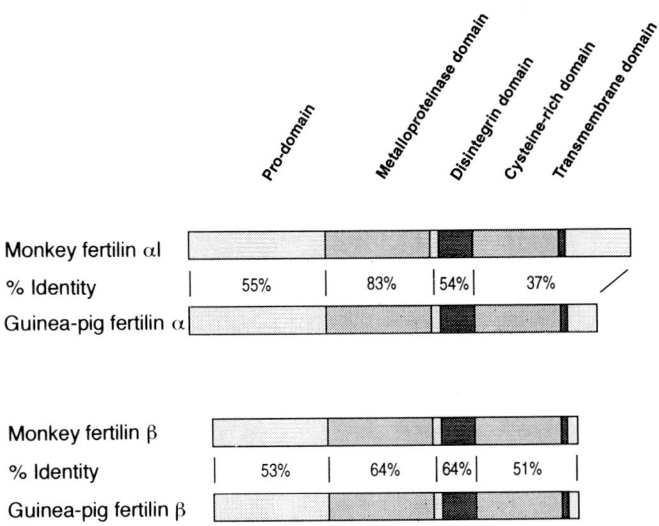

Figure 18.3. Sequence comparison of monkey and guinea-pig fertilin α and β precursors. The % nucleotide sequence identity of various regions is shown. Monkey fertilin α is unique amongst currently sequenced fertilins in possessing two closely related, but distinct, isoforms designated αI and αII (Perry *et al.*, 1995a). Monkey fertilin αI is represented here; the αII isoform gives similar identity values. Note the particularly high level of sequence conservation within the fertilin α metalloproteinase-like domain, consistent with this domain being catalytically active.

This high level of sequence conservation within the metalloproteinase-like domain of the fertilin α precursor, taken together with the presence of a conserved 'catalytic centre' motif, strongly suggest that the fertilin α precursor will prove to be an active zinc metalloproteinase. In the guinea-pig, fertilin α has been proposed to form a heterodimeric complex with fertilin β on spermatozoa, where it has been suggested to play a role in sperm–egg membrane fusion during fertilization (Blobel *et al.*, 1992). However, unlike fertilin β, it is expressed in a wide range of tissues (albeit at a much lower level than in the testis), implying that fertilin α may possess a more general or additional function. The metalloproteinase-like domain is absent from the mature form of fertilin α (in the reproductive tract at least), following endoproteolytic processing of the fertilin α precursor in the testis. It is therefore not inconceivable that the metalloproteinase-like domain might be involved in the autocatalytic processing of the fertilin α precursor, similar to the self-processing of the pro-Kex2 endoproteinase. In this respect, all sequenced fertilin α precursors possess a positionally conserved di-, tri- or tetrabasic site (monkey: RKRR; guinea-pig: RRRR; mouse: FKKK; rabbit: RTRR; rat: HKRR) between the metalloproteinase-like and disintegrin-like domains. Alternatively, the fertilin α precursor may play a role in the processing of other proteins or activation of other endoproteinases; possibly other MDC family members.

Recent work on human fertilin (Jury *et al.,* 1997) has indicated that the human genome apparently contains a single, non-functional fertilin α gene, which is transcribed but which does not encode a functional protein (due to the presence of a number of deletions and in-frame termination codons). This implies that the proposed role of fertilin α in sperm–egg recognition, as well as any additional functions in non-reproductive tissues, must be a facilitatory rather than an essential one, or that another MDC protein may substitute for fertilin α in the human.

Meltrin α is an MDC family member which is implicated in skeletal muscle development. During this process myoblast cells aggregate and their plasma membranes fuse to form multinucleated myotubules although the molecular mechanisms involved are poorly understood. Meltrin α, like fertilin α possesses a region which resembles the fusion peptides associated with some viruses (e.g. rubella virus). This has led Yagami-Hiromasa *et al.* (1995) to propose that gamete and myoblast fusion may proceed by a common mechanism involving the potential fusogenic peptides associated with these MDC proteins. Meltrin α is expressed predominantly in neonatal muscle and in differentiation-induced cultured C2 cells, consistent with its proposed role in myoblast fusion. Furthermore, when the level of meltrin α produced by C2 cells in culture was reduced by expression of an antisense RNA, there was a concomitant suppression of myotube formation, strongly implicating meltrin α in this process. Meltrin α is synthesised as a 115 kDa precursor which undergoes processing to produce an 86 kDa form which lacks the pro-domain and to smaller forms which lack the metalloproteinase-like domain (similar to fertilin α). Preliminary data (Yagami-Hiromasa *et al.*, 1995) using truncated constructs in C2 cells suggest that processed forms of meltrin α facilitate myoblast fusion, whereas over-expression of the full-length precursor suppresses fusion, suggesting that proteolytic processing of the meltrin α precursor may be important in governing its activity. As in the case of fertilin α, the highly conserved 'catalytic centre' of the metalloproteinase-like domain of the meltrin α precursor suggests that it may be an active zinc metalloproteinase, possibly involved in its own processing to produce the mature fusogenic form.

Other members of the MDC family of mammalian membrane proteins which, on the basis of their conserved 'catalytic centre' motif, may possess zinc metalloproteinase activity include metargidin (Krätzschmar *et al.*, 1996) and MDC9 (Weskamp *et al.*, 1996), both of which are expressed in a wide range of tissues. Both of these MDC proteins have been localized to the plasma membrane of the cell and both are of a size which indicates that the mature form possesses both the disintegrin-like and the metalloproteinase-like domains. Whilst the functions of these two proteins have yet to be established, it is interesting to note that they both possess proline-rich motifs within their cytoplasmic tails which closely resemble Src homology 3 (SH3) ligand domains (consensus sequence: RPLPXXP), a feature found in some cytoskeletal proteins and signalling molecules, raising the possibility that MDC9 and metargidin may interact with cytoplasmic proteins. Similar putative SH3 ligand domains are also to be found in some other MDC proteins including meltrin α and MS2.

Finally, when attempting to establish functions for the MDC family of proteins it should be remembered that they all possess a conserved disintegrin-like domain. In common with other proteins which interact with integrins (e.g. fibronectin) many, but not all, snake venom disintegrins contain a consensus binding sequence, RGD, and alterations of this sequence affect either the affinity or the specificity of binding. In the majority of MDC proteins, this RGD binding motif is absent but an XCD (often ECD) motif is found at a similar position. Interestingly, ECD is probably the most common alternative tripeptide in non-RGD-containing snake venom disintegrins or the disintegrin domains of mosaic haemorrhagic proteins (e.g. *T. flavoviridis* HR1B, *B. jararaca* jararhagin, *Vipera russelli* RVV-X, *Echis pyramidum leakeyi* EcH-II). There is therefore every reason to believe that the disintegrin-like domain of many, if not all, of the mammalian MDC proteins will be capable of interacting with cognate integrin receptors and such an interaction has already been established between the mouse fertilin β-subunit on the sperm surface and an integrin $\alpha_6\beta_1$ receptor on the oolemma plasma membrane (Almeida *et al.*, 1995). Many mammalian MDCs therefore have the potential to form specific cell–cell or cell–matrix interactions, thereby affording a means of precisely targeting their putative metalloproteinase activity.

In conclusion, the mammalian MDCs, together with the closely related snake venom reprolysins form a fascinating, but as yet poorly understood, family of proteins with diverse functions ranging from sperm–egg recognition and myoblast fusion in mammals to haemorrhagic activity in snake venoms.

References

Almeida EAC, Huovila A-PJ, Sutherland AE *et al.* (1995) Mouse egg integrin $\alpha_6\beta_1$ functions as a sperm receptor. *Cell* **81:** 1095–1104.

Barker HL, Perry ACF, Jones R, Hall L. (1994) Sequence and expression of a monkey testicular transcript encoding tMDC I, a novel member of the metalloproteinase-like, disintegrin-like, cysteine-rich (MDC) protein family. *Biochim. Biophys. Acta* **1218:** 429–431.

Birkedal-Hansen H, Moore WGI, Bodden, MK, Windsor LJ, Birkedal-Hansen B, De Carlo A, Engler JA. (1993) Matrix metalloproteinases: a review. *Crit. Rev. Oral Biol. Med.* **4:** 197–250.

Blobel CP, Myles DG, Primakoff P, White JM. (1990) Proteolytic processing of a protein involved in sperm–egg fusion correlates with acquisition of fertilization competence. *J. Cell Biol.* **111:** 69–78.

Blobel CP, Wolfsberg TG, Turck CW, Myles DG, Primakoff P, White JM. (1992) A potential fusion peptide and an integrin ligand domain in a protein active in sperm–egg fusion. *Nature* **356:** 248–252.

Chantry A, Glynn P. (1990) A novel metalloproteinase originally isolated from brain myelin membranes is present in many tissues. *Biochem. J.* **268:** 245–248.

Chantry A, Earl C, Groome N, Glynn P. (1988) Metalloprotease cleavage of 18.2- and 14.1-kilodalton basic proteins dissociating from rodent myelin membranes generates 10.0- and 5.9-kilodalton C-terminal fragments. *J. Neurochem.* **50:** 688–694.

Fox JW, Campbell R, Beggerly L, Bjarnasson JB. (1986) Substrate specificities and inhibition of two hemorrhagic zinc proteases Ht-c and Ht-d from *Crotalus atrox* venom. *Eur. J. Biochem.* **156:** 65–72.

Gomis-Rüth F-X, Kress LF, Bode W. (1993) First structure of a snake venom metalloproteinase: a prototype for matrix metalloproteinases/collagenases. *EMBO J.* **12:** 4151–4157.

Heinlein UAO, Wallat S, Senftleben A, Lemaire L. (1994) Male germ cell-expressed mouse gene TAZ83 encodes a putative, cysteine rich transmembrane protein (cyritestin) sharing homologies with snake venom toxins and sperm–egg fusion proteins. *Dev. Growth Differen.* **36:** 49–58.

Hite LA, Jia L-G, Bjarnason, JB, Fox JW. (1994) cDNA sequences for four snake venom metalloproteinases: structure, classification, and their relationship to mammalian reproductive proteins. *Arch. Biochem. Biophys.* **308:** 182–191.

Howard L, Lu X, Mitchell S, Griffiths S, Glynn P. (1996) Molecular cloning of MADM: a catalytically active mammalian disintegrin-metalloprotease expressed in various cell types. *Biochem. J.* **317:** 45–50.

Jury JA, Frayne J, Hall L. (1997) The human fertilin α gene is non-functional; implications for its proposed role in fertilization. *Biochem. J.*, in press.

Krätzschmar J, Lum L, Blobel CP. (1996) Metargidin, a membrane-anchored metalloprotease-disintegrin protein with an RGD integrin binding sequence. *J. Biol. Chem.* **271:** 4593–4596.

Kress LF. (1986) Inactivation of human plasma serine proteinase inhibitors (serpins) by limited proteolysis of the reactive site loop with snake venom and bacterial metalloproteinases. *J. Cell. Biochem.* **32:** 51–58.

Nakahama K, Yoshimura K, Marumoto R, Kikuchi M, Lee IS, Hase T, Matsubara H. (1986) Cloning and sequencing of *Serratia* protease gene. *Nucleic. Acids Res.* **14:** 5843–5855.

Paine MJI, Desmond HP, Theakston RDG, Crampton JM. (1992) Jararhagin: purification, cloning, and molecular characterization of a high molecular weight haemorrhagic metalloprotease from *Bothrops jararaca* venom – insights into the disintegrin gene family. *J. Biol. Chem.* **267:** 22869–22876.

Pei D, Weiss SJ. (1995) Furin-dependent intracellular activation of the human stromelysin-3 zymogen. *Nature* **375:** 244–247.

Perry ACF, Jones R, Barker PJ, Hall L. (1992) A mammalian epididymal protein with remarkable sequence similarity to snake venom haemorrhagic peptides. *Biochem. J.* **286:** 671–675.

Perry ACF, Barker HL, Jones R, Hall L. (1994) Genetic evidence for an additional member of the metalloproteinase-like, disintegrin-like, cysteine-rich (MDC) family of mammalian proteins and its abundant expression in the testis. *Biochim. Biophys. Acta* **1207:** 134–137.

Perry ACF, Gichuhi PM, Jones R, Hall L. (1995a) Cloning and analysis of monkey fertilin reveals novel α subunit isoforms. *Biochem. J.* **307:** 843–850.

Perry ACF, Jones, R, Hall L. (1995b) Analysis of transcripts encoding novel members of the mammalian metalloprotease-like, disintegrin-like, cysteine-rich (MDC) protein family and their expression in reproductive and non-reproductive monkey tissues. *Biochem. J.* **312:** 239–244.

Primakoff P, Hyatt H, Tredick-Kline J. (1987) Identification and purification of a sperm surface protein with a potential role in sperm–egg membrane fusion. *J. Cell Biol.* **104:** 141–149.

Rawlings ND, Barrett AJ. (1995) In: *Proteolytic Enzymes: Aspartic and Metallopeptidases: Evolutionary Families of Metallopeptidases* (ed. A.J. Barrett), pp. 183–228. Academic Press, New York.

Rooke J, Pan D, Xu T, Rubin GM. (1996) KUZ, a conserved metalloprotease-disintegrin protein with two roles in *Drosophila* neurogenesis. *Science* **273:** 1227–1231.

Titani K, Torff H-J, Hormel S, Kumar S, Walsh KA, Rödl J, Neurath H, Zwilling R. (1987) Amino acid sequence of a unique protease from the crayfish *Astacus fluviatilis*. *Biochemistry* **26:** 222–226.

Weskamp G, Krätzschmar J, Reid MS, Blobel CP. (1996) MDC9, a widely expressed cellular disintegrin containing cytoplasmic SH3 ligand domains. *J. Cell Biol.* **132:** 717–726.

Wolfsberg TG, Bazan JF, Blobel CP, Myles DG, Primakoff P, White JM. (1993) The precursor region of a protein active in sperm–egg fusion contains a metalloprotease and a disintegrin domain: structural, functional, and evolutionary implications. *Proc. Natl Acad. Sci. USA* **90:** 10783–10787.

Wolfsberg TG, Straight PD, Gerena RL, Huovila AP, Primakoff P, Myles DG, White JM. (1995a) ADAM, a widely distributed and developmentally regulated gene family encoding membrane proteins with A Disintegrin And Metalloprotease domain. *Devel. Biol.* **169:** 378–383.

Wolfsberg TG, Primakoff P, Myles DG, White JM. (1995b) ADAM, a novel family of membrane proteins containing A Disintegrin And Metalloprotease domain: multipotential functions in cell–cell and cell–matrix interactions. *J. Cell Biol.* **131:** 275–278.

Yagami-Hiromasa T, Sato T, Kurisaki T, Kamijo K, Nabeshima Y-I, Fujisawa-Sehara A. (1995) A metalloprotease-disintegrin participating in myoblast fusion. *Nature* **377:** 652–656.

Chapter 19

Neuropeptide inactivation in the central nervous system

Anthony J. Turner

19.1 Introduction

The discovery of the opioid peptides, the enkephalins, in the early 1970s led to a proliferation of studies on the mechanisms by which their synaptic action was terminated. In general, two principal mechanisms have been identified by which neurotransmitters are inactivated. In the case of catecholamines or amino acid neurotransmitters, uptake into neurons or glia via specific membrane transport proteins is the predominant method for removal of the signal molecule. For acetylcholine, enzymic hydrolysis by synaptic membrane acetylcholinesterase predominates. The latter mechanism, enzymic degradation in the synaptic cleft, is now established as the principal means of terminating neuropeptide action in the nervous system. Such metabolism could occur through the action of membrane-bound or soluble secreted peptidases but, for the majority of cases studied to date, it is the action of membrane ectoenzymes that is of primary importance. Thus, in characterizing any novel peptidase as a potential neuropeptide-degrading enzyme, establishing its integral membrane character and ectoenzyme status is paramount. All too often, insufficiently rigorous criteria have been applied to the characterization of such activities. Subcellular fractionation of brain tissue can provide useful indications of the localization of specific peptidases and detection of a particular activity in intact synaptosomes, which is not significantly increased upon hypotonic or detergent lysis, is a good indicator of ectoenzyme status. Likewise, the use of intact, primary neuronal (or glial) cells in culture, to examine the metabolism of an individual neuropeptide can be helpful. It should be noted, however, that caution is needed in interpreting data obtained with transformed cells (e.g. neuroblastoma or glioma cell lines), which may not reflect the normal situation. Brain slices can be used to monitor the metabolism both of exogenously added neuropeptide or the endogenously released peptide and, coupled with the application of appropriate inhibitors, can aid in the identification of the physiologically relevant peptidase(s). Such strategies have successfully been used in a number of cases, particularly in delineating the pathways of metabolism of enkephalins, substance P and cholecystokinin (see below and de la Baume *et al.*, 1983; Davis *et al.*, 1992; Konkoy and Davis, 1995; Rose *et al.*, 1989). Such studies, however, rely exquisitely on the specificity of the applied inhibitor and, unless carefully controlled, can provide misleading conclusions about relevant peptidases. Examples of such problems are listed in subsequent sections.

Cell-Surface Peptidases in Health and Disease, edited by A.J. Kenny and C.M. Boustead.

For subcellular fractionation studies of nervous tissue, the Percoll procedure devised by Dunkley and colleagues (Dunkley *et al.*, 1986, 1988; Thorne *et al.*, 1991), is particularly useful, providing a relatively homogeneous population of synaptosomes in less than 1 h and allowing distinction between pre- and post-synaptic elements. Its application to the localization of membrane peptidases has aided in the location of endopeptidase-24.11 (E-24.11), angiotensin-converting enzyme (ACE), aminopeptidase N (AP-N) and dipeptidyl peptidase IV (DPP-IV) in brain (Barnes *et al.*, 1992, 1994). The identification of individual peptidases as integral membrane proteins has been discussed in Chapters 1 and 3, but a simple and reliable technique is the procedure of phase separation in Triton X-114 (Bordier, 1981). The purification of the protein and its immunolocalization at the light and electron microscopy levels in the nervous system provides the ultimate test of its amphipathic character, and its membrane and cellular location. *Table 19.1* provides a list of criteria for identification of a peptidase as a physiologically relevant neuropeptide-degrading enzyme. In relatively few cases have these criteria been rigorously applied, by far the best studied example being E-24.11.

A recent review of brain peptidases, particularly in relation to peptide metabolism at the blood–brain barrier is to be found in Brownlees and Williams (1993) and a more detailed discussion of ectoenzymes and neuropeptide metabolism is provided in Kenny and Turner (1987).

Table 19.1. Criteria for identification of a neuropeptide-inactivating enzyme

1.	Enzyme hydrolysis produces physiologically inactive products
2.	Enzyme is strategically located on neuronal or glial cell plasma membranes as an ectoenzyme
3.	Specific enzyme inhibition protects released neuropeptides from degradation
4.	Co-localization of peptidase and its substrate and/or neuropeptide receptor
5.	Stabilized peptide analogues show enhanced or prolonged biological activity

19.2 Endopeptidase-24.11 in the central nervous system

The general properties of E-24.11 have been described in Chapter 6 and this section will focus only on its roles in the central nervous system (CNS) in relation to the metabolism of neuropeptides. It was a year after the first purification of E-24.11 (Kerr and Kenny, 1974) that Hughes and colleagues identified the first two opioid peptides in the brain, [Met]- and [Leu]-enkephalin (Hughes *et al.*, 1975). It was not until several years later that these two fields of research converged with the report that the opioid peptides were inactivated principally through hydrolysis of the Gly3–Phe4 bond by an activity in brain (termed 'enkephalinase' at that time; Craves *et al.*, 1978; Malfroy *et al.*, 1978). The subsequent observation that phosphoramidon inhibited the hydrolysis of this bond led us to infer that brain 'enkephalinase' was identical with E-24.11 in renal microvilli (Fulcher *et al.*, 1982) and this was subsequently confirmed immunologically (Matsas *et al.*, 1983) and by purification and characterization of the brain enzyme (Relton *et al.*, 1983). In the intervening years many other peptide neurotransmitters were being identified, the majority of which appeared to be potential substrates for E-24.11, based on its known specificity.

Several pieces of evidence support a physiological role for E-24.11 in the inactivation of enkephalins. Thus, inhibition of E-24.11 by the synthetic inhibitor thiorphan *in vivo* produced naloxone-reversible, antinociceptive responses in mice (Roques *et al.*, 1980; see also Chapter 5). Thiorphan has also been shown to enhance the recovery of enkephalins released endogenously from brain slices (de la Baume *et al.*, 1983) although complete protection against metabolism also required the inclusion of an aminopeptidase inhibitor in the perfusion medium. A close correspondence has also been observed between the distribution in brain of E-24.11 and of opioid receptors detected autoradiographically (Waksman *et al.*, 1986a). Immunohistochemical and autoradiographic studies in the pig and rat nervous system, respectively, have shown that E-24.11 is particularly concentrated in a striatonigral pathway (Matsas *et al.*, 1986; Pollard *et al.*, 1987), co-localizing with enkephalin and with substance P immunoreactivity in striosomes (Barnes *et al.*, 1988a). Immunoperoxidase electron microscopy of the globus pallidus (Barnes *et al.*, 1988b) has revealed E-24.11 to be localized to synaptic terminals as well as to axonal membranes, but the enzyme is not present on glial cells in the CNS. A combination of lesioning and electron microscopic studies have also confirmed a predominantly neuronal localization (Marcel *et al.*, 1990; Waksman *et al.*, 1986b). Application of the Percoll gradient fractionation procedure of Dunkley *et al.* (1986, 1988) to homogenates of pig substantia nigra have suggested that E-24.11 is present on both pre- and post-synaptic membranes where it can efficiently hydrolyse peptides released into the synaptic cleft (Barnes *et al.*, 1992). Further pharmacological and ultrastructural studies have strongly suggested that E-24.11 plays a role in the synaptic inactivation of substance P. For example, inhibitors of E-24.11 can increase the recovery of substance P immunoreactivity released from slices of rat substantia nigra (Littlewood *et al.*, 1988; Mauborgne *et al.*, 1991) and inhibit completely the hydrolysis of the peptide by synaptic membranes prepared from pig striatum (Matsas *et al.*, 1983) or rat substantia nigra (Oblin *et al.*, 1988). Finally, double-labelling studies, employing immunogold and immunoperoxidase methods at the electron microscopic level, have established the presence of substance P within some terminals in the substantia nigra that also reveal positive staining for E-24.11 at the plasma membrane (Barnes *et al.*, 1993). This study provided the first direct evidence for the co-localization of a neuropeptide with its putative degrading enzyme. Other tachykinins (neurokinins A and B) as well as the natriuretic peptides are also hydrolysed efficiently by E-24.11 *in vitro* (Hooper and Turner, 1985; Hooper *et al.*, 1985; Kenny *et al.*, 1993) and may be target substrates *in vivo*. More recently, E-24.11 has been localized to the stratum oriens and stratum radiatum of the hippocampus where it is postulated to play a role in the inactivation of somatostatin (Barnes *et al.*, 1995). Other potential substrates for E-24.11 in the nervous system, at least in some brain regions, are cholecystokinin (CCK), neurotensin, vasoactive intestinal polypeptide, calcitonin gene-related peptide, neuropeptide Y and the endothelins. Thus E-24.11 appears to function at numerous peptidergic synapses in a manner analogous to that of acetylcholinesterase at cholinergic terminals. Inhibitors of E-24.11, either alone or in combination with other peptidase inhibitors, may therefore have therapeutic value by potentiating the synaptic levels of specific neuropeptides.

In addition to its localization on neuronal membranes in the CNS, E-24.11 is abundant on the brush border membranes of the choroid plexus and in pial membranes in the brain and spinal cord (Bourne *et al.*, 1989). In the choroid plexus, E-24.11, as well as other peptidases located there (e.g. ACE, AP-N), may play a role in regulating neuropeptide concentration in the cerebrospinal fluid. Although E-24.11 is not detectable on glial cells in the CNS, it is associated with Schwann cell membranes in the peripheral nervous system (PNS) in all types of peripheral nerve examined (Barnes *et al.*, 1991a). At these locations E-24.11 may play a role in the development and regeneration of axons rather than in terminating neuropeptide signals. Wallerian degeneration of pig peripheral nerve produces a striking cellular reorganization of certain membrane peptidases, including E-24.11 (Kenny and Bourne, 1991). Furthermore, phosphoramidon has been reported to affect mammalian embryonic development in a dose-dependent manner, producing craniofacial and at higher inhibitor doses, some cardiac defects, further suggesting a role for E-24.11 in developmental events in the nervous system and elsewhere (Spencer-Dene *et al.*, 1994).

19.3 Aminopeptidases in neuropeptide metabolism

19.3.1 Aminopeptidase N

The work of de la Baume *et al.* (1983) had shown that, in intact brain slices, a combination of both thiorphan and the general aminopeptidase inhibitor bestatin was required to inhibit maximally the metabolism of endogenously released enkephalins, implicating one or more aminopeptidases in enkephalin metabolism in addition to E-24.11. Bestatin had also been shown to exhibit antinociceptive activity consistent with this conclusion (Chaillet *et al.*, 1983). A potential candidate as the enkephalin-degrading aminopeptidase was AP-N which, like E-24.11, is an abundant renal brush border membrane activity. By using phase separation of brain synaptic membranes in Triton X-114 and immunological titration, it was possible to show that AP-N was the major integral membrane aminopeptidase hydrolysing enkephalins (Giros *et al.*, 1986; Gros *et al.*, 1985; Matsas *et al.*, 1985). Some subsequent studies localized AP-N mainly to brain microvessels (Hersh *et al.*, 1987; Solhonne *et al.*, 1987) and the microvascular location of AP-N is now well documented in both the CNS and PNS (Barnes *et al.*, 1988a, 1991a; Schnabel *et al.*, 1992). A more detailed study of AP-N localization in brain (Barnes *et al.*, 1994) confirmed that the enzyme was heterogeneously distributed on synaptic membranes, being present on both pre- and post-synaptic elements where it could play a role in neuropeptide metabolism. AP-N is additionally located on astrocytes and pericytes in the perivascular neuropil and is also abundant, as might be expected, in the choroid plexus and on pial membranes (Barnes *et al.*, 1994; Bourne *et al.*, 1989; Kunz *et al.*, 1994) where, like E-24.11, it may play a protective role. AP-N has a broad substrate specificity, removing N-terminal amino acids from almost all unsubstituted oligopeptides, although neutral amino acids such as Ala, Phe, Tyr and Leu are particularly favoured. Thus, potentially, AP-N could degrade most neuropeptides that are unprotected at the N terminus, examples of which include CCK and the neurokinins. Protection by aminopeptidase inhibitors of the metabolism of CCK-8 and the neurokinins by membrane preparations *in vitro* has been demonstrated (Hooper and Turner, 1985; Hooper *et al.*, 1985; Matsas *et al.*, 1984b; Nau *et al.*, 1986). However, to date, only the enkephalins have been established as substrates for AP-N *in vivo* in the CNS. In particular, by using *in vivo* microdialysis techniques, the administration of the combined E-24.11/AP-N inhibitor RB101 has been shown to maintain Met-enkephalin levels in the brain of freely moving rats (Daugé *et al.,* 1996).

19.3.2 Puromycin-sensitive aminopeptidase

Although AP-N has now been implicated as the major aminopeptidase hydrolysing the enkephalins and possibly other neuropeptides, other brain aminopeptidase activities with activity on enkephalins have been characterized in addition. These particularly include a puromycin-sensitive enzyme (K_i for puromycin = 1 μM) which has been referred to as aminopeptidase M II, aminoenkephalinase or, simply, puromycin-sensitive aminopeptidase (Hersh, 1981, 1985; Hui *et al.*, 1983). This was distinct from AP-N which is relatively insensitive to puromycin (K_i = 78 μM). The significance of the puromycin-sensitive aminopeptidase to brain enkephalin metabolism was principally based on three arguments: the enzyme existed in an apparently membrane-associated form, although the bulk of the activity was cytosolic; it was the major enkephalin-hydrolysing activity in a crude membrane preparation from brain and released Tyr from the pentapeptide but did not hydrolyse the peptide any further; puromycin produced a dose-related, naloxone-sensitive, analgesia in mice (Ehrenpreis, 1985; Herman *et al.*, 1985), suggesting that the enzyme may function in enkephalin metabolism *in vivo*. More careful analysis of the properties of this activity have, however, led to a re-assessment of its likely involvement in neuropeptide metabolism. Indeed the early studies on brain slices of de la Baume *et al.* (1983) had already suggested that puromycin was ineffective in protecting endogenously released enkephalin from degradation, emphasising the value of this type of study for identifying physiologically relevant enzymes. Examination of the membrane association and other properties of the purified puromycin-sensitive aminopeptidase revealed it to be an extrinsic membrane protein, highly similar in sequence (and perhaps identical) to the cytosolic enzyme (Dyer *et al.*, 1990; Matsas *et al.*, 1985; McLellan *et al.*, 1988). These studies led Hersh and colleagues to conclude that it was bound to intracellular membranes with its active site within the cytosol and therefore unlikely to play any role in neuropeptide inactivation (Dyer *et al.*, 1990). The recent cloning of the puromycin-sensitive aminopeptidase from a mouse neuroblastoma cell line (Constam *et al.*, 1995) has shown the protein to be a typical zinc peptidase, with 27–40% identity with other known aminopeptidases, including AP-N, with highest similarity around the HEXXH motif. Surprisingly, the sequence contained a nuclear localization signal and immunohistochemical analysis of the subcellular localization of the enzyme revealed it to be localized to the cytoplasm and the nucleus and to associate with microtubules of the spindle apparatus during mitosis (Constam *et al.*, 1995). Similarly, human puromycin-sensitive aminopeptidase, when expressed in HeLa cells, was localized in the cytoplasm and in a perinuclear region (Tobler *et al.*, 1997). It was suggested that the puromycin-sensitive aminopeptidase may participate in proteolytic events leading to cell growth and viability. No localization at the plasma membrane was seen, eliminating any role for this activity in neuropeptide inactivation at the cell surface. However, Northern blot analysis of different human tissues has shown that the enzyme is expressed at highest levels in the brain (Tobler *et al.*, 1997) with expression detected by *in situ* hybridization in cortical and cerebellar neurons, but not in glial cells or blood vessels.

19.3.3 Dipeptidyl peptidase IV

DPP-IV (CD26) is the only serine ectopeptidase of the renal brush border, and its distribution and properties have been described in Chapter 13. The specificity of DPP-IV is to remove the N-terminal dipeptide from polypeptides containing an unsubstituted N terminus and a penultimate prolyl residue in *trans* conformation. The Xaa–Pro sequence is found at the N terminus of more than 100 naturally occurring peptides and proteins, including peptide hormones, kinins, neurotransmitters, toxins and mediators of the immune response (Vanhoof *et al.*, 1995) and DPP-IV therefore has the potential to modulate the activity of such peptides. For example, substance P (RPKPNNFFGLMamide) is susceptible to hydrolysis by DPP-IV (Kato *et al.*, 1978), although as discussed above, E-24.11 appears to play the major role in synaptic inactivation of this peptide. Bradykinin (RPPGFSPFR), however, is resistant to the action of DPP-IV which cannot cleave the -Pro–Pro- bond (Kato *et al.*, 1978). An important group of neuropeptides susceptible to DPP-IV are members of the pancreatic polypeptide family which include neuropeptide Y and peptide YY. Processing of these peptides at their N terminus by DPP-IV generates receptor-selective forms of the peptides altering their biological effects, whereas E-24.11 serves to inactivate them (Medeiros and Turner, 1994, 1996). Another recently described substrate for DPP-IV is the naturally occurring peptide enterostatin (VPDPR) which is a potent inhibitor of food intake. The hydrolysis and inactivation of enterostatin by brain membranes appears to involve the combined actions of DPP-IV and carboxypeptidase P (Bouras *et al.*, 1995).

In the CNS, DPP-IV has a predominantly microvascular location, although it is less widely distributed than AP-N or ACE, and it does not appear to have a synaptosomal location (Barnes *et al.*, 1994). In the PNS, DPP-IV has been shown to be associated predominantly with microvessels (Barnes *et al.*, 1991a). Some reports have claimed a localization of DPP-IV with the membranes of Schwann cells in contact with the axonal membrane at sensory nerve terminals in various mammals (Dubovy, 1987; Dubovy and Malinovsky, 1984; Dubovy and Soukup, 1985), although we failed to detect DPP-IV by immunostaining in Schwann cell membranes in sections of peripheral nerves of the pig (Barnes *et al.*, 1991a). DPP IV has also been localized to the luminal surface of ependymal cells in the ventricles and central canal of the spinal cord in the pig (Bourne *et al.*, 1989).

19.3.4 Pyroglutamyl aminopeptidase II: a neuropeptide-specific peptidase

When first reviewing the field of neuropeptide metabolism (Turner *et al.*, 1985) we had advanced the view that the peptidases involved would be unlikely to be brain-specific or peptide-specific, based on the information that was accumulating at the time on E-24.11 and its broad substrate specificity (Matsas *et al.*, 1984a). Indeed, with the number of recognized neuropeptides exceeding 50, it was unreasonable to assume that each peptide would have an associated and individual peptidase. This hypothesis is still broadly accepted today and helped to direct research away from seeking such neuropeptide-specific enzymes. There are, however, exceptional examples where a highly specific enzyme may be appropriate, for example where the neuropeptide itself has a structure that is not amenable to hydrolysis by conventional peptidases. Such an example is provided by the hypothalamic peptide, thyrotrophin-releasing hormone (TRH), which is a tripeptide, blocked at both termini (Glp–His–ProNH$_2$) and consequently resistant to the usual exopeptidases. It also contains no internal hydrophobic residues and is therefore resistant to the action of E-24.11; ACE is also unable to hydrolyse TRH. A number of cytosolic brain enzymes had been shown to degrade TRH, for example proline endopeptidase (Browne and O'Cuinn, 1983; Hersh and McKelvy, 1979; Wilk, 1983) and a pyroglutamyl aminopeptidase (PGA I) (Bauer and Kleinkauf, 1980; Browne and O'Cuinn, 1983) but their relevance to TRH metabolism *in vivo* was questionable. Attention therefore shifted to seeking a membrane-bound TRH-degrading activity that might exist as an ectoenzyme, leading to the discovery of PGA II in brain and adenohypophysis (Bauer, 1987; Browne *et al.*, 1981; O'Connor and O'Cuinn, 1984). The activity of the adenohypophyseal enzyme was regulated by thyroid hormones, emphasizing its relevance to TRH metabolism (Bauer, 1987). Brain PGA II was shown to be associated with synaptosomal membranes as an ectoenzyme (Charli *et al.*, 1988; Garat *et al.*, 1985; O'Connor and O'Cuinn, 1984) from which it could be solubilized by papain treatment. The enzyme was subsequently purified to homogeneity and shown to be a metallopeptidase of M_r approximately 230 000 and with very highly restricted substrate specificity, hydrolysing only TRH or very closely related peptides (Bauer, 1994; O'Connor and O'Cuinn, 1985; Wilk and Wilk, 1989). It was thus confirmed as the first example of a truly neuropeptide-specific peptidase and, to date, no other natural peptide substrate has been identified. The recent cloning and expression of rat PGA II (Schauder *et al.*, 1994) has established it to be a type II integral membrane glycoprotein with a typical zinc consensus sequence (HEXXH) in the putative extracellular domain. The PGA II sequence revealed significant homology with AP-N and AP-A. A fuller account of this neuropeptide-specific peptidase is provided in Chapter 15 and an earlier review of TRH metabolism is to be found in O'Cuinn *et al.* (1990). A number of potent and specific inhibitors of PGA II have been developed (Wilk, 1989) and their therapeutic potential as modulators of TRH levels has been discussed in Kelly (1995).

19.4 Angiotensin-converting enzyme

ACE displays a wide distribution in mammalian tissues but is particularly abundant at the endothelial surface of the lung, in the renal brush border membranes and in the testes. Although the primary roles of ACE are regarded as the conversion of angiotensin I (AI) to angiotensin II (AII) and the inactivation of bradykinin (BK), a number of other neuropeptide substrates are susceptible to hydrolysis by ACE *in vitro*. Among these are neurotensin (Skidgel *et al.*, 1984), enkephalins, and C-terminally extended enkephalin precursors (Skidgel and Erdös, 1987), on all of which ACE acts as a peptidyl dipeptidase. ACE has also been shown to act as an endopeptidase on some neuropeptide substrates [e.g. substance P (Skidgel *et al.*, 1984; Yokosawa *et al.*, 1983), luteinizing hormone-releasing hormone (LHRH; Skidgel and Erdös, 1985), as well as cholecystokinin (CCK) and various gastrin analogues (Dubreuil *et al.*, 1989)]. Convincing evidence for a physiological role in the metabolism of any of these peptides other than AI and BK is, however, lacking and, in general, E-24.11 exhibits a much higher k_{cat}/K_m value for these neuropeptides than ACE (*Table 19.2*). Aspects of the biology of ACE are discussed in detail elsewhere (Chapters 7 and 8) and this section will focus on the localization and possible functions of ACE in the CNS.

Brain ACE was first detected in the rat, where the specific activity of the enzyme was found to be greatest in the striatum and pituitary (Yang and Neff, 1972). The distribution of ACE in brain and pituitary was first examined in detail by an autoradiographic study employing the tritiated ACE inhibitor, captopril (Strittmatter *et al.*, 1984). A similar method of *in situ* autoradiography, but using a [^{125}I]-labelled derivative of lisinopril (MK351A), has also been employed (Chai *et al.*, 1987; Correa *et al.*, 1985; Mendelsohn, 1984) and there is good correlation on ACE distribution in brain between autoradiographic data using labelled inhibitors and enzymic assay (Correa *et al.*, 1991). Some species differences have been observed in the density of ACE labelling in brain regions. For example, there are some notable differences between the distribution of ACE in rat, rabbit and monkey brain (Chai *et al.*, 1987, 1991; Rogerson *et al.*, 1995). Moderate to high levels of ACE are detectable throughout the rabbit and monkey cerebral cortex, and in the interpeduncular and suprachiasmatic nuclei but not in the rat. Thus, caution should be exercised in extrapolating data on the effects of ACE inhibitors in the rat brain to the human situation. Immunohistochemical studies of ACE distribution in the brain have generally agreed with the autoradiographic data (Strittmatter and Snyder, 1987) and have revealed a dual location of ACE, being present on all microvessels, but with a heterogeneous distribution in the neuropil (Barnes *et al.*, 1988a). In the PNS, ACE is exclusively localized on microvessels (Barnes *et al.*, 1991a). ACE has been mapped immunohistochemically to the choroid plexus, ependyma, subfornical organ, basal ganglia (caudate putamen and globus pallidus) but most notably, the

Table 19.2. Comparison of kinetic constants of E-24.11 and ACE for certain neuropeptides

	E-24.11			ACE		
Peptide	K_m (μM)	k_{cat} (min^{-1})	k_{cat}/K_m (min^{-1}.μM^{-1})	K_m (μM)	k_{cat} (min^{-1})	k_{cat}/K_m (min^{-1}.μM^{-1})
Bradykinin	92	6364	69	0.4	500	1250
[Leu^5]enkephalin	86	3781	43.9	1000	700	0.7
[Met^5]enkephalin	62	2597	41.9	1000	3500	3.5
Neurotensin	78	1280	16	14	34	2.4
Substance P	32	5062	159	25	225	9.0

The data are assembled from Kenny and Turner (1987).

enzyme was found to be highly localized to a striatonigral pathway (Barnes *et al.*, 1988a; Strittmatter *et al.*, 1984) but, unlike E-24.11, was not striosomally organized (Barnes *et al.*, 1988a). Since there is no evidence for AI, AII receptors or BK in the striatonigral region, this suggests a role for ACE in the metabolism of a distinct neuropeptide present in this neuronal pathway. ACE was found to be present on the epithelium of the choroid plexus and on the brain pial membranes, but was not significant in the cord pia (Bourne *et al.*, 1989). The distribution of ACE mRNA in rat brain has also been explored (Whiting *et al.*, 1991) with high levels being detectable in choroid plexus, caudate putamen and cerebellum, generally correlating well with localization of ACE protein.

Candidate peptides for brain ACE other than AI or BK might be enkephalins or tachykinins, especially in the striatonigral region, and the initial studies on enkephalin degradation implied incorrectly that ACE was the brain membrane activity hydrolysing the Gly3–Phe4 bond of the enkephalins (Benuck and Marks, 1979; Swerts *et al.*, 1979). Subsequent studies using selective inhibitors have generally established that E-24.11 is the enzyme responsible for this hydrolysis (see Section 19.2). One more recent study (Konkoy and Davis, 1995), however, suggests that ACE may participate in the metabolism of Met-enkephalin in intact brain slices. Captopril was observed to inhibit Met-enkephalin degradation by between 18 and 30%, depending on brain region. However, this observation may, in part, reflect the relatively high concentration of the substrate (400 μM) used in these experiments. ACE may, however, contribute to the hydrolysis of extended opioid peptides (Kase *et al.*, 1986), particularly since captopril can potentiate the analgesic activity of Met-enkephalin-Arg-Phe in mouse brain (Norman *et al.*, 1985). Other suggested substrates for ACE in the brain include neo-kyotorphin (Thr–Ser–Lys–Tyr–Arg; Kase *et al.*, 1986) and, in combination with aminopeptidases, δ sleep-inducing peptide (Augustijns *et al.*, 1995). One study has indicated a partial involvement of ACE in the metabolism of brain natriuretic peptide (BNP) *in vivo* in the rat (Vanneste *et al.*, 1990), although in a separate study of BNP metabolism by pig brain membranes and purified enzymes, E-24.11 was reported as the principal degrading activity (Bourne and Kenny, 1990).

In most mammalian tissues ACE is present as a single form of M_r 140 000–180 000 depending on species, although the testicular form of the enzyme is much smaller (M_r 90 000–100 000) and arises by alternative splicing of the primary transcript. In rat brain, however, two species of the enzyme of M_r 165 000 and 175 000 have been detected (Strittmatter *et al.*, 1985) and similarly in pig brain striatum two forms of ACE are observed of M_r approx 170 000 and 180 000 (Hooper and Turner, 1987), which have been shown to represent endothelial and neuronal forms of ACE respectively (Williams *et al.*, 1991) and which arise by differential glycosylation. It was suggested that the smaller isozyme of brain ACE may participate in the metabolism of substance P and neurokinin A (substance K) (Strittmatter *et al.*, 1985; Thiele *et al.*, 1985). However, Hooper *et al.* (1985) failed to detect hydrolysis of neurokinin A by purified ACE and concluded that E-24.11 was the principal inactivating enzyme for this neuropeptide in pig brain striatal membranes. Similar conclusions were reached with neurokinin B as substrate, although a bestatin-sensitive aminopeptidase participated in the degradation of this peptide in synaptic membranes in addition to E-24.11 (Hooper and Turner, 1985). No differences in specificity or inhibitor sensitivity were observed between the endothelial and neuronal forms of ACE (Hooper and Turner, 1987). By using a combination of subcellular fractionation and immunoblotting, ACE has been shown to reside on post-synaptic, but not pre-synaptic, membranes in pig brain (Barnes *et al.*, 1992). Thus, although ACE is appropriately located in the nervous system to facilitate neuropeptide processing or metabolism, specific substrates other than AI and BK have yet to be established.

19.5 The thimet oligopeptidase family

19.5.1 Thimet oligopeptidase/endopeptidase-24.15

Thimet oligopeptidase, as it is now referred to, has had a chequered history and, like E-24.11, has been discovered in various guises depending on the tissue examined and the peptide substrates employed for assay purposes. It was originally detected as a possible mammalian collagen-degrading enzyme (Morales and Woessner, 1977) by using a synthetic substrate referred to as Pz-peptide, and hence the enzyme became known as Pz-peptidase (Lukac and Koren, 1977). A potential for the enzyme in neuropeptide metabolism came with the report of an apparent cysteine peptidase hydrolysing the Phe5–Ser6 bond of bradykinin and it was referred to as kininase A (Oliveira *et al.*, 1976) and subsequently as endo-oligopeptidase A (EC 3.4.22.19; Coelho *et al.*, 1981). This activity hydrolysed a number of other biologically active peptides, including the Arg8–Arg9 bond of neurotensin (Camargo *et al.*, 1983) and acted to generate Met-enkephalin from several enkephalin-containing peptides such as the 12-residue opioid peptide from bovine adrenal medulla (BAM-12P), by cleavage at the Met5–Arg6 position (Camargo *et al.*, 1985, 1987). Thus, the enzyme was proposed to play a role in the processing and metabolism of certain biologically active peptides, including LHRH (Lasdun *et al.*, 1989; Medeiros *et al.*, 1992). Independently, an apparently unrelated, soluble metallopeptidase was characterized in rat brain and pituitary which was capable of hydrolysing similar peptides and which was subsequently referred to as endopeptidase-24.15 (E-24.15; EC 3.4.24.15; Orlowski *et al.*, 1983). Endo-oligopeptidase A and E-24.15 were, for some time, regarded as distinct enzymes because of their allocation to different classes of peptidases (thiol and metallo). However, it was subsequently shown that both activities were due to the action of a single enzyme which was a thiol-activated metallopeptidase, which hydrolysed internal bonds in peptides up to about 17-amino-acid residues in length. Hence it was then referred to as 'thimet' oligopeptidase (Barrett and Brown, 1990; Tisljar and Barrett, 1990). The rules governing the specificity of the enzyme are still unclear but hydrophobic residues tend to be preferred in positions P1 and P3′, and Pro in P2′ and there is a clear preference for cleavage near the C terminus (Dando *et al.*, 1993; Knight *et al.*, 1995; Orlowski *et al.*, 1988).

Thimet oligopeptidase is widely distributed as a soluble, intracellular enzyme in mammalian cells and tissues and it has a high activity in brain and testis (reviewed in

Tisljar, 1993). In rat brain, a synaptosomal membrane-associated form of the enzyme has been reported (Acker *et al.*, 1987; Oliveira *et al.*, 1990) although evidence that it exists as an ectoenzyme is lacking (Millican *et al.*, 1991). Immunohistochemical studies have suggested E-24.15 may also be located in endosome-like structures (Serizawa *et al.*, 1995) and in the nucleus (Healy and Orlowski, 1992). The latter observation may, however, be artefactual and reflect the specificity of the antibody, in that subcellular localization studies have failed to detect significant amounts of enzyme activity in the nuclear fraction from rat brain (Oliveira *et al.*, 1990). The enzyme has been purified from a number of sources as a 78 kDa polypeptide and was first cloned from rat testis (Pierotti *et al.*, 1990). The predicted protein sequence of the enzyme has subsequently been slightly corrected (McKie *et al.*, 1993) due to a cloning error in the original work of Pierotti *et al.* (1990) but neither reported sequence possesses any hydrophobic membrane-spanning region consistent with its predominant cytosolic location. The enzyme is a typical zinc endopeptidase with an HEXXH motif. The thiol dependence of the enzyme has recently been explored by the use of site-directed mutagenesis (Shrimpton *et al.*, 1996). It appears that, in the absence of thiol reagents, E-24.15 normally exists as an oligomer due to interchain disulphide bond formation, and disruption of the oligomer by thiol reagents enhances access to the active site and hence activates the enzyme. Thus, in the normal intracellular reducing environment, E-24.15 would be expected to be fully active. Three cysteine residues in the polypeptide have been identified by mutagenesis as being responsible for this effect.

Attempts have been made to establish a role for thimet oligopeptidase in neuropeptide inactivation *in vivo*, especially of LHRH, through the use of apparently selective inhibitors of the enzyme such as Cpp–Ala–Ala–Phe–pAb [in which Cpp = *N*-(1-(*RS*)-carboxy-3-phenylpropyl)-, and pAb = *p*-aminobenzoate; Lasdun *et al.*, 1989; Molineaux *et al.*, 1988]. However, it was discovered that E-24.11 efficiently cleaves the Ala–Phe bond in the inhibitor, producing Cpp–Ala–Ala, which inhibits ACE potently (Williams *et al.*, 1993). Thus, some of the effects of the inhibitor *in vivo* may be due to indirect inhibition of ACE (Williams *et al.*, 1993). More recent studies have implicated thimet oligopeptidase in the metabolism of somatostatin (Dahms and Mentlein, 1992; Mentlein and Dahms, 1994), neurotensin (Davis *et al.*, 1992; Mentlein and Dahms, 1994; Vincent *et al.*, 1995) and of nociceptin/orphanin FQ (Montiel *et al.*, 1997). All these observations are difficult to equate with an enzyme that appears to be predominantly cytosolic but with a weakly hydrophobic domain which could cause a minor, membrane-associated component (Acker *et al.*, 1987; Orlowski *et al.*, 1983). Attention has also been directed towards a possible role of E-24.15 in the metabolism of the Alzheimer's β-amyloid precursor protein (β-APP). Thus, E-24.15 is able to hydrolyse oligopeptide substrates, mimicking one of the amyloidogenic cleavage sites (McDermott *et al.*, 1992; Thompson *et al.*, 1995) and it has been suggested that the enzyme may contribute to the pathogenesis of the disease (Papastoitsis *et al.*, 1994). Furthermore, human E-24.15 is located at chromosomal locus 19q13.3, which is within the linkage region for late-onset Alzheimer's disease (Meckelein *et al.*, 1996). The predominantly cytosolic location of E-24.15 and its strict preference for oligopeptides, however, makes it difficult to reconcile with a functional role in the metabolism of β-APP (Hooper *et al.*, 1997). The enzyme responsible for the proteolytic processing of farnesylated peptides also has characteristics resembling E-24.15 (Akopyan *et al.*, 1994).

19.5.2 Endopeptidase-24.16 (neurolysin)

Independent studies on the mode of degradation of the neuropeptide, neurotensin, led to the discovery of another metallopeptidase, originally termed 'neurotensin-degrading peptidase' (Checler *et al.*, 1986) and, following its classification as EC 3.4.24.16, often referred to as endopeptidase-24.16 (E-24.16). The name 'neurolysin' was recommended by IUBMB in 1992. The enzyme was originally identified in rat brain synaptic membrane preparations as one of several activities contributing to neurotensin degradation (Checler *et al.*, 1986). E-24.11 hydrolysed the peptide at several sites (*Figure 19.1*), and E-24.15 was identified as being responsible for hydrolysis of the Arg8–Arg9 bond but when both these activities were inhibited, hydrolysis of the Pro10–Tyr11 bond was still detectable. This activity was first purified from detergent-solubilized brain synaptic membranes and a similar activity subsequently isolated from rat ileum and kidney. Several lines of evidence suggested that E-24.16 may contribute to neurotensin degradation *in vivo* both in the brain and in the gastrointestinal tract (reviewed in Checler *et al.*, 1995). In particular, the use of inhibitors such as Pro–Ile and phosphodiepryl 03 (Checler *et al.*, 1995; Dauch *et al.*, 1991), have proved useful in this context, although conclusions based on such studies are completely dependent on the specificity of the inhibitors. E-24.16 is heterogeneously distributed in rat brain (Checler *et al.*, 1995), with highest concentrations being found in the olfactory bulb and tubercle, cingulate cortex, medial striatum and globus pallidus. There appeared to be a positive correlation between the distribution of enzyme activity and the density of neurotensin receptors.

Like E-24.15, E-24.16 exists predominantly as a soluble, cytosolic enzyme (Checler *et al.*, 1995) although there is undoubtedly a small membrane-associated component. Phase-separation experiments with Triton X-114 have failed to reveal an amphipathic form of the enzyme in a pig brain membrane fraction and the ectoenzyme nature of E-24.16 has been questioned (Millican *et al.*, 1991). A major advance in our understanding of the biology of E-24.16 came with the report of Serizawa *et al.* (1995) of a rat liver metallopeptidase that exhibited many of the characteristics of E-24.16 although it was primarily located in the mitochondrial inner membrane space where it could interact non-covalently with the inner membrane. The authors concluded that this activity,

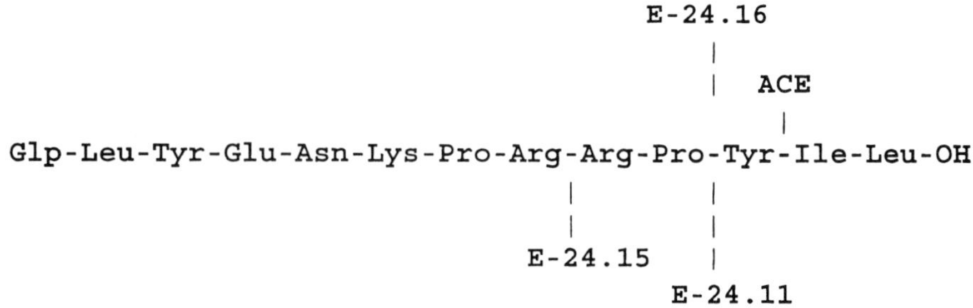

Figure 19.1. Sites of hydrolysis of neurotensin by various peptidases.

which they termed oligopeptidase M, was none other than E-24.16 and was also identical with two other proteins reported in the literature: 'rabbit microsomal endopeptidase' and pig 'soluble angiotensin II-binding protein'. Yet again, it appeared that a single protein had appeared under a number of different guises in the literature. The soluble AII-binding protein had previously been cloned and the deduced amino acid sequence had identified it as a homologue of thimet oligopeptidase, therefore adding E-24.16/neurolysin to this family. We had previously recognized the biochemical similarities between E-24.15 and E-24.16 in our study of the pig enzymes (Millican *et al.*, 1991). Shortly afterwards, Dauch *et al.* (1995) reported the cloning and expression in COS cells of rat brain E-24.16. It comprised a 704-amino-acid polypeptide with no potential membrane-spanning segments nor glycosylation sites, excluding its occurrence as an ectoenzyme. The predicted sequence of rat brain E-24.16 showed 90% homology with rabbit liver microsomal endopeptidase, suggesting that it is the species homologue, as predicted by Serizawa *et al.* (1995). The sequence of rat E-24.16 also showed 60% homology with that of rat E-24.15, assigning them as members of the same metallopeptidase family (now referred to as family M3; Rawlings and Barrett, 1993). Other eukaryotic members of this family include mitochondrial intermediate peptidase (EC 3.4.24.59), which is a matrix protein in the mitochondria of rat and yeast, and yeast peptidase *yscD* (saccharolysin; EC 3.4.24.37). Thus, E-24.15 and E-24.16 are closely related proteins, predominantly cytosolic but for which a membrane-associated component may exist in some tissues. There is marked heterogeneity in the 5′ end of E-24.16 mRNA and six types of mRNA were demonstrated by 5′-RACE (rapid amplification of 5′ cDNA ends) which could be classified into two categories, with or without a mitochondrial targeting sequence (Sugiura *et al.*, 1992). It seems that the cytosolic and mitochondrial neurolysins derive from a single gene by alternative choice of promoter and translation initiation codon. To date, no plasma membrane targeting sequence has been identified for isoforms of E-24.15 or E-24.16. Although very similar, there are a number of biochemical characteristics that distinguish E-24.15 from E-24.16. These include distinct sites of hydrolysis of neurotensin, sensitivity to dipeptide inhibitors (e.g. Pro–Ile), and by the lack of activation of E-24.16 by thiol reagents such as dithiothreitol. In this context, two of the three cysteines in E-24.15 required for the thiol activation phenomenon are not conserved in E-24.16.

All these studies apparently call into question a role for this family of enzymes in terminating neuropeptide action *in vivo*. In an attempt to resolve this dilemma, Vincent *et al.* (1996) have compared the subcellular distribution of E-24.16 in neurons and astrocytes. Striking differences were observed between the two cell types, especially during differentiation. Thus, astrocytes were found to contain only a soluble form of the enzyme which could apparently be secreted from the cells despite the lack of the requisite signal sequence for secretion. This secretory process was unaffected by pre-treatment with monensin, brefeldin A or forskolin and resembled the reported secretion of certain other proteins lacking a classical signal sequence (Halban and Irminger, 1994; Rubartelli *et al.*, 1990). There is one report that E-24.15 may also undergo such a secretory event from a glioma cell line (Ferro *et al.*, 1993). In contrast, primary neurons expressing neurolysin show the presence of increasing amounts of a membrane-associated form of the enzyme during differentiation (Vincent *et al.*, 1996). This membrane-associated form, by several criteria, appeared to behave as an ectoenzyme and no secretion of neuronal activity was observed. Thus, it may be that a novel form of neurolysin can be expressed on neurons, perhaps through alternative splicing, that could serve as an ectopeptidase, but this remains to be demonstrated unequivocally.

In conclusion, evidence that members of the thimet oligopeptidase family participate in the metabolism of some neuropeptides is compelling but difficult to reconcile with their predominant subcellular locations. That a proportion of these activities may exist as cell-surface ectoenzymes is still a matter for debate and further experimentation. In this context, human kidney cells stably transfected with rat brain E-24.16 cDNA are reported to exhibit a membrane-associated form of the enzyme detectable at the cell surface, similar to the 'neuronal' form of the enzyme (Vincent *et al.*, 1997). Approximately 50% of the total activity was membrane-associated and resistant to solubilization by digitonin and Na_2CO_3. This transfected cell model may prove useful for establishing the structural requirements, by site-directed mutagenesis, of this neuronal form of E-24.16. Finally, both E-24.15 and E-24.16 hydrolyse a number of neuropeptides and therefore cannot be considered as neuropeptide-specific peptidases.

19.6 Metabolism of CCK by a serine endopeptidase: tripeptidyl peptidase II

CCK is one of the most abundant neuropeptides present in the brain and has been implicated in the control of food intake, in anxiety states and in psychotic disorders. Like gastrin, CCK exists in a number of molecular forms, the major form in the brain being the sulphated octapeptide (CCK-8; *Figure 19.2*) which has all the characteristics of a neurotransmitter (Bunnett, 1987). CCK is also present in the gut, mainly as an extended 33-amino-acid peptide where it has been implicated in intestinal motility, gastric emptying and the control of gall bladder contraction. CCK is rapidly metabolized and inactivated following its secretion from neurons, although there has been some controversy surrounding the enzyme(s) responsible. It has been shown that E-24.11 efficiently hydrolyses CCK-8 *in vitro* at two distinct sites, Gly4–Trp5 and Asp7–Phe8NH_2 (Deschodt-Lanckman *et al.*, 1983, 1984; Matsas *et al.*, 1984b) and that a combination of E-24.11 and aminopeptidase activities fully accounts for the hydrolysis of CCK-8 in pig striatal synaptic membranes (Matsas *et al.*, 1984b) and rat synaptic membranes

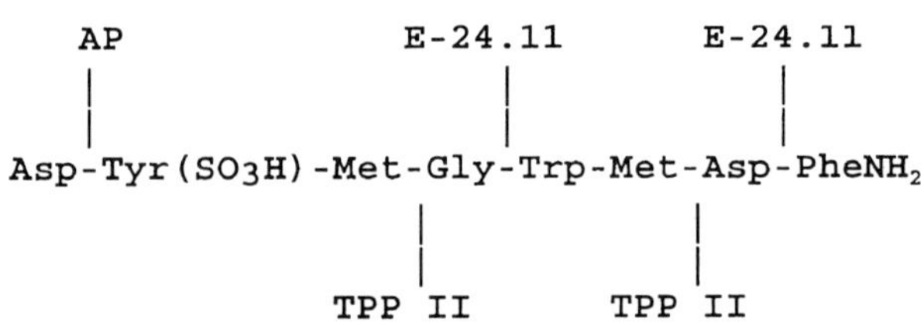

Figure 19.2. Sites of hydrolysis of CCK-8 by aminopeptidase (AP), E-24.11 and TPP II activities. Hydrolysis by TPP II occurs sequentially from the N terminus.

(Deschodt-Lanckman and Strosberg, 1983). ACE can also hydrolyse CCK *in vitro* (Dubreuil *et al.*, 1989) but ACE inhibitors do not protect CCK from degradation by synaptic membrane preparations (Matsas *et al.*, 1984b) or brain slices (Konkoy and Davis, 1995). Attempts to examine the effects of peptidase inhibitors on the recovery of CCK released endogenously from brain slices have produced equivocal results. Thus, Zuzel *et al.* (1985) reported that the E-24.11 inhibitor thiorphan did not significantly affect the recovery of the peptide from slices of cerebral cortex (a CCK-rich brain region but deficient in E-24.11), and had only a small effect in the case of striatal slices (an E-24.11-rich region). In cerebral cortex, the release experiments have implicated a novel elastase-like serine endopeptidase in brain that cleaves the two peptide bonds of CCK-8 where the P_1 residue is methionine (Camus *et al.*, 1989; Rose *et al.*, 1988, 1989) (*Figure 19.2*). This activity was claimed to constitute a major inactivation ectoenzyme for the neuropeptide cleaving initially at the Met3–Gly4 bond and then at the Met6–Asp7 bond. A contribution of aminopeptidases to CCK inactivation was also reported. More recently, Konkoy and Davis (1995) have re-examined the effects of peptidase inhibitors on the metabolism of CCK by intact brain microslices from a number of different brain regions (nucleus accumbens, caudate putamen, frontal cortex). In general, the degradation of exogenously added CCK-8 (sulphated) (50 μM final concentration) was much less sensitive to peptidase inhibitors than was Met-enkephalin. The metabolic fragments identified were generally consistent with those reported previously by Camus *et al.* (1989) with the CCK-(2–8) and CCK-(4–8) fragments as major metabolites and the formation of the latter fragment was partially inhibited (13–24% depending on brain region) by a cocktail of general serine protease inhibitors. Neither phosphoramidon nor captopril had any significant effect on CCK metabolism in these brain regions.

Rose *et al.* (1996) have now identified unequivocally the putative serine peptidase involved in CCK metabolism as tripeptidyl peptidase II (TPP II; EC 3.4.14.10). Their strategy was to isolate the activity from a membrane preparation from rat cerebral cortex and they succeeded in purifying a protein of subunit M_r 135 000 which was irreversibly labelled by [^{3}H]di-isopropyl fluorophosphate. Microsequencing of three internal peptides and comparisons with protein sequence databases led to the identification of the CCK-degrading enzyme as TPP II, a subtilisin-like peptidase originally isolated from rat liver and human erythrocytes and subsequently cloned by Tomkinson and Jonsson (1991). The specificity of TPP II is to remove tripeptides sequentially from the N terminus of longer peptides. An enigma was that TPP II had previously been characterized solely as a cytosolic enzyme (Balow *et al.*, 1983) and the hydropathy profile of the cloned TPP II does not contain any potential membrane-spanning segment. However, Rose *et al.* (1996) claim that TPP II-like activity is present in cerebral membranes as an integral membrane protein which can be released by treatment with phosphatidylinositol-specific phospholipase C, although no data were presented to support this. Nevertheless, it remains possible that a glycosyl phosphatidylinositol-anchored form of TPP II could arise in brain by alternative splicing and serve as an ectoenzyme inactivating CCK-8 and possibly some other neuropeptides. Although the highest specificity constants for hydrolysis were obtained with CCK as substrate, cleavage of neurokinin A, somatostatin, vasopressin, calcitonin gene-related peptide, dynorphin A and vasoactive intestinal polypeptide occurred at a lower but detectable rate (Rose *et al.*, 1996). Further evidence for a role of TPP II in CCK metabolism came from the rational development of a potent (K_i = 7 nM) and reportedly selective inhibitor, butabindide (*Figure 19.3*), which was able to enhance substantially the recovery of CCK released from brain slices (Rose *et al.*, 1996). *In vivo*, in mice and in rats, butabindide was also shown to reduce food intake significantly, consistent with the known pro-satiating properties of CCK. Thus, this membrane-bound form of TPP II appears to be a novel neuropeptide-degrading enzyme which is of relevance to CCK metabolism in particular, but possibly of other neuropeptides in at least some brain regions.

19.7 NAALA dipeptidase: a neuropeptidase and prostate-specific tumour antigen

The dipeptide *N*-acetyl-L-aspartyl-L-glutamate (NAAG) is extremely abundant in the human brain (Curatolo *et al.*, 1965). It exceeds the levels of most other brain peptides by three orders of magnitude and, indeed, approaches the levels of the amino acids glutamate and aspartate themselves. There has been much speculation over the years as to the physiological role of NAAG which has ranged from its functioning as an acidic counterion through to a role as a neuropeptide in the CNS. The latter role was emphasized by its independent re-isolation as a convulsant peptide which potently inhibited chloride-dependent glutamate binding to brain membranes ((Zaczek *et al.*, 1983), suggesting that it may particularly serve as an excitatory neurotransmitter at glutamate synapses. NAAG has also been shown to be released from neurons in a Ca^{2+}-dependent manner (Tsai *et al.*, 1990), again implicating it as a neurotransmitter. A detailed immunocytochemical study of NAAG has shown it to be localized in most of the forebrain glutamate pathways and it also localizes to cholinergic systems of the forebrain (Tsai *et al.*, 1993).

A membrane-bound peptidase has been shown to hydrolyse NAAG to *N*-acetylaspartate and glutamate, thereby possibly also serving as a source of the excitatory neurotransmitter, glutamate. This enzyme, usually referred to as '*N*-acetylated α-linked

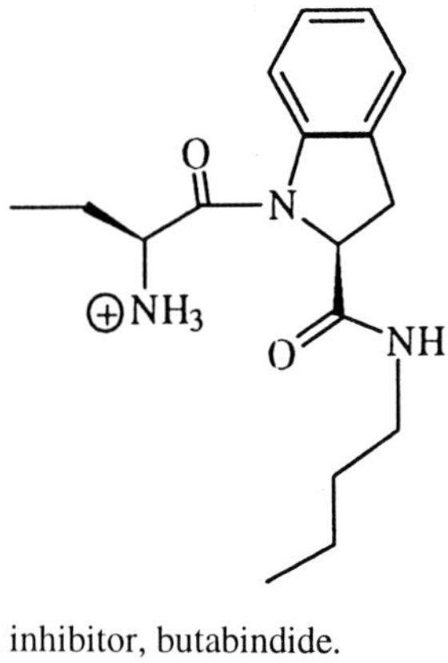

Figure 19.3. Structure of the TPP II inhibitor, butabindide.

acidic dipeptidase' (NAALA dipeptidase), therefore has the potential to serve as a neuropeptidase, inactivating the intrinsic activity of NAAG, or to function as a processing enzyme regulating glutamate availability. Initial characterization of NAALA dipeptidase revealed it to be a membrane-bound neutral metallopeptidase enriched in synaptosomal membranes (Robinson *et al.*, 1987). It was also shown to be abundant in the kidney cortex and a distinctive feature of the activity was its sensitivity to inhibition by quisqualate (I_{50} = 0.48 μM), distinguishing it from other known exo- or endopeptidases (Robinson *et al.*, 1987). The enzyme was subsequently purified to homogeneity from rat brain as a 94 kDa polypeptide (Slusher *et al.*, 1990) and shown to have a high affinity for its substrate (K_m = 0.14 μM). By using a poylclonal antibody raised to the purified enzyme, the distribution of the dipeptidase was revealed in brain and kidney (Slusher *et al.*, 1990, 1992). NAALA dipeptidase was found to be widely yet heterogeneously distributed in brain, being abundant in neuropil. It was found to be selectively enriched in areas containing NAAG immunoreactivity, for example amygdala, caudate putamen, substantia nigra, globus pallidus, hippocampus, hypothalamus, olfactory bulb and thalamus. In the cerebellar cortex it was detected in the molecular and granule cell layers but was absent from Purkinje cells. In the rat kidney cortex, NAALA dipeptidase immunoreactivity was revealed both in the proximal tubules and glomeruli. In this respect its abundance in kidney resembles that seen for certain other cell-surface peptidases such as E-24.11, AP-N and ACE. The overall distribution of the dipeptidase was, however, broader than that of its substrate, suggesting that the enzyme may have roles additional to hydrolysis of NAAG. In the PNS, NAALA dipeptidase was shown both by enzyme assay and by immunoblotting to be present in sciatic and phrenic nerve, cervical dorsal root ganglion and the superior cervical ganglion (Berger *et al.*, 1995a). It was exclusively co-localized in peripheral nerves with glial fibrillary acidic protein. By using electron microscopic immunocytochemistry, the enzyme was revealed in the plasma membranes of non-myelin-forming Schwann cells in the superior cervical ganglion. These studies suggested that NAALA dipeptidase may be involved in signalling between axons and Schwann cells, for example during development or regeneration, a role analogous to that proposed for E-24.11 (see Section 19.2). NAAG is also found in concentrations exceeding millimolar in mammalian motor neurons and in the ventral roots of the spinal cord (Ory-Lavollée *et al.*, 1987) where, on release, it can function as an agonist at the *N*-methyl-D-aspartate (NMDA) class of glutamate receptors (Sekiguchi *et al.*, 1992) and also be cleaved to release glutamate by NAALA dipeptidase. A detailed immunocytochemical study of the rat neuromuscular junction has revealed the presence of NAAG in motor neuron terminals, NAALA dipeptidase in perisynaptic Schwann cells, and the NMDAR-1 glutamate receptor on post-synaptic muscle membranes, suggesting a role for the peptide in the modulation of neuromuscular transmission (Berger *et al.*, 1995b).

By using the antibody raised against purified NAALA dipeptidase to screen a λgt11 rat brain cDNA library, it proved possible to isolate several immunopositive clones (Carter *et al.*, 1996). Sequencing of one of these partial length clones revealed a protein with 86% identity to a human prostatic carcinoma antigen of previously unknown function but which had been identified as a useful clinical marker for prostate cancer (Israeli *et al.*, 1993). Expression of the human prostate-specific membrane antigen cDNA confirmed that it did indeed possess NAALA dipeptidase enzyme activity (Carter *et al.*, 1996). The sequence of the prostate antigen revealed it to be a type II integral membrane protein with a short (19-amino-acid) cytoplasmic domain containing a protein kinase C consensus site. Surprisingly, no homologies with other cell-surface peptidases were revealed, indicating NAALA dipeptidase to be a novel enzyme. The precise role of NAALA dipeptidase in the reproductive system, and its substrate at that location, remains unknown.

19.8 Conclusions

To date, only a small group of CNS peptidases have been characterized as neuropeptide-inactivating ectoenzymes. Most of these have counterparts in the renal brush border which has much facilitated their characterization. Some novel activities have, however, been identified in the nervous system, for example PGA II and NAALA dipeptidase which had not previously been characterized in kidney or other tissues. E-24.11 has the most general role in neuropeptide metabolism but AP-N is also an important contributor. The specific role of ACE in the striatonigral pathway remains an enigma. Studies of changes in levels of neuropeptide-degrading enzymes in neurological disease have been limited in extent and no consistent pattern yet appears. For example, levels of ACE in the cortex have been reported to be reduced (Ichai *et al.*, 1994) or unchanged or elevated (Barnes *et al.*, 1991b) in post-mortem brains from Alzheimer's patients when compared with controls. Levels of DPP-IV and AP-N may also be decreased in the frontal cortex of Alzheimer's patients (Ichai *et al.*, 1994). Neuroleptic drug treatment has also been shown to affect brain neuropeptide concentrations and the levels of some peptidases (Waters *et al.*, 1996). For example, administration of haloperidol or apomorphine to rats has been shown to produce differential effects on the levels of AP-N, E-24.11 and E-24.15, but not ACE, depending on brain region. The mechanism underlying these effects was not explored. There is clearly much scope for examination of factors that may regulate the expression of neuropeptide-degrading enzymes both in normal and in the pathological state in the nervous system. Finally, it cannot be excluded that there are yet further neuropeptide-degrading enzymes to be discovered, although they are likely to be very limited in number.

Acknowledgements

I thank the Medical Research Council (UK) and the Wellcome Trust who have supported our work in this area.

References

Acker GR, Molineaux C, Orlowski M. (1987) Synaptosomal membrane-bound form of endopeptidase-24.15 generates Leu-enkephalin from dynorphin$_{1-8}$, α- and β-neoendorphin, and Met-enkephalin from Met-enkephalin-Arg^6-Gly^7-Leu^8. *J. Neurochem.* **48**: 284–292.

Akopyan TN, Couedel Y, Orlowski M, Fournié-Zaluski MC, Roques BP. (1994) Proteolytic processing of farnesylated peptides: assay and partial purification from pig brain membranes of an endopeptidase which has the characteristics of E.C. 3.4.24.15. *Biochem. Biophys. Res. Commun.* **198**: 787–794.

Augustijns PF, Ng KY, Williams TM, Borchardt RT. (1995) Peptidyl dipeptidase A-catalyzed metabolism of delta sleep-inducing peptide in bovine brain microvessel endothelial cells: a cell culture model of the blood–brain barrier. *Biochem. Biophys. Res. Commun.* **210**: 987–994.

Balow R-M, Ragnarsson U, Zetterqvist O. (1983) Tripeptidyl aminopeptidase in the extralysosomal fraction of rat liver. *J. Biol. Chem.* **258**: 11622–11628.

Barnes K, Matsas R, Hooper NM, Turner AJ, Kenny AJ. (1988a) Endopeptidase-24.11 is striosomally ordered in pig brain and, in contrast to aminopeptidase N and peptidyl dipeptidase A ('angiotensin converting enzyme'), is a marker for a set of striatal efferent fibres. *Neuroscience* **27**: 799–817.

Barnes K, Turner AJ, Kenny AJ. (1988b) Electron microscopic immunocytochemistry of pig brain shows that endopeptidase-24.11 is localized in neuronal membranes. *Neurosci. Lett.* **94**: 64–69.

Barnes K, Bourne A, Cook PA, Turner AJ, Kenny AJ. (1991a) Membrane peptidases in the peripheral nervous system of the pig: their localization by immunohistochemistry at light and electron microscopic levels. *Neuroscience* **44**: 245–261.

Barnes NM, Cheng CH, Costall B, Naylor RJ, Williams TJ, Wischik CM. (1991b) Angiotensin converting enzyme density is increased in temporal cortex from patients with Alzheimer's disease. *Eur. J. Pharmacol.* **200**: 289–292.

Barnes K, Turner AJ, Kenny AJ. (1992) Membrane localization of endopeptidase-24.11 and peptidyl dipeptidase A (angiotensin converting enzyme) in the pig brain: a study using subcellular fractionation and electron microscopic immunocytochemistry. *J. Neurochem.* **58**: 2088–2096.

Barnes K, Turner AJ, Kenny AJ. (1993) An immunoelectron microscopic study of pig substantia nigra shows co-localisation of endopeptidase-24.11 with substance P. *Neuroscience* **53**: 1073–1082.

Barnes K, Kenny AJ, Turner AJ. (1994) Localization of aminopeptidase N and dipeptidyl peptidase IV in pig striatum and in neuronal and glial cell cultures. *Eur. J. Neurosci.* **6**: 531–537.

Barnes K, Doherty S, Turner AJ. (1995) Endopeptidase-24.11 is the integral membrane peptidase initiating degradation of somatostatin in the hippocampus. *J. Neurochem.* **64**: 1826–1832.

Barrett AJ, Brown MA. (1990) Chicken liver Pz-peptidase, a thiol-dependent metallo-endopeptidase. *Biochem. J.* **271**: 701–706.

Bauer K. (1987) Adenohypophyseal degradation of thyrotropin releasing hormone regulated by thyroid hormones. *Nature* **330**: 375–377.

Bauer K. (1994) Purification and characterization of the thyrotropin-releasing-hormone-degrading-ectoenzyme. *Eur. J. Biochem.* **224**: 387–396.

Bauer K, Kleinkauf H. (1980) Catabolism of thyroliberin by rat adenohypophyseal tissue extract. *Eur. J. Biochem.* **106**: 107–117.

de la Baume S, Yi CC, Schwartz J-C, Chaillet P, Marçais-Collado HM, Costentin J. (1983) Participation of both 'enkephalinase' and aminopeptidase activities in the metabolism of endogenous enkephalins. *Neuroscience* **8**: 143–151.

Benuck M, Marks N. (1979) Co-identity of brain angiotensin converting enzyme with a membrane bound dipeptidyl carboxypeptidase inactivating met-enkephalin. *Biochem. Biophys. Res. Commun.* **88**: 215–221.

Berger UV, Carter RE, McKee M, Coyle JT. (1995a) *N*-acetylated alpha-linked acidic dipeptidase is expressed by non-myelinating Schwann cells in the peripheral nervous system. *J. Neurocytol.* **24**: 99–109.

Berger UV, Carter RE, Coyle JT. (1995b) The immunocytochemical localization of *N*-acetylaspartyl glutamate, its hydrolysing enzyme NAALADase, and the NMDAR-1 receptor at a vertebrate neuromuscular junction. *Neuroscience* **64**: 847–850.

Bordier C. (1981) Phase separation of integral membrane proteins in Triton X-114 solution. *J. Biol. Chem.* **256**: 1604–1607.

Bouras M, Huneau JF, Luengo C, Erlanson-Albertsson C, Tome D. (1995) Metabolism of enterostatin in rat intestine, brain membranes, and serum: differential involvement of proline-specific peptidases. *Peptides* **16**: 399–405.

Bourne A, Kenny AJ. (1990) The hydrolysis of brain and atrial natriuretic peptides by porcine choroid plexus is attributable to endopeptidase-24.11. *Biochem. J.* **271**: 381–385.

Bourne A, Barnes K, Taylor BA, Turner AJ, Kenny AJ. (1989) Membrane peptidases in the pig choroid plexus and on other cell surfaces in contact with the cerebrospinal fluid. *Biochem. J.* **259**: 69–80.

Browne P, O'Cuinn G. (1983) An evaluation of the role of a pyroglutamyl peptidase, a post-proline cleaving enzyme and a post-proline dipeptidyl aminopeptidase, each purified from the soluble fraction of guinea-pig brain, in the degradation of thyroliberin *in vitro*. *Eur. J. Biochem.* **137**: 75–87.

Browne P, Phelan J, O'Cuinn G. (1981) TRH pyroglutamate aminopeptidase activity in two different subcellular fractions of guinea pig brain. *Ir. J. Med. Sci.* 150: 348–359.

Brownlees J, Williams CH. (1993) Peptidases, peptides and the mammalian blood-brain barrier. *J. Neurochem.* **60**: 793–803.

Bunnett NW. (1987) Gastrin, cholecystokinin and gastrin releasing peptide: physiological roles in the alimentary tract and the brain. In: *Neuropeptides and their Peptidases* (ed A.J. Turner), pp. 107–133. Ellis Horwood, Chichester.

Camargo ACM, Caldo H, Emson PC. (1983) Degradation of neurotensin by rabbit brain endo-oligopeptidase A and endo-oligopeptidase B (proline-endopeptidase). *Biochem. Biophys. Res. Commun.* **166**: 1151–1159.

Camargo ACM, Ribeiro MJVF, Schwartz WN. (1985) Conversion and inactivation of opioid peptides by rabbit brain endo-oligopeptidase A. *Biochem. Biophys. Res. Commun.* **130**: 932–938.

Camargo ACM, Oliveira EB, Toffoletto O, Metters KM, Rossier J. (1987) Brain endo-oligopeptidase A, a putative enkephalin converting enzyme. *J. Neurochem.* **48**: 1258–1263.

Camus A, Rose C, Schwartz JC. (1989) Role of a serine endopeptidase in the hydrolysis of exogenous cholecystokinin by brain slices. *Neuroscience* **29**: 595–602.

Carter RE, Feldman AR, Coyle JT. (1996) Prostate-specific membrane antigen is a hydrolase with substrate and pharmacologic characteristics of a neuropeptidase. *Proc. Natl Acad. Sci. USA* **93**: 749–753.

Chai SY, Mendelsohn FAO, Paxinos G. (1987) Angiotensin converting enzyme in rat brain visualized by quantitative *in vitro* autoradiography. *Neuroscience* **20**: 615–627.

Chai SY, McKinley MJ, Paxinos G, Mendelsohn FA. (1991) Angiotensin converting enzyme in the monkey (*Macaca fascicularis*) brain visualized by *in vitro* autoradiography. *Neuroscience* **42**: 483–495.

Chaillet P, Marçais-Collado L, Costentin J, Li CC, de la Baume S, Schwartz J-C. (1983) Inhibition of enkephalin metabolism by, and antinociceptive action of bestatin, an aminopeptidase inhibitor. *Eur. J. Pharmacol.* **86**: 329–336.

Charli J-L, Cruz C, Vargas M-A, Joseph-Bravo P. (1988) The narrow specificity pyroglutamate aminopeptidase degrading TRH in rat brain is an ectoenzyme. *Neurochem. Int.* **13**: 237–242.

Checler F, Vincent JP, Kitabgi P. (1986) Purification and characterization of a novel neurotensin-degrading peptidase from rat brain synaptic membranes. *J. Biol. Chem.* **261**: 11274–11281.

Checler F, Dauch P, Barelli H, Dive V, Masuo Y, Vincent B, Vincent JP. (1995) Identification and distribution of endopeptidase-24.16 in central nervous system. *Meth. Neurosci.* **23**: 363–382.

Coelho HLL, Cicilini MA, Carvalho KM, Carvalho IF, Camargo ACM. (1981) Inhibition of rabbit tissue kininase by anti-(endo-oligopeptidase A) antibodies. *Biochem. J.* **197**: 85–93.

Constam DB, Tobler AR, Rensing-Ehl A, Kemler I, Hersh LB, Fontana A. (1995) Puromycin-sensitive aminopeptidase. Sequence analysis, expression, and functional characterization. *J. Biol. Chem.* **270**: 26931–26939.

Correa FM, Plunkett LM, Saavedra JM, Hichens M. (1985) Quantitative autoradiographic determination of angiotensin-converting enzyme (kininase II) binding in individual rat brain nuclei with ^{125}I-351A, a specific enzyme inhibitor. *Brain Res.* **347**: 192–195.

Correa FM, Guilhaume SS, Saavedra JM. (1991) Comparative quantification of rat brain and pituitary angiotensin-converting enzyme with autoradiographic and enzymatic methods. *Brain Res.* **545**: 215–222.

Craves FB, Law PY, Hunt CA, Loh HH. (1978) The metabolic disposition of radiolabeled enkephalins *in vitro* and *in situ*. *J. Pharmacol. Exp. Ther.* **206**: 492–506.

Curatolo A, D'Arcangelo P, Lino A, Brancati A. (1965) Distribution of *N*-acetyl-aspartic and *N*-acetyl-aspartyl-glutamic acids in nervous tissue. *J. Neurochem.* **12**: 339–342.

Dahms P, Mentlein R. (1992) Purification of the main somatostatin-degrading proteases from rat and pig brains, their action on other neuropeptides, and their identification as endopeptidases 24.15 and 24.16. *Eur. J. Biochem.* **208**: 145–154.

Dando PM, Brown MA, Barrett AJ. (1993) Human thimet oligopeptidase. *Biochem. J.* **294**: 451–457.

Dauch P, Vincent JP, Checler F. (1991) Specific inhibition of endopeptidase 24.16 by dipeptides. *Eur. J. Biochem.* **202**: 269–276.

Dauch P, Vincent J-P, Checler F. (1995) Molecular cloning and expression of rat brain endopeptidase 3.4.24.16. *J. Biol. Chem.* **270**: 27266–27271.

Daugé V, Mauborgne A, Cesselin F, Fournié-Zaluski M-C, Roques BP. (1996) The dual peptidase inhibitor RB101 induces a long-lasting increase in the extracellular level of Met-enkephalin-like material in the nucleus accumbens in freely moving rats. *J. Neurochem.* **67**: 1301–1308.

Davis TP, Gillespie TJ, Konings PNM. (1992) Specificity of neurotensin metabolism by regional rat brain slices. *J. Neurochem.* **58**: 608–617.

Deschodt-Lanckman M, Strosberg AD. (1983) *In vitro* degradation of the C-terminal octapeptide of cholecystokinin by 'enkephalinase A'. *FEBS Lett.* **152**: 109–113.

Deschodt-Lanckman M, Bui ND, Koulischer D, Paroutaud P, Strosberg AD. (1983) Cholecystokinin octa- and tetrapeptide degradation by synaptic membranes. II. Solubilization and separation of membrane-bound CCK-8 cleaving enzymes. *Peptides* **4**: 71–78.

Deschodt-Lanckman M, Koulischer D, Przedborski S, Lauwereys M. (1984) Cholecystokinin octa- and tetrapeptide degradation by synaptic membranes. III. Inactivation of CCK-8 by a phosphoramidon-sensitive endopeptidase. *Peptides* **5**: 649–651.

Dubovy P. (1987) Histochemical evidence for the presence of dipeptidyl peptidase IV in the Schwann cells of skin unmyelinated axons. *Experientia* **43**: 883–884.

Dubovy P, Malinovsky L. (1984) Localization of DPPIV in sensory corpuscles by means of light and electron microscopy. *Histochem. J.* **16**: 473–475.

Dubovy P, Soukup T. (1985) Histochemical localization of dipeptidyl peptidase IV in muscle spindles of the rat. *Histochem. J.* **17**: 582–584.

Dubreuil P, Fulcrand P, Rodriguez M, Fulcrand H, Laur J, Martinez M. (1989) Novel activity of angiotensin-converting enzyme. Hydrolysis of cholecystokinin and gastrin analogues with release of the amidated C-terminal dipeptide. *Biochem. J.* **262**: 125–130.

Dunkley PR, Jarvie PE, Heath JW, Kidd GJ, Rostas JA. (1986) A rapid method for isolation of synaptosomes on Percoll gradients. *Brain Res.* **372**: 115–129.

Dunkley PR, Heath JW, Harrison SM, Jarvie PE, Glenfield PJ, Rostas JA. (1988) A rapid Percoll gradient procedure for isolation of synaptosomes directly from an S1 fraction: homogeneity and morphology of subcellular fractions. *Brain Res.* **441**: 59–71.

Dyer SH, Slaughter CA, Orth K, Moomaw CR, Hersh LB. (1990) Comparison of the soluble and membrane-bound forms of the puromycin-sensitive enkephalin-degrading aminopeptidases from the rat. *J. Neurochem.* **54**: 547–554.

Ehrenpreis S. (1985) Analgesic properties of enkephalinase inhibitors: animal and human studies. *Progr. Clin. Biol. Res.* **192**: 363–370.

Ferro ES, Tambourgi DV, Gobersztejn F, Gomes MD, Sucupira M, Armelin MC, Kipnis TL, Camargo AC. (1993) Secretion of a neuropeptide-metabolizing enzyme similar to endopeptidase 22.19 by glioma C6 cells. *Biochem. Biophys. Res. Commun.* **191**: 275–281.

Fulcher IS, Matsas R, Turner AJ, Kenny AJ. (1982) Kidney neutral endopeptidase and the hydrolysis of enkephalin by synaptic membranes show similar sensitivity to inhibitors. *Biochem. J.* **203**: 519–522.

Garat B, Miranda J, Charli J-L, Joseph-Bravo P. (1985) Presence of a membrane-bound pyroglutamyl aminopeptidase degrading thyroliberin releasing hormone in rat brain. *Neuropeptides* **6**: 27–40.

Giros B, Gros C, Solhonne B, Schwartz JC. (1986) Characterization of aminopeptidases responsible for inactivating endogenous (Met5) enkephalin in brain slices using peptidase inhibitors and anti-aminopeptidase M antibodies. *Mol. Pharmacol.* **29**: 281–287.

Gros C, Giros B, Schwartz J-C. (1985) Identification of aminopeptidase M as an enkephalin-inactivating enzyme in rat cerebral membranes. *Biochemistry* **24**: 2179–2185.

Halban PA, Irminger JC. (1994) Sorting and processing of secretory proteins. *Biochem. J.* **299**: 1–18.

Healy DP, Orlowski M. (1992) Immunocytochemical localization of endopeptidase 24.15 in rat brain. *Brain Res.* **571**: 121–128.

Herman ZS, Stachura Z, Laskawiec G, Kowalski J, Obuchowicz E. (1985) Antinociceptive effects of puromycin and bacitracin. *Polish J. Pharmacol. Pharm.* **37**: 133–140.

Hersh LB. (1981) Solubilization and characterization of two rat brain membrane-bound aminopeptidases active on Met-enkephalin. *Biochemistry* **20**: 2345–2350.

Hersh LB. (1985) Characterization of membrane-bound aminopeptidases from rat brain: identification of the enkephalin-degrading aminopeptidase. *J. Neurochem.* **44**: 1427–1435.

Hersh LB, McKelvy JF. (1979) Enzymes involved in the degradation of thyrotropin releasing hormone (TRH) and luteinising hormone releasing hormone (LHRH) in bovine brain. *Brain Res.* **168**: 553–564.

Hersh LB, Aboukhair N, Watson S. (1987) Immunohistochemical localization of aminopeptidase M in rat brain and periphery: relationship of enzyme localization and enkephalin metabolism. *Peptides* **8**: 523–532.

Hooper NM, Turner AJ. (1985) Neurokinin B is hydrolysed by synaptic membranes and by endopeptidase-24.11 ('enkephalinase') but not by angiotensin converting enzyme). *FEBS Lett.* **190**: 133–136.

Hooper NM, Turner AJ. (1987) Isolation of two differentially glycosylated forms of peptidyl dipeptidase A (angiotensin converting enzyme) from pig brain: a re-evaluation of their role in neuropeptide metabolism. *Biochem. J.* **241**: 625–633.

Hooper NM, Kenny AJ, Turner AJ. (1985) Neurokinin A (substance K) is a substrate for endopeptidase-24.11 but not for peptidyl dipeptidase A (angiotensin converting enzyme). *Biochem. J.* **231**: 357–361.

Hooper NM, Karran E, Turner AJ. (1997) Membrane protein secretases. *Biochem. J.* **321**: 265–279.

Hughes J, Smith TW, Kosterlitz HW, Fothergill LA, Morgan BA, Morris HR. (1975) Identification of two related pentapeptides from the brain with potent opiate agonist activity. *Nature* **258**: 577–580.

Hui KS, Wang YJ, Lajtha A. (1983) Purification and characterization of an enkephalin aminopeptidase from rat brain membranes. *Biochemistry* **22**: 1062–1067.

Ichai C, Chevallier N, Delaere P, Dournaud P, Epelbaum J, Hauw JJ, Vincent JP, Checler F. (1994) Influence of region-specific alterations of neuropeptidase content on the catabolic fates of neuropeptides in Alzheimer's disease. *J. Neurochem.* **62**: 645–655.

Israeli, RS, Powell CT, Fair WR, Heston WDW. (1993) Molecular cloning of a complementary DNA encoding a prostate-specific membrane antigen. *Cancer Res.* **53**: 227–230.

Kase R, Sekine R, Katayama T, Takagi H, Hazato T. (1986) Hydrolysis of neo-kyotorphin (Thr-Ser-Lys-Tyr-Arg) and [Met]enkephalin-Arg^6-Phe^7 by angiotensin converting enzyme from monkey brain. *Biochem. Pharmacol.* **35**: 4499–4503.

Kato T, Nagatsu T, Fukasawa K, Harada M, Nagatsu I, Sakakibura S. (1978). Successive cleavage of Arg-Pro and Lys-Pro from substance P, but no release of Arg-Pro from bradykinin, by X-Pro dipeptidyl aminopeptidase. *Biochim. Biophys. Acta* **525**: 417–422.

Kelly JA. (1995) Thyrotropin-releasing hormone: basis and potential for its therapeutic use. *Essays Biochem.* **30**: 133–149.

Kenny AJ, Bourne A. (1991) Cellular reorganisation of membrane peptidases in Wallerian degeneration of pig peripheral nerve. *J. Neurocytol.* **20**: 875–885.

Kenny AJ, Turner AJ. (eds) (1987) *Mammalian Ectoenzymes,* p. 359. Elsevier, Amsterdam.

Kenny AJ, Bourne A, Ingram J. (1993) Hydrolysis of human and pig brain natriuretic peptides, urodilatin, C-type natriuretic peptide and some C-receptor ligands by endopeptidase-24.11. *Biochem. J.* **291**: 83–88.

Kerr MA, Kenny AJ. (1974) The purification and specificities of a neutral endopeptidase from rabbit kidney brush border. *Biochem. J.* **137**: 477–488.

Knight CG, Dando PM, Barrett AJ. (1995) Thimet oligopeptidase specificity: evidence of preferential cleavage near the C-terminus and product inhibition from kinetic analysis of peptide hydrolysis. *Biochem. J.* **308**: 145–150.

Konkoy CS, Davis TP. (1995) Regional metabolism of Met-enkephalin and cholecystokinin on intact rat brain slices: characterization of specific peptidases. *J. Neurochem.* **65**: 2773–2782.

Kunz J, Krause D, Kremer M, Dermietzel R. (1994) The 140-kDa protein of blood–brain barrier-associated pericytes is identical to aminopeptidase N. *J. Neurochem.* **62**: 2375–2386.

Lasdun A, Reznik S, Molineaux CJ, Orlowski M. (1989) Inhibition of endopeptidase 24.15 slows the *in vivo* degradation of luteinizing hormone-releasing hormone. *J. Pharmacol. Exp. Therap.* **251**: 439–447.

Littlewood GM, Iversen LL, Turner AJ. (1988) Neuropeptides and their peptidases: functional considerations. *Neurochem. Int.* **12**: 383–389.

Lukac J, Koren E. (1977) The metalloenzymic nature of collagenase-like peptidase of the rat testis. *J. Reprod. Fertil.* **49**: 95–99.

McDermott JR, Biggins JA, Gibson AM. (1992) Human brain peptidase activity with the specificity to generate the N-terminus of the Alzheimer beta-amyloid protein from its precursor. *Biochem. Biophys. Res. Commun.* **185**: 746–752.

McKie N, Dando PM, Rawlings ND, Barrett AJ. (1993) Thimet oligopeptidase: similarity to 'soluble angiotensin II-binding protein' and some corrections to the published amino acid sequence of the rat testis enzyme. *Biochem. J.* **295**: 57–60.

McLellan S, Dyer SH, Rodriguez G, Hersh LB. (1988) Studies on the tissue distribution of the puromycin-sensitive enkephalin-degrading aminopeptidases. *J. Neurochem.* **51**: 1552–1559.

Malfroy B, Swerts JP, Guyon A, Roques BP. (1978) High-affinity enkephalin-degrading peptidase in brain is increased after morphine. *Nature* **276**: 523–526.

Marcel D, Pollard H, Verroust P, Schwartz JC, Beaudet A. (1990) Electron microscopic localization of immunoreactive enkephalinase (EC 3.4.24.11) in the neostriatum of the rat. *J. Neurosci.* **10**: 2804–2817.

Matsas R, Fulcher IS, Kenny AJ, Turner AJ. (1983) Substance P and[Leu]enkephalin are hydrolyzed by an enzyme in pig caudate synaptic membranes that is identical with the endopeptidase of kidney microvilli. *Proc. Natl Acad. Sci. USA* **80**: 3111–3115.

Matsas R, Kenny AJ, Turner AJ. (1984a) The metabolism of neuropeptides. The hydrolysis of peptides, including enkephalins, tachykinins and their analogues, by endopeptidase-24.11. *Biochem. J.* **223**: 433–440.

Matsas R, Kenny AJ, Turner AJ. (1984b) Endopeptidase-24.11 and aminopeptidase activity in brain synaptic membranes are jointly responsible for the hydrolysis of cholecystokinin octapeptide (CCK-8). *FEBS Lett.* **175**: 124–128.

Matsas R, Stephenson SL, Hryszko J, Kenny AJ, Turner AJ. (1985) The metabolism of neuropeptides: phase separation of synaptic membrane preparations with Triton X-114 reveals the presence of aminopeptidase N. *Biochem. J.* **231**: 445–449.

Matsas R, Kenny AJ, Turner AJ. (1986) An immunohistochemical study of endopeptidase-24.11 ('enkephalinase') in the pig nervous system. *Neuroscience* **18**: 991–1012.

Mauborgne A, Bourgoin S, Benoliel JJ, Hamon M, Cesselin F. (1991) Is substance P released from slices of the rat spinal cord inactivated by peptidase(s) distinct from both 'enkephalinase' and 'angiotensin-converting enzyme'? *Neurosci. Lett.* **123**: 221–225.

Meckelein B, De Silva HAR, Roses AD, Rao PN, Pettenati MJ, Xu PT, Hodge R, Glucksman MJ, Abraham CR. (1996) Human endopeptidase (THOP1) is localized on chromosome 19 within the linkage region for the late-onset Alzheimer disease AD2 locus. *Genomics* **31**: 246–249.

Medeiros MS, Turner AJ. (1994) Processing and metabolism of peptide YY: pivotal roles for dipeptidyl peptidase IV, aminopeptidase P and endopeptidase-24.11. *Endocrinology* **134**: 2088–2094.

Medeiros MS, Turner AJ. (1996) Metabolism and functions of neuropeptide Y. *Neurochem. Res.* **21**: 1131–1138.

Medeiros MS, Iazigi N, Camargo ACM, Oliveira EB. (1992) Distribution and properties of endo-oligopeptidases A and B in the human neuroendocrine system. *J. Endocrinol.* **135**: 579–588.

Mendelsohn FA. (1984) Location of angiotensin convering enyme in rat forebrain and other tissues by *in vitro* autoradiography using ^{125}I-labelled MK351A *Clin. Exp. Pharmacol. Physiol.* **11**: 431–435.

Mentlein R, Dahms P. (1994) Endopeptidases 24.16 and 24.15 are responsible for the degradation of somatostatin, neurotensin, and other neuropeptides by cultivated rat cortical astrocytes. *J. Neurochem.* **62**: 27–36.

Millican PE, Kenny AJ, Turner AJ. (1991) Purification and properties of a neurotensin-degrading endopeptidase from pig brain. *Biochem. J.* **276**: 583–591.

Molineaux CJ, Lasdun A, Michaud C, Orlowski M. (1988) Endopeptidase-24.15 is the primary enzyme that degrades luteinizing hormone releasing hormone both *in vitro* and *in vivo*. *J. Neurochem.* **51**: 624–633.

Montiel J-L, Cornille F, Roques BP, Noble F. (1997) Nociceptin/orphanin FQ metabolism. Role of aminopeptidase and endopeptidase 24.15. *J. Neurochem.* **68**: 354–361.

Morales TI, Woessner JF Jr. (1977) PZ-Peptidase from chick embryos. Purification, properties, and action on collagen peptides. *J. Biol. Chem.* **252**: 4855–4860.

Nau R, Schäfer G, Deacon CF, Cole T, Agoston DV, Conlon JM. (1986) Proteolytic inactivation of substance P and neurokinin A in the longitudinal muscle layer of guinea pig small intestine. *J. Neurochem.* **47**: 856–864.

Norman JA, Autry WL, Barbaz BS. (1985) Angiotensin converting enzyme inhibitors potentiate the analgesic activity of [Met]-enkephalin-Arg^6-Phe^7 by inhibiting its degradation in mouse brain. *Mol. Pharmacol.* **28**: 521–526.

Oblin A, Danse MJ, Zivkovic B. (1988) Degradation of substance P by membrane peptidases in the rat substantia nigra: effect of selective inhibitors. *Neurosci. Lett.* **84**: 91–96.

O'Connor B, O'Cuinn G. (1984) Localisation of a narrow specificity thyroliberin hydrolysing pyroglutamate aminopeptidase in synaptosomal membranes of guinea-pig brain. *Eur. J. Biochem.* **144**: 271–278.

O'Connor B, O'Cuinn G. (1985) Purification of and kinetic studies on a narrow specificity synaptosomal membrane pyroglutamate aminopeptidase from guinea-pig brain. *Eur. J. Biochem.* **150**: 47–52.

O'Cuinn G, O'Connor B, Elmore M. (1990) Degradation of thyrotropin-releasing hormone and luteinising hormone-releasing hormone by enzymes of brain tissue. *J. Neurochem.* **54**: 1–13.

Oliveira EB, Martins AR, Camargo ACM. (1976) Isolation of brain endopeptidases: influence of size and sequence of substrates structurally related to bradykinin. *Biochemistry* **15**: 1967–1974.

Oliveira ES, Leite PEP, Spillatini MG, Camargo ACM, Hunt SP. (1990) Localization of endo-oligopeptidase (EC 3.4.22.19) in the rat nervous tissue. *J. Neurochem.* **55**: 1114–1121.

Orlowski M, Michaud C, Chu TG. (1983) A soluble metalloendopeptidase from rat brain. Purification of the enzyme and determination of specificity with synthetic and natural substrates. *Eur. J. Biochem.* **135**: 81–88.

Orlowski M, Michaud C, Molineaux CJ. (1988) Substrate-related potent inhibitors of brain metalloendopeptidase. *Biochemistry* **27**: 597–602.

Ory-Lavollée L, Blakely RD, Coyle JT. (1987) Neurochemical and immunocytochemical satudies on the distribution of *N*-acetyl-aspartylglutamate and *N*-acetylaspartate in rat spinal cord and some peripheral nervous tissues. *J. Neurochem.* **48**: 895–899.

Papastoitsis G, Siman R, Scott R, Abraham CR. (1994) Identification of a metalloprotease from Alzheimer's disease brain able to degrade the β-amyloid precursor protein and generate amyloidogenic fragments. *Biochemistry* **33**: 192–199.

Pierotti AR, Dong K-W, Glucksman MJ, Orlowski M, Roberts JL. (1990) Molecular cloning and primary structure of rat testes metalloendopeptidase EC 3.4.24.15. *Biochemistry* **29**: 10323–10329.

Pollard H, Llorens-Cortes C, Couraud JY, Ronco P, Verroust P, Schwartz JC. (1987) Enkephalinase (EC 3.4.24.11) is highly localized to a striatonigral pathway in rat brain. *Neurosci. Lett.* **77**: 267–271.

Rawlings ND, Barrett AJ. (1993) Evolutionary families of peptidases. *Biochem. J.* **290**: 205–218.

Relton JM, Gee NS, Matsas R, Turner AJ, Kenny AJ. (1983) Purification of endopeptidase-24.11 ('enkephalinase') from pig brain by immunoadsorbent chromatography. *Biochem. J.* **215**: 519–523.

Robinson MB, Blakely RD, Couto R, Coyle JT. (1987) Hydrolysis of the brain dipeptide *N*-acetyl-L-aspartyl-L-glutamate. Identification and characterization of a novel *N*-acetylated α-linked acidic dipeptidase activity from rat brain. *J. Biol. Chem.* **262**: 14498–14506.

Rogerson FM, Schlawe I, Paxinos G, Chai SY, McKinley MJ, Mendelsohn FA. (1995) Localization of angiotensin converting enzyme by *in vitro* autoradiography in the rabbit brain. *J. Chem. Neuroanat.* **8**: 227–243.

Roques BP, Fournié-Zaluski MC, Soroca E, Leconmte JM, Malfroy B, Llorens C, Schwartz J-C. (1980) The enkephalinase inhibitor thiorphan shows antinociceptive activity in mice. *Nature* **288**: 286–288.

Rose C, Camus A, Schwartz J-C. (1988) A serine peptidase is responsible for the inactivation of endogenous cholecystokinin in brain. *Proc. Natl Acad. Sci. USA* **85**: 8326–8330.

Rose C, Camus A, Schwartz J-C. (1989) Protection by serine protease inhibitors of endogenous cholecystokinin released from brain slices. *Neuroscience* **29**: 583–594.

Rose C, Vargas F, Facchinetti P *et al.* (1996) Characterization and inhibition of a cholecystokinin-inactivating serine peptidase. *Nature* **380**: 403–409.

Rubartelli A, Cozzolino F, Talio M, Sitia R. (1990) A novel secretory pathway for interleukin-1b, a protein lacking a signal sequence. *EMBO J.* **9**: 1503–1510.

Schauder B, Schomburg L, Köhrle J, Bauer K. (1994) Cloning of a cDNA encoding an ectoenzyme that degrades thyrotropin-releasing hormone. *Proc. Natl Acad. Sci. USA* **91**: 9534–9538.

Schnabel R, Bernstein H-G, Luppa H, Lojda Z, Barth A. (1992) Aminopeptidase in the circumventricular organs of the mouse brain: a histochemical study. *Neuroscience* **47**: 431–438.

Sekiguchi M, Wada K, Wenthold RJ. (1992) *N*-acetylaspartylglutamate acts as an agonist upon homomeric NMDA receptor (NMDAR1) expressed in *Xenopus* oocytes. *FEBS Lett.* **311**: 285–289.

Serizawa A, Dando PM, Barrett AJ. (1995) Characterization of a mitochondrial metallopeptidase reveals neurolysin as a homologue of thimet oligopeptidase. *J. Biol. Chem.* **270**: 2092–2098.

Shrimpton CN, Glucksman MJ, Lew RA, Tullai JW, Margulies EH, Roberts JL, Smith AI. (1997) Thiol activation of endopeptidase EC 3.4.24.15; a novel mechanism for the regulation of catalytic activity. *J. Biol. Chem.,* in press.

Skidgel RA, Erdös EG. (1985) Novel activity of angiotensin converting enzyme: release of the NH_2- and COOH-terminal tripeptides from the luteinising hormone-releasing hormone. *Proc. Natl. Acad. Sci. USA.* **82**: 1025–1029.

Skidgel RA, Erdös EG. (1987) The broad substrate specificity of human angiotensin I converting enzyme. *Clin. Exp. Hypertension* **9**: 243–259.

Skidgel RA, Engelbrecht S, Johnson AR, Erdös EG. (1984) Hydrolysis of substance P and neurotensin by converting enzyme and neutral endopeptidase. *Peptides* **5**: 769–776.

Slusher BS, Robinson MB, Tsai G, Simmons ML, Richards SS, Coyle JT. (1990) Rat brain N-acetylated α-linked acidic dipeptidase activity. Purification and immunologic characterization. *J. Biol. Chem.* **265**: 21297–21301.

Slusher BS, Tsai G, Yoo G, Coyle JT. (1992) Immunocytochemical localization of the *N*-acetylaspartylglutamate (NAAG) hydrolyzing enzyme *N*-acetylated alpha-linked acidic dipeptidase (NAALADase). *J. Comp. Neurol.* **315**: 217–229.

Solhonne B, Gros C, Pollard H, Schwartz J-C. (1987) Major localization of aminopeptidase M in rat brain microvessels. *Neuroscience* **22**: 225–232.

Spencer-Dene B, Thorogood P, Nair S, Kenny AJ, Harris M, Henderson B. (1994) Distribution of, and a putative role for, the cell-surface neutral metallo-endopeptidases during mammalian craniofacial develpment. *Development* **120**: 3213–3226.

Strittmatter SM, Snyder SH. (1987) Angiotensin converting enzyme immunohistochemistry in rat brain and pituitary gland: correlation of isozyme type with cellular localization. *Neuroscience* **21**: 407–420.

Strittmatter SM, Lo MMS, Javitch JA, Snyder SH. (1984) Autoradiographic localization of angiotensin-converting enzyme in rat brain with [^{3}H]captopril. Localization to a striatonigral pathway. *Proc. Natl Acad. Sci. USA* **81**: 1599–1603.

Strittmatter SM, Thiele EA, Kapiloff MS, Snyder SH. (1985) A rat brain isozyme of angiotensin converting enzyme. Unique specificity for amidated peptide substrates. *J. Biol. Chem.* **260**: 9825–9832.

Sugiura N, Hagiwara H, Hirose S. (1992) Molecular cloning of porcine soluble angiotensin-binding protein. *J. Biol. Chem.* **267**: 18067–18072.

Swerts JP, Perrisot R, Malfroy B, Schwartz JC. (1979) Is 'enkephalinase' identical with angiotensin converting enzyme? *Eur. J. Pharmacol.* **53**: 209–210.

Thiele EA, Strittmatter SM, Snyder SH. (1985) Substance K and substance P as possible endogenous substrates of angiotensin converting enzyme in the brain. *Biochem. Biophys. Res. Commun.* **128:** 317–324.

Thompson A, Huber G, Malherbe P. (1995) Cloning and functional expression of a metalloendopeptidase from human brain with the ability to cleave a β-APP substrate peptide. *Biochem. Biophys. Res. Commun.* **213**: 66–73.

Thorne B, Wonnacott S, Dunkley PR. (1991) Isolation of hippocampal synaptosomes on Percoll gradients: cholinergic markers and ligand binding sites. *J. Neurochem.* **56**: 479–484.

Tisljar U. (1993) Thimet oligopeptidase – a review of a thiol-dependent metallo-endopeptidase also known as Pz-peptidase, endopeptidase-24.15 and endo-oligopeptidase. *Biol. Chem. Hoppe-Seyler* **374**: 91–100.

Tisljar U, Barrett AJ. (1990) Thiol-dependent metalloendopeptidase characteristics of Pz-peptidase in rat and rabbit. *Biochem. J.* **267**: 531–533.

Tobler AR, Constam DB, Schmitt-Graff A, Malipiero U, Schalpbach R, Fontana A. (1997) Cloning of the human puromycin-sensitive aminopeptidase and evidence for expression in neurons. *J. Neurochem.* **68:** 889–897.

Tomkinson B, Jonsson A-K. (1991) Characterization of cDNA for human tripeptidyl peptidase II: the N-terminal part of the enzyme is similar to subtilisin. *Biochemistry* **30**: 168–174.

Tsai G, Stauch BL, Vornov JJ, Deshpande JK, Coyle JT. (1990) Selective release of *N*-acetylaspartylglutamate from rat optic nerve terminals *in vivo*. *Brain Res.* **518**: 313–316.

Tsai G, Slusher BS, Sim L, Coyle JT. (1993) Immunocytochemical distribution of *N*-acetylaspartylglutamate in the rat forebrain and glutamatergic pathways. *J. Chem. Neuroanat.* **6**: 277–292.

Turner AJ, Matsas R, Kenny AJ. (1985) Are there neuropeptide-specific peptidases? *Biochem. Pharmacol.* **34**: 1347–1356.

Vanhoof G, Goossens F, De Meester I, Hendriks D, Scharpé S. (1995) Proline motifs in peptides and their biological processing. *FASEB J.* **9**: 736–744.

Vanneste Y, Pauwels S, Lambotte L, Deschodt-Lanckman M. (1990) *In vivo* metabolism of brain natriuretic peptide in the rat involves endopeptidase 24.11 and angiotensin converting enzyme. *Biochem. Biophys. Res. Commun.* **173**: 265–271.

Vincent B, Dive V, Yiotakis A, Smajda C, Maldonado R, Vincent JP, Checler F. (1995) Phosphorus-containing peptides as mixed inhibitors of endopeptidase 3.4.24.15 and 3.4.24.16: effect on neurotensin degradation *in vitro* and *in vivo*. *Br. J. Pharmacol.* **115**: 1053–1063.

Vincent B, Beaudet A, Dauch P, Vincent J-P, Checler F. (1996) Distinct properties of neuronal and astrocytic endopeptidase 3.4.24.16: a study on differentiation, subcellular distribution, and secretion processes. *J. Neurosci.* **16**: 5049–5059.

Vincent B, Dauch P, Vincent JP, Checler F. (1997) Stably transfected human cells overexpressing rat brain endopeptidase 3.4.24.16: biochemical characterization of the activity and expression of a membrane-bound counterpart. *J. Neurochem.* **68:** 837–845.

Waksman G, Hamel E, Fournié-Zaluski MC, Roques BP. (1986a) Autoradiographic comparison of the distribution of the neutral endopeptidase 'enkephalinase' and of μ and δ opioid receptors in rat brain. *Proc. Natl Acad. Sci. USA* **83**: 1523–1527.

Waksman G, Hamel E, Delay-Goyet P, Roques BP. (1986b) Neuronal localization of the neutral endopeptidase 'enkephalinase' in rat brain revealed by lesions and autoradiography. *EMBO J.* **5**: 3163–3166.

Waters SM, Konkoy CS, Davis TP. (1996) Haloperidol and apomorphine differentially affect neuropeptidase content. *J. Pharmacol. Exp. Ther.* **277**: 113–120.

Whiting P, Nava S, Mozley L, Eastham H, Poat J. (1991) Expression of angiotensin converting enzyme mRNA in rat brain. *Brain Res. (Mol. Brain Res.)* **11**: 93–96.

Wilk S. (1983) Prolyl endopeptidase. *Life Sci.* **33**: 2149–2157.

Wilk S. (1989) Inhibitors of TRH-degrading enzyme. *Ann. NY Acad. Sci.* **553**: 252–264.

Wilk S, Wilk EK. (1989) Pyroglutamyl peptidase II, a thyrotropin releasing hormone degrading enzyme: purification and specificity studies of the rabbit brain enzyme. *Neurochem. Int.* **15**: 81–90.

Williams TA, Hooper NM, Turner AJ. (1991) Characterization of neuronal and endothelial forms of angiotensin converting enzyme in pig brain. *J. Neurochem.* **57**: 193–199.

Williams CH, Yamamoto T, Walsh DM, Allsop D. (1993) Endopeptidase 3.4.24.11 converts *N*-1-(*R,S*)carboxy-3-phenylpropyl-Ala-Ala-Phe-*p*-carboxyanilide into a potent inhibitor of angiotensin-converting enzyme. *Biochem. J.* **294**: 681–684.

Yang HYT, Neff NH. (1972) Distribution and properties of angiotensin-converting enzyme of rat brain. *J. Neurochem.* **19**: 2443–2450.

Yokosawa H, Ogura Y, Ishii S. (1983) Purification and inhibition by neuropeptides of angiotensin converting enzyme from rat brain. *J. Neurochem.* **41**: 403–410.

Zaczek R, Koller K, Cotter R, Heller D, Coyle JT. (1983) *N*-Acetylaspartylglutamate: an endogenous peptide with high affinity for a brain "glutamate- receptor. *Proc. Natl Acad. Sci. USA* **80**: 1116–1119.

Zuzel KA, Rose C, Schwartz J-C. (1985) Assessment of the role of 'enkephalinase' in cholecystokinin inactivation. *Neuroscience* **15**: 149–158.

Chapter 20

Natriuretic peptide metabolism: inhibitors of endopeptidase-24.11 as possible therapeutic agents for cardiovascular disease

Martin R. Wilkins and A. John Kenny

20.1 The natriuretic peptides

20.1.1 The natriuretic peptide family

This is a group of peptides possessing natriuretic, diuretic and hypotensive activities. The first to be recognized, originally referred to as atrial natriuretic *factor*, was identified in storage granules in atrial muscle. The existence of these granules, with a presumed secretory role, was noted nearly 40 years ago by morphologists and an early insight into their role was gained by the observation that their abundance could be influenced by changes in the intake of sodium and water and that injection of extracts of atria caused diuresis and natriuresis (De Bold *et al.*, 1981). The major circulating form in the human is a 28-amino-acid residue peptide, α-human atrial natriuretic peptide (α-hANP, Kangawa and Matsuo, 1984); see *Figure 20.1* for amino acid sequences. It contains (as do all the natriuretic peptides) a disulphide-linked loop of 17 residues, the integrity of which is essential for biological activity (Misono *et al.*, 1984). Other members of the family are brain natriuretic peptide (BNP) and C-type natriuretic peptide (CNP). In the human the main forms of these are hBNP-32 (32 residues) (Sudoh *et al.*, 1988) and hCNP-22 (22 residues) (Komatsu *et al.*, 1991; Sudoh *et al.*, 1990). Another peptide, urodilatin, is identical to α-hANP with a four-residue extension at the N terminus (Schulz-Knappe *et al.*, 1988). Pig BNP-32 differs in sequence from hBNP-32 and is also found in a truncated form, pBNP-26, while in pig brain there exists a larger form of CNP of 53 residues (Minamino *et al.*, 1990).

20.1.2 Locations of natriuretic peptides

ANP is most abundant in cardiac atria, but is also found in several other tissues including the brain, adrenals and testis. BNP is present in brain and heart, the highest concentration being in the atria, but a greater quantity (because of their greater size) is present in the ventricles (Ogawa *et al.*, 1991); in cardiac failure, the proportional increase in

Cell-Surface Peptidases in Health and Disease, edited by A.J. Kenny and C.M. Boustead.

Peptide	Sequence
α-hANP-28	SLRRSSCFGGRMDRIGAQSGLGCNSFRY
Urodilatin-32	APRLSRRSSCFGGRMDRIGAQSGLGCNSFRY
hBNP-32	SPKMVQGSGCFGRKMDRISSSSGLGCKVLRRH
pBNP-32	SPKTMRDSGCFGRRLDRIGSLSGLGCNVLRRY
pBNP-26	DSGCFGRRLDRIGSLSGLGCNVLRRY
hCNP-22	GLSKGCFGLKLDRIGSMSGLGCamide
SC46542	SSCGRMDRIGSGLGCNSFRY
C-ANP$_{4\text{-}23}$	RSSCFGGRIDRIGAC

Figure 20.1. Amino acid sequences of some natriuretic peptides and two synthetic peptides. The lines above the sequences define the cystine-linked loop. The arrows indicate the initial site, where known, of attack by E-24.11 (Bourne and Kenny, 1990; Kenny *et al.,* 1993; Stephenson and Kenny, 1987).

plasma BNP levels, relative to basal levels, is larger than that of ANP (Hosoda *et al.*, 1991; Mukoyama *et al.*, 1990, 1991). CNP is present in the brain in much higher concentrations than ANP or BNP (Komatsu *et al.*, 1991; Ueda *et al.*, 1991). CNP is also produced in endothelial cells, where its release may be stimulated by ANP and BNP (Nazario *et al.*, 1995; Suga *et al.*, 1993). Urodilatin was first isolated from human urine and has been postulated to be produced and act locally within the kidney (Schulz-Knappe *et al.,* 1988).

20.1.3 Pathophysiological role of natriuretic peptides

The natriuretic peptides participate in the regulation of blood volume and blood pressure (Ruskoaho, 1992). Cardiac secretion of ANP and BNP increases in response to stretching of the myocardial wall, such as occurs with volume or pressure overload of the heart, and these peptides act to reduce cardiac work by increasing urinary sodium and water excretion and reducing blood pressure. CNP acts a local regulator of vascular tone (Stingo *et al.,* 1992), although its role is not well defined. ANP and BNP have been shown to inhibit renin and aldosterone secretion and collectively the natriuretic peptide family act as physiological antagonists of the renin–angiotensin system. In addition, ANP and CNP have been shown to exert anti-trophic effects in vascular endothelial and smooth muscle cells (Furuya *et al.,* 1991; Itoh *et al.,* 1990).

This profile of biological activity is attractive to pharmacologists seeking new treatments for the management of cardiovascular diseases, such as heart failure and systemic and pulmonary hypertension, and for other disorders associated with sodium and water retention, such as chronic renal and liver disease. Attempts to synthesise natriuretic peptide analogues that are more resistant to degradation in the gut have not been very successful. Attention has turned, therefore, to maximizing the biological activity of the endogenous peptides by inhibiting their degradation.

20.2 Metabolism of natriuretic peptides *in vitro*

20.2.1 Role of peptidases

Renal microvillus membrane preparations contain a battery of ectopeptidases and provide a useful model system which can degrade rapidly a wide range of simple peptides. Moreover, it is possible, by the use of appropriate inhibitors, to define the contributions of the different peptidases (Stephenson and Kenny, 1987b). For many peptides (including the angiotensins, bradykinin, substance P) endopeptidase-24.11 (E-24.11) was shown to be the enzyme initiating the attack. α-hANP is also metabolized rapidly by microvillus membranes, at rates comparable to angiotensins; again E-24.11 initiates the attack (Stephenson and Kenny, 1987a). Many of the same ectopeptidases are present on choroid plexus membranes. Here, too, the hydrolysis of α-hANP and BNP-26 is initiated by E-24.11 (Bourne and Kenny, 1990). Neither ANP nor BNP is susceptible to attack by angiotensin-converting enzyme (ACE; Bourne and Kenny, 1990; Kenny *et al.*, 1993; Stephenson and Kenny, 1987a).

The first bond in α-hANP to be hydrolysed is Cys7–Phe8 (Kenny and Stephenson, 1988; Stephenson and Kenny, 1987a) the cleavage of which opens the ring and generates a linear, inactive form. This point may be important in interpreting the results of immunoassays, since some of the antisera to α-hANP may not distinguish the intact and linear forms. Other susceptible bonds within the ring and in the tail are cleaved subsequently (Stephenson and Kenny, 1987a; Vanneste *et al.*, 1988). Although urodilatin has been reported to be resistant to hydrolysis by dog kidney cortex membranes (Gagelmann *et al.*, 1988), it was hydrolysed by purified E-24.11 at about half the rate [determined by high performance liquid chromatography (HPLC) analysis of the products] observed for ANP (Kenny *et al.,* 1993). In another report (Abassi *et al.*, 1994), a very much lower rate of hydrolysis was observed, but these authors used iodine-labelled ANP, for which the analysis depends on the release of the iodine-labelled C-terminal tripeptide (Phe–Arg–Tyr), which might not be representative of the rate of inactivation.

Within the family of natriuretic peptides there are remarkable differences in the rates of hydrolysis by E-24.11 and in the identity of the bond first cleaved (Kenny *et al.,*

1993). When the peptides were incubated in identical experiments with purified E-24.11, hBNP-32 was hydrolysed very slowly, with a half-life in these conditions of 2 h, compared with a half-life of 8 min for α-hANP. CNP was hydrolysed more rapidly than α-hANP, as were the synthetic NPR-C (see Section 20.2.2) inhibitory peptides, SC46542 and C-ANP$_{4-23}$. Moreover, hBNP-32 was hydrolysed differently from any of the other peptides. In particular, the initial cleavage was not within the ring but in the head peptide at Met4–Val5. It is not known whether loss of the N-terminal tetrapeptide has any effect on activity, but since the sequences of the head peptides in this family of peptides vary greatly, it is probable that inactivation of hBNP-32 also requires subsequent hydrolysis at bonds within the ring. The pattern of attack on bonds within the disulphide-linked ring of hBNP-32 differs substantially from that of pBNP-26 (Bourne and Kenny, 1990).

Kinetic studies also reveal major differences in the hydrolysis of the natriuretic peptides by E-24.11 (Kenny *et al.*, 1993) see *Table 20.1*. For α-hANP, pBNP-26 and pBNP-32 the kinetics were determined in detail by conventional methods. For others a short-cut was employed, whereby the peptide was used as an inhibitor in the hydrolysis of radiolabelled substrates, ANP and B-chain of insulin. In such experiments K_i approximates to K_m. Among the natural peptides CNP stands out as having the highest affinity for E-24.11. hBNP-32 appeared to be in a different category, for which the K_i was much higher. Thus hBNP-32 is unusually resistant to hydrolysis by E-24.11. In contrast, the rapidity of hydrolysis and low K_m value for CNP, suggest that it may be an important target for E-24.11.

Table 20.1. Kinetic constants for the hydrolysis of some natriuretic peptides and C-receptor ligands by pure E-24.11[a]

Peptide	K_m (or K_i for those marked*) (μM)	V_{max} (μmol min^{-1} mg^{-1})	K_{cat} (min^{-1})	k_{cat}/K_m (μmol^{-1} min^{-1})
CNP*	2.5, 13			
SC 46542*	2.7, 25			
pBNP-26	15.3±3.7	4.2±0.1	379	25
Urodilatin*	31, 62			
α-hANP*	26, 34			
C-ANP$_{4-23}$*	31, 82			
pBNP-32	111±11	9.4±0.5	611	8
α-hANP	121±14	19±1	1690	14
hBNP-32*	151,172			

[a] The peptides are ranked in ascending order of their K_m values. Those marked * were used as inhibitors in assays of the hydrolysis of two radio-iodine labelled substrates, namely insulin B-chain and α-hANP, the K_i values then approximating to K_m values, the two values being those with each substrate. SC46542 and C-ANP$_{4-23}$ are synthetic C-receptor inhibitors.
Data derived from Kenny *et al.* (1993).

20.2.2 Role of receptors

Natriuretic peptides may also be metabolized by receptor-mediated endocytosis. ANP receptors have been identified in many tissues, including vascular smooth muscle, renal glomerulus, adrenal zona glomerulosa cells, choroid plexus and brain (Bianchi *et al.*, 1985,1986a, b; Lynch *et al.*, 1986; Quirion *et al.*, 1986). One receptor type is a transmembrane protein capable of binding ANP or BNP at the extracytoplasmic domain and of generating cyclic GMP at the cytoplasmic domain (Chinkers *et al.*, 1989; Kuno *et al.*, 1986). This class (referred to as NPR-GC) is therefore biologically active and exists as two subtypes, NPR-A and NPR-B. ANP is the most potent peptide for the stimulation of NPR-A; BNP also activates this receptor subtype, but a 10-fold higher concentration is required (Schulz *et al.*, 1989). CNP selectively activates NPR-B, with an affinity at least 50-fold greater than either ANP or BNP (Koller *et al.*, 1991).

Another type, NPR-C, lacks the cytoplasmic guanylate cyclase domain and was therefore thought to be biologically 'silent' (Fuller *et al.*, 1988; Maack *et al.*, 1987). However, it has been suggested that it may be coupled to adenylate cyclase, since ANP (at nanomolar concentrations) inhibited cyclic AMP formation in rat platelets (which lack NPR-A and -B) (Anand-Srivastava *et al.*, 1991). While NPR-A and NPR-B display rigorous criteria requiring intact peptides for ligand binding, the NPR-C subtype is much less demanding, tolerating truncated forms and D-amino acid substitutions (Scarborough *et al.*, 1988). NPR-C, which seems to out-number the other types (Leitman *et al.*, 1986; Scarborough *et al.*, 1986), has been shown to internalize ANP (Johnson *et al.*, 1990; Rathinavelu and Isom, 1991) and thus perform a clearance role. In rat kidney two-thirds of the receptors are of the NPR-C subtype in glomeruli, while all of the receptors in the papillae are guanylate cyclase-linked. Unlike ANP, the sequences of BNP from various species reveal considerable differences (see *Figure 20.1*) and this is reflected in marked species specificity for receptors (Kambayashi *et al.*, 1990). But this apart, ANP and BNP seem to be approximately equipotent in various model systems (Brown and Czarnecki, 1990; Iwata *et al.*, 1989) and to bind to similar sites in the kidney (Maeda *et al.*, 1990). An analogue of ANP with five residues (Gln18–Gly22) deleted from within the ring (C-ANP_{4-23}, see *Figure 20.1*) selectively binds with high efficiency to NPR-C on vascular smooth muscle cells and *in vivo* causes an increase in plasma ANP levels (Maack *et al.*, 1987). In attempting to assess the roles of NPR-C and E-24.11 in the metabolism of these peptides, it should be noted that the K_d for receptor binding of ANP has been reported to be in the range 60 pM–2 nM (Napier *et al.*, 1986; Takayanagi *et al.*, 1987), while estimates of the K_m values for hydrolysis of natriuretic peptides by E-24.11 are in the micromolar range (see *Table 20.1*) (Kenny *et al.*, 1993; Koehn *et al.*, 1987). However, the relative capacities of the two clearance systems are also relevant. It seems likely that both NPR-C and E-24.11 have roles in modulating plasma ANP in the physiological range (Chevalier *et al.*, 1991). It has been shown that ANP bound to the NPR-A receptors is readily dissociable, leading to repeated cycles of binding and dissociation, unaccompanied by internalization of the ligand–receptor complex (Abe *et al.*, 1995). This phenomenon (which incidentally makes it impossible to determine K_d for NPR-A at 37°C) would make the dissociated peptide available to attack by E-24.11. In contrast, the binding to NPR-C of both intact and various truncated forms of ANP is not followed by dissociation, leading, in this case, to internalization of the complex and degradation of the ligand intracellularly.

20.3 Clearance of natriuretic peptides *in vivo*

The contribution of E-24.11 and NPR-C to the regulation of natriuretic peptide levels *in vivo* has been addressed by examining the effect of inhibitors of the two clearance pathways on the pharmacokinetics of radiolabelled ANP in the rat. The disappearance of an IV bolus dose of [^{125}I]-ANP from plasma in intact rats is biphasic, with an initial half-life of around 16 sec and a terminal half-life of about 90 sec: the initial phase is thought to reflect distribution of peptide around the body and the later phase its clearance. Pretreatment with an E-24.11 inhibitor increases the terminal half-life of [^{125}I]-ANP but does not effect the volume of distribution of the peptide; the net effect is a modest (approximately two-fold) rise in area-under-curve (AUC) (Barclay *et al.,* 1991; Chiu *et al.,* 1991). The NPR-C ligand, C-ANP$_{4-23}$, reduces the initial half-life of ANP together with the volume of distribution and increases AUC by two- to seven-fold in different studies (Barclay *et al.,* 1992; Chiu *et al.,* 1991). These data have to be interpreted with some caution as C-ANP$_{4-23}$ is also a substrate for E-24.11, and therefore some of its effect on ANP pharmacokinetics may be mediated by competition with ANP for the enzyme rather than NPR-C. Nonetheless the studies indicate that both E-24.11 and NPR-C act to clear ANP *in vivo* and it can be argued that NPR-C makes the larger contribution. Indeed NPR-C may compensate for and so attenuate the effects of E-24.11 inhibition when ANP levels are in the 'physiological range' and the NPR-C pathway is not saturated. Consistent with this notion, combined administration of E-24.11 inhibitor and NPR-C ligand has been shown to produce an increase in AUC (up to 14-fold) that is greater than the sum of either agent given alone (Chiu *et al.,* 1991).

The effect of E-24.11 inhibitors and C-ANP$_{4-23}$ on the pharmacokinetics of ANP is not influenced by nephrectomy, indicating that E-24.11 and NPR-C in the kidney contribute little to the regulation of plasma ANP levels (Barclay *et al.,* 1992; Chiu *et al.,* 1991). Although E-24.11 is present in glomeruli, it is found predominantly in the brush border of the proximal tubule where it acts to degrade ANP filtered at the glomerulus. E-24.11 inhibition is accompanied by a marked increase in immunoreactive ANP in the urine (Burnier *et al.,* 1991) which co-elutes with the biologically active peptide on HPLC analysis (Sybertz *et al.,* 1989). This is further evidence that E-24.11 participates in the metabolism of ANP *in vivo* and although renal E-24.11 may not contribute to ANP clearance, inhibition of the enzyme in the brush border and exposure of the nephron distal to the proximal tubule to an increased load of filtered intact ANP may be important to the renal actions of E-24.11 inhibitors.

E-24.11 inhibitors have been reported to have little effect on the metabolic clearance of [^{125}I]-urodilatin *in vivo* (Abassi *et al.,* 1992) (see Section 20.2.1).

20.4 Cardiovascular and renal effects of E-24.11 inhibitors in animals

20.4.1 Co-administration of E-24.11 inhibitors and exogenous natriuretic peptides

The observation that ANP is a substrate for E-24.11 *in vitro* led rapidly to studies of the effect of E-24.11 inhibition on ANP-induced sodium and water excretion and hypotension in animals. These experiments demonstrated marked dose-dependent potentiation of ANP-induced natriuresis and diuresis. The earliest studies were conducted with thiorphan and phosphoramidon in rodents (Trapani *et al.*, 1989) but have since been reproduced with other more potent and specific E-24.11 inhibitors in a wider range of species including dogs (Margulies *et al.*, 1990) and monkeys (Seymour *et al.*, 1994a). E-24.11 inhibition has a less pronounced effect on the hypotensive response to ANP which is best observed in hypertensive models where it prolongs rather than increases the maximum fall in blood pressure (Seymour *et al.*, 1989). These actions are accompanied by a significant increase in the magnitude and duration of the rise in urinary cyclic GMP excretion following ANP administration.

Similarly the renal and hypotensive actions of BNP can be enhanced in rats and monkeys by concomitant E-24.11 inhibition, providing care is taken to use the BNP appropriate for the test species (Kirk and Wilkins, 1993; Seymour *et al.*, 1992). Thus SQ 28 603 potentiates the natriuresis and prolongs the hypotension produced by rBNP-32 in the spontaneously hypertensive rat (SHR) and to a lesser extent pBNP-26 but has no apparent effect on hBNP-32 in this animal; however, the E-24.11 inhibitor augments effectively the renal and hypotensive actions of hBNP-32 in cynomolgus monkeys (Seymour *et al.*, 1992). There are limited data available on the response to CNP and urodilatin in the presence of E-24.11 inhibitors. Co-administration of candoxatrilat and CNP had no significant effect on blood pressure in normal rats; a small natriuresis was observed but this may be attributable to competition between CNP and other natriuretic peptides for NPR-C leading to increased ANP/BNP activity (J.E. Kirk and M.R. Wilkins, unpublished observations).

20.4.2 Effects of E-24.11 inhibitors in animal models of disease

Given alone to healthy conscious or anaesthetized rats, E-24.11 inhibitors produce a small natriuresis and diuresis but have no effect on blood pressure (Wilkins *et al.*, 1990, 1992). The renal response is enhanced when rats are volume expanded from fluid infusion or partial nephrectomy (Lafferty *et al.*, 1989; Shepperson *et al.*, 1991), stimuli known to increase circulating ANP and BNP levels.

Endogenous circulating ANP and BNP levels are elevated in animal models of heart failure. Several models have been studied in some detail, including the aorto-venocaval fistula rat and the coronary artery ligated rat (which develop high- and low-output cardiac failure respectively), the cardiomyopathic hamster and the rapid atrial/ventricular paced dog. In all models dose-dependent increases in urinary sodium and water excretion (up to four-fold above baseline) have been demonstrated following administration of an E-24.11 inhibitor (Margulies *et al.*, 1990; Seymour *et al.*, 1993, 1996; Trippodo *et al.*, 1991; Wilkins *et al.*, 1990). The renal response is accompanied by an increase in urinary cyclic GMP excretion but often a comparatively small rise (usually less than two-fold) in immunoreactive ANP levels and blood pressure is unaffected. There is some evidence for continued efficacy with chronic dosing. SCH 39370 30 mg kg^{-1} twice daily subcutaneously for 6 days produced similar increases in plasma ANP and cyclic GMP after the first and twelfth injections in the coronary artery ligated rat (Helin *et al.*, 1991). Oral administration of candoxatril 10 mg kg^{-1} twice daily for 6 days preserved the renal response to ANP and attenuated the development of cardiac failure in rapid ventricular paced dogs (Margulies *et al.*, 1995). SQ 28,603 30 mg kg^{-1} twice daily by oral gavage for 30 days has been observed to reduce the cardiac hypertrophy that normally follows introduction of an aorto-venocaval fistula in the rat and improve the renal response of this animal to acute volume expansion, although this was achieved independent of changes in plasma ANP levels (Willenbrock *et al.*, 1996).

Hypotensive responses to E-24.11 inhibition have been observed in rat models of hypertension but not all models are equally responsive. The deoxycorticosterone acetate (DOCA)-salt rat, in which hypertension is volume-dependent and relatively resistant to ACE inhibitors and angiotensin II receptor antagonists, responds particularly well to E-24.11 inhibitors (Pham *et al.*, 1993; Seymour *et al.*, 1991; Sybertz *et al.*, 1990a). For example, 10 and 90 mg kg^{-1} orally produced maximum falls in mean arterial blood pressure of around 30 and 45 mmHg respectively within 1–2 h of dosing and lasting for at least 4 h (Sybertz *et al.*, 1990a). In contrast, similar doses given subcutaneously twice daily had no acute effect on blood pressure in the SHR but produced a modest (~15 mmHg) fall from day 3 to day 5 of a chronic dosing schedule (Sybertz *et al.*, 1990a). E-24.11 inhibition has no effect on blood pressure in the one-clip, two-kidney renovascular rat, a renin-dependent model of hypertension (Pham *et al.*, 1993), or sheep made hypertensive by infusion of angiotensin II in a dose sufficient to double plasma levels of this vasoconstrictor (Charles *et al.*, 1995). Chronic administration of SCH 34 826 100 mg kg^{-1} orally twice daily, while having no long-term effect on blood pressure in the SHR, has been shown to attenuate the development of cardiac hypertrophy and decrease the extent of cardiac fibrosis (Monopoli *et al.*, 1992).

The pulmonary circulation is a primary target for ANP and BNP released from the right ventricle and there has been interest in the potential for enhancing the vasorelaxant effects of these peptides in pulmonary hypertension. A commonly used animal model has been chronic hypoxia-induced pulmonary hypertension, where the elevated pressure is due to a combination of vasoconstriction and vascular remodelling. Pretreatment with candoxatrilat or SCH 34 826 has been shown to reduce pulmonary artery pressure and the attendant right ventricular hypertrophy and vascular structural changes in this model (Klinger *et al.*, 1993; Winter *et al.*, 1991).

20.5 Cardiovascular and renal effects of E-24.11 inhibitors in humans

20.5.1 Studies in healthy volunteers

Acute E-24.11 inhibition produces a small (less than two-fold) increase in urinary sodium and water excretion in the 8–12 h following dosing in healthy volunteers on a normal salt diet (Gros *et al.*, 1989; Richards *et al.*, 1990; Schmitt *et al.*, 1994). The natriuretic effects of E-24.11 inhibitors are increased when subjects are placed on a high salt diet (Sagnella *et al.*, 1995) or by co-infusion of ANP or BNP (Richards *et al.*, 1991a). Plasma ANP levels increase up to two-fold above baseline and urinary cyclic GMP excretion becomes elevated. Urinary phosphate excretion also rises but potassium, magnesium, calcium and uric acid excretion are unaffected (Burnier *et al.*, 1991; Richards *et al.*, 1990). Despite the natriuresis–diuresis, plasma renin activity and aldosterone levels tend to fall. A transient increase in glomerular filtration rate together with a fall in renal plasma flow has been reported in some studies (Schmitt *et al.*, 1994). Blood pressure is usually unaffected although single 200 mg doses of candoxatril have been observed to reduce central venous pressure and increase systolic blood pressure with no apparent increase in sympathetic drive (Ando *et al.*, 1995). Continued dosing with candoxatril for 4–10 days produces a persistent elevation of urinary cyclic GMP excretion without affecting plasma ANP levels. Sodium excretion is increased on the first day only; data on whether a negative sodium balance is sustained are conflicting (O'Connell *et al.*, 1993; Richards *et al.*, 1991b).

20.5.2 Studies in patients with cardiac failure

Plasma ANP and BNP levels are elevated in patients with cardiac failure in proportion to the severity of the disease. These patients also exhibit resistance to the renal effects of ANP infusion, mediated, at least in part, by increased activity of the renin–angiotensin and sympathetic nervous systems and perhaps down-regulation of ANP receptors. There are no data on the effect of cardiac failure on E-24.11 activity in humans but several studies have now examined the effect of E-24.11 inhibitors in this patient group.

In an early study, 10–400 mg of UK69578 (a racemic mixture of candoxatrilat) was given by intravenous infusion to six patients with mild [New York Heart Association (NYHA) grade 2] heart failure (Northridge *et al.*, 1989). A three-fold increase in plasma ANP and urinary sodium excretion was recorded in the first 4 h period following dosing, together with significant falls in right atrial and pulmonary capillary wedge pressures. Plasma renin activity and systemic blood pressure were not affected by E-24.11 inhibition.

Further studies in patients with heart failure of similar or greater severity have confirmed these rather modest renal and haemodynamic effects. Candoxatrilat (150 mg) given intravenously 8-hourly over a 24 h period to patients with heart failure (NYHA grades 3 and 4) increased plasma ANP level 2.5-fold and urinary sodium excretion six-fold 2 h after the first dose and both remained elevated above baseline for the 24 h study period (Munzel *et al.*, 1992). There was a close correlation between the natriuretic response and basal cardiac output, such that those patients with the poorest output responded least well to the inhibitor. Right atrial and pulmonary capillary wedge pressures were reduced for the duration of the study while systemic blood pressure was not altered. Candoxatril 200–400 mg by mouth increased plasma ANP levels about 2.5-fold but had no effect on left ventricular hydraulic load when haemodynamic measurements were made 90 min after dosing with heart failure (NYHA grades 2 and 3) (Kromer *et al.*, 1991). In a similar group of patients, candoxatril 150 mg orally twice daily for 10 days produced a sustained elevation of plasma ANP and reduction in plasma aldosterone levels and reduced body weight (Elsner *et al.*, 1992). Pulmonary capillary wedge pressure was lower, and cardiac index increased, on active compared with placebo treatment but the changes did not reach statistical significance.

Single oral doses of sinorphan (30–60 mg) have been reported to reduce pulmonary capillary wedge pressure but data on the response to chronic dosing are not available.

20.5.3 Studies in patients with hypertension

ANP and BNP levels are elevated in some patients with essential hypertension, particularly those with target-organ damage such as left ventricular hypertrophy and renal impairment. There have been several reports that E-24.11 inhibition reduces blood pressure in patients with essential hypertension but this action is not common to all inhibitors and care must be taken in interpreting open-label studies (Lefrancois *et al.*, 1990; Ogihara *et al.*, 1994). In a double-blind cross-over study, SCH 34 826 400 mg 6-hourly for five doses had no acute antihypertensive effect, despite increasing plasma ANP levels and urinary sodium excretion (Richards *et al.*, 1992a). However, in a separate study in 24 black patients with mild essential hypertension, treatment for 14 days reduced mean supine systolic and diastolic blood pressure by 16 and 12 mmHg, compared with 3 and 2.6 mmHg in the placebo group (Kosoglou *et al.*, 1990). Similarly SCH 42 495 200 mg twice daily for 4 days had no effect on blood pressure in the first 12 h but reduced mean 24 h systolic and diastolic blood pressure by around 9 and 3 mmHg, respectively, once E-24.11 inhibition was established (Richards *et al.*, 1993). Although single oral doses of candoxatril 200 mg daily increased plasma ANP levels two-fold in patients with essential hypertension and stimulated a natriuresis (Singer *et al.*, 1991), it did not lower blood pressure when given twice daily for 28 days (Bevan *et al.*, 1992).

20.5.4 Studies in patients with other disorders

Although patients with hepatic cirrhosis and ascites retain sodium and water, blood volume may be normal or reduced and plasma ANP levels are not necessarily elevated. Their renal response to infusion of ANP is blunted compared to healthy individuals, largely because of the associated fall in blood pressure. Nonetheless, sinorphan 30–100 mg has been observed to produce a transient two-fold increase in urinary sodium excretion in this patient group in the first 2 h following oral dosing which was not seen with placebo treatment (Dussaule *et al.*, 1991). Aldosterone levels decreased during active treatment. The same investigators have also reported that sinorphan 100 mg increased sodium excretion 1.9-fold in the second hour after ingestion in patients with chronic renal failure (Dussaule *et al.*, 1993).

The natriuretic peptides are synthesised locally within the eye, which also exhibits receptors for these peptides (Fernandez-Durango *et al.*, 1995). This suggests a role for ANP in the regulation of intraocular pressure. In a small study in subjects with normal pressures, candoxatril 200 mg twice daily for 4 weeks lowered intraocular pressure by 11% and 16% in the right and left eye, respectively (Wolfensberger *et al.*, 1994).

20.6 Mechanism of the cardiovascular and renal response to E-24.11 inhibition *in vivo*

Since a variety of endogenous peptides are substrates for E-24.11, the notion that the cardiovascular and renal effects of E-24.11 inhibition are due to inhibition of the degradation of the natriuretic peptides should be examined with some care. Several observations support this claim: (i) E-24.11 inhibitors increase plasma ANP and BNP levels and potentiate the biological effects of infusions of exogenous ANP and BNP; (ii) the renal and hypotensive responses to E-24.11 inhibitors are more pronounced when endogenous ANP and BNP levels are elevated; (iii) the effects of E-24.11 inhibitors are accompanied by an increase in plasma and urinary cyclic GMP levels; and (iv) the cardiovascular and renal response to E-24.11 inhibitors can be blocked or attenuated significantly by antibodies raised against ANP (Bralet *et al.,* 1990; Shepperson *et al.,* 1991) and by an ANP receptor antagonist (Hirata *et al.,* 1994).

It is clear, however, that the response to E-24.11 inhibitors cannot be explained solely by the effect of these agents on circulating natriuretic peptide levels. In the aorto-venocaval fistula rat, thiorphan produced twice the rise in urinary sodium excretion produced by pharmacological doses of exogenous ANP, but with only a fraction (<20%) of the increase in plasma ANP concentration (Wilkins *et al.,* 1990). In the same model, chronic E-24.11 inhibition has been reported to attenuate the increase in cardiac mass without altering plasma ANP levels (Willenbrock *et al.,* 1996). A similar dissociation between the change in natriuretic peptide levels and fall in blood pressure has been noted in the DOCA-salt rat. One explanation is that E-24.11 inhibition enhances the local effects of natriuretic peptides. Inhibition of the enzyme expressed by endothelial cells would increase the local concentration of natriuretic peptides to which vascular smooth muscle was exposed. In the kidney, inhibition of the enzyme in the renal brush border allows an increased filtered load of intact ANP to reach more distal segments of the nephron and may unmask or augment actions of these peptides on the renal tubule.

This does not exclude the possibility that factors other than the natriuretic peptides contribute to or, equally important, modify the response to E-24.11 inhibitors. Indeed differences in the haemodynamic and hormonal response to ANP and candoxatril have been noted in studies in healthy subjects (Ando *et al.,* 1995); specifically, although both reduce central venous pressure, candoxatril increased systolic pressure while ANP had no effect on this parameter. Candidate peptides for this modifying role are the kinins, angiotensins and endothelins. There is no evidence that bradykinin influences the hypotensive or vascular actions of selective E-24.11 inhibitors (Seymour *et al.,* 1994b; Sybertz *et al.,* 1989, 1990b). Furthermore, retrothiorphan has been reported to induce a natriuresis in kininogen-deficient Brown-Norway rats with and without DOCA-salt hypertension (Pham *et al.,* 1992). However, SQ 28,603 potentiated the peak natriuretic response to intrarenal bradykinin in anaesthetized dogs (Seymour *et al.,* 1994b). Conversely, bradykinin receptor blockade inhibited the increase in renal papillary blood flow produced by phosphoramidon (Roman *et al.,* 1988), abolished the natriuretic response to thiorphan and low-dose ANP in anaesthetized normotensive rats (Smits *et al.,* 1990) and the renal effects of candoxatrilat in Sprague-Dawley (but not DOCA-salt) rats (Nomura *et al.,* 1995). Thus kinins, perhaps released locally in the kidney rather than circulating, may play a permissive role in the effect of E-24.11 inhibitors on natriuretic peptide-induced natriuresis.

Treatment with candoxatril 25 mg twice daily for 5 days has been reported to increase basal angiotensin II levels and the pressor response to a short infusion of the peptide in humans (Richards *et al.,* 1992b). SCH 39 370 also increases angiotensin levels in sheep infused with the peptide (Charles *et al.,* 1995). Single 200 mg doses of candoxatril have been shown to increase plasma endothelin and adrenaline levels in healthy volunteers (Ando *et al.,* 1995). These actions would be expected to offset in part the hypotensive effect of E-24.11 inhibitors *in vivo.* Infusion of thiorphan and candoxatrilat has been reported to cause vasoconstriction in human forearm vessels (Ferro *et al.*, 1996); however, no change in plasma angiotensin II levels was observed and the effect was attributed to inhibition of endothelin degradation (although this was not established in this study). In contrast, infusion of phosphoramidon has been shown to dilate human forearm vessels (Haynes and Webb, 1994), an action assigned to inhibition of the conversion of big endothelin to its more active (21-amino-acid) species. This property is not shared by thiorphan or candoxatrilat. Thus E-24.11 inhibitors vary in their effect on vasomotor tone depending on their ability to co-inhibit other enzymes.

20.7 Cardiovascular effects of dual E-24.11/ACE inhibitors

From the foregoing discussion it can be appreciated that one approach to enhancing the efficacy of E-24.11 inhibitors is to co-inhibit systems that oppose their action, such as the renin–angiotensin–aldosterone system and the endothelins. To date most work has been done with combined E-24.11 and ACE inhibition. This has been given impetus by the development of compounds which inhibit both enzymes. Several of these dual inhibitors have now been studied in animal models of hypertension and heart failure.

A consistent finding is that dual E-24.11/ACE inhibitors exhibit antihypertensive effects over a wider range of animal models of hypertension than inhibitors of either enzyme alone. Whereas E-24.11 inhibitors are most effective in low-renin hypertension (e.g. DOCA-salt rats), a group resistant to the actions of ACE inhibitors, and least effective in the SHR and renovascular hypertensive animals, both responsive to ACE inhibitors, inhibition of both enzymes lowers blood pressure in all three models (French *et al.,* 1995; Gonzales Vera *et al.,* 1995; Pham *et al.,* 1993).

Similarly, combined E-24.11 and ACE inhibition reduces left ventricular end-diastolic volume, peripheral vascular resistance, blood pressure and cardiac hypertrophy and increases cardiac output in a variety of models of heart failure (Bralet *et al.,* 1994; Marie *et al.,* 1995; Seymour *et al.,* 1993; Trippodo *et al.,* 1993, 1995b). As in hypertensive animals, the haemodynamic effects of combined inhibition were often greater than those produced by an E-24.11 or ACE inhibitor alone (Trippodo *et al.,* 1995b). In contrast, ACE inhibition appears to reduce the acute natriuretic effect of E-24.11 inhibition, most likely because of the greater fall in systemic blood pressure and consequent reduction in renal perfusion pressure (Kirk and Wilkins, 1996). There is evidence, at least in heart failure, that dual inhibitors produce less stimulation of renin secretion (Bralet *et al.,* 1994) and that bradykinin makes a greater contribution to the haemodynamic effects of these agents than it does to E-24.11 or ACE inhibitors alone (Trippodo *et al.,* 1995a).

Data in humans are few. Co-treatment with an ACE inhibitor has been reported to reduce the early natriuretic response to candoxatrilat in healthy volunteers (Motwani *et al.,* 1995); a similar interaction has been reported with ACE inhibitors and ANP infusions and probably relates to the accompanying fall in systemic blood pressure. In a study in hypertensive patients, the combination of captopril and sinorphan for 4 weeks had a significantly greater effect on 24 h blood pressure reduction than either agent alone (Favrat *et al.*, 1995).

20.8 Overview and future prospects

Cardiovascular disease is a major cause of death in Western society and naturally an important focus of activity of the pharmaceutical industry. The efficacy and safety of diuretics, β-blockers, calcium antagonists and ACE inhibitors in the treatment of hypertension and (particularly ACE inhibitors) cardiac failure has made this a highly competitive market. Not all patients are able to tolerate these drugs, however, and current therapeutic regimens still fall short of the desired impact on disease progression and cardiovascular mortality.

E-24.11 inhibitors have a number of attractive properties. In contrast to thiazide and loop-diuretics, E-24.11 inhibitors are relatively potassium-sparing and do not affect adversely uric acid excretion. Given the dependence of these inhibitors on circulating natriuretic peptides for their activity, the risks of dehydration should be negligible; ANP and BNP levels would be expected to fall with successful diuresis, checking the action of the drugs as the need for them diminishes. Another property of E-24.11 inhibitors that distinguishes them from currently available diuretics is that the former antagonize rather than stimulate the renin–aldosterone system.

It must be accepted that as monotherapy, E-24.11 inhibitors are not as potent as loop diuretics in inducing natriuresis and appear to be relatively ineffective for the treatment of hypertension. Furthermore, although these drugs have been shown to have beneficial haemodynamic effects in patients with heart failure, it is unlikely that E-24.11 inhibitors alone can compete with ACE inhibitors which have established a secure role in the treatment of this condition. The most likely future clinical application of E-24.11 inhibition in the treatment of cardiovascular disease will be in conjunction with inhibition of other enzymes, notably through the development of dual E-24.11/ACE and E-24.11/endothelin-converting enzyme (ECE) inhibitors. As discussed above, the cardiovascular effects of ACE and ECE inhibition are not only complementary to those of E-24.11 inhibition, but reducing the synthesis of angiotensin II and the endothelins also attenuates the 'adverse' effect of E-24.11 inhibitors on the clearance of these factors. One caveat is that the haemodynamic actions of ACE and ECE inhibition would also be expected to reduce the stimulus to natriuretic peptide secretion and so attenuate to some extent the benefit of dual inhibitors. Thus careful consideration may have to be given to the balance of E-24.11 and ACE or ECE inhibition in these dual inhibitors. Nonetheless, studies to date in animal models have been encouraging. Studies of this new class of pharmaceutical agents in humans are underway and the results are awaited with interest.

References

Abassi ZA, Tate J, Hunsberger S, Klein H, Trachewsky D, Keiser HR. (1992) Pharmacokinetics of ANF and urodilatin during cANF receptor blockade and neutral endopeptidase inhibition. *Am. J. Physiol.* **263:** E870–E876.

Abassi ZA, Golomb E, Agbaria R, Roller PP, Tate JE, Keiser HR. (1994) Hydrolysis of iodine labelled urodilatin and ANP by recombinant neutral endopeptidase EC 3.4.24.11. *Br. J. Pharmacol.* **113:** 204–208.

Abe T, Arakawa Y, Rajasekaran AK, Yu TK, Wada O. (1995) Interaction of atrial natriuretic peptide with its receptors in bovine lung membranes. *J. Biol. Chem.* **270:** 7672–7678.

Anand-Srivastava MB, Gutkowska J, Cantin M. (1991) The presence of atrial natriuretic factor receptors of ANF-R2 subtype in rat platelets. Coupling to adenylate cyclase/cyclic AMP signal-transduction. *Biochem. J.* **278:** 211–217.

Ando S, Rahman MA, Butler GC, Senn BL, Floras JS. (1995) Comparison of candoxatril and atrial natriuretic factor in healthy men. Effects on hemodynamics, sympathetic activity, heart rate variability and endothelin. *Hypertension* **26:** 1160–1166.

Barclay PL, Bennett JA, Samuels GMR, Shepperson NB. (1991) The atriopeptidase inhibitor (±) candoxatrilat reduces the clearance of atrial natriuretic factor in both intact and nephrectomized rats: evidence for an extrarenal site of action. *Biochem. Pharmacol.* **41:** 841–844.

Barclay PL, Bennett JA, Greengrass PM, Griffin A, Samuel GMR, Shepperson NB. (1992) The pharmacokinetics of ^{125}I-atrial natriuretic factor in anaesthetized rats. Effects of neutral endopeptidase inhibition with candoxatrilat and of ANF-C receptor blockade. *Biochem. Pharmacol.* **44:** 1013–1022.

Bevan EG, Connell JMC, Doyle J, Carmichael HA, Davies DL, Lorimer AL, McInnes GT. (1992) Candoxatril, a neutral endopeptidase inhibitor: efficacy and tolerability in essential hypertension. *J. Hypertens.* **10:** 607–613.

Bianchi C, Gutowska J, Thibault G, Garcia R, Genest J, Cantin M. (1985) Radioautographic localisation of ^{125}I-atrial natriuretic factor (ANF) in rat tissues. *Histochemistry* **82:** 441–452.

Bianchi C, Anand-Srivastava MB, Lean AD, Gutowska J. (1986a) Localization and characterisation of specific receptors for atrial natriuretic factor in the ciliary processes of the eye. *Curr. Eye Res.* **5:** 283–293.

Bianchi C, Gutowska J, Ballak M, Thibault G, Garcia R, Genest J, Cantin M. (1986b) Radioautographic localisation of ^{125}I-atrial natriuretic factor binding sites in the brain. *Neuroendocrinology* **44:** 365–372.

Bourne A, Kenny AJ. (1990) The hydrolysis of brain and atrial natriuretic peptides by porcine choroid plexus is attributable to endopeptidase-24.11. *Biochem. J.* **271:** 381–385.

Bralet J, Mossiat C, Lecomte J, Charpentier S, Gros C, Schwartz J. (1990) Diuretic and natriuretic responses in rats treated with enkephalinase inhibitors. *Eur. J. Pharmacol.* **179**: 57–64.

Bralet J, Marie C, Mossiat C, Lecomte J, Gros C, Schwartz J. (1994) Effects of alatriopril, a mixed inhibitor of atriopeptidase and angiotensin I-converting enzyme, on cardiac hypertrophy and hormonal responses in rats with myocardial infarction. Comparison with captopril. *J. Pharmacol. Exp. Ther.* **270**: 8–14.

Brown J, Czarnecki A. (1990) Binding of atrial and brain natriuretic peptides in brains of hypertensive rats. *Brain Res.* **512:** 132–137.

Burnier M, Ganslmayer M, Perret F, Porchet M, Kosoglou T, Gould A, Nussberger J, Waeber B, Brunner HR. (1991) Effects of SCH 34826, an orally active inhibitor of atrial natriuretic peptide degradation, in healthy volunteers. *Clin. Pharmacol. Ther.* **50**: 181–191.

Charles CJ, Espiner EA, Richards AM, Sybertz EJ. (1995) Endopeptidase inhibition in angiotensin-induced hypertension. Effect of SCH 39370 in sheep. *Hypertension* **26**: 89–94.

Chevalier RL, Garmey M, Scarborough RM, Linden J, Gomez RA, Peach MJ, Carey RM. (1991) Inhibition of ANP clearance receptors and endopeptidase 24.11 in maturing rats. *Am. J. Physiol.* **260:** R1218–R1228.

Chinkers M, Garbers DL, Chang M-S, Lowe DG, Chin H, Goeddel DV, Schulz S. (1989) A membrane form of guanylate cyclase is an atrial natriuretic peptide receptor. *Nature* **338:** 78–83.

Chiu PJS, Tetzloff G, Romano MT, Foster CJ, Sybertz EJ. (1991) Influence of C-ANF receptor and neutral endopeptidase on pharmacokinetics of ANF in rats. *Am. J. Physiol.* **260:** R208–R216.

De Bold AJ, Borenstein HB, Veress AT, Sonnenberg H. (1981) A rapid and potent natriuretic response to intravenous injection of atrial myocardial extracts in rats. *Life Sci* **28:** 859–865.

Dussaule J, Grange J, Wolf J, Lecomte J, Gros C, Schwartz J, Bodin F, Ardaillou R. (1991) Effect of sinorphan, an enkephalinase inhibitor, on plasma atrial natriuretic factor and sodium urinary excretion in cirrhotic patients with ascites. *J. Clin. Endocrinol. Metab.* **72**, 653–659.

Dussaule JC, Michel C, Peraldi MN, Lecomte JM, Gros C, Mignon F, Ardaillou R. (1993) Inhibition of neutral endopeptidase stimulates renal sodium excretion in patients with chronic renal failure. *Clin. Sci.* **84**, 31–39.

Elsner D, Muntze A, Kromer EP, Riegger GAJ. (1992) Effectiveness of endopeptidase inhibition (candoxatril) in congestive heart failure. *Am. J. Cardiol.* **70:** 494–498.

Favrat B, Burnier M, Nussberger J, Lecomte JM, Brouard R, Waeber B, Brunner HR. (1995) Neutral endopeptidase versus angiotensin converting enzyme inhibition in essential hypertension. *J. Hypertens.* **13:** 797–804.

Fernandez-Durango R, Nunez DJR, Brown MJ. (1995) Messenger RNAs encoding the natriuretic peptides and their receptors are expressed in the eye. *Exp. Eye Res.* **61:** 723–729.

Ferro CJ, Haynes WG, Webb DJ. (1996) Neutral endopeptidase inhibitors cause vasoconstriction in man. *J. Hypertens.* **14** (Suppl 1): S8.

French JF, Anderson BAA, Downs TR, Dage RC. (1995) Dual inhibition of angiotensin-converting enzyme and neutral endopeptidase in rats with hypertension. *J. Cardiovasc. Pharmacol.* **26:** 107–113.

Fuller F, Porter JG, Arfsten AE, Miller J, Schilling JW, Scarborough RM, Lewicki JA, Schenk DB. (1988) Atrial natriuretic peptide clearance receptor. Complete sequence and functional expression of cDNA clones. *J. Biol. Chem.* **263:** 9395–9401.

Furuya M, Yoshida M, Hayashi Y, Ohnuma N, Minamino N, Kangawa K, Matsuo H. (1991) C-type natriuretic peptide is a growth inhibitor of rat vascular smooth muscle cells. *Biochem. Biophys. Res. Commun.* **177:** 927–931.

Gagelmann M, Hock D, Forssmann WG. (1988) Urodilatin (CDD/ANP-95–126) is not biologically inactivated by a peptidase from dog kidney cortex membranes in contrast to atrial natriuretic peptide/cardiodilatin (α-hANP/CDD-99–126). *FEBS Lett.* **233:** 249–254.

Gonzales Vera W, Fournie-Zaluski M, Pham I, Laboulandine I, Roques BP, Michel J. (1995) Hypotensive and natriuretic effects of RB 105, a new dual inhibitor of angiotensin converting enzyme and neutral endopeptidase in hypertensive rats. *J. Pharmacol. Exp. Ther.* **272:** 343–351.

Gros C, Souque A, Schwartz J, Duchier J, Cournot A, Baumer P, Lecomte J. (1989) Protection of atrial natriuretic factor against degradation: diuretic and natriuretic responses after *in vivo* inhibition of enkephalinase (EC 3.4.24.11) by acetorphan. *Proc. Natl Acad. Sci. USA* **86:** 7580–7584.

Haynes WG, Webb DJ. (1994) Contribution of endogenous generation of endothelin-1 to basal vascular tone. *Lancet* **344:** 832–833.

Helin K, Tikkanen I, Tikkanen T, Saijonmaa O, Sybertz EJ, Vemulapalli S, Sariola H, Fyhrquist F. (1991) Prolonged neutral endopeptidase inhibition in heart failure. *Eur. J. Pharmacol.* **198:** 23–30.

Hirata Y, Suzuki Y, Suzuki E, *et al.* (1994) Mechanisms underlying the augmented responses of deoxycorticosterone acetate-salt hypertensive rats to neutral endopeptidase inhibtors. *J. Hypertens.* **12:** 367–374.

Hosoda K, Nakao K, Mukoyama M, *et al.* (1991) Expression of brain natriuretic peptide gene in human heart. Production in the ventricle. *Hypertension* **17:** 1152–1155.

Itoh H, Pratt RE, Dzau VJ. (1990) Atrial natriuretic polypeptide inhibits hypertrophy of vascular smooth muscle. *J. Clin. Invest.* **867:** 1690–1697.

Iwata T, Inui K-i, Nakao K, Imura H, Matsuo H, Hori R. (1989) Effect of brain natriuretic peptide on cyclic GMP accumulation in a kidney epithelial cell line (LLC-PK1). *Eur. J. Pharmacol.* **159:** 321–323.

Johnson GR, Arik L, Pitts BJR, Forster CJ. (1990) Rapid receptor-mediated catabolism of ^{125}I-atrial natriuretic factor by vascular cells. *Biochem. J.* **268:** 771–776.

Kambayashi Y, Nakao K, Kimura H, Kawabata T, Nakamura M, Inouye K, Yoshida N, Imura H. (1990) Biological characterization of human brain natriuretic peptide (BNP) and rat BNP; species specific actions. *Biochem. Biophys. Res. Commun.* **173:** 599–605.

Kangawa K, Matsuo H. (1984) Purification and complete amino acid sequence of α-human atrial natriuretic polypeptide (α-hANP). *Biochem. Biophys. Res. Commun.* **118:** 131–139.

Kenny AJ, Stephenson SL. (1988) Role of endopeptidase-24.11 in the inactivation of atrial natriuretic peptide. *FEBS Letts.* **232:** 1–8.

Kenny AJ, Bourne A, Ingram J. (1993) Hydrolysis of natriuretic peptides and some C-receptor ligands by endopeptidase-24.11. *Biochem. J.* **191:** 83–88.

Kirk JE, Wilkins MR. (1993) Effect of endopeptidase-24.11 inhibition and of atrial natriuretic peptide clearance receptor ligand on the response to rat brain natriuretic peptide in the conscious rat. *Br. J. Pharmacol.* **110:** 350–354.

Kirk JE, Wilkins MR. (1996) Renal effects of concurrent E-24.11 and ACE inhibition in the aortovenocaval fistula rat. *Br. J. Pharmacol.* **119:** 943–948.

Klinger JR, Petit RD, Warburton RR, Wrenn DS, Arnal F, Hill NS. (1993) Neutral endopeptidase inhibition attenuates development of hypoxic pulmonary hypertension in rats. *J. Appl. Physiol.* **75:** 1615–1623.

Koehn JA, Norman JA, Jones BN, LeSueur L, Sakane Y, Ghai RD. (1987) Degradation of atrial natriuretic factor by kidney cortex membranes. Isolation and characterization of the primary proteolytic product. *J. Biol. Chem.* **262:** 11623–11627.

Koller KJ, Lowe DG, Bennett GL, Minamino N, Kangawa K, Matsuo H, Goeddel DV. (1991) Selective activation of the B natriuretic peptide receptor by C-type natriuretic peptide (CNP). *Science* **252:** 120–123.

Komatsu Y, Nakao K, Suga S, *et al.* (1991) C-type natriuretic peptide (CNP) in rats and humans. *Endocrinology* **129:** 1104–1106.

Kosoglou T, Herron JM, Chen R, Given BD, Sybertz EJ, Affrime MB. (1990) Antihypertensive effects of the atriopeptidase inhibitor SCH 34826 in essential hypertension. *Circulation* **82** (Suppl.3): 554 (abstract 2201).

Kromer EP, Elsner D, Kahles HW, Riegger GAJ. (1991) Effects of atriopeptidase inhibitor UK 79300 on left ventricular hydraulic load in patients with congestive cardiac heart failure. *Am. J. Hypertens.* **4:** 460–463.

Kuno T, Andresen JW, Kamisaki Y, Waldman SA, Chang LY, Saheki S, Leitman DC, Naane M, Murad F. (1986) Copurification of an atrial natriuretic factor receptor and particulate guanylate cyclase from rat lung. *J. Biol. Chem.* **261:** 5817–5823.

Lafferty HM, Gunning M, Silva P, Zimmerman MB, Brenner BM, Anderson S. (1989) Enkephalinase inhibition increases plasma atrial natriuretic peptide levels, glomerular filtration rate, and urinary sodium excretion in rats with reduced renal mass. *Circ. Res.* **65:** 640–646.

Lefrancois P, Clerc G, Duchier J, Lim C, Lecomte JM, Gros C, Schwartz J. (1990) Antihypertensive activity of sinorphan. *Lancet* **336:** 307–308.

Leitman DC, Andresen JW, Kuno T, Kamisaki Y, Chang JK, Murad F. (1986) Identification of multiple binding sites for atrial natriuretic factor by affinity cross-linking in cultured endothelial cells. *J. Biol. Chem.* **261:** 11650–11655.

Lynch DR, Braas KM, Snyder SH. (1986) Atrial natriuretic factor receptors in rat kidney, adrenal gland and brain: autoradiographic localisation and fluid balance dependent changes. *Proc. Natl Acad. Sci. USA* **83:** 3357–3361.

Maack T, Suzuki M, Almeida FA, Nussenzveig D, Scarborough RM, McEnroe GA, Lewicki JA. (1987) Physiological role of silent receptors of atrial natriuretic factor. *Science* **238:** 675–678.

Maeda T, Niwa M, Shigematsu K, *et al.* (1990) Specific [^{125}I]brain natriuretic peptide-26 binding sites in rat and pig kidneys. *Eur. J. Pharmacol.* **176:** 341–350.

Margulies KB, Cavero PG, Seymour AA, Delaney NG, Burnett JC. (1990) Neutral endopeptidase inhibition potentiates the renal actions of atrial natriuretic factor. *Kidney Int.* **38:** 67–72.

Margulies KB, Barclay PL, Burnett JC. (1995) The role of neutral endopeptidase in dogs with evolving congestive heart failure. *Circulation* **91:** 2036–2042.

Marie C, Mossiat C, Lecomte J, Schwartz J, Bralet J. (1995) Hemodynamic effects of acute and chronic treatment with aladotril, a mixed inhibitor of neutral endopeptidase and angiotensin 1-converting enzyme, in conscious rats with myocardial infarction. *J. Pharmacol. Exp. Ther.* **275:** 1324–1331.

Minamino N, Kangawa K, Matsuo H. (1990) N-terminally extended form of C-type natriuretic peptide (CNP-53) identified in porcine brain. *Biochem. Biophys. Res. Commun.* **170:** 973–979.

Misono KS, Fukumi H, Grammer RT, Inagami T. (1984) Rat atrial natriuretic factor: complete amino acid sequence and disulfide linkage essential for biological activity. *Biochem. Biophys. Res. Commun.* **119:** 524–529.

Monopoli A, Ongini E, Cigola E, Olivetti G. (1992) The neutral endopeptidase inhibitor, SCH 34826, reduces left ventricular hypertrophy in spontaneously hypertensive rats. *J. Cardiovasc. Pharmacol.* **20:** 496–504.

Motwani JG, Lang CC, Cramb G, Struthers AD. (1995) Natriuretic response to neutral endopeptidase inhibition is blunted by enalapril in healthy men. *Hypertension* **25:** 637–642.

Mukoyama M, Nakao K, Saito Y, Ogawa Y, Hosoda K, Suga S, Shirakami G, Jougasaki M, Imura H. (1990) Increased brain natriuretic peptide in congestive heart failure. *New Engl. J. Med.* **323:** 757–758.

Mukoyama M, Nakao K, Hosoda K, *et al.* (1991) Brain natriuretic peptide as a novel cardiac hormone in humans. Evidence for an exquisite dual natriuretic peptide system, atrial natriuretic peptide and brain natriuretic peptide. *J. Clin. Invest.* **87:** 1402–1412.

Munzel T, Kurz S, Holtz J, Busse R, Steinhauer H, Just H, Drexler H. (1992) Neurohormonal inhibition and hemodynamic unloading during prolonged inhibition of ANF degradation in patients with severe chronic heart failure. *Circulation* **86:** 1089–1098.

Napier MA, Arcuri KE, Vandlen RL. (1986) Binding and internalization of atrial natriuretic factor by high-affinity receptors in A10 smooth muscle cells. *Arch. Biochem. Biophys.* **248:** 516–522.

Nazario B, Hu RM, Pedram A, Prins B, Levin ER. (1995) Atrial and brain natriuretic peptides stimulate the production and secretion of C-type natriuretic peptide from bovine aortic endothelial cells. *J. Clin. Invest.* **95:** 1151–1157.

Nomura N, Shimamoto K, Ura N, Iwata M, Aoyama T, Takagawa Y, Iimura O. (1995) The role of kinins and atrial natriuretic peptide on the renal effects of neutral endopeptidase inhibitor in normotensive and hypertensive rats. *Clin. Exp. Hypertens.* **17:** 1219–1231.

Northridge DB, Alabaster CT, Connell JMC, *et al.* (1989) Effects of UK 69 578: a novel atriopeptidase inhibitor. *Lancet* **ii:** 591–593.

O'Connell JE, Jardine AG, Davies DL, McQueen J, Connell JM. (1993) Renal and hormonal effects of chronic inhibition of neutral endopeptidase (EC 3.4.24.11) in normal man. *Clin. Sci.* **85:** 19–26.

Ogawa Y, Nakao K, Mukoyama M, *et al.* (1991) Natriuretic peptides as cardiac hormones in normotensive and spontaneously hypertensive rats. The ventricle is a major site of synthesis and secretion of brain natriuretic peptide. *Circ. Res.* **69:** 491–500.

Ogihara T, Rakugi H, Masuo K, Yu H, Nagano M, Mikami H. (1994) Antihypertensive effects of the neutral endopeptidase inhibitor SCH 42495 in essential hypertension. *Am. J. Hypertens.* **7:** 943–947.

Pham I, el Amrani AI, Fournie-Zaluski MC, Corvol P, Roques BP, Michel JB. (1992) Effects of the selective neutral endopeptidase inhibitor, retrothiorphan, on renal function and blood pressure in conscious normtensive Wistar and hypertensive DOCA-salt rats. *J. Cardiovasc. Pharmacol.* **20:** 847–857.

Pham I, Gonzalez W, Amrani A-IK, Fournie-Zaluski MC, Philippe M, Laboulandine I, Roques BP, Michel J-B. (1993) Effects of converting enzyme inhibitor and neutral endopeptidase inhibitor on blood pressure and renal function in experimental hypertension. *J. Pharmacol. Exp. Ther.* **265:** 1339–1347.

Quirion R, Dalpe M, Dam TV. (1986) Characterization and distribution of receptors for the atrial natriuretic peptides in mammalian brain. *Proc. Natl Acad. Sci. USA* **83:** 174–178.

Rathinavelu A, Isom GE. (1991) Differential internalization and processing of atrial-natriuretic factor B and C receptors in PC12 cells. *Biochem. J.* **276:** 493–497.

Richards M, Espiner E, Frampton C, Ikram H, Yandle T, Sopwith M, Cussans N. (1990) Inhibition of endopeptidase EC 24.11 in humans. *Hypertension* **16:** 269–276.

Richards AM, Wittert G, Espiner EA, Yandle TG, Frampton C, Ikram H. (1991a) EC 24.11 inhibition in man alters clearance of atrial natriuretic peptide. *J. Clin. Endocrinol. Metab.* **72:** 1317–1322.

Richards AM, Wittert G, Espiner EA, Yandle TG, Frampton C, Ikram H. (1991b) Prolonged inhibition of endopeptidase 24.11 in normal man: renal, endocrine and haemodynamic effects. *J. Hypertens.* **9:** 955–962.

Richards AM, Crozier IG, Espiner EA, Ikram H, Yandle TG, Kosoglou T, Rallings M, Frampton C. (1992a) Acute inhibition of endopeptidase 24.11 in essential hypertension: SCH34826 enhances atrial natriuretic peptide and natriuresis without lowering blood pressure. *J. Cardiovasc. Pharmacol.* **20:** 735–741.

Richards AM, Wittert GA, Espiner EA, Yandle TG, Ikram H, Frampton C. (1992b) Effect of inhibition of endopeptidase 24.11 on responses to angiotensin II in human volunteers. *Circ. Res.* **71:** 1501–1507.

Richards AM, Crozier IG, Kosoglou T, Rallings M, Espiner EA, Nicholls MG, Yandle TG, Ikram H, Frampton C. (1993) Endopeptidase 24.11 inhibition by SCH 42495 in essential hypertension. *Hypertension* **22:** 119–126.

Roman RJ, Kaldunski ML, Scicli AG, Carretero OA. (1988) Influence of kinins and angiotensin II on the regulation of papillary blood flow. *Am. J. Physiol.* **255:** F690–F698.

Ruskoaho R. (1992) Atrial natriuretic peptide: synthesis, release and metabolism. *Pharmacol. Rev.* **44:** 479–602.

Sagnella GA, Markandu ND, Buckley MG, Miller MA, Blackwood A, Singer DR, MacGregor GA. (1995) Hormonal and renal responses to neutral endopeptidase inhibition in normal humans on a low and on a high sodium intake. *Eur. J. Clin. Invest.* **25:** 165–170.

Scarborough RM, Schenk DB, McEnroe GA, Arfsten A, Kang L-L, Schwartz K, Lewicki JA. (1986) Truncated atrial natriuretic peptide analogs. Comparison between receptor binding and stimulation of cyclic GMP accumulation in cultured vascular smooth muscle cells. *J. Biol. Chem.* **261:** 12960–12964.

Scarborough RM, McEnroe GA, Arfsten A, Kang LL, Schwartz K, Lewicki JA. (1988) D-amino acid-substituted atrial natriuretic peptide analogs reveal novel receptor recognition requirements. *J. Biol. Chem.* **263:** 16818–16822.

Schmitt F, Martinez F, Ikeni A, Savoiu C, Natov S, Laborde K, Lacour B, Grunfeld J, Hannedouche T. (1994) Acute renal effects of neutral endopeptidase inhibition in humans. *Am. J. Physiol.* **267:** F20–F27.

Schulz M-S, Singh S, Bellet RA, Singh G, Tubb DJ, Chin H, Garbers DL. (1989) The primary structure of a plasma membrane guanylate cyclase demonstrates diversity within this new receptor family. *Cell* **58:** 1155–1162.

Schulz-Knappe P, Forssmann K, Herbst F, Hock D, Pipkorn R, Forssmann WG. (1988) Isolation and structural analysis of 'urodilatin', a new peptide of the cardiodilatin- (ANP)-family, extracted from human urine. *Klin. Wochenschr.* **66:** 752–759.

Seymour AA, Swerdel JN, Fennell SA, Druckman SP, Neubeck R, Delaney NG. (1989) Potentiation of the depressor responses to atrial natriuretic peptides in conscious SHR by an inhibitor of neutral endopeptidase. *J. Cardiovasc. Pharmacol.* **14:** 194–204.

Seymour AA, Norman JA, Asaad MM, Fennell SA, Little DK, Kratunis VJ, Rogers WL. (1991) Antihypertensive and renal activity of SQ 28,603, an inhibitor of neutral endopeptidase. *J. Cardiovasc. Pharmacol.* **17:** 296–304.

Seymour AA, Asaad MM, Abboa-Offei BE, Rovnyak PL, Fennell S, Rogers WL. (1992) Potentiation of brain natriuretic peptides by SQ 28,603, an inhibitor of neutral endopeptidase 3.4.24.11, in monkeys and rats. *J. Pharmacol. Exp. Ther.* **262:** 60–70.

Seymour AA, Asaad MM, Lanoce VM, Langenbacher KM, Fennell SA, Rogers WL. (1993) Systemic hemodynamics, renal function and hormonal levels during inhibition of neutral endopeptidase 3.4.24.11 and angiotensin-converting enzyme in conscious dogs with pacing-induced heart failure. *J. Pharmacol. Exp. Ther.* **266:** 872–883.

Seymour AA, Asaad MM, Abboa-Offei B, Smith PL, Rogers WL. (1994a) Sodium loads enhance the natriuretic responses to atrial natriuretic peptide and neutral endopeptidase inhibitors in conscious cynomolgus monkeys. *Clin. Exp. Pharmacol. Physiol.* **21:** 845–856.

Seymour AA, Sheldon JH, Smith PL, Asaad MM, Rogers WL. (1994b) Potentiation of the renal responses to bradykinin by inhibition of neutral endopeptidase 3.4.24.11 and angiotensin-converting enzyme in anesthetized dogs. *J. Pharmacol. Exp. Ther.* **269:** 263–270.

Seymour AA, Asaad MM, Lanoce VM, Fennell SA, Cheung HS, Rogers WL. (1996) Inhibition of neutral endopeptidase 3.4.24.11 in conscious dogs with pacing induced heart failure. *Cardiovasc. Res.* **27:** 1015–1023.

Shepperson NB, Barclay PL, Bennett JA, Samuels GMR. (1991) Inhibition of neutral endopeptidase (EC 3.4.24.11) leads to an atrial natriuretic factor-mediated natriuretic, diuretic and antihypertensive response in rodents. *Clin. Sci.* **80:** 265–269.

Singer DRJ, Markandu ND, Buckley MG, Miller MA, Sagnella GA, MacGregor GA. (1991) Dietary sodium and inhibition of neutral endopeptidase 24.11 in essential hypertension. *Hypertension* **18:** 798–804.

Smits GJ, McGraw DE, Trapani AJ. (1990) Interaction of ANP and bradykinin during endopeptidase 24.11 inhibition: renal effects. *Am. J. Physiol.* **258:** F1417–F1424.

Stephenson SL, Kenny AJ. (1987a) The hydrolysis of α-human atrial natriuretic peptide by pig kidney microvillar membranes is initiated by endopeptidase-24.11. *Biochem. J.* **243:** 183–187.

Stephenson SL, Kenny AJ. (1987b) Metabolism of neuropeptides. Hydrolysis of angiotensins, bradykinin, substance P and oxytocin by pig microvillar membranes. *Biochem. J.* **241:** 237–247.

Stingo AJ, Clavell AL, Heublein DM, Wei C-M, Pittelkow MR, Burnett JC. (1992) Presence of C-type natriuretic peptide in cultured human endothelial cells and plasma. *Am. J. Physiol.* **263:** H1318–H1321.

Sudoh T, Kangawa K, Minamino N, Matsuo H. (1988) A new natriuretic peptide in porcine brain. *Nature* **332:** 78–81.

Sudoh T, Minamino N, Kangawa K, Matsuo H. (1990) C-type natriuretic peptide (CNP): a new member of natriuretic peptide family identified in porcine brain. *Biochem. Biophys. Res. Commun.*, **168:** 863–870.

Suga S, Itoh H, Komatsu Y, Ogawa N, Hama N, Yoshima T, Nakao K. (1993) Cytokine-induced C-type natriuretic peptide (CNP) secretion from vascular endothelial cells – evidence for CNP as a novel autocrine/paracrine regulator from endothelial cells. *Endocrinology* **133:** 3038–3041.

Sybertz EJ, Chiu PJS, Vemulapalli S, Pitts B, Foster CJ, Watkins RW, Barnett A, Haslanger MF. (1989) SCH 39370, a neutral metalloendopeptidase inhibitor, potentiates biological responses to atrial natriuretic factor and lowers blood pressure in desoxycorticosterone acetate–sodium hypertensive rats. *J. Pharmacol. Exp. Ther.* **250:** 624–631.

Sybertz EJ, Chiu PJS, Vemulapalli S, Watkins R, Haslanger MF. (1990a) Atrial natriuretic factor – potentiating and antihypertensive activity of SCH 34826, an orally active neutral metalloendopeptidase inhibitor. *Hypertension* **15:** 152–161.

Sybertz EJ, Chiu PJS, Watkins RW, Vemulapalli S. (1990b) Neutral metalloendopeptidase inhibition: a novel means of circulatory modulation. *J. Hypertens.* **8** (Suppl.7): S161–S167.

Takayanagi R, Inagami T, Snajdar RM, Imada T, Tamura M, Misono KS. (1987) Two distinct forms of receptors for atrial natriuretic factor in bovine adrenocortical cells. Purification, ligand binding and peptide mapping. *J. Biol. Chem.* **262:** 12104–12113.

Trapani AJ, Smits GL, McGraw DE, Spear KL, Koepke JP, Olins GM, Blaine EH. (1989) Thiorphan, an inhibitor of endopeptidase 24.11, potentiates the natriuretic activity of atrial natriuretic peptide. *J. Cardiovasc. Pharmacol.* **14:** 419–424.

Trippodo NC, Gabel RA, Harvey CM, Asaad MM, Rogers WL. (1991) Heart failure augments the cardiovascular and renal effects of neutral endopeptidase inhibition in rats. *J. Cardiovasc. Pharmacol.* **18:** 308–316.

Trippodo NC, Fox M, Natarajan V, Panchal BC, Dorso CR, Asaad MM. (1993) Combined inhibition of neutral endopeptidase and angiotensin converting enzyme in cardiomyopathic hamsters with compensated heart failure. *J. Pharmacol. Exp. Ther.* **267:** 108–116.

Trippodo NC, Panchal BC, Fox M. (1995a) Repression of angiotensin II and potentiation of bradykinin contribute to the synergistic effects of dual metalloprotease inhibition in heart failure. *J. Pharmacol. Exp. Ther.* **272:** 619–627.

Trippodo NC, Robl JA, Asaad MM, Bird JE, Panchal BC, Schaeffer TR, Fox M, Giancarli MR, Cheung HS. (1995b) Cardiovascular effects of the novel dual inhibitor of neutral endopeptidase and angiotensin-converting enzyme BMS-182657 in experimental hypertension and heart failure. *J. Pharmacol. Exp. Ther.* **275:** 745–752.

Ueda S, Minamino N, Aburaya M, Kangawa K, Matsukura S, Matsuo H. (1991) Distribution and characterization of immunoreactive porcine C-type natriuretic peptide. *Biochem. Biophys. Res. Commun.* **175:** 759–767.

Vanneste Y, Michel A, Dimaline R, Najdovski T, Deschodt-Lanckman M. (1988) Hydrolysis of α-human atrial natriuretic peptide *in vitro* by human kidney membranes and purified endopeptidase-24.11. *Biochem. J.* **254:** 531–537.

Wilkins MR, Settle SL, Stockmann PT, Needleman P. (1990) Maximizing the natriuretic effect of endogenous atriopeptin in a rat model of heart failure. *Proc. Natl Acad. Sci. USA* **87:** 6465–6469.

Wilkins MR, Settle SL, Kirk JE, Taylor SA, Moore KP, Unwin RJ. (1992) Response to atrial natriuretic peptide, endopeptidase-24.11 inhibitor and C-ANP receptor ligand in the rat. *Br. J. Pharmacol.* **107:** 50–57.

Willenbrock R, Scheuermann M, Hohnel K, Luft FC, Dietz R. (1996) Acute and chronic neutral endopeptidase inhibition in rats with aortocaval shunt. *Hypertension* **27:** 1259–1266.

Winter RJD, Zhao L, Krauz T, Hughes JMB. (1991) Neutral endopeptidase 24.11 inhibition reduces pulmonary vascular remodelling in rats exposed to chronic hypoxia. *Am. Rev. Respir. Dis.* **144:** 1342–1346.

Wolfensberger TJ, Singer DR, Freegard T, Markandu ND, Buckley MG, MacGregor GA. (1994) Evidence for a new role of natriuretic peptides: control of intraoccular pressure. *Br. J. Ophthalmol.* **78:** 446–448.

Chapter 21

Cell-surface peptidases and the immune system

Didier Marguet, Anne-Marie Bernard and Michel Pierres

21.1 Introduction

Most of our knowledge of lymphocyte physiology (differentiation, maturation, migration, effector functions, cell–cell and cell–matrix interactions, induction of long-term memory and death) stems from *in vivo* and *in vitro* experimental systems using reagents directed at differentiation antigens selectively expressed on the lymphocyte surface at defined developmental and effector stages. It is of interest that among the amazingly large family of cluster differentiation (CD) antigens of cells of the immune system, a significant number are now recognized as members of the ectoenzyme family. These molecules have been shown to be associated with rather diverse enzymic activities, such as alkaline phosphatase, 5′-nucleotidase, glycosyl transferase, nicotinamide adenine dinucleotide (NAD) ribosyl cyclase, etc. (Naquet and Pierres, 1991). A major group of ectoenzymes expressed by haematopoietic cells belong to the ectopeptidase group. This family of molecules was initially recognized as being conventional cell-surface markers by means of monoclonal antibodies (mAbs), although early studies had indeed identified specific peptidase activities on haematopoietic cell surfaces. It is clear, however, that cDNA cloning technology has permitted the identification and characterization of the vast majority of haematopoietic cell-surface peptidases on the basis of their structural relationships with the ectopeptidases expressed by cells from non-lymphoid tissues.

This chapter will review some of the recent advances in the structural characterization and biological functions of cell-surface peptidases that belong to this particular class of haematopoietic cell differentiation antigens.

21.2 The ectopeptidases characterized in the immune system

The common features shared by this group of cell-surface peptidases include an extracellular domain associated with defined enzymic activities, often exhibiting a broad spectrum of substrate specificity, a wide distribution on the surface of a variety of cell types and a precise window of expression during the specific stages of haematopoietic cell development. Based on these structural properties, specificity and expression pattern, one

Cell-Surface Peptidases in Health and Disease, edited by A.J. Kenny and C.M. Boustead.

can envisage that the ectopeptidases of the immune system are primarily concerned with the modulation or termination of the actions of regulatory peptides. Almost all peptidases share the general function of reducing cellular responses to peptide hormones. The latter, which are produced in a variety of tissues, play a central role in regulating immune responses. The activity of most of the peptide hormones appears to be controlled locally by surface ectopeptidases, some of which act in concert to degrade common substrates. Many ectopeptidases have been identified and characterized in kidney, lung and brush borders of small intestinal cells. Within this group of molecules, some are expressed at specific stages of lymphoid or myeloid cell differentiation. This precise expression pattern clearly suggests that these molecules play a specific role in cell growth, differentiation and maturation in the haematopoietic system.

21.2.1 Endopeptidase-24.11 (E-24.11, CD10; EC 3.4.24.11)

This molecule was initially identified as the common acute lymphoblastic leukaemia antigen (CALLA), a marker of the most frequent form of childhood leukaemia. This antigen is present on leukaemic cells from approximately 80% of leukaemic patients, but is not detected on peripheral blood mononuclear cells. Its expression is precisely regulated during bone marrow haematopoiesis (CD10 is a marker of the pre-B stage of B-cell differentiation). Interestingly, CD10 is also expressed by bone marrow stromal cells and mature granulocytes, and at low levels on peripheral mature B and T cells (Mari *et al.,* 1992). cDNA cloning has revealed that CD10 is identical to E-24.11 (Letarte *et al.,* 1988; Shipp *et al.,* 1989). Structurally, E-24.11 is a type II integral glycoprotein that is highly conserved among human, rat, mouse and rabbit species (> 90% identity). It belongs to the zinc metallopeptidase family and exerts a catalytic activity which is closely related to that of the bacterial enzyme thermolysin, cleaving small peptides adjacent to hydrophobic amino acid residues. Only a brief summary of E-24.11 will be given here; for more detailed accounts, see Chapters 6, 19, 20, 22 and 24.

The overall function of E-24.11 is concerned with the reduction of cellular responses to peptide hormones. It should be stressed that a given target cell expresses on its surface both this ectopeptidase and the receptor for the relevant peptide substrate. E-24.11 has been shown to limit the neutrophil inflammatory cascade, as shown by its ability to reduce peptide-mediated inflammatory responses. This is consistent with the observation that the concentrations of enkephalin needed for generating a neutrophil inflammatory burst differ by several orders of magnitude depending on the presence of E-24.11 on the cell surface (Shipp *et al.,* 1991a). Similarly, E-24.11 can modulate the required concentrations of other inflammatory peptides such as substance P and fMet–Leu–Phe (Painter *et al.,* 1988; Shipp *et al.,* 1991a). The latter chemotactic peptide is initially cleaved by E-24.11 before a further hydrolysis by aminopeptidase N (AP-N, CD13). It is of particular interest that such a concerted action between E-24.11 and AP-N results in a limitation of the neutrophil inflammatory responses.

Another important property of E-24.11 is its ability to regulate peptide-mediated cellular proliferation. Thus, E-24.11 hydrolyses bombesin-like peptides which are potent inhibitors for fibroblasts and normal bronchial epithelial cells, and essential autocrine growth factors for most small-cell lung carcinomas. This is shown by the ability of E-

24.11 to inhibit the growth and proliferation of bombesin-like peptide-dependent small-cell lung carcinomas. A reciprocal effect is observed by the inhibition of E-24.11 enzymic activity (Shipp *et al.*, 1991b).

E-24.11 is also involved in the regulation of stromal cell-dependent B-cell lymphopoiesis. This is consistent with the expression of this ectopeptidase on pre-B cells as well as on stromal cells. More specifically, E-24.11 is expressed at low levels on the surface of Thy-1^{low}B220^{+} large cycling lymphoid progenitor cells. Mouse bone marrow stromal cells that support B-cell differentiation of haematopoietic cells also express high levels of E-24.11 (Salles *et al.*, 1992). Using the Whitlock–Witte culture system, in which uncommitted haematopoietic progenitors are plated on stromal cells that support the earliest stages of B-cell differentiation, the inhibition of the E-24.11 activity, by means of specific inhibitors (e.g., thiorphan or phosphoramidon), induced a significant increase of early lymphoid colony formation (Salles *et al.*, 1992). Similar results were also obtained *in vivo* by assaying the reconstitution of lethally irradiated recipients by bone marrow cells in the presence of E-24.11 inhibitors (Salles *et al.*, 1993). Altogether, these observations suggest that E-24.11 functionally regulates B-cell development by inactivating growth factors that are critical in the generation of bone marrow microenvironments.

Recent studies based on genetically modified mouse strains, in which the E-24.11 gene was disrupted by homologous gene recombination, indicate that E-24.11 is also involved in the modulation of septic shock *in vivo* (Lu *et al.*, 1995). E-24.11$^{-/-}$ mice (constructed by replacing E-24.11 exon 13 and intron 13 by a gene encoding neomycin resistance) exhibit a 10-fold increased sensitivity to endotoxin, as compared to wild-type animals.

21.2.2 Angiotensin-converting enzyme (ACE; EC 3.4.15.1)

This ectoenzyme has been characterized at the surface of pre-B and myelomonocytic leukaemic cell lines and T lymphocytes (Costerousse *et al.*, 1993). It exists as both a membrane-bound and soluble metallopeptidase of 170 kDa (see Chapter 7). ACE cleaves substrates, such as angiotensin I and substance P, and its inhibitors are therefore of major therapeutic interest. In CD8^{+} cytolytic T lymphocytes, ACE participates in the processing of endogenous antigenic peptides and increases antigen presentation to MHC class I restricted lymphocytes (Einserlohr *et al.*, 1992).

21.2.3 Aminopeptidase N (AP-N, CD13; EC 3.4.11.2)

CD13 has been described as a marker of precursor and differentiated cells of the granulocyte and monocyte cell lineages (see Chapters 11 and 23). It has also been detected on acute myeloid and lymphoid leukaemias as well as on bone marrow stromal cells (Griffin *et al.*, 1981). Molecular cloning of CD13 has revealed its identity with AP-N (Look *et al.*, 1989). AP-N is a 130–150 kDa surface protein containing a large extracellular domain with the pentapeptide consensus sequence of the zinc metallopeptidase family. AP-N collaborates with E-24.11 to achieve the processing of a variety of substrates including fMet–Leu–Phe involved in the neutrophil-mediated inflammatory responses (Connelly *et al.*, 1985). AP-N has also been recognized as the receptor for the two transmissible gastroenteritis viruses and HCV-229E coronavirus (see Chapter 25). Its functional role on cells of the myelomonocytic lineages remains to be explored.

21.2.4 Aminopeptidase A (AP-A, BP-1/6C3; EC 3.4.11.7)

BP1/6C3 has been characterized as a differentiation antigen of pre-B cells (Cooper *et al.*, 1986; Pillemer *et al.*, 1984; Wu *et al.*, 1989). It is also detected on a subset of cortical thymic epithelial cells and bone marrow stromal cells. This murine marker has been identified as AP-A (Wu *et al.*, 1990, 1991; and see Chapter 12). It cleaves acidic amino acids from the N terminus of peptides, including angiotensin II. Its expression pattern suggests that AP-A participates in the regulation of early stages of B-cell development. It is of interest that IL-7, which promotes B-cell proliferation, is capable of significantly enhancing AP-A expression on B-cell precursors in bone marrow. This ectopeptidase is supposed to degrade a small peptide which serves as a natural inhibitor of B-cell proliferation (Welch, 1995). Although the AP-A gene has been cloned in man, its functional role in the regulation of human B-cell differentiation remains to be determined.

21.2.5 Dipeptidyl peptidase IV (DPP-IV, CD26; EC 3.4.14.5)

CD26 has been extensively characterized in cells of the immune system in mouse, rat and human species (Fleischer, 1994; Marguet *et al.*, 1993). Molecular cloning of this surface peptidase (Marguet *et al.*, 1992; Tanaka *et al.*, 1992) has revealed that CD26 is identical to DPP-IV, a serine peptidase which is a non-classical member of this large class of peptidases, having a catalytic triad organized in a different topological order (Ser, Asp, His). DPP-IV cleaves X–Pro and X–Ala from peptides such as substance P and growth hormone-releasing hormone. DPP-IV has been characterized as a dimeric or multimeric molecule composed of monomers of 110 kDa. (For a detailed account of DPP-IV, see Chapter 13.) CD13 is widely distributed on cells of the immune system, mainly immature thymocytes in mouse and rat, and medullary thymocytes in human. It is also expressed by peripheral T and B cells but is absent from cells in splenic and lymph node germinal centres. A subset of bone marrow cells also express CD26 in both mouse and rat. Interestingly, CD26 participates in cell proliferation, as shown by the ability of a CD26-specific mAb to drive T cells to proliferate. This activation pathway involves the CD3–TCR complex. Mitogenic anti-CD26 mAbs induce early events of T-cell activation (CD3 zeta phosphorylation and increase of P56*l*lck activity). CD26 has been shown to be associated with the protein tyrosine phosphatase CD45 and, at the level of its extracellular domain, with adenosine deaminase, the lack of which is associated with severe combined immunodeficiency. The role of DPP-IV in the immune system is not yet clearly defined. It seems very likely that it participates in the generation of microenvironments by participating in cell–cell or cell–matrix interactions, during which a variety of growth factors or small peptides can potentially engage specific cell-surface receptors. Its expression in bone marrow and thymus strongly suggests that CD26 participates in the homeostasis of haematopoietic cell production. Finally, the possibility exists that DPP-IV could play a specific role in the control of the communication between the neuroendocrine and the immune systems. The pleiotropic functions of DPP-IV and its various molecular interactions make it difficult to precisely define the structural bases of each of its functional features. For example, experimental evidence indicates that the DPP-IV activity of CD26 is not associated with the T-cell triggering capacity of this molecule. A distinction can be also made between the enzymic and the putative receptor functions of this molecule.

21.3 Conclusions

The diverse ectoenzymic activities so far characterized on cells of the immune system are likely to play a role in the control of differentiation, maturation and proliferation of haematopoietic cells. Most of these molecules seem to interact with biologically active peptides that can be degraded in a cooperative fashion so as to limit their cellular responses. There are now several approaches for gaining insights into the role of ectopeptidases in immunophysiology. One approach is to use specific inhibitors to block the function(s) of a given ectoenzyme, by assaying their effects in *in vitro* (for example in fetal thymic organ culture) or *in vivo* experimental systems. Complementary approaches make it possible to explore the functional effects of overexpressing either a wild-type, or a mutant form of a given ectoenzyme in transgenic animals, as well as to analyse the phenotype of appropriate single or multiple gene inactivations in ectoenzyme-deficient mice. These approaches are ongoing in several laboratories and should give specific information on the role of these ectopeptidases in haematopoietic cell biology. Finally, a better definition of the role of these molecules in the immune system will undoubtedly generate new fields of investigation, leading to the identification of agonist or antagonist compounds of potential clinical use.

Acknowledgements

This work was supported by Institutional funds from INSERM and CNRS, and by grants from ARC and LNFCC.

References

Connelly JC, Skidgel RA, Schulz WW, Johnson AR, Erdös EG. (1985) Neutral endopeptidase 24.11 in human neutrophils: cleavage of chemotactic peptide *Proc. Natl Acad. Sci. USA* **82**: 8737–8742.

Cooper MD, Mulvaney D, Coutinho A, Cazenave PA. (1986) A novel cell surface molecule on early B-lineage cells. *Nature* **321**: 616–618

Costerousse O, Allegrini J, Alhenc-Gelas F. (1993) Angiotensin I-converting enzyme in human mononuclear cells: genetic polymorphism of expression in T-lymphocytes. *Biochem. J.* **290**: 33–40.

Einsenlohr LC, Bacik I, Bennink JR, Bernstein K, Yewdell JW. (1992) Expression of a membrane protease enhances presentation of endogenous antigens to MHC class-I-restricted T lymphocytes. *Cell* **71**: 963–972.

Fleischer B. (1994) CD26: a surface protease involved in T cell activation. *Immunol. Today* **15**: 180–184.

Griffin JD, Ritz J, Nadler LM, Schlossman SF. (1981) Expression of myeloid differentiation antigens on normal and malignant myeloid cells. *J. Clin. Invest.* **68***:* 932–937.

Letartre M, Vera S, Tran R, Addis JB, Omizuka RJ, Quackenbush EJ, Jongeneel CV, McInnis RR. (1988) The common acute lymphoblastic leukemia antigen is identical to neutral endopeptidase. *J. Exp. Med.* **168**: 1247–1253.

Look AT, Ashmun RA, Shapiro LH, Peiper SC. (1989) Human myeloid plasma membrane glycoprotein CD13 (gp150) is identical to aminopeptidase N. *J. Clin. Invest.* **83**: 1299–1307.

Lu B, Gerard NP, Kolakowski LF Jr, Bozza, M, Zurakowski D, Finco O, Carroll ML, Gerard G. (1995) Neutral endopeptidase modulation of septic shock. *J. Exp. Med.* **181:** 2271–2275.

Marguet D, Bernard A-M, Vivier I, Darmoul D, Naquet, P, Pierres M. (1992) cDNA cloning for mouse-thymocyte-activating molecule: a multifunctional ectodipeptidyl peptidase IV (CD26) included in a subgroup of serine proteases. *J. Biol. Chem.* **267**: 2200–2208.

Marguet D, David F, Vivier I, Bernard A-M, Naquet P, Pierres M. (1993) Dipeptidyl peptidase IV (CD26): expression, structure and functions in lymphocytes. *Adv. Neuroimmunol.* **3:** 209–215.

Mari B, Checler F, Ponzio G, Peyron JF, Manie S, Farahifar D, Rossi B, Auberger P. (1992) Jurkat T cells express a functional neutral endopeptidase activity (CALLA) involved in T cell activation *EMBO J.* **11**: 3875–3895.

Naquet P, Pierres M. (1991) Cell-surface enzymes and lymphocyte functions. *Curr. Opin. Immunol.* **3**: 326–329.

Painter RG, Dukes R, Sullivan J, Carter R, Erdös EG, Johnson AR. (1988) Function of neutral endopeptidase on the cell membrane of human neutrophils. *J. Biol. Chem.* **263**: 9456–9461.

Pillemer E, Whitlock C, Weissman I. (1984) Transformation-associated proteins in murine B-cell lymphomas that are distinct from Abelson virus gene products. *Proc. Natl Acad. Sci. USA* **81**: 4434–4438

Salles G, Chen C-Y, Reinherz EL, Shipp, MA. (1992) CD10/NEP is expressed on Thy-1low B220+ murine B cell progenitors and functions to regulate stromal cell-dependent lymphopoiesis. *Blood* **80**: 2021–2029.

Salles G, Rodewald H-R, Chin BS, Reinherz EL, Shipp MA. (1993) Inhibition of CD10/neutral endopeptidase 24.11 promotes B cell reconstitution and maturation *in vivo. Proc. Natl Acad. Sci. USA* **90**: 7618–7622

Shipp MA, Vijayaraghaven J, Schmidt EV, Masteller EL, D'Adamio L, Hersh LB, Reinherz EL (1989) Common acute lymphoblastic leukemia antigen (CALLA) is active neutral endopeptidase 24.11 ('enkephalinase'): direct evidence by cDNA transfection analysis. *Proc. Natl. Acad. Sci. USA* **86**: 297–301.

Shipp MA, Stefano GB, Switzer SN, Griffin JD, Reinherz EL. (1991a) CD10 (CALLA)/neutral endopeptidase-24.11 modulates inflammatory peptide-induced changes in neutrophil morphology, migration, and adhesion proteins and is itself regulated by neutrophil activation. *Blood* **78**: 1834–1841.

Shipp MA, Tarr GE, Chen C-Y, Switzer SN, Hersh LB, Stein H, Reinherz EL. (1991b) CD10/NEP hydrolyses bombesin-like peptides and regulates growth of small cell carcinomas of the lung. *Proc. Natl Acad. Sci. USA* **88**: 10662–10667.

Tanaka T, Camerini D, Seed B, Torimoto Y, Dang NH, Kameoka J, Dahlberg HN, Schlossman SF, Morimoto C. (1992) Cloning and functional expression of the T cell activation antigen CD26. *J. Immunol.* **149**: 481–486.

Welch PA. (1995) Regulation of B cell precursor proliferation by aminopeptidase A. *Int. Immunol.* **7**: 737–746.

Wu Q, Tidmarsh GF, Welch PA, Pierce JH, Weissman IL, Cooper MD. (1989) The early B lineage antigen BP-1 and the transformation-associated antigen 6C3 are on the same molecule. *J. Immunol.* **143**: 3303–3308.

Wu Q, Lahti JM, Air GM, Burrows PD. Cooper MD. (1990) Molecular cloning of the murine BP-1/6C3 antigen: a member of the zinc-dependent metallopeptidase family. *Proc. Natl Acad. Sci. USA* **87:** 993–997.

Wu Q, Li L, Cooper MD, Pierres M, Gorvel JP. (1991) Aminopeptidase A activity of the murine B-lymphocyte differentiation antigen BP-1/6C3. *Proc. Natl Acad. Sci. USA* **88:** 676–680.

Chapter 22

The biological functions of CD10/endopeptidase-24.11 in normal and malignant cells

Michelle Letarte, Adonna Greaves and Eiichi Ishii

22.1 Introduction

This review will focus on the potential roles of CD10 with a particular emphasis on haematopoietic cells, childhood leukaemia and melanoma cells. CD10 was demonstrated by molecular cloning to be identical to endopeptidase-24.11 (E-24.11; Letarte *et al.*, 1988) and to be expressed as a normal active enzyme in malignant cells (Jongeneel *et al.*, 1989). Within the haematopoietic cells, CD10 is primarily associated with precursors of B lymphocytes found in human bone marrow and fetal liver. In children, acute lymphoblastic leukaemia (ALL) arises in cells which have a phenotype characteristic of early stages of B-cell lineage in 75–80% of cases and consequently express CD10. This antigen was originally described as CALLA (common acute lymphoblastic leukaemia antigen) because of its association with this disease and its usefulness in diagnosis (Greaves *et al.*, 1975). CD10 is observed on more than 90% of B-lineage ALL and on 40% of T-lineage ALL; however in children, T-ALL account for less than 20% of leukaemia cases. In the large group of patients with B-lineage ALL, which in general respond well to chemotherapy, CD10 has no independent prognostic significance; in T-ALL it appears to be associated with a favourable prognosis (Pui *et al.*, 1993).

During haematopoiesis, the expression of CD10 is tightly regulated and is transient in the B-lineage; it is more stable on the stromal elements of bone marrow. B and T cells and monocytes are devoid of significant CD10, while mature granulocytes are the only cells in peripheral blood which express significant levels of CD10, presumed to regulate peptide-induced chemotaxis and inflammatory responses (LeBien and McCormack, 1989; Painter *et al.*, 1988; Shipp *et al.*, 1991). Expression of CD10 on melanoma cells is variable and suggests a complex regulation. It is present on 30–50% of cases of melanoma, whether they be primary tumours, advanced tumours or metastases; however the proportion of cells within a given lesion which express CD10 appears to increase with progression of the disease (Carrel *et al.*, 1993).

In this chapter we summarize results obtained with human malignant cells growing in immunodeficient mice. We observed transient induction of CD10 in a pre-B-leukaemia line derived from a relapse case of ALL and the stable induction of CD10 in

Cell-Surface Peptidases in Health and Disease, edited by A.J. Kenny and C.M. Boustead.

a line derived from an early benign radial growth phase (RGP) of melanoma. Our results suggest that the expression of CD10 is associated with increased tumour growth and dissemination of the disease. The microenvironment, as observed in various murine tissues, can induce CD10 which in turn modifies interactions with adjacent cell types by hydrolysing the peptide substrates present at the points of contacts between cells.

Any discussion of the role of CD10 in normal and malignant cells remains speculative because of the lack of information on the relevant substrates and the complexity of the biological systems under study. This chapter is not meant to be comprehensive and the reader should refer to earlier review articles (LeBien and McCormack, 1989; Letarte and Ishii, 1993; Roques *et al.*, 1993; Shipp and Look, 1993) and to Chapters 6, 19 and 20 of this book for additional information on the potential functions of CD10.

22.2 Potential role of CD10 in haematopoiesis

Differentiation of haematopoietic precursors into specific lineages is a complex process which requires their close contact with stromal cells and extracellular matrix, the so-called microenvironment of bone marrow and other haematopoietic organs, such as fetal liver and thymus. At the surface of stromal cells, defined areas of contact with, for example, a B-lineage versus a myeloid progenitor, will generate foci of high concentrations of specific growth factors, produced by the interacting cell types and the adjacent milieu. These foci could give specific differentiation signals to the progenitor cells. The expression of CD10 on both human and murine bone marrow stromal cells (Indig *et al.*, 1990; Keating *et al.*, 1983; Kee *et al.*, 1992), suggests that the apposition of neutral E-24.11 at the stromal surface with the progenitor cells might hydrolyse specific substrates and trigger a differentiation event. For example, hydrolysis of mitogenic peptides which normally maintain the B-lineage precursor in a self-renewal mode might shift the balance towards a commitment to differentiation and exit from the bone marrow. The CD10$^+$ stromal cells, which include fibroblasts, endothelial and macrophage-like cells, can in fact support the growth of B-cell precursors *in vitro* (Indig *et al.*, 1990; Keating et al., 1983; Saeland *et al.*, 1991; Wolf *et al.*, 1991). In the human, the transition to a pre-B lymphocyte stage is also accompanied by the acquisition of CD10 on the surface of the precursor itself.

While mature B and T cells are generally devoid of CD10, human B-cell colonies can be generated *in vitro* from CD10$^+$ precursors and prior to immunoglobulin light chain gene rearrangement (McGinnes *et al.*, 1991). In fact, the CD10$^+$CD34$^+$CD19$^+$ subset of bone marrow lymphoid cells is the most adherent to the stromal layer (Ryan *et al.*, 1990). Thus commitment to B-cell lineage occurs in a microenvironment where CD10 is expressed on both stromal cells and interacting pre-B-cells.

In murine bone marrow, CD10 is expressed on a small subset of large adherent cells, which are most likely stromal. Several cell lines of stromal origin are able to support early B-lineage commitment which do express CD10 (Kalled *et al.*, 1995; Kee *et al.*, 1992). However lymphoid progenitors and a variety of Abelson virus transformed pre-B cell lines do not express CD10 (Kalled *et al.*, 1995; Kee *et al.*, 1992). This suggests that if this enzyme plays a role in early lymphoid differentiation in mouse, it must be as the result of the hydrolysis of regulatory peptides by the stromal cells alone. The earlier report (Salles *et al.*, 1992) that CD10 was expressed on the more immature pro-B cells [based on polymerase chain reaction (PCR) analysis and small effects of enzymic inhibitors on generation of colonies] was not confirmed; preliminary data suggest that irradiation of the animals might have led to induction of CD10 on murine B cell progenitors (Kalled *et al.*, 1995).

E-24.11 cleaves at the amino side of hydrophobic residues in small peptides and will not hydrolyse the known cytokines and growth factors capable of supporting haematopoiesis. The observation that phosphoramidon can potentiate the effects of IL-1β on the proliferation of a bone marrow stromal cell line (Delikat *et al.*, 1994) might be due to an indirect effect as IL-1β is an extremely poor substrate for this enzyme. However there exists a multiplicity of small peptides which could be potential substrates of this enzyme in bone marrow. Their identification is a very difficult task and little progress has been made in that direction.

22.3 Transient up-regulation of CD10/E-24.11 expression in the G2 pre-B leukaemia cell line growing in *scid* mice

Human ALL transplanted into immunodeficient *scid* mice proliferate in haematopoietic tissues, invade several peripheral organs and lead to death of the mice. The ability of cells to grow in these mice correlated with prognosis of the human disease: bone marrow cells from pre-B ALL at relapse (and cell lines derived from the leukaemic blasts) disseminated widely in *scid* mice while those from pre-B ALL patients at diagnosis proliferated poorly, if at all (Kamel-Reid *et al.*, 1989, 1991). The G2 line, derived from a young child who relapsed early in the course of the disease, grew rapidly in *scid* mice. The injection of 10^4–10^7 cells intravenously into irradiated (400 cGy) *scid* mice led to massive infiltration in several tissues and death of the animals at 12 weeks, while injection of 100 cells gave similar effects at 15 weeks (Kamel-Reid *et al.*, 1992). This cell line, although originally derived from CD10$^+$ leukaemic blasts, is CD10$^-$ in culture, but positive when recovered from the thymus of *scid* mice, 10–12 weeks after injection. All subclones selected *in vitro* from individual colonies were CD10$^-$ and showed a similar pattern of expression in the *scid* mice. This indicated that the haematopoietic microenvironment could induce CD10 in G2 cells, which are probably arrested at a very early stage of pre-B-cell differentiation, a stage at which induction of CD10 must occur in normal bone marrow.

The temporal pattern of G2 cell infiltration and their CD10 induction in *scid* mice was studied by quantitative reverse transcriptase PCR (Ishii *et al.*, 1993). Two weeks after intravenous injection of 10^6 G2 cells into irradiated *scid* mice, leukaemic cells were found almost exclusively in bone marrow. At 6–7 weeks, spleen and lung contained 30% human RNA, while peripheral blood, liver and kidney showed 2–3%; infiltration of brain and thymus was observed only at 8–9 weeks. CD10 induction (10-fold) was observed at 6–7 weeks, corresponding to the onset of dissemination. It was seen in bone marrow, spleen, peripheral blood and liver, but not in lung despite 30% infiltration. Increased levels of CD10 were seen in all tissues between 8 and 10 weeks; the highest levels were observed in leukaemic cells in thymus (113-fold) and in the circulation (Ishii *et al.*, 1993). These findings suggest that the initial induction of CD10 occurs in haematopoietic tissues at the time of rapid proliferation of the leukaemic cells and their infiltration of peripheral tissues. At later time points, CD10 is expressed on the leukaemic cells found in all organs suggesting that a general up-regulation of the enzyme is associated with advanced disease state.

To further ascertain whether expression of CD10 was correlated with increased tumorigenicity and/or ability to invade peripheral tissues, we co-injected subcutaneously into non-irradiated *scid* mice G2 cells and Matrigel (a reconstituted extracellular matrix, shown previously to increase tumour formation by several malignant cell types) (Fridman *et al.*, 1990; Kath *et al.*, 1991). Lymphatic tumours grew and infiltrated the dermis; they were observed at 10–13 weeks in eight Matrigel-treated mice, but in only two control mice receiving leukaemic cells alone (Ishii *et al.*, 1995). Infiltration of bone marrow, spleen, thymus, lung and liver was observed earlier and was also more extensive in the Matrigel-treated group. Tumour cells derived from Matrigel-treated mice could be passaged *in vitro* and their colony-forming ability was higher than that of the original G2 line. When re-injected intravenously into non-irradiated *scid* mice, these tumour cells invaded the thymus earlier than did the G2 cells. However the expression of CD10 was induced at high levels in all tumours growing in *scid* mice, irrespective of the use of Matrigel as shown in *Table 22.1*. So Matrigel *per se* did not stimulate CD10 expression, but accelerated the process of

Table 22.1. Modulation of CD10 expression in G2 human pre-B-leukaemic cells growing in *scid* mice or cultured *in vitro*

Cells injected *in vivo* or cultured *in vitro*	Time in weeks	CD10 level[a] on leukaemic cells derived from: Inoculum	Tumour	Bone marrow	Spleen	Thymus
G2 cells subcutaneously						
+ Matrigel	10	0.01	24±1	7±2	6±1	None
– Matrigel	10	0.01	None	9±1	6±1	None
+ Matrigel	12	0.01	31±5	12±1	9±1	27±9
– Matrigel	12	0.01	41±5	13±4	10±2	17±1
Cells intravenously						
Tumour variant 64[b]	6	24	None	4±1	6±1	9±1
	8	24	None	7±1	6±1	13±1
G2	6	0.01	None	5±2	4±1	None
	8	0.01	None	5±1	5±1	14±5
Cells *in vitro*						
Tumour variant 64	0	24±1				
	1	11±2				
	3	5±1				
	5	2±1				
G2		0.01				

[a] CD10 level is expressed as the CD10/CD44 ratio calculated from the mean fluorescence intensity values of CD10 and CD44 as determined by flow cytometry, assuming constant expression of CD44 and equivalent proportions of cells labelled with both antibodies.

[b] Tumour variant 64 was derived from a subcutaneous tumour, 10 weeks after injection of G2 cells in Matrigel and was verified intraveneously, or grown *in vitro*.

Adapted from Ishii *et al.* (1995).

tumour formation by G2 cells. The lymphoid tumour microenvironment was in turn capable of inducing CD10. This is a transient process since the tumour variants grown *in vitro* or re-injected *in vivo* lost expression of CD10. However, 6–8 weeks later, CD10 was observed on both tumour variants and parental G2 cells growing in the thymus, bone marrow and spleen (see *Table 22.1* and Ishii *et al.*, 1995).

Culturing G2 cells *in vitro* at high density or in the presence of known growth-promoting cytokines such as IL-3, IL-6, IL-7 and granulocyte-macrophage colony-stimulating factor did not stimulate the expression of CD10 mRNA (Ishii *et al.*, 1995). The induction of this surface endopeptidase was thus associated with growth of leukaemic cells in the specific microenvironments provided particularly by lymphoid tumours and thymus in *scid* mice. It also appears to require a critical cell density and, at later stages of the disease, is observed in most tissues. The function of CD10 is related to the hydrolysis of peptides which are critical in regulating interactions between adjacent pre-B cells, the stromal microenvironment and the transduction of growth and/or differentiation signals. At later stages of the disease, proliferation is rapid and dysregulated; expression of CD10 might contribute to this process of growth autonomy by degrading peptides which normally inhibit proliferation. Alternatively, CD10 expression might be a consequence of the advanced malignancy and represent a desperate attempt at degrading peptides which stimulate the uncontrolled growth.

22.4 Induction of stable expression of CD10/E-24.11 in the WM35 melanoma line proliferating in nude mice

Human melanomas become more malignant by progression through defined clinical and pathological stages and cell lines can be established from these different stages, facilitating a comparative study of steps leading to malignancy. Melanocytes are well differentiated in the epidermis where they interact with keratinocytes which provide them with paracrine growth factors. In response to ultra-violet irradiation, increased production of endothelins by keratinocytes leads to increased proliferation of melanocytes (Imokawa *et al.*, 1992). Expression of CD10 on cells at that stage of development, would represent a way of regulating an environmentally induced mitogenic substrate and the rate of proliferation of melanocytes. The first stage of melanoma is the radial growth phase (RGP) where cells grow in the epidermis in a horizontal plaque-like manner; these lesions are curable and cells do not metastasize, despite the fact that a few cells can be seen invading the dermis (Herlyn, 1993). The next stage is the vertical growth phase (VGP) when cells leave the adjacent keratinocytes and begin to grow downwards into the underlying dermis. VGP lesions less than 0.76 mm in thickness (thin VGP) are usually curable by surgery and are metastatically incompetent whereas those thicker than 0.76 mm (thick VGP) have a worse prognosis and metastasize (Herlyn, 1993). Thus the extent of invasion of cells within the dermis determines the outcome of the disease and it is critical to understand the mechanisms regulating this process.

CD10 was reported to be present on 30–50% of cases of melanoma, whether they be primary tumours, advanced tumours or metastases; however the proportion of cells within a given lesion which express CD10 and consequently the level of enzymic activity present in the tumour appears to increase with progression of the disease (Carrel *et al.*, 1993). We observed that the early melanoma cell line WM35 was CD10⁻ while two other early lines WM1341B and WM793 were CD10⁺. All three lines were derived from RGP melanomas and did not metastasize in the patients (Herlyn, 1993; Kobayashi *et al.*, 1994). However, co-injection of these cells with Matrigel into athymic nude mice led to the rapid formation of progressively growing tumours from which permanent cell lines could be derived. Repeated passage of these tumour variants *in vivo*, and subsequent derivation of cell lines from the tumours, led to the generation of additional variants characterized by a step-wise increase in tumorigenic properties, such as acquisition of cytokine resistance (Kobayashi *et al.*, 1994).

The WM35 cell line showed a gradual increase in CD10 as cells were serially passaged *in vivo*, as shown in *Figure 22.1*. Two populations of cells can be identified at passages 1 and 2, which express low and high levels of CD10 respectively. At passage 3, and in a lung metastatic clone isolated at passage 2, a single population of cells with high levels of CD10 was observed. The expression of CD10 was stable in culture for more than 1 year with the various clones. With the CD10⁺ melanoma lines, WM1341B and WM793, no changes in the level of this endopeptidase occurred with passage *in vivo*. The rate of growth of the sublines established from the variants of the three melanoma lines was higher than that of the parental lines, and they formed tumours

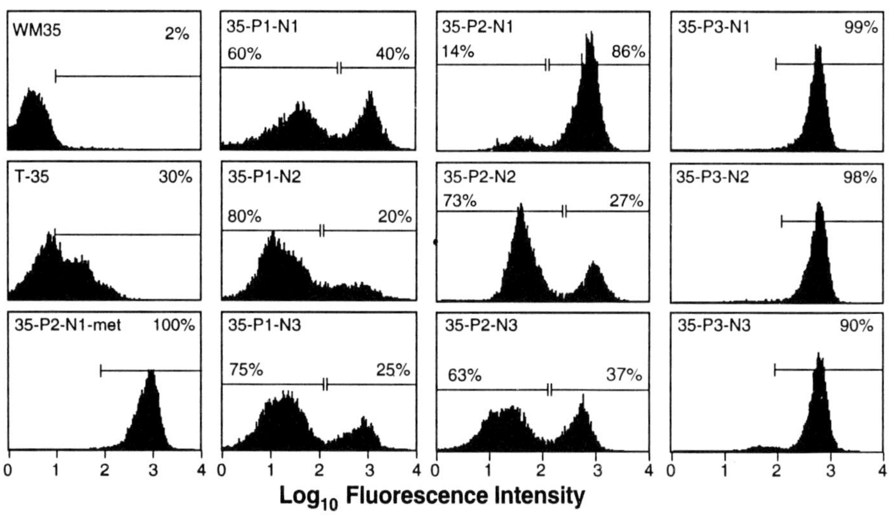

Figure 22.1. Induction of CD10 on melanoma tumorigenic variants. WM35, an early melanoma parental cell, and its sublines, derived by *in vivo* passage in Matrigel in nude mice (Kobayashi *et al.*, 1994), were analysed by flow cytometry for their expression of CD10 using MAb 44C10 and FITC-goat anti-mouse IgG. P represents the passage number and N the mouse number. WM35 cells express no detectable CD10 when compared to the IgG2b negative control (2% cells within the positive gate as indicated). Thirty per cent of T-35 cells, derived from the tumour that grew in one of 10 mice after one passage without Matrigel, showed weak fluorescence. After one and two passages *in vivo*, two populations of cells expressing low and high levels of CD10 could be identified within the sublines. After three passages *in vivo*, all cells expressed high levels of CD10, as did the 35-P2-N1-met cell line, established from a lung metastasis.

more readily *in vivo*. We can postulate that induction of CD10 is an early event in the transformation of melanocytes, suggesting that the WM35 line is derived from a less well differentiated melanocyte. Increased tumorigenicity in all three RGP melanoma lines was associated with an increase in a3β1 integrin but to no major changes in other integrins. The CD10⁺ variants of WM35 were not more adhesive to fibronectin, laminin, collagen type IV and to monolayers of endothelial cells, and their migration or invasion of various matrix proteins was not altered. Thus no direct correlation could be established between changes in CD10 and in integrin-related functions.

However, when thiorphan (1 μM) was added to cultures of the tumorigenic variant 35-P3-N1, dramatic changes in morphology were observed (see *Figure 22.2*). While the $CD10^-$ WM35 parental line grew as small clusters with or without thiorpan, the $CD10^+$ tumorigenic variants formed large clusters of interacting cells in the presence of the specific E-24.11 inhibitor. This would suggest that CD10 might be preventing the close apposition of cells by degrading peptides involved in this contact-dependent proliferation signal. CD10 could then contribute to growth autonomy of tumour cells and facilitate their escape from the tumour mass.

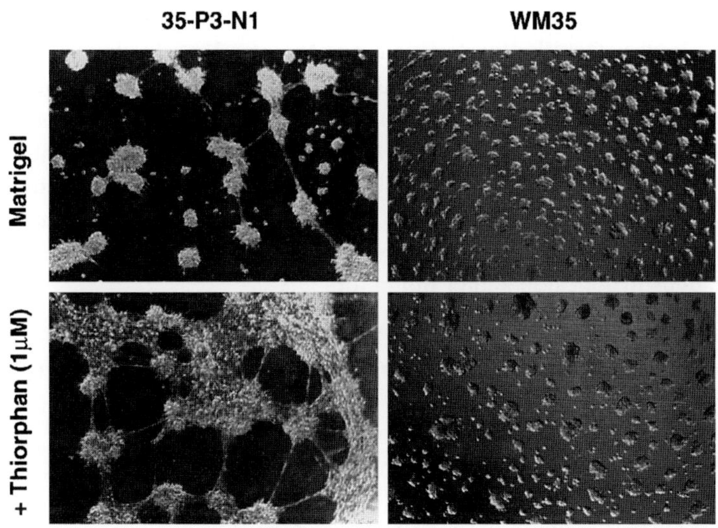

Figure 22.2. Effects of thiorphan on WM35 melanoma tumorigenic variants. The tumorigenic variant 35-P3-N1 and the parental WM35 early melanoma cell line were seeded on Matrigel-coated wells with or without 1 μM thiorphan. The presence of this specific inhibitor of CD10/E-24.11 induced a significant increase in cell clustering of the $CD10^+$ variant as observed here after 48 h in culture but had no effect on the parental $CD10^-$ line.

22.5 Role of CD10/E-24.11 in regulating mitogenic neuropeptides and potential link to focal adhesion kinase, integrins and malignancy

CD10/E-24.11 is abundant in membranes of the brush border of intestine where it can degrade several peptides such as gastrin and neuropeptide Y implicated in the control of the growth of gastrointestinal epithelial cells (Goodlad and Wright, 1987). In this system, E-24.11 would decrease cell proliferation by degrading growth-promoting peptides. In fetal, normal and neoplastic respiratory tissues, bombesin-like peptides, which act as potent growth factors, are also degraded by E-24.11 expressed in bronchial epithelial cells; furthermore, administration to mice of another specific E-24.11 inhibitor, SCH32615, promoted the development of fetal lung (King *et al.*, 1993). However, the expression of CD10 was highest at the time of maximal lung development, suggesting that regulation of the levels of bombesin is very critical. In the case of the bombesin-responsive small cell lung carcinomas, the more malignant cells had lost CD10 expression whereas $CD10^+$ cells could be stimulated by the addition of inhibitors (Shipp and Look, 1993; Shipp *et al.*, 1991).

Focal adhesion kinase ($p125^{FAK}$) has now been implicated in the action of mitogenic neuropeptides such as bombesin, vasopressin and endothelin, which act through G protein-coupled receptors (Leeb-Lundberg and Song, 1991; Zachary and Rozengurt, 1992). This protein was independently identified as a cytoplasmic substrate of the $pp60^{src}$ protein kinase and as a kinase specifically localized to points of focal contact (Schaller *et al.*, 1992). Phosphorylation of $p125^{FAK}$ can be induced by ligation of integrins to extracellular matrix, or with antibodies, leading to growth and/or differentiation signals (Juliano and Haskill, 1993). This led to the idea that this kinase might act as a point of convergence between neuropeptide receptor occupancy, activation of integrins and oncogenic transformation (Zachary and Rozengurt, 1992). We propose that expression of CD10/E-24.11 (and potentially other cell-surface peptidases acting in cascade), could release cells from the control of mitogenic peptides and facilitate integrin-mediated changes in cell–cell or cell–matrix interactions. Alternatively, cells might be released from the influence of regulatory peptides to undergo a differentiation step requiring changes in adhesion and/or migration.

22.6 Role of CD10/E-24.11 in modulation of inflammatory responses and septic shock

In all cell types where CD10/E-24.11 function has been studied, it appears to reduce cellular responses to these peptides, such as enkephalin-mediated analgesia, and substance P-, fMLP- or enkephalin-mediated inflammatory responses (Roques *et al.*, 1993; Shipp and Look, 1993). For example, inhibition of CD10 leads to an increase in substance P-mediated neurogenic inflammation of the lung which is associated with cough, bronchial constriction, vasodilation and neutrophil migration (Shipp and Look, 1993). The enzyme is, in fact, regulating the level of peptide hormones by controlled (normal situation) or uncontrolled (disease situation) degradation. This in turn leads to a reduction in signalling via the specific peptide hormone receptor(s) and in subsequent cellular responses.

Targeted disruption of the CD10/E-24.11 locus in mice resulted in increased lethality in response to endotoxin-induced septic shock (Lu *et al.*, 1995). Galactosamine-sensitized lipopolysaccharide (LPS) gives a complex inflammatory stimulus which activates several pro-inflammatory pathways and triggers macrophages to release IL-1 and TNF-α. CD10-

deficient animals were more sensitive to LPS (LD_{50} 1.2 μg versus > 16.7 μg) and were 100 times more likely to die from the treatment. The effect was dependent on gene dosage and animals were protected from lethal shock by injection of recombinant CD10. The null mice also demonstrated increased sensitivity to IL-1 and TNF-α, suggesting that a mixture of peptides, such as endothelin, bradykinin, tachykinins, atriopeptin and enkephalins, is likely to be released, which potentiates the lethal shock in response to these mediators (Lu *et al.*, 1995).

22.7 Conclusions

The results observed with the null mice reveal that the function of CD10/E-24.11 is not essential to development, nor to life in general, as long as adversity is not encountered. However the null mice were used to illustrate one example of adversity, namely septic shock, which in the absence of E-24.11, leads to death. It is also very likely that stress and various diseases which generate peptide substrates of E-24.11 which can be detrimental to health, if present in excess, will lead to severe and potentially lethal conditions in the CD10 null mice. Thus, this animal model should be adequate to establish the role of this enzyme in regulating such complex biological processes as hypertension, pain control and inflammation. The model might also be suitable to study the contribution of CD10 to the development of a fully functional immune system and to the progression of malignancies.

Acknowledgements

M. Letarte is a Terry Fox Research Scientist of the National Cancer Institute of Canada. This research was funded by the Medical Research Council of Canada, the National Cancer Institute of Canada and the Leukemia Research Fund of Canada. We thank Dr H. Kobayashi and Dr R. Kerbel (Sunnybrook Hospital, Toronto) and Dr M. Herlyn (Wistar Institute, Philadelphia) for providing us with the melanoma cell lines.

References

Carrel S, Zografos L, Schreyer M, Rimoldi D. (1993) Expression of CALLA/CD10 on human melanoma cells. *Melan. Res.* **3:** 319–323.

Delikat SE, Galvani DW, Zuzel M. (1994) A function of CD10 on bone marrow stroma. *Br. J. Haemat.* **87:** 655–657.

Fridman R, Giaccone G, Kanemoto T, Martin GR, Gazdar AF, Mulshine JL. (1990) Reconstituted basement membrane (matrigel) and laminin can enhance the tumorigenicity and the drug resistance of small cell lung cell lines. *Proc. Natl Acad. Sci. USA* **87:** 6698–6702.

Goodlad RA, Wright NA. (1987) Peptides and epithelial growth regulation. *Experientia* **43:** 780–784.

Greaves MF, Brown G, Rapson N, Lister TA. (1975) Antisera to acute lymphoblastic leukaemia cells. *Clin. Immunol. Immunopathol.* **4:** 67–84.

Herlyn M. (1993) *Molecular and Cellular Biology of Melanoma* (ed. R.G. Landes). CRC Press, Austin/Georgetown, TX.

Imokawa G, Yada Y, Miyagishi M. (1992) Endothelins secreted from human keratinocytes are intrinsic mitogens for human melanocytes. *J. Biol. Chem.* **267:** 24675–24680.

Indig FE, Benayahu D, Fried A, Wientroub S, Blumberg S. (1990) Neutral endopeptidase (EC 3.4.24.11) is highly expressed on osteoblastic cells and other marrow stromal cell types. *Biochem. Biophys. Res. Commun.* **172:** 620–626.

Ishii E, Greaves A, Grunberger T, Freedman MH, Letarte M. (1993) The induction of CD10 on a pre-B leukemia cell line occurs with progression of the disease in *scid* mice. *Leukemia* **7:** 1592–1601.

Ishii E, Greaves A, Grunberger T, Freedman MH, Letarte M. (1995) Tumour formation by a human pre-B leukemia cell line in *Scid* mice is enhanced by Matrigel and is associated with induction of CD10 expression. *Leukemia* **9:** 175–184.

Jongeneel CV, Quackenbush EJ, Ronco P, Verroust P, Carrel S, Letarte M. (1989) Common acute lymphoblastic leukemia antigen expressed on leukemia and melanoma cell lines has neutral endopeptidase activity. *J. Clin. Invest.* **83:** 713–717.

Juliano RL, Haskill S. (1993) Signal transduction from extracellular matrix. *J. Cell Biol.* **120:** 577–585.

Kalled SL, Siva N, Stein H, Reinherz EL. (1995) The distribution of CD10 (NEP 24.11, CALLA) in humans and mice is similar in non-lymphoid organs but differs within the hematopoeitic system: absence on murine T and B lymphoid progenitors. *Eur. J. Immunol.* **25:** 677–687.

Kamel-Reid S, Letarte M, Sirard C, Doedens M, Grunberger T, Fulop G, Freedman MH, Phillips RA, Dick JE. (1989) A model of human acute lymphoblastic leukemia in immune-deficient SCID mice. *Science* **246:** 1597–1600.

Kamel-Reid S, Letarte M, Doedens M *et al.* (1991) Bone marrow from children in relapse with pre-B acute lymphoblastic leukemia proliferates and disseminates rapidly in *scid* mice. *Blood* **78:** 2973–2981.

Kamel-Reid S, Dick JE, Greaves A *et al.* (1992) Differential kinetics of engraftment and induction of CD10 on human pre-B leukemia cell lines in immune deficient *scid* mice. *Leukemia* **6:** 8–17.

Kath R, Jambrosic JA, Holland L, Rodeck U, Herlyn M. (1991) Development of invasive and growth factor-independent cell variants from primary melanomas. *Cancer Res.* **51:** 2205–2211.

Keating A, Whalen CK, Singer JW. (1983) Cultured marrow stromal cells express common acute lymphoblastic leukaemia antigen (CALLA): implications for marrow transplantation. *Br. J. Haematol.* **55:** 623–628.

Kee BL, Paige CJ, Letarte M. (1992) Characterization of murine CD10, an endopeptidase expressed on bone marrow adherent cells. *Intern. Immunol.* **4:** 1041–1047.

King KA, Hua J, Torday JS, Drazen JM, Graham SA, Shipp MA, Sunday ME. (1993) CD10/neutral endopeptidase-24.11 regulates fetal lung growth and maturation by potentiating endogenous bombesin-like peptides. *J. Clin. Invest.* **91:** 1969–1973.

Kobayashi H, Man S, MacDougall JR, Graham CH, Lu C, Kerbel RS. (1994) Variant sublines of early-stage human melanomas selected for tumorigenicity in nude mide express a multicytokine-resistant phenotype. *Am. J. Pathol.* **144:** 776–786.

LeBien TW, McCormack RT. (1989) The common acute lymphoblastic leukemia antigen (CD10)-emancipation from a functional enigma. *Blood* **73:** 625–635.

Leeb-Lundberg LMF, Song X-H. (1991) Bradykinin and bombesin rapidly stimulate tyrosine phosphorylation of a 120-kDa group of proteins in Swiss 3T3 cells. *J. Biol. Chem.* **266:** 7746–7749.

Letarte M, Ishii E. (1993) The potential role of CD10/neutral endopeptidase-24.11 in the immune system. *Adv. Neuroimmunol.* **3:** 183–194.

Letarte M, Vera S, Tran R, Addis JBL, Onizuka RJ, Quackenbush EJ, Jongeneel CV, McInnes RR. (1988) Common acute lymphocytic leukemia antigen is identical to neutral endopeptidase. *J. Exp. Med.* **168:** 1247–1253.

Lu B, Gerard NP, Kolakowski LFJ, Bozza M, Zurakowski D, Finco O, Carroll MC, Gerard C. (1995) Neutral endopeptidase modulation of septic shock. *J. Exp. Med.* **181:** 2271–2275.

McGinnes K, Letarte M, Paige CJ. (1991) B-lineage colonies from normal, human bone marrow are initiated by B-cells and their progenitors. *Blood* **77:** 961–970.

Painter RG, Dukes R, Sullivan J, Carter R, Erdos EG, Johnson AR. (1988) Function of neutral endopeptidase on the cell membrane of human neutrophils. *J. Biol. Chem.* **263:** 9456–9461.

Pui C-H, Rivera GK, Hancock ML *et al.* (1993) Clinical significance of CD10 expression in childhood acute lymphoblastic leukemia. *Leukemia* **7:** 35–40.

Roques BP, Noble F, Daugé V, Fournié-Zaluski MC, Beaumont A. (1993) Neutral endopeptidase-24.11: structure, inhibition, and experimental and clinical pharmacology. *Pharmacol. Rev.* **45:** 87–146.

Ryan DH, Nuccie BL, Abboud CN, Liesveld JL. (1990) Maturation-dependent adhesion of human B-cell precursors to the bone marrow microenvironment. *J. Immunol.* **145:** 477–484.

Saeland S, Duvert V, Pandrau D, Caux C, Durand I, Wrighton N, Wideman J, Lee F, Banchereau J. (1991) Interleukin-7 induces the proliferation of normal human B-cell precursors. *Blood* **78:** 2229–2238.

Salles G, Chen C, Reinherz EL, Shipp MA. (1992) CD10/NEP is expressed on Thy-1^{low}B220^{+} murine B-cell progenitors and functions to regulate stromal cell-dependent lymphopoiesis. *Blood* **80:** 2021–2029.

Schaller MD, Borgman CA, Cobb BS, Vines RV, Reynolds AB, Parsons JT. (1992) pp125 FAK, a structurally distinctive protein-tyrosine kinase associated with focal adhesions. *Proc. Natl Acad. Sci. USA* **89:** 5192–5196.

Shipp MA, Look AT. (1993) Hematopoietic differentiation antigens that are membrane-associated enzymes: cutting is the key! *Blood* **82:** 1052–1070.

Shipp MA, Tarr GE, Chen C, Switzer SN, Hersh LB, Stein H, Sunday ME, Reinherz EL. (1991) CD10/neutral endopeptidase-24.11 hydrolyses bombesin-like peptides and regulates the growth of small cell carcinomas of the lung. *Proc. Natl Acad. Sci. USA* **88:** 10662–10666.

Wolf ML, Buckley JA, Goldfarb A, Law C-H, LeBien TW. (1991) Development of a bone marrow culture for maintenance and growth of normal human B-cell precursors. *J. Immunol.* **147:** 3324–3330.

Zachary I, Rozengurt E. (1992) Focal adhesion kinase (p125FAK): a point of convergence in the action of neuropeptides, integrins, and oncogenes. *Cell* **71:** 891–894.

Chapter 23

Molecular and functional aspects of human myelomonocytic ectopeptidases

Brigitte Bauvois

23.1 Introduction

During the past 10 years, the application of modern immunological and molecular techniques has resulted in the characterization of an increasing number of ectopeptidases, thereby providing the means to study these enzymes in molecular detail. This chapter describes our contribution on ectopeptidases from the human myelomonocytic system and on their potential roles in immune–inflammatory responses.

We and others have shown that human myelomonocytic lineage cells possess several types of ectopeptidases: aminopeptidases, γ-glutamyl transpeptidase (γ-GT; EC 2.3.2.2) and dipeptidyl peptidase IV (DPP-IV; EC 3.4.14.5). Their distinctive patterns of expression that are seen in differentiating neutrophils and macrophages appeared to be related to differences in the stage of myeloid maturation. Interestingly, such up- and down-regulation phenomena appear to occur for other surface proteins crucially involved in myeloid activation and maturation, such as cytokine receptors and adhesion receptors. Thus, all these controlling mechanisms deserve attention because they are likely to be critical determinants in the activation and differentiation of inflammatory cells. In addition, dysregulation in blood cell ectopeptidase levels may be important in malignant pathology.

We have also focused on the emerging evidence indicating a critical role of ectopeptidases as leukocyte stimulatory molecules and their relation to the development of the immune and inflammatory responses. Monocyte/macrophage lineage cells release a variety of diffusible mediators, which can recruit and activate cells involved in tissue injury and repair. As assessed by polyacrylamide gel electophoresis and bioassay, we have shown that the cell-surface-associated peptidases, cathepsin G, DPP-IV and γ-GT, are important in the control of certain cytokine activities: tumour necrosis factor α (TNF-α) was degraded into biologically inactive fragments of 2 kDa or less, whereas interleukin 6 (IL-6) and insulin-like growth factor 1 (IGF-1) were processed by a combination of degradation and transpeptidation into dimeric IL-6 and IGF-1 species. Thus, the regulatory role played by such enzymes may be of importance in altering the magnitude and quality of cellular responses induced by these cytokines.

Cell-Surface Peptidases in Health and Disease, edited by A.J. Kenny and C.M. Boustead.

Apart from their role in the regulation of cytokine levels, γ-GT, present at the surface of human myeloid precursor cells, seems to be functionally involved in macrophage maturation and transforming growth factor β (TGF-β) production. Hence, as our understanding of the role of ectopeptidases in the human myeloid system increases, the future challenge will be to define the relationship between ectopeptidases and signal transduction pathways leading to cell activation.

23.2 Expression and regulation of human myelomonocytic ectopeptidases

23.2.1 Expression

Several cell-surface peptidases have been described on human blood granulocytes and monocytes. Endopeptidase-24.11 (CD10; EC 3.4.24.11) is found at low density on granulocytes while monocytes and granulocytes express aminopeptidase-N (CD13; EC 3.4.11.2) (Look *et al.,* 1987; Painter *et al.,* 1988). Moreover, γ-GT was detected at the surface of human monocytes (Grisk *et al.,* 1993). γ-GT is a widely distributed enzyme that acts as a transferase in the transfer of the γ-glutamyl group from a wide variety of peptide donors to other amino acids and peptide acceptors and as a hydrolase in removing the γ-glutamyl residue from such peptides (Tate and Meister, 1981). It was not clear, however, whether cell-surface peptidase expression was regulated on haematopoietic cells along the granulocyte and macrophage maturation pathways. In our laboratory, we have used human promyelocytic leukaemia (HL-60) and monoblastic leukaemia (U937) cells which represent precursors of myeloid lineage. Using a spectrophotometric method, we showed it to be a sensitive assay of aminopeptidase and DPP-IV activities detected on the surface of intact cells, as well as in highly purified cell membrane fractions. Aminopeptidase activity was demonstrated by the bestatin inhibitor profile and specificity towards the synthetic substrates Ala-, Leu-, Lys- and Arg-*p*-nitroanilide (Arg-pNA) (Bauvois *et al.,* 1992; Laouar *et al.,* 1993b) (see *Figure 23.1a).* Cell-surface γ-GT (assayed with γ-Glu-pNA as substrate and Gly–Gly as acceptor and abolished by the specific inhibitor of γ-GT, i.e. acivicin) was confirmed in the two leukaemia cell lines (Bauvois *et al.,* 1995) (see *Figure 23.1a*). Interestingly, under our experimental conditions, a serine peptidase, identified as DPP-IV by its hydrolysis of Gly–Pro-pNA and its sensitivity to inhibition by diisopropyl phosphofluoridate and diprotin A was also observed (Bauvois *et al.,* 1992; Laouar *et al.,* 1993b) (see *Figure 23.1a*). Peripheral blood monocytes, granulocytes as well as macrophages developed *in vitro* from monocytes, were found to exhibit the same three ectopeptidase activities detected on HL-60 and U937 cells (Bauvois and Laouar, 1992; Bauvois *et al.,* 1995; Laouar *et al.,* 1993b) (see *Figure 23.1b*).

Another line of evidence that human myeloid cells express γ-GT was obtained from Northern blot analysis. In humans, there are at least four potential genes for γ-GT located on chromosome 22 (Guellaën, 1989). Using a liver γ-GT probe, we have detected an mRNA species of 2.4 kb, corresponding to that previously described for the HepG2 cell line (Bulle *et al.,* 1987), in both cell lines and in granulocytes and macrophages (Bauvois *et al.,* 1995) (see *Figure 23.2*). CD26 has been previously shown

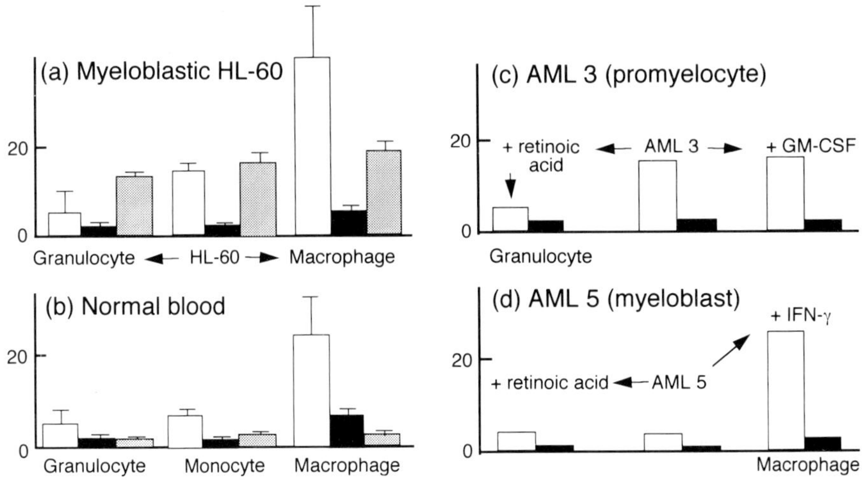

Figure 23.1. Ectopeptidase activities during differentiation into granulocytes or macrophages, by HL-60 cells, human blood cells and by cells from patients with acute leukaemias. (a) HL-60 cells were induced to differentiate into granulocytes with 1 μM retinoic acid, or macrophages with 3 ng ml^{-1} GM-CSF. (b) Granulocytes and monocytes were isolated from normal blood by Ficoll Hypaque density gradient centrifugation; monocytes were cultured for 5 days in the presence of 3 ng ml^{-1} GM-CSF to develop into macrophages. (c) Human cells from a patient with acute promyelocytic leukaemia (AML3) were stimulated for 5 days with 1 μM retinoic acid or 3 ng ml^{-1} GM-CSF. (d) Human cells from a patient with acute myeloblastic/ monocytic leukaemia (AML5) were stimulated for 5 days with 1 μM retinoic acid or 100 U ml^{-1} IFN-γ. γ-GT activity (shaded) was determined with 1 mg ml^{-1} γ-Glu-pNA and 20 mM Gly–Gly as acceptor, Ala aminopeptidase activity (open) with 1 mg ml^{-1} Ala-pNA and DPP-IV-like activity (solid) with 1 ng ml^{-1} Gly–Pro-pNA. The columns show peptidase activities (nmol/30 min/10^5 cells).

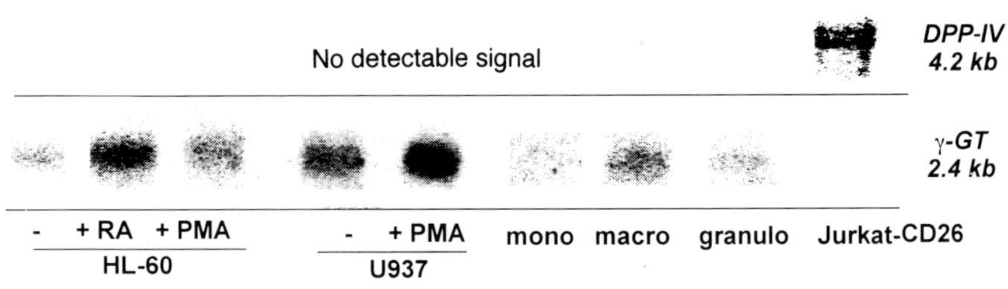

Figure 23.2. DPP-IV and γ-GT mRNA expression in HL-60 cells, U937 cells, monocytes, granulocytes, macrophages and CD26-transfected Jurkat T cells. Cells were cultured in the absence or in the presence of 1 μM retinoic acid (for HL-60 to differentiate into granulocytes) or 1 ng ml^{-1} phorbol myristate acetate (for HL-60 cells, U937 cells and monocytes, to differentiate into macrophages). Total RNAs extracted were prepared by acid guanidinium thiocyanate–phenol–chloroform extraction. Northern blots were hybridized with [^{32}P]-labelled DPP-IV or γ-GT cDNAs.

to possess DPP-IV activity (Hegen *et al.,* 1990). In our hands, HL-60 and U937 cells, monocytes and macrophages, differentiated *in vitro* from monocytes, displayed surface expression of CD26 molecules when detected with the monoclonal antibodies Ta1 (from Coulter Clones) and TA5.9 (from Eurogenetics). However, these antibodies neither inhibited nor immunoprecipitated U937/monocyte DPP-IV activities.

Moreover, no CD26 transcripts were detected with a CD26 cDNA encoding human T CD26 in any cells of the myelomonocytic lineage (see *Figure 23.2*). Other authors have shown that freshly isolated peripheral blood cells do not express CD26 mRNA, although some cells have low levels of surface CD26 (Tanaka *et al.,* 1992). Recently, Böhm *et al.* (1995) have shown that the DPP-IV promoter possesses the ability to initiate transcription in a tissue-specific fashion. Together, these data suggest that myeloid DPP-IV gene expression is tightly regulated. However, our data cannot exclude the possibility that myeloid cells possess a plasma membrane-associated ectopeptidase related to, but distinct from DPP-IV; this possibility needs to be investigated.

23.2.2 Regulation

Phorbol esters, granulocyte–macrophage colony stimulating factor (GM-CSF), interferon-γ and TGF-β1 are potent inducers of the maturation of monocyte precursors towards a macrophage phenotype (De Benedetti *et al.,* 1990; Harris and Ralph, 1985; Kamijo *et al.,* 1990). Differentiation of HL-60 and U937 cells into macrophages by such inducers was accompanied by a marked increase in cell-surface concentration and affinity of the enzymes (Laouar *et al.,* 1992, 1993b) (see *Figure 23.1a*). In contrast, HL-60 cells, induced to differentiate into a mature neutrophil-like phenotype by retinoic acid or dimethyl sulphoxide (DMSO) (Breitman *et al.,* 1980; Harris and Ralph, 1985), down-regulated cell-surface aminopeptidase activity, whereas DPP-IV activity was only minimally affected (Laouar *et al.*, 1992, 1993b) (see *Figure 23.1a*). Down-regulation of aminopeptidase activity was caused by a reduction in surface enzyme concentration associated with a decrease in the apparent affinity of the enzyme(s) for the substrates. Neither retinol nor TNF-α, which both inhibit cell proliferation without inducing HL-60 cell differentiation, modified cell-surface peptidase activities. Moreover, this down-regulation of cell-surface aminopeptidases was not correlated with an increased release into the culture medium, thus suggesting that this decrease could be caused by reduced rates of enzyme synthesis, or by endocytosis of the ectopeptidases, as reported for CD10 on activated neutrophils (Ërdos *et al.,* 1989).

With regard to γ-GT, we observed that the levels of 2.4 kb γ-GT mRNA species increased in cells undergoing granulocyte or macrophage differentiation upon addition of any of the inducers of granulocyte–macrophage differentiation tested (Baùvois *et al.,* 1995) (see *Figure 23.2*) and this was accompanied by a significant enhancement of γ-GT activity in the cytosol of stimulated cells (Bauvois *et al.,* 1995). However, no up-modulation of γ-GT expression was observed at the surface of stimulated cells (see *Figure 23.1a*), nor were detectable levels of soluble γ-GT released into the culture supernatants of differentiating cells (Bauvois *et al.,* 1995).

We observed comparable results with normal blood cells: the mature granulocytes expressed at their surface low levels of aminopeptidase, γ-GT and DPP-IV activities, whereas monocytes, differentiated *in vitro* to mature macrophages, showed significantly enhanced aminopeptidase and DPP-IV activities (Bauvois *et al.,* 1995) (see *Figure 23.1b*).

In conclusion, our data indicated that the differentiation process observed in the two immature HL-60 and U937 cell lines, reflects molecular events in normal myeloid differentiation: up- or down-expression of ectopeptidase correlates with the acquisition of a differentiated phenotype (granulocyte or monocyte/macrophage). The exact biochemical mechanisms by which ectopeptidase expression is altered remain to be determined. In particular, we do not know whether these ectopeptidases relate only to functions of differentiating cells, or whether they are also involved in mature cells.

23.2.3 Pathology

Dysregulation peptidase levels in blood cells may be important in malignant pathology. DPP-IV/CD26, aminopeptidase-N/CD13 and γ-GT have been shown to be overexpressed in a number of leukaemias, including acute myelomonocytic leukaemias (Ashmun *et al.,* 1990; Morell *et al.,* 1986; Scott *et al.,* 1988). Our results on cell-surface aminopeptidase and DPP-IV activities determined in cells from two patients with acute myeloblastic leukaemia (AML3) and monoblastic leukaemia (AML5) are shown in *Figures 23.1c* and *d* respectively. Aminopeptidase on AML3 cells was markedly higher than in normal granulocytes. *In vitro* differentiation of AML3 cells into granulocytes by retinoic acid was accompanied by a down-regulation of aminopeptidase expression (see *Figure 23.1c*). AML5 cells, induced to differentiate into macrophages in the presence of γ-interferon (IFN-γ), up-regulated both aminopeptidase and DPP-IV activities (see *Figure 23.1d*). These data suggest that such ectoenzyme activities may serve as diagnostic predictors in diseases before and after therapy.

23.3 Proteolytic processing of TNF-α, IL-6 and IGF-1 by myelomonocytic ectopeptidases: involvement of γ-GT and DPP-IV

Proteases secreted by stimulated granulocytes and monocytes are thought to be responsible for the breakdown of components of the extracellular environment at sites of inflammation. It has become important to determine whether ectopeptidases might play the same role as secreted enzymes. We have incubated intact U937 or HL-60 cells (or purified plasma membrane fractions) with a panel of radiolabelled molecules. No processing of acid and basic fibroblast growth factors, epidermal growth factor, GM-CSF, IFN-α and IFN-γ, interleukin-1β (IL-1β) and transforming growth factor-β1 (TGF-β1) was observed. However, we did find that intact U937 cells as well as their purified plasma membranes were capable of degrading labelled recombinant human TNF-α ([^{125}I]-TNF-α) into biologically inactive fragments of 2 kDa or less (inactive for cytotoxic activity towards murine L929 cells) (Bauvois *et al.,* 1992) (see *Figure 23.3a*). Proteolysis of TNF-α sequentially produced fragments of 8 kDa and 2 kDa or less. It has been reported that cathepsin G and elastase, purified from neutrophils, could degrade TNF-α into 11 kDa and 7.6 kDa fragments (Van Kessel *et al.,* 1991). However, under our experimental condi-

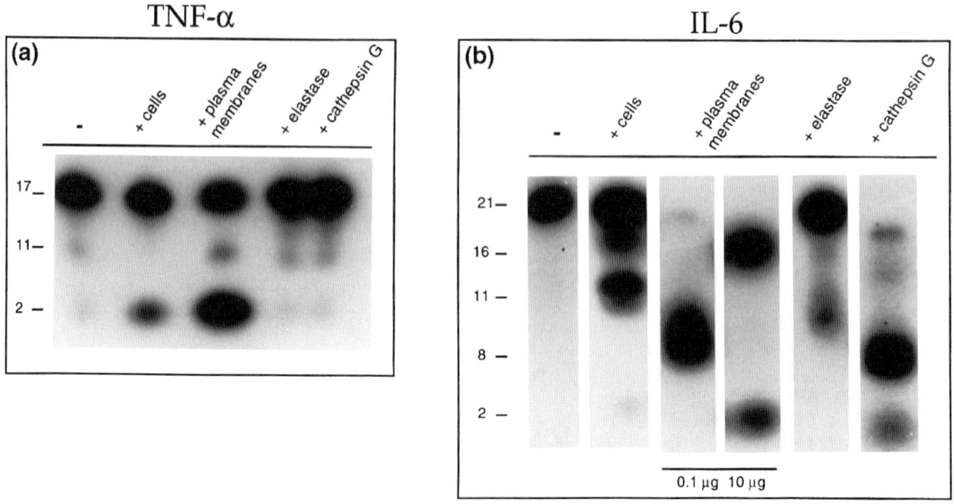

Figure 23.3. Processing of TNF-α and IL-6 (analysed by SDS-PAGE) by intact U937 cells, by plasma membranes prepared from U937 cells and by commercial proteases. [^{125}I]-TNF-α (5 ng) or [^{125}I]-IL-6 (10 ng) were incubated at 37°C for 2 h in the absence or presence (2×10^5) of intact U937 cells, or for 18 h with plasma membranes (10 μg), cathepsin G (2 mU) or elastase (2 mU). The samples were analysed by SDS-PAGE under reducing conditions, using 20% acrylamide monomer and a Pharmacia Minigel system, and followed by autoradiography of reaction products. (a) TNF-α processing; (b) IL-6 processing. Position of markers are shown in kDa.

tions, active elastase and cathepsin G, purified from U937 cells, were incapable of hydrolysing TNF-α (Bauvois *et al.*, 1992) (see *Figure 23.3a*). Degradation of [^{125}I]-TNF-α by plasma membranes was completely prevented by diisopropyl phosphofluoridate and partially prevented by diprotin A, a specific inhibitor of DPP-IV (Bauvois *et al.*, 1992). Thus, a serine ectopeptidase (distinct from elastase and cathepsin G) and including DPP-IV appeared to participate in TNF-α degradation and inactivation.

In contrast to TNF-α, recombinant human IL-6 was degraded by cathepsin G, purified from U937 cells, into 8 kDa fragments (Laouar *et al.*, 1993, 1994) whereas U937 cell elastase was ineffective (Laouar *et al.*, 1994) (see *Figure 23.3b*). Purified plasma membranes, as well as intact U937 cells, degraded [^{125}I]-IL-6 into 8–11 kDa fragments (see *Figure 23.3b*) and cathepsin G activity associated with the U937 cell-surface was indeed involved in this degradation. Moreover, a more detailed study of the processing of [^{125}I]-IL-6 indicated that the 8 kDa fragments were obtained by the degradation of the 11 kDa intermediates (Laouar *et al.*, 1993a) (see *Figure 23.3b*). The most interesting finding was that prolonged incubation of the 8 kDa [^{125}I]-IL-6 fragments with U937 cells or plasma membranes led to their dimerization through a specific cell-surface transpeptidation reaction. This was insensitive to group inhibitors of metallo-, aspartate- and serine-peptidases (Laouar *et al.*, 1993a) (see *Figure 23.3b*), thus excluding the potential participation of DPP-IV, elastase and cathepsin G activities. The dimerization process was, however, inhibited by acivicin and glutathione, reagents which are specific inhibitors of γ-GT on the myeloid cell surface (Laouar *et al.*, 1994). Surprisingly, solubilized plasma membranes from U937 cells which were cleared of γ-GT activity by immunoprecipitation with anti-γ-GT and adsorption on Protein-G–Sepharose, were still able to convert the intermediate forms of IL-6 into dimers. Our observations thus indicated that the ectotranspeptidase involved in the formation of the dimeric species of IL-6 belonged to the family of γ-GT. Of note, γ-GT activity was demonstrated to be present on cells of all haematopoietic lineages. Although the 8 and 16 kDa intermediates did not retain growth-stimulating activity for 7TD1 hybridoma cells, the possibility that they displayed other biological properties remains an open question.

It is known that carbohydrate side chains can stabilize and protect proteins from proteolytic degradation. Our work indicated that natural N or O glycosylated IL-6 was as sensitive to cell-surface-associated cathepsin G and γ-GT activities as the recombinant unglycosylated form of IL-6 from *Escherichia coli*. This latter property may be of significance in relation to the pharmacokinetics of the recombinant IL-6 which has been used in a number of clinical studies.

Natural human IGF-1 is a 7.6 kDa single-chain, non-glycosylated peptide. Like IL-6, it was processed, by a combination of hydrolysis and transpeptidation, into a 14 kDa peptide which was at least equipotent with native IGF-1 in the BALB/c 3T3 fibroblast DNA synthesis response (Laouar *et al.*, 1994).

Together, our observations may be relevant to the reported presence of high molecular mass species of IL-6 (17–85 kDa) and IGF-1 (21.5–26 kDa) found in human serum and in some cell culture supernatants (Clemmons and Shaw, 1986; Helfgott *et al.*, 1989; Kestler *et al.*, 1995; Revel, 1989; Sancéau *et al.*, 1991).

23.4 γ-GT and its signalling functions in human myelomonocytic cells

Several recent *in vitro* studies, employing monoclonal antibodies and/or synthetic inhibitors of DPP-IV (CD26), endopeptidase-24.11 (CD10) or AP-N (CD13), have indicated that these ectopeptidases are involved in regulating cell activation, proliferation and cytokine secretion (Marguet *et al.*, 1993; Mari *et al.*, 1992; Shipp and Look, 1993). Nichols *et al.* (1989) initially reported that HL-60 cells treated with the specific inhibitor of γ-GT, acivicin, were induced to become monocyte-like cells, as assessed by increased naphthyl butyrate esterase and H_2O_2 production. Our results have confirmed these data and have revealed several new findings. When incubated with HL-60 cells, acivicin produced a dose- and time-dependent inhibitory growth effect (Bauvois *et al.*, 1995) (see *Figure 23.4a*), associated with the withdrawal of most cells from the cell cycle, which accumulated in the G_0–G_1 phase. Acivicin modulated the mRNA levels encoded by the retinoblastoma (Rb) tumour suppressor gene and the *c*-myc gene (see *Figure 23.4b*), whose nuclear proteins are involved in the control of the cell cycle and transcription. According to Boyd and Metcalf (1984), the G_1 phase is implicated as a control point for HL-60 differentiation. We have found that acivicin-stimulated HL-60 and U937 cells exhibited a marked shift towards a macrophage morphology, associated with an enhancement of the maturation markers, CD71 and CD11b (Bauvois *et al.*, 1995) (see *Figure 23.5a* and *b*). When cultured

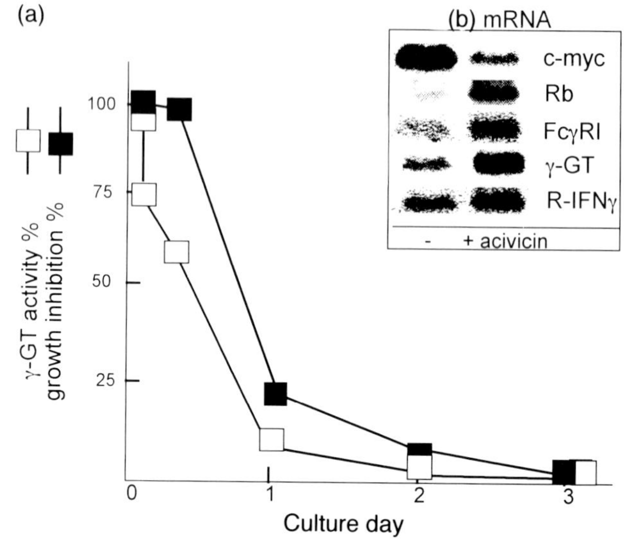

Figure 23.4. Effects of acivicin on growth, cell-surface γ-GT activity and mRNA expression in HL-60 cells. HL-60 cells were cultured in the absence (control) or in the presence of 10 μg ml^{-1} acivicin for up to 3 days. After each day of incubation, HL-60 cells were assayed for cell number and γ-GT activity, which was assayed by incubating cells (10^6/ml) with 1 mg ml^{-1} γ-Glu-pNA and 20 mM Gly–Gly for 30 min at 37°C. Results are expressed relative to the controls (=100). Total RNAs from unstimulated (–) or stimulated cells for 2 days with 10 μg ml^{-1} acivicin (+) were extracted by acid guanidinium thiocyanate–phenol–chloroform extraction. Northern blots were hybridized with [^{32}P]-labelled c-myc, Rb, γ-GT, FcγRI, R-IFN-γ or GAPDH cDNAs. GAPDH mRNA (not shown) was taken as a control because it did not change upon treatment with acivicin.

in the presence of acivicin, freshly isolated monocytes also underwent characteristic changes in morphology and antigenic phenotype consistent with their differentiation into macrophages (Bauvois *et al.,* 1995) (see *Figure 23.5c* and *d*). As expected, macrophage maturation correlated with the up-regulation of aminopeptidase and DPP-IV activities at the surface of acivicin-treated cells. The acquisition of macrophage phenotype in acivicin-treated cells prompted us to search for modulation of expression of some myeloid genes. Differentiation of acivicin-treated HL-60 cells along the macrophage pathway was accompanied by an increase in γ-GT, FcγRI and IFN-γ receptor (R-IFN-γ) mRNA levels (see *Figure 23.4b*) whereas mRNAs specific for TNF-α, 2′5′ oligoA synthetase and IFN-responsive factor (IRF-1) were not induced.

We have investigated the possibility of an indirect effect of acivicin on myeloid cells that could involve growth factors or cytokines operating by an autocrine mechanism. Unstimulated HL-60 cells, U937 cells and monocytes, as well as cells stimulated by acivicin, did not secrete detectable levels of IL-1β, IL-10, TNF-α, GM-CSF or IL-6. In contrast, significant levels of TGF-β (neutralized by an antibody against TGF-β1/TGF-β2) were released by HL-60 and U937 cells (up to 1200 pg ml^{-1}) and by monocytes (up to 280 pg ml^{-1}) (Bauvois *et al.,* 1995). Although steady-state levels of TGF-β1 mRNA

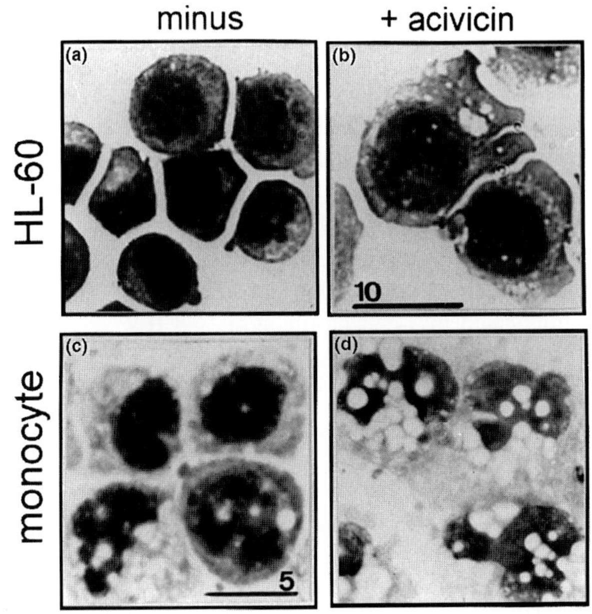

Figure 23.5. Light microscopic morphology of HL-60 cells and monocytes exposed to acivicin. Cultured cells were exposed to 25 μg ml^{-1} acivicin for 1 day (monocytes) or 3 days (HL-60). Bar, 5 or 10 μm. (a,c) Control cells; (b,d) acivicin-treated cells.

species were apparently not affected by acivicin treatment, our data did not rule out the possibility that cell activation via acivicin induces TGF-β synthesis. Moreover, the neutralizing antibody, anti-TGF-β1/β2 was unable to prevent cell differentiation induced by acivicin (Bauvois *et al.*, 1995).

Together, our data indicate that acivicin induced two separate events in human myeloid cells: macrophage maturation and cytokine secretion (latent TGF-β).

23.5 Prospects for future development

Haematopoietic cells communicate with their environment through several kinds of cell-surface receptors. Ectopeptidases are now viewed as physiological receptors. In this chapter, we have shown that certain ectopeptidases present on human myelomonocytic cells are differentially regulated in the two maturation pathways (granulocyte and monocyte/macrophage), during which critical regulatory events are likely to occur. Thus, one could expect that one aspect of these ectopeptidases might involve changes in their activity depending on the state of cell activation in response to environmental signals. In this context, we need further identification of factors controlling the regulation of myeloid cell ectopeptidases. A 'protective' mechanism in normal development may have implications for preventing the process of cell transformation. Indeed, any mechanism for reducing or increasing the levels of ectopeptidases could result in substantial pathological problems.

A positive correlation has already been observed between over-expression of an ectopeptidase and hyperproliferative properties of leukaemia cells, as in acute myeloblastic and myelomonocytic leukaemias, which express higher γ-GT activity than do normal monocytes (Morell *et al.,* 1986). Thus, the inhibition of enzymes such as γ-GT, could be a new approach to suppressing the growth or metastasis of cancer. Our studies using a γ-GT inhibitor, acivicin, suggested that the blockade of γ-GT activity at the surface of myelomonocytic precursors induced cell activation leading to arrest of growth, macrophage maturation and TGF-β secretion.

Some other ectopeptidases, including endopeptidase-24.11 (CD10) and DPP-IV (CD26), have been implicated in T-cell proliferation and IL-2 secretion (Fleischer 1992; Marguet *et al.,* 1993; Mari *et al.,* 1992). Although the effects of peptidase inhibitors are still difficult to evaluate, these data do suggest that such ectopeptidases participate in some critical steps of cellular responses. Type I and type II cell-surface peptidases have very short intracytoplasmic tails and it is unclear whether these enzymes are connected to functional molecules through which might be channelled activation signals. It will become essential to define the molecular mechanism(s) underlying ectopeptidase-mediated cell activation. Attempts to specifically control the action of ectopeptidases are hampered by our ignorance of the physiological substrates as well as by the lack of effective inhibitors.

In an effort to identify the potential substrates of myeloid ectopeptidases, we showed that ectopeptidases, by degrading and inactivating cytokines, may account for the regulation of the extracellular cytokine concentration. Through the sequential actions of surface-associated cathepsin G and γ-GT, human myeloid cells produced *in vitro* dimeric species of IL-6 and IGF-1. IL-6 and IGF-1 are cytokines influencing both cell proliferation and differentiation, by binding to their respective surface receptors. With regard to high molecular mass species of IGF-1 and IL-6 found in human fluids, our observations may be of physiological relevance for the synthesis of IL-6 and IGF-1-derived molecules. Determinations of the chemical structures of these dimers and investigation of their binding capacity to cell-surface receptors will help to establish their possible biological roles.

Acknowledgements

This work was supported by grants from Institut National de la Santé et de la Recherche Médicale (INSERM) and Association pour la Recherche sur la Cancer (ARC No. 6689). The author gratefully acknowledges E. Barthélémy, Dr A. Laouar, D. Rouillard, Dr C. Villiers and Dr J. Wietzerbin for their participation in this work and A. Birot for her secretarial expertise. The author thanks Dr G. Guellaën for the gift of human liver γ-GT cDNA, Dr B. Seed for the gift of human T CD26 cDNA and Dr M. Hegen for the gift of CD26-transfected Jurkat T cells. The author is indebted to Dr C. Chomienne for providing AML3 cells and Dr C. Mathiot for providing AML5 cells.

References

Ashmun RA, Look AT. (1990) Metalloprotease activity of CD13/aminopeptidase-N on the surface of human myeloid cells. *Blood* **75**: 462–469.

Bauvois B, Sancéau J, Wietzerbin J. (1992) Human U937 cell-surface peptidase activities : characterization and degradative effect on tumour necrosis factor-α. *Eur. J. Immunol.* **22**: 923–930.

Bauvois B, Laouar A, Rouillard D, Wietzerbin J. (1995) Inhibition of γ-glutamyl transpeptidase activity at the surface of human myeloid cells is correlated with macrophage maturation and transforming growth factor-β production. *Cell Growth Differen.* **6**: 1163–1170.

Böhm SK, Gum JR. Jr, Erickson RH, Hicks JW, Kim YS. (1995) Human dipeptidyl peptidase IV gene promoter: tissue-specific regulation from a TAT-less GC-rich sequence characteristic of a housekeeping gene promoter. *Biochem. J.* **311:** 835–843.

Boyd AW, Metcalf D. (1984) Induction of differentiation in HL-60 leukemic cells: a cell cycle dependent all-or-none event. *Leuk. Res.* **8:** 27–32.

Breitman TR, Selonick SE, Collins SJ. (1980) Induction of differentiation of the human promyelocytic leukemia cell line (HL-60) by retinoic acid. *Proc. Natl Acad. Sci. USA* **77**: 2936–2940.

Bulle F, Matte MG, Siegrist S, Pawlak A, Passage E, Chobert MN, Laperche Y, Guellaën G. (1987) Assignment of the human γ-glutamyl transferase gene to the long arm of chromosome 22. *Hum. Genet.* **76**: 283–286.

Clemmons DR, Shaw DS. (1986) Purification and biologic properties of fibroblast somatomedin. *J. Biol. Chem.* **261**: 10293–10298.

De Benedetti F, Falk LA, Ellingsworth LR, Ruscetti FW, Flatyneck CR. (1990) Synergy between transforming growth factor-β and tumor necrosis factor-α in the induction of monocytic differentiation of human leukemic cell lines. *Blood* **75**: 626–632.

Erdös EG, Wagner B, Harbury CB, Painter RG, Skidgel RA, Fa XG. (1989) Down-regulation and inactivation of neutral endopeptidase 24.11 (enkephalinase) in human neutrophils. *J. Biol. Chem.* **264:** 14519–14523.

Fleischer B. (1992) CD26: a surface protease involved in T-cell activation. *Immunol. Today* **15**: 180–184.

Grisk O, Küster U, Ansörge S. (1993) The activity of γ-glutamyl transpeptidase (γ-GT) in populations of mononuclear cells from human peripheral blood. *Biol. Chem. Hoppe-Seyler* **374:** 287–290.

Guellaën C. (1989) Rôle physiologique de la γ-glutamyl transpeptidase. *Med. Sci.* **5:** 637–644.

Harris P, Ralph P. (1985) Human leukemic models for myelomonocytic development: a review of the HL-60 and U937 cell lines. *Leuk. Biol.* **37**: 407–422.

Hegen M, Niedobitek G, Klein CB, Stein H, Fleischer B. (1990) The T cell triggering molecule Tp103 is associated with dipeptidyl aminopeptidase IV activity. *J. Immunol.* **144:** 2908–2914.

Helfgott DC, Tatter SB, Santhanam U, Clarick RH, Bhardwaj N, May LT, Seghal PB. (1989) Multiple forms of IFN-β2/IL-6 in serum and body fluids during acute bacterial infection. *J. Immunol.* **142**: 948–953.

Kamijo R, Takeda K, Nagumo M, Konno K. (1990) Effects of combinations of transforming growth factor-β1 and tumour necrosis factor on induction of differentiation of human myelogenous leukemic cell lines. *J. Immunol.* **144**: 1311–1316.

Kestler DP, Agarwal S, Cobb J, Goldstein KM, Hall RE. (1995) Detection and analysis of an alternatively spliced isoform of interleukin-6 mRNA in peripheral blood mononuclear cells. *Blood* **86:** 4559–4567.

Laouar A, Bauvois B. (1992) Characterization and modulation of cell-surface proteases on human myeloblastic (HL-60) cells and comparison to normal myeloid cells. *Immunol. Lett.* **34:** 257–266.

Laouar A, Villiers C, Sancéau J, Maison C, Colomb M, Wietzerbin J, Bauvois B. (1993a) Inactivation of interleukin-6 *in vitro* by monoblastic U937 cell plasma membranes involves both protease and peptidyl-transferase activities. *Eur. J. Biochem.* **215**: 825–831.

Laouar A, Wietzerbin J, Bauvois B. (1993b) Divergent regulation of cell-surface protease expression in HL-60 cells differentiated into macrophages with granulocyte macrophage colony stimulating factor or neutrophils with retinoic acid. *Int. Immunol.* **5**: 965–973.

Laouar A, Wietzerbin J, Bauvois B. (1994) Protease-catalyzed conversion of insulin-like growth factor-1 and interleukin-6 into high-molecular-mass species through the sequential action of hematopoietic surface-associated cathepsin G and γ-glutamyl transpeptidase-related activities. *Eur. J. Biochem.* **223:** 617–624.

Look AT, Ashmun RA, Shapiro LH, Peiper SC. (1987) Human myeloid plasma membrane glycoprotein CD13 (gp 150) is identical to aminopeptidase N. *J. Clin. Invest.* **83**: 1299–1307.

Marguet D, David F, Vivier I, Bernard AM, Naquet P, Pierres M. (1993) Dipeptidyl peptidase IV (CD26): expression, structure and functions in lymphocytes. *Adv. Neuroimmunol.* **3:** 209–215.

Mari B, Checler F, Ponzio G, Peyron JF, Manie S, Farahifar D, Rossi B, Auberger P. (1992) Jurkat T cells express a functional neutral endopeptidase activity (CALLA) involved in T cell activation. *EMBO J.* **11:** 3875–3885.

Morell A, Losa G, Carrel S, Heumann D, Von Fliedner VE. (1986) Determination of ectoenzyme activities in leukemic cells and in established hematopoietic cell lines. *Am. J. Hematol.* **21:** 289–298.

Nichols KE, Chitneni SR, Moore JO, Weinberg JB. (1989) Monocytoid differentiation of freshly isolated human myeloid leukemia cells and HL-60 cells by the glutamine antagonist acivicin. *Blood* **74**: 1728–1737.

Painter RG, Dukes R, Sullivan J, Carter R, Erdös EG, Johnson AR. (1988) Function of neutral endopeptidase on the cell membrane of human neutrophils. *J. Biol. Chem.* **263**: 9456–9461.

Revel M. (1989) Host defense against infection and inflammation: role of the multifunctional IL-6/IFN-β2 cytokine. *Experientia* **45**: 549–557.

Rom WN, Basset P, Fells GA, Nukiwa T, Trapnell BC, Crystal RG. (1988) Alveolar macrophages release an insulin like growth factor I-type molecule. *J. Clin. Invest.* **82**: 1685–1693.

Sancéau J, Wijdenes J, Revel M, Wieztzerbin J. (1991) IL-6 and IL-6 receptor modulation by IFN-γ and tumour necrosis factor-α in human monocytic cell line THP-1. *J. Immunol.* **147**: 630–637.

Scott CS, Stark AN, Minowada J, Drexler HG. (1988) Quantitative and qualitative studies of leukaemic cell dipeptidyl peptidases II and IV. *Leuk. Res.* **12**: 129–134.

Shipp M, Look AT. (1993) Hematopoietic differentiation antigens that are membrane-associated enzymes: cutting is the key. *Blood* **68**: 1052–1070.

Tanaka T, Camerini D, Seed B, Torimoto Y, Dang NH, Kameoka J, Dahlberg HN, Schlossman SF, Morimoto C. (1992) Cloning and functional expression of the T cell activation antigen CD26. *J. Immunol.* **149:** 481–486.

Tate SS, Meister A. (1981) γ-glutamyl transpeptidase: catalytic, structural and functional aspects. *Mol. Cell Biol.* **39**: 357–368.

Van Kessel KPM, Van Strijp JAG, Verhoef J. (1991) Inactivation of recombinant human tumour necrosis factor-α by proteolytic enzymes released from stimulated human neutrophils. *J. Immunol.* **147:** 3862–3868.

Chapter 24

Expression of cell-surface peptidases in neoplastic cells

David M. Nanus, Christos N. Papandreou and Anthony P. Albino

24.1 Introduction

In 1989, Kenny and colleagues proposed that cell-surface peptidases play a key role in the control of growth and differentiation by modulating the activity of peptide factors and regulating their access to adjacent cells (Kenny *et al.*, 1989). They suggested that cell-surface peptidases might either activate or inactivate peptide growth factors. This idea has direct implications for the development of neoplasia. Two basic mechanisms of cell-surface peptidase involvement in the malignant process can be predicted: loss of function, resulting in an inability of the transformed cell to inactivate a mitogenic peptide or activate an inhibitory peptide; and gain of function, resulting in the activation of a mitogenic peptide or the inactivation of an inhibitory peptide. The molecular and genetic events which could lead to alterations in cell-surface peptidase function in a cancer cell are many, and include gene amplification, deletion, mutation, and loss or expression of a cell-surface peptidase and/or its inhibitor which upsets the balance between an active and inactive peptidase. These alterations in cell-surface peptidase activity may disrupt normal cellular homeostasis and contribute to neoplastic transformation and progression. Data which implicate specific cell-surface peptidases in the pathogenesis of human cancers are now beginning to emerge. This chapter will review the current understanding of the involvement of cell-surface peptidases in human tumours.

24.2 Haematological malignancies

The classification of leukaemias and lymphomas is based in part on the expression of cell-surface antigens which are also present on normal precursor haematopoietic cells. Many of these haematopoietic differentiation antigens, which have been assigned cluster differentiation (CD) numbers, are cell-surface peptidases (CD10: endopeptidase-24.11, E-24.11; CD13: aminopeptidase N, AP-N; CD26: dipeptidyl peptidase IV, DPP-IV). These ectoenzymes are involved in regulating normal haematopoiesis through regulation of cellular responses to peptides (see also Chapters 22 and 23), and appear to also play a role in the development and maintenance of certain haematopoietic malignancies.

Cell-Surface Peptidases in Health and Disease, edited by A.J. Kenny and C.M. Boustead.

24.2.1 E-24.11/CALLA/CD10

The common acute lymphoblastic leukaemia antigen (CALLA or CD10), which was initially identified as a diagnostic marker for acute lymphoblastic leukaemias (ALL), is identical to E-24.11. It is normally expressed by pre-B and early-B lymphocytes during normal B-cell differentiation, prior to immunoglobulin light chain commitment (McGinnes *et al.,* 1991), and by mature neutrophils, but not by mature lymphocytes, eosinophils, monocytes or basophils (Iwamoto *et al.,* 1991). CD10 is also expressed on bone marrow stromal cells which support B-lymphoid progenitor development and is involved in regulating the earliest stages of stromal cell-dependent B lymphopoiesis (Salles *et al.,* 1992).

Expression of CD10 by leukaemic cells is indicative of the differentiation stage of normal lymphopoiesis from which the malignant cell derives (see *Figure 24.1*). Thus, CD10 is found on non-T-cell ALL cells which derive from pre-B lymphocytes, but not on leukaemia cells or lymphomas which originate in more mature B or T cells. Approximately 90% of paediatric and 65–75% of adult cases of non-T-cell ALL are CD10 positive, and patients with CD10-positive ALL have a more favourable outcome in both children and adults (Bradstock, 1993). The CD10 gene in ALL cells is not mutated (Kee *et al.,* 1991) and the protein is enzymically active (Jongeneel *et al.,* 1989). The function of CD10 in normal and malignant lymphopoiesis is unclear, but may involve the regulation of B-cell development through inactivation of a peptide substrate which stimulates B-cell proliferation and differentiation; or the activation of a peptide precursor whose cleavage product inhibits early lymphoma development (Salles *et al.*, 1993).

Expression of CD10 is also found on chronic myelogenous leukaemia cells in lymphoid blast crisis, occurring in 35–40% of patients and representing transformation of a pluripotent myeloid stem cell (Hokland *et al.,* 1983). In addition, CD10 is expressed by virtually all Burkitt's lymphomas and nodular poorly differentiated lymphocytic lymphomas, which derive from the mature B cell, and it is found on 10–20% of T-cell ALLs and 40% of T-cell lymphoblastic lymphomas (Ritz *et al.,* 1981), indicating that these malignancies arise from an early-B-cell progenitor cell which still expresses T-cell antigens. In contrast with ALL, no correlation has been established between CD10 positivity in lymphomas and clinical outcome. However, CD10 expression is occasionally present on multiple myeloma cells with an immature histological appearance and

---

Figure 24.1. Haematopoietic differentiation, neoplastic transformation and expression of cell-surface peptidases. Lymphoid and myeloid stem cells originate from a pluripotent stem cell. The lymphoid stem cell further differentiates into either B- or T-lymphocyte progenitors, which further differentiate into mature B and T cells. CD10 (E-24.11) is normally expressed by pre-B cells; murine AP-A (BP-1/6C3) is also expressed at this stage of differentiation. CD26 (DPP-IV) is normally expressed during the final stages of T-cell differentiation. Haematological malignancies that arise from pre-B cells, or mature thymocytes/mature T cells, will express CD10, or CD26, respectively. The myeloid stem cell gives rise to two major lines of differentiation, maturing into either neutrophils or monocytes. Both lineages express CD13 (AP-N) at all stages of maturation. As a result, most acute and chronic myeloid leukaemia cells express CD13. ATL, acute T-cell lymphocytic leukaemia; CFU-GM, colony-forming unit granulocyte-monocyte; CTCL, cutaneous T-cell lymphoma; T-CLL, T-cell chronic lymphocytic leukaemia; Tγ-LPD, Tγ-cell lymphoproliferative disease.

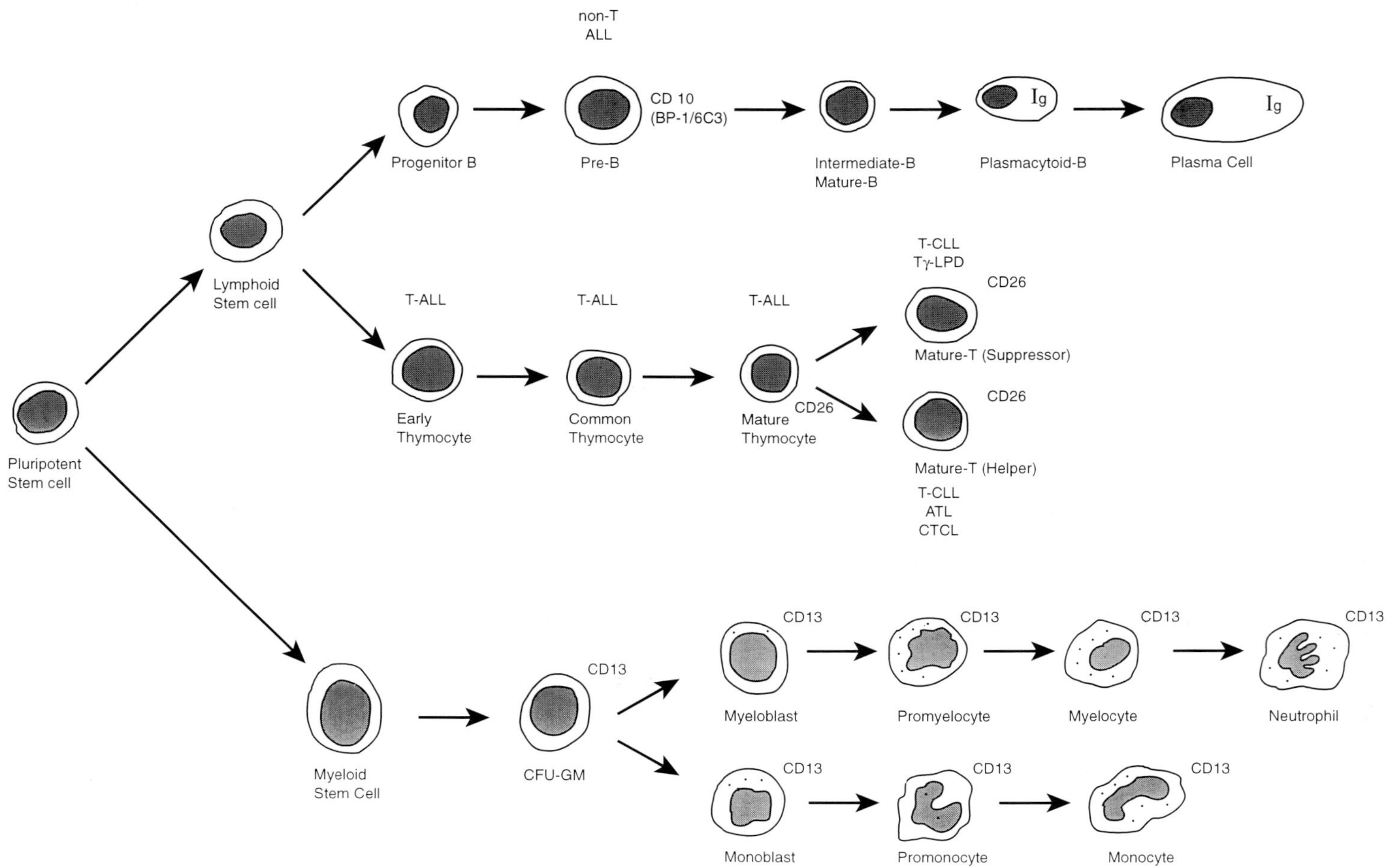

non-T
ALL
CD 10
(BP-1/6C3)
Progenitor B
Pre-B
Intermediate-B
Mature-B
Plasmacytoid-B
Plasma Cell
Ig
Ig
Lymphoid
Stem cell
T-ALL
T-ALL
T-ALL
T-CLL
Tγ-LPD
CD26
Mature-T (Suppressor)
CD26
CD26
Early
Thymocyte
Common
Thymocyte
Mature
Thymocyte
Mature-T (Helper)
T-CLL
ATL
CTCL
Pluripotent
Stem cell
CD13
Myeloid
Stem Cell
CFU-GM
CD13
CD13
CD13
CD13
Myeloblast
Promyelocyte
Myelocyte
Neutrophil
CD13
CD13
CD13
Monoblast
Promonocyte
Monocyte

in this disease, CD10 positivity correlates with aggressive clinical disease (Durie and Grogan, 1985).

24.2.2 AP-N/CD13

CD13, which is identical to AP-N, was initially identified as a glycoprotein expressed by committed granulocyte-monocyte progenitors and by cells of the granulocytic and monocytic lineages at distinct stages of differentiation (Griffin *et al.*, 1981, 1983). CD13 is expressed by 76% of acute myelogenous leukaemias (AML), more frequently in those of a myelomonocytic subtype (M4), by 92% of chronic myelogenous leukaemias, but by only 1% of ALLs (Foon and Todd, 1986). Although CD13 and CD10 are jointly expressed on non-haematopoietic tissues, expression of these two enzymes on leukocytes is mutually exclusive, such that their expression is used to discriminate myeloid and lymphoblastic leukaemias.

Adult patients with CD13-positive AML have a lower probability of responding to chemotherapy and an inferior survival (Schwarzinger *et al.,* 1990). CD13 expression is also associated with a lower response rate and shorter survival time in adult patients with non-T-cell ALL, but not in adults with T-cell ALL (Sobol *et al.,* 1987). The function of CD13 in leukaemogenesis is unclear but ALL cells with high CD13 expression are highly mobile and capable of migration through bone marrow stroma and extracellular matrix proteins (Makrynikola *et al.,* 1994), suggesting that AP-N activity may facilitate tumour cell invasion as has been proposed in some non-haematopoietic cells.

24.2.3 Aminopeptidase-A/BP-1/C63

Aminopeptidase A (AP-A) expression in murine lymphoid development is restricted to cytoplasmic immunoglobulin-positive pre-B lymphoid cells, but it is not expressed by mature B-lineage cells in blood, spleen, Peyer's patches or lymph nodes (Cooper *et al.,* 1986). Enzymically active AP-A is present on neoplastic murine pre-B-cell lines (Wu *et al.,* 1989) and on virally transformed murine pre-B cells (Ramakrishnan *et al.,* 1990). Low levels of AP-A have been detected on some immortal human pre-B-cell lines but not on B cells or T cells (Li *et al.,* 1993), suggesting that this cell-surface peptidase is only expressed during a very early stage of human precursor B-cell development, and on transformed cells which derive from cells at that stage of B-cell maturation. The function of AP-A in early lymphoid haematopoiesis and transformation is unknown. AP-A expression can be regulated by bone marrow stromal cell-derived interleukin-7 (IL-7) (Welch *et al.,* 1990), and IL-7 can induce early B-cell proliferation. IL-7 is a theoretical substrate for AP-A (Li *et al.,* 1993), suggesting a regulatory function by AP-A, but a direct relationship has not been established.

24.2.4 DPP-IV/CD26

CD26 was initially characterized as a murine T-cell differentiation antigen and is predominantly expressed in humans by CD3-positive medullary thymocytes (Dang *et al.,* 1991) (see *Figure 24.1*). CD26 is also present on subsets of peripheral T cells (Fleischer, 1987), and its expression increases with T-cell activation by interleukin-2

(IL-2) and other mitogens (Mattern *et al.*, 1991). Studies suggest that CD26 can mediate antigen-induced T-cell proliferation (Dang *et al.*, 1990; Fleischer, 1987).

CD26 is expressed by most T-cell prolymphocytic leukaemias and T-cell ALLs (Crockard *et al.*, 1984). However, immature forms of T-cell leukaemias and Sezary syndrome lack CD26 reactivity, and progression of T-cell derived mycosis fungoides to a more malignant stage is associated with a decrease in CD26 expression, which may have prognostic significance (Sterry *et al.*, 1982). CD26 is occasionally expressed by B-cell lymphomas and surface immunoglobulin-positive B-cell ALLs, and in a subgroup of T-cell lymphoblastic lymphomas (Carbone *et al.*, 1995). Expression of CD26 in lymphoblastic lymphomas and in T-cell ALL is associated with inferior survival (Carbone *et al.*, 1995). CD26 expression is also found in most anaplastic large-cell lymphomas and approximately one half of T-cell-derived non-Hodgkin's lymphomas, but not in B-cell non-Hodgkin's lymphomas (Cozzi *et al.*, 1990) or in Reed–Sternberg cells of Hodgkin's lymphoma (Carbone *et al.*, 1994).

The mechanism by which CD26 confers a proliferative advantage to malignant T cells is unknown, but may involve a direct stimulatory effect on T-cell proliferation through the ability of CD26 to transduce mitogenic signals in T cells (Fleischer, 1994; Morimoto *et al.*, 1995). CD26-positive T lymphocytes, and T-cell lymphoblastic lymphoma cells into which CD26 cDNA has been stably introduced, produce large amounts of IL-2 (Morimoto *et al.*, 1995; Ulmer *et al.*, 1992), and T-cell proliferation is inhibited by CD26 enzyme inhibitors concomitant with a decrease in IL-2 production (Schon *et al.*, 1987). In addition, CD26 is also a receptor for fibronectin and collagen (Shipp and Look, 1993) and therefore may enhance the invasive properties of malignant T cells.

24.3 Solid tumour malignancies

The expression pattern of individual cell-surface peptidases in malignant cells and their normal counterparts has been characterized in many different solid tumour neoplasms (see *Table 24.1*). Recent studies have also begun to examine the mechanism and biological effect of activation or loss of specific ectoenzymes. In various tumour types, different cell-surface peptidases have been linked to the regulation of peptide growth factors, differentiation, tumour initiation, tumour cell invasion and the development of metastases.

24.3.1 Gastrointestinal neoplasms

Colon cancers frequently originate as adenomas which become invasive and progress to carcinomas. As AP-N is associated with tumour invasion in other tumour types (Fujii *et al.*, 1995), the pattern of AP-N expression in pre-malignant polyps and adenomas, and malignant colon carcinomas was assessed, but no relationship with the invasive phenotype could be demonstrated. AP-N is expressed in normal colonic epithelial cells at the brush border along the basal membrane, while in specimens from patients with colon polyps or cancers AP-N is found in normal mucosa adjacent to the tumours on the luminal membrane (Quaroni *et al.*, 1992). AP-N is not usually found in the tumour cells of adenomas or adenocarcinomas (Dixon *et al.*, 1994; Quaroni *et al.*, 1992; Young *et al.*,

Table 24.1. Expression of cell-surface peptidases in solid tumor malignancies

Tumour type	Cell-surface peptidase	Expression pattern	
		Normal	Neoplastic
Bladder cancer	E-24.11	Epithelium	Decreased with invasion
Basal cell carcinoma	AP-N	Not expressed	Tumour-associated stroma
	DPP-IV	Occ. keratinocytes	Carcinoma and stroma
Breast cancer	E-24.11	Myoepithelium	Tumour-associated stroma
	AP-N	Stromal fibroblasts	Tumour-associated stroma
		Epithelium	Occ. tumor cells
	DPP-IV	Stroma	Stroma
Colon cancer carcinomas	AP-N	Epithelium	Well differentiated adenomas
Oesophageal cancer	AP-N	ND	Metaplasia, carcinoma
Kidney cancer	AP-A	Proximal tubule	Absent 30% tumours
	AP-N	Proximal tubule	Absent 35% tumours
	DPP-IV	Proximal tubule	>90% tumours
	E-24.11	Proximal tubule	Weak expression
	ACE	Proximal tubule	Strong expression 100% tumours
Liver cancer	E-24.11	Hepatocytes	Increased in carcinomas
Lung cancer	DPP-IV	ND	Adenocarcinoma histology only
	E-24.11	Epithelium	Absent most carcinomas
Melanoma	AP-N	Not expressed	Dysplastic naevi (25%) Melanoma *in situ* (44%) Invasive melanoma (67%) Metastatic melanoma (100%)
	DPP-IV	Melanocyte	Absent in melanomas
	E-24.11	ND	All stages melanoma (30–40%)
Prostate cancer	AP-A	Not expressed	Tumour-associated stroma
	AP-N	Epithelium	Absent 75% tumours
	DPP-IV	Epithelium	Primary tumours, absent 50% metastases
	E-24.11	Epithelium	Decreased hormone-resistant
Stomach cancer	AP-N	Not expressed	Metaplasia, carcinoma
Squamous cell carcinoma	AP-N	Not expressed	Tumour-associated stroma
	DPP-IV	Occ. keratinocytes	Invasive tumour cells
Testicular cancer	ACE	Immature germ cells	Seminoma histologies
Thyroid cancer	DPP-IV	Not expressed	Carcinomas

ND: not done; occ.: occasionally.
Expression pattern of normal cellular counterpart and neoplastic cells unless otherwise indicated.

1992), and when present, the expression pattern demonstrates a loss of polarity and dense cytoplasmic staining (Young *et al.*, 1992). AP-N-expressing tumours are typically moderately or well-differentiated histologically (Quaroni *et al.*, 1992), in agreement with a study of colon cancer cell lines in which differentiated cancers express AP-N (Chantret *et al.*, 1988). Similar studies of colon cancer cell lines which undergo spontaneous enterocytic differentiation *in vitro*, have also demonstrated an increase in E-24.11 and DPP-IV expression concomitant with cell differentiation (Jalal *et al.*, 1992; Yoshioka *et al.*, 1991).

Analyses of cell-surface peptidase expression in upper gastrointestinal tract cancers (gastric, oesophagus, pancreas and liver) is predicated by the observation that numerous neuropeptides are growth factors for cells of these types of tumour (Townsend *et al.*,

1994), although few studies of peptidase expression in these tumours have been done. AP-N is not expressed by normal stomach epithelium, but can be detected in intestinal metaplasia and some gastric adenocarcinomas (Matsukura *et al.,* 1980; Osborn *et al.,* 1988). In oesophageal tumours, AP-N is present in 84% of pre-malignant oesophageal metaplasia specimens (Barret's oesophagus) and in 71% of oesophageal adenocarcinomas (Moore *et al.,* 1994). AP-N is normally expressed by the epithelium of pancreatic ducts and the epithelium of bile ducts and bile canaliculi, but its expression pattern in pancreatic or hepatocellular carcinoma has not been determined.

DPP-IV levels are significantly decreased in gastric cancer tissue and in the serum of patients with gastric cancer, and increase to normal levels following gastrectomy suggesting it could serve as a tumour marker (Yoshii *et al.,* 1981). In hepatocellular carcinomas, an increase in E-24.11 expression is present in comparison with normal liver (Dragovic *et al.,* 1994).

24.3.2 Genitourinary neoplasms

Renal cell carcinomas (RCCs) express a panel of kidney differentiation antigens, many of which are cell-surface peptidases [AP-A (URO-2), AP-N (URO-3), DPP-IV (URO-4)] (Bander *et al.,* 1989; Nanus *et al.,* 1993) (see *Figure 24.2*). While all of these antigens, in addition to E-24.11 and ACE, are expressed by proximal tubular cells of the nephron, the normal cellular counterpart of renal cancer cells, expression of one or more of these antigens is often absent in both primary and metastatic RCCs (Borowitz *et al.,* 1986; Ebert *et al.,* 1990; Finstad *et al.,* 1985). Loss of AP-A expression occurs in 30–40% of cultured RCCs (Ebert *et al.,* 1990), as a result of decreased transcription of the AP-A gene. AP-A protein is detected more frequently in RCC specimens, but may represent a catalytically inactive form of the AP-A protein (D.M. Nanus *et al.,* unpublished observations).

AP-N expression, like that of AP-A, is absent in about one-third of RCCs (Ebert *et al.,* 1990), and is due to lack of AP-N mRNA. Invasion of a highly metastatic AP-N-expressing renal cancer cell line into reconstituted basement membrane can be inhibited by a monoclonal antibody to AP-N (Saiki *et al.,* 1993) or by the AP-N inhibitor, bestatin (Yoneda *et al.,* 1992). AP-N expression and enzymic activity on RCC cells can be up-regulated by T-cell-derived cytokines, interleukin-4 (IL-4) and interleukin-13 (IL-13) (Riemann *et al.,* 1995), and the percentage of AP-N-expressing tumour-infiltrating T-cells in RCCs is increased (Riemann *et al.,* 1994), suggesting that AP-N may play an important, but undefined role in the renal cancer phenotype.

DPP-IV is expressed by virtually all RCCs (Ebert *et al.,* 1990), and its expression can be increased by IL-4 and IL-13 (Riemann *et al.,* 1995). E-24.11 is also present in most RCCs (Droz *et al.,* 1990) but displays weak immunoreactivity in a large proportion (Borowitz *et al.,* 1986). Angiotensin-converting enzyme (ACE) is expressed by RCCs (Takada *et al.,* 1985), as are both angiotensin II receptor subtypes (Goldfarb *et al.,* 1994), suggesting that angiotensin II may act as a growth factor for RCCs.

We examined the expression pattern of AP-A, AP-N and DPP-IV in benign and malignant prostate tissues, including primary and metastatic cancers (Bogenrieder *et al.,* 1997). While benign prostatic stroma exhibited no AP-A, AP-N or DPP-IV immunoreactivity, stromal cells surrounding prostatic carcinoma cells demonstrated

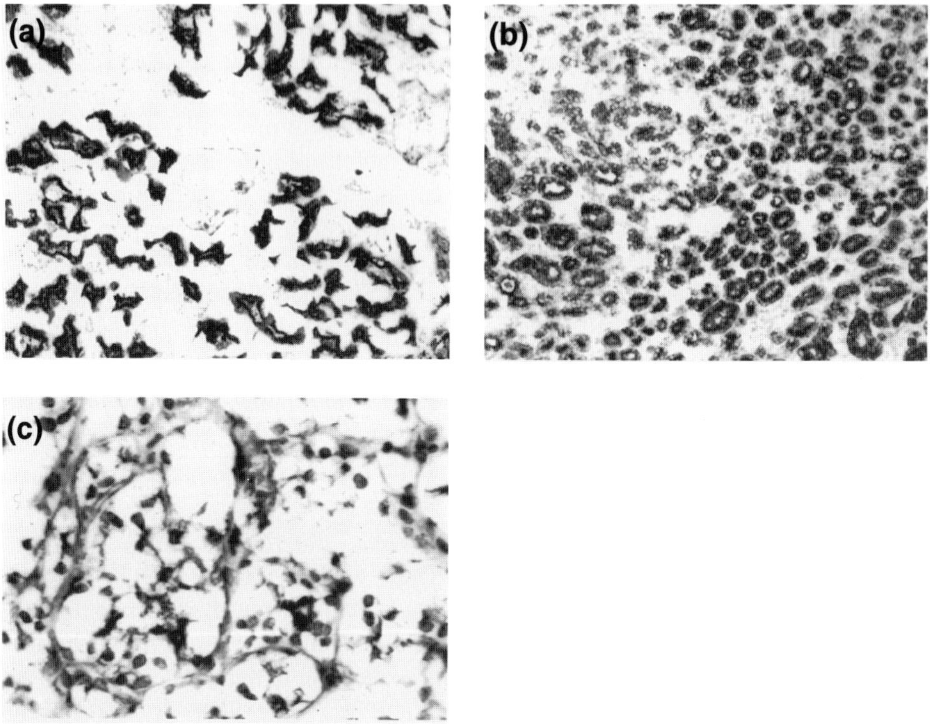

Figure 24.2. Aminopeptidase A expression in normal human kidney and renal cancer. (a) Normal renal cortex: AP-A immunoreactivity is present in the proximal tubule cells of the nephron. (b) Renal cell carcinoma, papillary growth pattern, chromophilic type: note strong staining and strong immunoreactivity in the cancer cells. (c) Renal cell carcinoma, acinar growth pattern, clear cell type (high power): note strong immunoreactivity along the cell-surface membrane of the cancer cells.

increased expression of AP-A in 73% of tumours. Benign prostatic epithelial cells strongly expressed AP-N and DPP-IV but not AP-A. In contrast, AP-N was minimally expressed or absent in three-quarters of tumour specimens. DPP-IV was strongly expressed in virtually all primary prostate cancers, but could be detected in only one-half of metastases. This result is similar to a previous study which also implicated loss of DPP-IV expression in metastases (Dinjens *et al.,* 1990).

E-24.11, which is also expressed by normal prostate epithelial cells (Erdos *et al.,* 1985), can hydrolyse a number of neuropeptide growth factors for prostate cancer cells, including bombesin, neurotensin and endothelin-1. E-24.11 is expressed by prostate cancer cells which are still dependent on androgen for growth, but lost in metastatic prostate cancers which continue to proliferate despite androgen withdrawal (Papandreou *et al.,* 1997). This occurs in part because E-24.11 is transcriptionally activated by androgen and transcriptionally repressed by androgen deprivation (Papandreou *et al.,* 1997). This potentially leads to an increased concentration of neuropeptides following androgen withdrawal, a standard treatment for metastatic prostate cancer, and propagation of an androgen-independent phenotype which is universally lethal.

In superficial bladder cancer, E-24.11 is highly expressed in contrast with invasive bladder cancer in which only low levels of E-24.11 expression and enzymic activity are present, suggesting that E-24.11 could serve as a marker for invasive bladder cancer (Koiso *et al.*, 1994). In testicular cancer, ACE is only expressed in seminoma and not in non-seminomatous histologies (Saint-Andre *et al.*, 1988).

24.3.3 Breast cancer

Cell-surface peptidases are expressed by distinct cell types in the intact mammary tissue. E-24.11 is present on myoepithelial cells (Gusterson *et al.*, 1986), AP-N on all stromal fibroblasts and on the apical membrane of luminal epithelial cells (Atherton *et al.*, 1992a), and DPP-IV on interlobular but not intralobular fibroblasts (Atherton *et al.*, 1992b). In breast tumours, E-24.11 shows significant reactivity in tumour-associated stromal cells (Atherton *et al.*, 1994). AP-N is expressed by the majority of associated stromal fibroblasts and occasional tumour cells (Atherton *et al.*, 1992a), and has been observed in the tumour cells in as many as 20% of primary breast cancers (Dixon *et al.*, 1994). DPP-IV is only found on stromal cells surrounding tumour cells, similar to benign breast tissue (Atherton *et al.*, 1992b).

The elevated levels of endothelin-1 in breast cancer cells, which promotes growth in concert with insulin-like growth factors, may result from the presence of endothelin-converting enzyme (ECE) in the tumour cells (Patel and Schrey, 1995). Serum ACE levels are significantly lower in women with breast cancer compared to healthy controls (Varela and Bosco Lopez Saez, 1993).

24.3.4 Lung cancer

E-24.11 is expressed by fetal distal airway epithelium and adult bronchial epithelial cells (Sunday *et al.*, 1992), where it modulates tachykinins such as substance P which mediate inflammation (Sunday *et al.*, 1992) and can regulate peptide-mediated cellular proliferation by hydrolysing bombesin-like peptides (BLPs) which are potent mitogens for fibroblasts and bronchial epithelial cells (King *et al.*, 1993). These peptides are also mitogenic for small cell (Ganju *et al.*, 1994) and non-small cell lung cancers (Cohen *et al.*, 1996). In many small cell lung cancers (SCLCs), an autocrine growth loop exists in which the tumour cells secrete BLPs, express BLP receptors, and exhibit a mitogenic response to BLPs (Sunday *et al.*, 1992). E-24.11 inactivates BLPs through hydrolysis of the conserved seven-amino-acid sequence at the C terminus, which is required for biological activity (Shipp *et al.*, 1991). BLP-stimulated growth of SCLCs, which possess reduced levels of E-24.11, is inhibited by recombinant E-24.11 and potentiated by E-24.11 inhibition (Cohen *et al.*, 1996; Shipp *et al.*, 1991). In addition, recombinant E-24.11 can inhibit SCLC xenografts (Brenner *et al.*, 1995). Furthermore, SCLCs occur exclusively in cigarette smokers; cigarette smoke inhibits E-24.11, and cigarette smokers have elevated levels of BLPs in their bronchoalveolar fluid (Shipp and Look, 1993). Thus, reduced E-24.11 enzymic activity which results in increased BLPs has been directly implicated in the development of SCLC.

DPP-IV expression has also been examined in different histological subtypes of lung cancer (Asada *et al.,* 1993). It is present in over 90% of adenocarcinomas, whereas small cell carcinomas, large cell carcinomas, squamous cell carcinomas, and carcinoid tumours all lack DPP-IV expression. AP-N could not be detected in any lung cancer, although it is present on the epithelium of secretory ducts in bronchial glands (Dixon *et al.,* 1994). Numerous studies have shown that serum ACE levels are lower in patients with lung cancers (Prochazka *et al.,* 1991; Varela and Bosco Lopez Saez, 1993).

24.3.5 Skin cancers

Cell-surface peptidases are speculated to play distinct roles in each of the clinical hallmarks associated with the transformation and progression of skin tumours (i.e. proliferation, invasion and metastatic potential).

Basal and squamous cell carcinomas. AP-N, which is not detectable in normal skin, is highly expressed by connective tissue (stromal cells) surrounding basal cell carcinomas and squamous cell carcinomas, but not the tumour cells (Moehrle *et al.,* 1995). DPP-IV, which is expressed at low levels in some basal and suprabasal keratinocytes, is strongly expressed by both basal cell carcinoma tumour cells and the surrounding connective tissue (Moehrle *et al.,* 1995). However, while well-differentiated squamous cell carcinomas exhibit focal DPP-IV activity, poorly differentiated squamous cell carcinomas show no detectable expression (Moehrle *et al.,* 1995). Retention of DPP-IV activity by non-invasive basal cell carcinomas and the loss of DPP-IV activity upon acquisition of invasive potential associated with squamous cell carcinomas, implicates this ectoenzyme in cell–matrix adhesion and motility functions. This suggestion is supported by the observation that essentially all invasive and metastatic malignant melanomas extinguish the expression of DPP-IV.

Malignant melanoma. Cell-surface peptidases show qualitative and quantitative alterations in expression on melanoma cells as compared with normal progenitor melanocytes (Albino *et al.,* 1996). DPP-IV is specifically regulated during melanocyte transformation and melanoma progression (Albino *et al.,* 1992; Houghton *et al.,* 1988; Morrison *et al.,* 1993), being consistently expressed by melanocytes *in vitro* and *in vivo,* but not by either primary or metastatic melanomas (Houghton *et al.,* 1988). Loss of DPP-IV occurs in naevi which display architectural disorder and cytological atypia (atypical or dysplastic naevi) (Houghton *et al.,* 1988). Such naevi may be a precursor of melanoma suggesting that DPP-IV loss occurs at an early stage of melanoma progression before cells acquire invasive and metastatic competence. The function of DPP-IV in the melanocyte lineage is unclear, but several possible substrates of DPP-IV are potential mitogens and may contribute to melanoma progression; such as interleukins-1 and -2, and basic fibroblast growth factor (Albino *et al.,* 1996). DPP-IV may regulate melanocyte growth by inactivating such factors (Morrison *et al.,* 1993) and DPP-IV loss may thus provide the dysplastic or malignant melanocyte with a selective growth advantage (Albino *et al.,* 1992; Houghton *et al.,* 1988). Alternatively, similar to the invasive keratinocyte, the invasive malignant melanocytes may require the loss of DPP-IV activity in order to digest and invade through the basement membrane and cell structure of the dermis.

AP-N, in contrast to DPP-IV, is not expressed by normal melanocytes but becomes increasingly prevalent as melanocytes undergo tumour progression from dysplastic naevocytes to malignant melanoma cells (Elder *et al.,* 1989; Menrad *et al.,* 1993). AP-N in melanomas shows prominent localization at the cell–cell contacts of melanoma cells (Menrad *et al.,* 1993), thereby implicating AP-N in adhesion to extracellular matrices. More recent studies show that AP-N has a direct role in melanoma tumour cell invasion and extracellular matrix degradation (Saiki *et al.,* 1993), which clinically correlates positively with the development of widespread metastases and increased mortality rates (Albino *et al.,* 1996). Bestatin, a competitive inhibitor of AP-N function, as well as antibodies to AP-N, inhibits the penetration of melanoma cells through an artificial basement membrane *in vitro* without affecting adhesion or cell growth (Menrad *et al.,* 1993; Saiki *et al.,* 1993). Thus, the expression of AP-N is thought to play a critical role as one member of a cascade of enzymes that hydrolyse extracellular membrane proteins (Menrad *et al.,* 1993; Saiki *et al.,* 1993). AP-N may serve as an activator of type IV collagenase and other matrix proteins by cleaving N-terminal amino acids, and thereby allowing the acquisition by melanoma cells of invasive and metastatic competence (Fujii *et al.,* 1995; Saiki *et al.,* 1993).

E-24.11 is expressed by both one-third to one-half of primary and metastatic melanomas, but the percentage of E-24.11-positive cells in a given lesion appears to increase with tumour progression (Carrel *et al.,* 1993). Thus, primary and metastatic melanoma cells do not fit with the paradigm that E-24.11 is lost upon tumour progression. Presumably, in melanoma cells, gain of E-24.11 function may be advantageous (see Chapter 22).

24.3.6 Other neoplasms

DPP-IV is homogeneously expressed by tumour cells in nearly all follicular and papillary thyroid carcinomas, in contrast to follicular adenomas where only 27% of tumours exhibit positive staining (Kotani *et al.,* 1992). Clinically, DPP-IV expression has been used to distinguish benign from malignant thyroid cancer in cytology specimens from needle biopsies of thyroid nodules from patients with suspicious lesions (Aratake *et al.,* 1991).

AP-N is expressed in varying histologies of malignant mesenchymal tumours, generally mirroring the pattern of expression found in the cell of origin (Mechtersheimer and Moller, 1990). E-24.11 is expressed by high grade malignant gliomas (Monod *et al.,* 1992). E-24.11 which regulates the potent vasoconstrictor endothelin-1 in the endometrium (Pekonen *et al.,* 1994) is lost in endometrial cancers (Pekonen *et al.,* 1995).

24.4 Conclusions

The limited data available to date hint at an important role for cell-surface peptidases in the pathogenesis of a variety of malignancies. Most studies have, as yet, only examined and compared the expression pattern of cell-surface peptidases in normal and malignant counterparts. It is apparent from these investigations, which have documented frequent alterations, that the role of a particular peptidase in various stages of the malignant process is tumour type-specific. However, few studies have addressed the molecular basis for these alterations, and the biological effects on the tumour cell of an increase or decrease in peptidase expression. Defining the precise roles of cell-surface peptidases in neoplasia will require further study, including the use of *in vitro* (gene transfer) and *in vivo* (transgenic and knockout mice) models in order to accurately determine the biological function as related to the neoplastic phenotype of each peptidase.

The use of cell-surface peptidase expression patterns in the clinic remains an active area of investigation. Ectoenzyme expression has been used for over 20 years to diagnose and to assess the prognosis of patients with leukaemias, and emerging data suggest that expression patterns may also be useful in defining clinical subgroups of patients with solid tumour malignancies. More important, a number of studies have now shown that cell-surface peptidases may be therapeutic targets for treating cancer patients. Bestatin, a potent aminopeptidase inhibitor, has been used to treat patients with acute and chronic leukaemias (Shibuya *et al.,* 1991; Usuka and Saito, 1991), and has demonstrated therapeutic efficacy in promoting graft vs. leukaemia effect in leukaemia patients following allogeneic bone marrow transplant (Hiraoka *et al.,* 1992). Preclinical studies suggest that bestatin may also be useful in inhibiting solid tumour cancer cells (Inoi *et al.,* 1994; Yoneda *et al.,* 1992). As data emerge over the next few years, undoubtedly the use of more potent and specific enzyme inhibitors, as well as gene therapy to re-express an absent peptidase, will take on an important role in the treatment of cancer patients.

References

Albino AP, Sozzi G, Nanus DM, Jhanwar SC, Houghton AN. (1992) Malignant transformation of human melanocytes: induction of a complete melanoma phenotype and genotype. *Oncogene* **7:** 2315–2321.

Albino AP, Reed JA, McNutt NS. (1997) Molecular biology of malignant melanoma. In: *Cancer: Principles and Practice of Oncology* (eds V.T. DeVita, S. Hellman and S.A. Rosenberg), pp. 1935–1946. J.B. Lippincott Co., New York.

Aratake Y, Kotani T, Tamura K, Araki Y, Kuribayashi T, Konoe K, Ohtaki S. (1991) Dipeptidyl aminopeptidase IV staining of cytologic preparations to distinguish benign from malignant thyroid diseases. *Am. J. Clin. Pathol.* **96:** 306–310.

Asada Y, Aratake Y, Kotani T, Marutsuka K, Araki Y, Ohtaki S, Sumiyoshi A. (1993) Expression of dipeptidyl aminopeptidase IV activity in human lung carcinoma. *Histopathology* **23:** 265–270.

Atherton AJ, Monaghan P, Warburton MJ, Gusterson BA. (1992a) Immunocytochemical localization of the ectoenzyme aminopeptidase N in the human breast. *J. Histochem. Cytochem.* **40:** 705–710.

Atherton AJ, Monaghan P, Warburton MJ, Robertson D, Kenny AJ, Gusterson BA. (1992b) Dipeptidyl peptidase IV expression identifies a functional sub-population of breast fibroblasts. *Int. J. Cancer* **50:** 15–19.

Atherton AJ, O'Hare MJ, Buluwela L, Titley J, Monaghan P, Paterson, HF, Warburton MJ, Gusterson BA. (1994) Ectoenzyme regulation by phenotypically distinct fibroblast sub-populations isolated from the human mammary gland. *J. Cell Sci.* **107:** 2931–2939.

Bander NH, Finstad CL, Cordon–Cardo C, Ramsawak RD, Vaughan ED, Jr, Whitmore WF, Jr, Oettgen HF, Melamed MR, Old LJ. (1989) Analysis of a mouse monoclonal antibody that reacts with a specific region of the human proximal tubule and subsets of renal cell carcinomas. *Cancer Res.* **49**: 6774–6780.

Bogenrieder T, Finstad CL, Freeman RH, Papandreou CN, Albino AP, Reuter VE, Nanus DM. (1997) Expression and localization of aminopeptidase A, aminopeptidase N, and dipeptidyl peptidase IV in benign and malignant human prostate tissue. *Prostate,* in press.

Borowitz MJ, Weiss MA, Bossen EH, Metzgar RS. (1986) Characterization of renal neoplasms with monoclonal antibodies to leucocyte differentiation antigens. *Cancer* **57:** 251–256.

Bradstock KF. (1993) The diagnostic and prognostic value of immunophenotyping in acute leukaemia. *Pathology* **25:** 367–374.

Brenner DG, Magill–Solc C, Spencer DR, Biroc SL, Warner NJ, Dykes DJ. (1995) Antitumour activity of neutral endopeptidase in nude mouse small cell lung cancer xenograft models. *Proc. Am. Assoc. Cancer Res.* **36:** 265.

Carbone A, Cozzi M, Gloghini A, Pinto A. (1994) CD26/dipeptidyl peptidase IV expression in human lymphomas is restricted to CD30-positive anaplastic large cell and a subset of T-cell non-Hodgkin's lymphomas. *Hum. Pathol.* **25:** 1360–1365.

Carbone A, Gloghini A, Zagonel V *et al.* (1995) The expression of CD26 and CD40 ligand is mutually exclusive in human T-cell non-Hodgkin's lymphomas/leukaemias. *Blood* **86:** 4617–4626.

Carrel S, Zografos L, Schreyer M, Rimoldi D. (1993) Expression of CALLA/CD10 on human melanoma cells. *Melanoma Res.* **3:** 319–323.

Chantret I, Barbat A, Dussaulx E, Brattain MG, Zweibaum A. (1988) Epithelial polarity, villin expression, and enterocytic differentiation of cultured human colon carcinoma cells: a survey of twenty cell lines. *Cancer Res.* **48:** 1936–1942.

Cohen AJ, Bunn PA, Franklin W *et al.* (1996) Neutral endopeptidase variable expression in human lung, inactivation in lung cancer, and modulation of peptide induced calcium flux. *Cancer Res.* **56:** 831–839.

Cooper MD, Mulvaney D, Coutinho A, Cazenave PA. (1986) A novel cell surface molecule on early B-lineage cells. *Nature* **321:** 616–618.

Cozzi M, Gloghini A, Volpe R, Carbone A. (1990) Immunohistocytochemical correlation of DAP IV-CD 26 reactivity with immunologic markers of lymphocyte activation in human lymphoid tissues. *Br. J. Haematol.* **75:** 325–332.

Crockard AD, MacFarlane E, Andrews C, Bridges JM, Catovsky D. (1984) Dipeptidylaminopeptidase IV activity in normal and leukemic T-cell subpopulations. *Am. J. Clin. Pathol.* **82:** 294–299.

Dang NH, Hafler DA, Schlossman SF, Breitmeyer JB. (1990) FcR-mediated crosslinking of Ta1 (CDw26) induces human T lymphocyte activation. *Cell. Immunol.* **125:** 42–57.

Dang NH, Torimoto Y, Shimamura K, Tanaka T, Daley JF, Schlossman SF, Morimoto C. (1991) 1F7 (CD26): a marker of thymic maturation involved in the differential regulation of the CD3 and CD2 pathways of human thymocyte activation. *J. Immunol.* **147:** 2825–2832.

Dinjens WNM, Kate JT, Kirch JAJM, Tanke HJ, Van Der Linden EPM, Van Den Ingh HFGM, Steenbrugge GJV, Khan PM, Bosman FT. (1990) Adenosine deaminase complexing protein (ADCP) expression and metastic potential in prostatic adenocarcinomas. *J. Pathol.* **160:** 195–201.

Dixon J, Kaklamanis L, Turley H, Hickson ID, Leek RD, Harris AL, Gatter KC. (1994) Expression of aminopeptidase–N (CD 13) in normal tissues and malignant neoplasms of epithelial and lymphoid origin. *J. Clin. Pathol.* **47:** 43–47.

Dragovic T, Deddish PA, Tan F, Weber G, Erdos EG. (1994) Increased expression of neprilysin (neutral endopeptidase 24.11) in rat and human hepatocellular carcinomas. *Lab. Invest.* **70:** 107–113.

Droz D, Zachar D, Charbit L, Gogusev J, Chretein Y, Iris L. (1990) Expression of the human nephron differentiation molecules in renal cell carcinomas. *Am. J. Pathol.* **137:** 895–905.

Durie BG, Grogan TM. (1985) CALLA-positive myeloma: an aggressive subtype with poor survival. *Blood* **66:** 229–232.

Ebert T, Bander NH, Finstad CL, Ramsawak RD, Old LJ. (1990) Establishment and characterization of human renal cancer and normal kidney cell lines. *Cancer Res.* **50:** 5531–5536.

Elder DE, Rodeck U, Thurin J, Cardillo F, Clark WH, Stewart R, Herlyn M. (1989) Antigenic profile of tumour progression stages in human melanocytic nevi and melanomas. *Cancer Res.* **49:** 5091–5096.

Erdos EG, Schulz WW, Gafford JT, Defendini R. (1985) Neutral metalloendopeptidase in human male genital tract. Comparison to angiotensin I-converting enzyme. *Lab. Invest.* **52:** 437–447.

Finstad CL, Cordon Cardo C, Bander NH, Whitmore WF, Melamed MR, Old LJ. (1985) Specificity analysis of mouse monoclonal antibodies defining cell surface antigens of human renal cancer. *Proc. Natl Acad. Sci. USA* **82:** 2955–2959.

Fleischer B. (1987) A novel pathway of human T cell activation via a 103 kD T cell activation antigen. *J. Immunol.* **138:** 1346–1350.

Fleischer B. (1994) CD26: a surface protease involved in T-cell activation. *Immunol. Today* **15:** 180–184.

Foon KA, Todd RF, III. (1986) Immunologic classification of leukaemia and lymphoma. *Blood* **68:** 1–31.

Fujii H, Nakajima M, Saiki I, Yoneda J, Azuma I, Tsuruo T. (1995) Human melanoma invasion and metastasis enhancement by high expression of aminopeptidase N/CD13. *Clin. Exp. Met.* **13:** 337–344.

Ganju RK, Sunday M, Tsarwhas DG, Card A, Shipp MA. (1994) CD10/NEP in non-small cell lung carcinomas. Relationship to cellular proliferation. *J. Clin. Invest.* **94:** 1784–1791.

Goldfarb DA, Diz DI, Tubbs RR, Ferrario CM, Novick AC. (1994) Angiotensin II receptor subtypes in the human renal cortex and renal cell carcinoma. *J. Urol.* **151:** 208–213.

Griffin JD, Ritz J, Nadler LM, Schlossman SF. (1981) Expression of myeloid differentiation antigens on normal and malignant myeloid cells. *J. Clin. Invest.* **68:** 932–941.

Griffin JD, Ritz J, Beveridge RP, Lipton JM, Daley JF, Schlossman SF. (1983) Expression of MY7 antigen on myeloid precursor cells. *Int. J. Cell Clon.* **1:** 33–48.

Gusterson BA, Brown G, Mahendran R, Ellis J, O'Hare MJ. (1986) Identification of myoepithelial cells in human and rat breasts by anti-common lymphoblastic leukaemia antigen antibody A12. *J. Natl Cancer Inst.* **77:** 343–349.

Hiraoka A, Shibata H, Masaoka T. (1992) Immunopotentiation with Ubenimex for prevention of leukaemia relapse after allogeneic BMT. The Study Group of Ubenimex for BMT. *Transplant. Proc.* **24:** 3047–3048.

Hokland P, Rosenthal P, Griffin JD, Nadler LM, Daley J, Hokland M, Schlossman SF, Ritz J. (1983) Purification and characterization of fetal haematopoietic cells that express the common acute lymphoblastic leukaemia antigen (CALLA). *J. Exp. Med.* **157:** 114–129.

Houghton AN, Albino AP, Cordon Cardo C, Davis LJ, Eisinger M. (1988) Cell surface antigens of human melanocytes and melanoma. Expression of adenosine deaminase binding protein is extinguished with melanocyte transformation. *J. Exp. Med.* **167:** 197–212.

Inoi K, Goto S, Okamoto T, Nomura S, Nawa A, Isobe K, Mizutani S, Tomoda Y. (1994) Expression of aminopeptidase N on human choriocarcinoma cells and cell growth suppression by the inhibition of aminopeptidase N activity. *Jpn. J. Cancer Res.* **85:** 927–933.

Iwamoto I, Kimura A, Ochiai K, Tomioka H, Yoshida S. (1991) Distribution of neutral endopeptidase activity in human blood leukocytes. *J. Leuk. Biol.* **49:** 116–125.

Jalal F, Jumarie C, Bawab W, Corbeil D, Malo C, Berteloot A, Crine P. (1992) Polarized distribution of neutral endopeptidase 24.11 at the cell surface of cultured human intestinal epithelial Caco-2 cells. *Biochem. J.* **288:** 945–951.

Jongeneel CV, Quackenbush EJ, Ronco P, Verroust P, Carrel S, Letarte M. (1989) Common acute lymphoblastic leukaemia antigen expressed on leukaemia and melanoma cell lines has neutral endopeptidase activity. *J. Clin. Invest.* **83:** 713–717.

Kee BL, Dadi HK, Tran-Paterson R, Quackenbush EJ, Andrulis IL, Letarte M. (1991) CD10 and CD44 genes of leukemic cells and malignant cell lines show no evidence of transformation-related alterations. *J. Cell. Physiol.* **148:** 414–420.

Kenny AJ, O'Hare MJ, Gusterson BA. (1989) Cell-surface peptidases as modulators of growth and differentiation. *Lancet* **2:** 785–787.

King KA, Hua J, Torday JS, Drazen JM, Graham SA, Shipp MA, Sunday, ME. (1993) CD10/neutral endopeptidase 24.11 regulates fetal lung growth and maturation *in utero* by potentiating endogenous bombesin-like peptides [see comments]. *J. Clin. Invest.* **91:** 1969–1973.

Koiso K, Akaza H, Ohtani M, Miyanaga N, Aoyagi K. (1994) A new tumour marker for bladder cancer. *Int. J. Urol.* **1:** 33–36.

Kotani T, Kawano J, Suganuma T, Hirai K, Umeki K, Aratake Y, Konoe, K, Ohtaki S. (1992) Immunohistochemical localization of dipeptidyl aminopeptidase IV in thyroid papillary carcinoma. *Int. J. Exp. Pathol.* **73:** 215–222.

Li L, Wang J, Cooper MD. (1993) cDNA cloning and expression of human glutamyl aminopeptidase (aminopeptidase A). *Genomics* **17:** 657–664.

Makrynikola V, Bianchi A, Bradstock K, Gottlieb D, Hewson J. (1994) Migration of acute lymphoblastic leukaemia cells into human bone marrow stroma. *Leukaemia* **8:** 1734–1743.

Matsukura N, Suzuki K, Kawachi T *et al.* (1980) Distribution of marker enzymes and mucin in intestinal metaplasia in human stomach and relation to complete and incomplete types of intestinal metaplasia to minute gastric carcinomas. *J. Natl Cancer Inst.* **65:** 231–240.

Mattern T, Scholz W, Feller AC, Flad HD, Ulmer AJ. (1991) Expression of CD26 (dipeptidyl peptidase IV) on resting and activated human T-lymphocytes. *Scand. J. Immunol.* **33:** 737–748.

McGinnes K, Letarte M, Paige CJ. (1991) B-lineage colonies from normal, human bone marrow are initiated by B cells and their progenitors. *Blood* **77:** 961–970.

Mechtersheimer G, Moller P. (1990) Expression of aminopeptidase N (CD13) in mesenchymal tumours. *Am. J. Pathol.* **137:** 1215–1222.

Menrad A, Speicher D, Wacker J, Herlyn M. (1993) Biochemical and functional characterization of aminopeptidase N expressed by human melanoma cells. *Cancer Res.* **53:** 1450–1455.

Moehrle MC, Schlagenhauff BE, Klessen C, Rassner G. (1995) Aminopeptidase M and dipeptidyl peptidase IV activity in epithelial skin tumours: a histochemical study. *J. Cutan. Pathol.* **22:** 241–247.

Monod L, Hamou MF, Ronco P, Verroust P, de Tribolet N. (1992) Expression of cALLa/NEP on gliomas: a possible marker of malignancy. *Acta Neurochir.* **114:** 3–7.

Moore JH, Lesser EJ, Erdody DH, Natale RB, Orringer MB, Beer DG. (1994) Intestinal differentiation and p53 gene alterations in Barrett's oesophagus and oesophageal adenocarcinoma. *Int. J. Cancer* **56:** 487–493.

Morimoto C, Kameoka J, Schlossman SF. (1995) Overview of CD26. In: *Leukocyte Typing V. White Cell Differentiation Antigens. Proceedings of the 5th International Workshop and Conferrence,* pp. 1105–1109. Oxford University Press, Oxford.

Morrison ME, Vijayasaradhi S, Engelstein D, Albino AP, Houghton AH. (1993) A marker for neoplastic progression of human melanocytes is a cell surface ectopeptidase. *J. Exp. Med.* **177:** 1135–1143.

Nanus DM, Engelstein D, Gastl GA, Gluck L, Vidal MJ, Morrison M, Finstad CL, Bander NH, Albino AP. (1993) Molecular cloning of the human kidney differentiation antigen gp160: human aminopeptidase A. *Proc. Natl Acad. Sci. USA* **90:** 7069–7073.

Osborn M, Mazzoleni G, Santini D, Marrano D, Martinelli G, Weber K. (1988) Villin, intestinal brush border hydrolases and keratin polypeptides in intestinal metaplasia and gastric cancer; an immunohistologic study emphasizing the different degrees of intestinal and gastric differentiation in signet ring cell carcinomas. *Virch. Archiv.* **413:** 303–312.

Papandreou CN, Usmani B, Bogenrieder T *et al.* (1997) Loss of neprilysin (neutral endopeptidase 24.11) in androgen independent prostate cancers. *Nature Med.,* in press.

Patel KV, Schrey MP. (1995) Human breast cancer cells contain a phosphoramidon-sensitive metalloproteinase which can process exogenous big endothelin-1 to endothelin-1: a proposed mitogen for human breast fibroblasts. *Br. J. Cancer* **71:** 442–447.

Pekonen F, Nyman T, Rutanen EM. (1994) Differential expression of mRNAs for endothelin-related proteins in human endometrium, myometrium and leiomyoma. *Mol. Cell. Endocrinol.* **103:** 165–170.

Pekonen F, Nyman T, Ammala M, Rutanen EM. (1995) Decreased expression of messenger RNAs encoding endothelin receptors and neutral endopeptidase 24.11 in endometrial cancer. *Br. J. Cancer* **71:** 59–63.

Prochazka J, Krepela E, Sedo A, Viklicky J, Fiala P. (1991) Aminopeptidases and angiotensin I-converting enzyme activities in primary human lung tumours and lung parenchyma. *Neoplasma* **38:** 501–508.

Quaroni A, Nichols BL, Quaroni E, Hurst K, Herrera L, Weiser MM, Hamilton SR. (1992) Expression and different polarity of aminopeptidase N in normal human colonic mucosa and colonic tumours. *Int. J. Cancer* **51:** 404–411.

Ramakrishnan L, Wu Q, Yue A, Cooper MD, Rosenberg N. (1990) BP-1/6C3 expression defines a differentiation stage of transformed pre–B cells and is not related to malignant potential. *J. Immunol.* **145:** 1603–1608.

Riemann D, Gohring B, Langner J. (1994) Expression of aminopeptidase N/CD13 in tumour-infiltrating lymphocytes from human renal cell carcinoma. *Immunol. Lett.* **42:** 19–23.

Riemann D, Kehlen A, Langner J. (1995) Stimulation of the expression and the enzyme activity of aminopeptidase N/CD13 and dipeptidylpeptidase IV/CD26 on human renal cell carcinoma cells and renal tubular epithelial cells by T cell-derived cytokines, such as IL-4 and IL-13. *Clin. Exp. Immunol.* **100:** 277–283.

Ritz J, Nadler LM, Bhan AK, Notis-McConarty J, Pesando JM, Schlossman SF. (1981) Expression of common acute lymphoblastic leukaemia antigen (CALLA) by lymphomas of B-cell and T-cell lineage. *Blood* **58:** 648–652.

Saiki I, Fujii J, Abe F, Nakajima M, Tsuruo T, Azuma I. (1993) Role of aminopeptidase N (CD 13) in tumour-cell invasion and extracellular matrix degradation. *Int. J. Cancer* **54:** 137–143.

Saint-Andre JP, Alhenc–Gelas F, Rohmer V, Chretien MF, Bigorgne JC, Corvol P. (1988) Angiotensin-I-converting enzyme in germinomas. *Hum. Pathol.* **19:** 208–213.

Salles G, Chen CY, Reinherz EL, Shipp MA. (1992) CD10/NEP is expressed on Thy-1low B220+ murine B-cell progenitors and functions to regulate stromal cell-dependent lymphopoiesis. *Blood* **80:** 2021–2029.

Salles G, Rodewald HR, Chin BS, Reinherz EL, Shipp MA. (1993) Inhibition of CD10/neutral endopeptidase 24.11 promotes B-cell reconstitution and maturation *in vivo. Proc. Natl Acad. Sci. USA* **90:** 7618–7622.

Schon E, Jahn S, Kiessig ST, Demuth HU, Neubert K, Barth A, Von, Baehr R, Ansorge S. (1987) The role of dipeptidyl peptidase IV in human T lymphocyte activation. Inhibitors and antibodies against dipeptidyl peptidase IV suppress lymphocyte proliferation and immunoglobulin synthesis *in vitro. Eur. J. Immunol.* **17:** 1821–1826.

Schwarzinger I, Valent P, Koller U, Marosi C, Schneider B, Haas O, Knapp W, Lechner K, Bettelheim P. (1990) Prognostic significance of surface marker expression on blasts of patients with *de novo* acute myeloblastic leukaemia. *J. Clin. Oncol.* **8:** 423–430.

Shibuya K, Chiba S, Hino M, Kitamura T, Miyagawa K, Takaku F, Miyazano K. (1991) Enhancing effect of ubenimex (bestatin) on proliferation and differentiation of haematopoietic progenitor cells, and the suppressive effect on proliferation of leukemic cell lines via peptidase regulation. *Biomed. Pharmacother.* **45:** 71–80.

Shipp MA, Look AT. (1993) Haematopoietic differentiation antigens that are membrane-associated enzymes: cutting is the key! *Blood* **84:** 1052–1070.

Shipp MA, Tarr GE, Chen CY, Switzer SN, Hersh LB, Stein H, Sunday, ME, Reinherz EL. (1991) CD10/neutral endopeptidase 24.11 hydrolyzes bombesin-like peptides and regulates the growth of small cell carcinomas of the lung. *Proc. Natl Acad. Sci. USA* **88:** 10662–10666.

Sobol RE, Mick R, Royston I. *et al* (1987) Clinical importance of myeloid antigen expression in adult acute lymphoblastic leukaemia. *N. Engl. J. Med.* **316:** 1111–1117.

Sterry W, Steigleder GK, Goos M, Meisterernst E. (1982) *In situ* identification and enumeration of helper T lymphocytes in cutaneous malignant lymphomas by dipeptidyl-peptidase IV (DPP IV). *J. Cutan. Pathol.* **9:** 119–120.

Sunday ME, Hua J, Torday JS, Reyer B, Shipp MA. (1992) CD/10 neutral endopeptidase 24.11 in developing human fetal lung. *J. Clin. Invest.* **90:** 2517–2525.

Takada Y, Hiwada K, Yokoyama M, Ochi K, Takeuchi M, Kokubu T. (1985) Angiotensin converting enzyme. A possible histologic indicator for human renal cell carcinoma. *Cancer* **56:** 130–133.

Townsend CM, Jr, Bold RJ, Ishizuka J. (1994) Gastrointestinal hormones and cell proliferation. *Surgery Today* **24:** 772–777.

Ulmer AJ, Mattern T, Flad HD. (1992) Expression of CD26 (dipeptidyl peptidase IV) on memory and naive T lymphocytes. *Scand. J. Immunol.* **35:** 551–559.

Usuka Y, Saito Y. (1991) Bestatin treatment of myelodysplastic syndromes and chronic myelogenous leukaemia. *Biomed. Pharmacother.* **45:** 87–93.

Varela AS, Bosco Lopez Saez JJ. (1993) Utility of serum activity of angiotensin-converting enzyme as a tumour marker. *Oncology* **50:** 430–435.

Welch PA, Burrows PD, Namen A, Gillis S, Cooper MD. (1990) Bone marrow stromal cells and interleukin-7 induce coordinate expression of the BP-1/6C3 antigen and pre-B cell growth. *Int. Immunol.* **2:** 697–705.

Wu Q, Tidmarsh GF, Welch PA, Pierce JH, Weissman IL, Cooper MD. (1989) The early B lineage antigen BP-1 and the transformation–associated antigen 6C3 are on the same molecule. *J. Immunol.* **143:** 3303–3308.

Yoneda J, Saiki I, Fujii H, Abe F, Kojima Y, Azuma I. (1992) Inhibition of tumour invasion and extracellular matrix degradation by ubenimex (bestatin). *Clin. Exp. Met.* **10:** 49–59.

Yoshii Y, Kasugai T, Kato T, Nagatsu T, Sakakibara S. (1981) Changes in serum dipeptidyl-aminopeptidase IV (glycylprolyl dipeptidyl-aminopeptidase) activity of patients with gastric carcinoma after surgical excision and the enzyme activity in the carcinoma tissue. *Biochem. Med.* **25:** 276–282.

Yoshioka M, Erickson RH, Matsumoto H, Gum E, Kim YS. (1991) Expression of dipeptidyl aminopeptidase IV during enterocytic differentiation of human colon cancer (Caco-2) cells. *Int. J. Cancer* **47:** 916–921.

Young GP, Macrae FA, Gibson PR, Alexeyeff M, Whitehead RH. (1992) Brush border hydrolases in normal and neoplastic colonic epithelium. *J. Gastroenterol. Hepatol.* **7:** 347–354.

Chapter 25

Cell-surface peptidases as receptors for pathogens and toxins

Bernard Delmas

25.1 Introduction

Infections usually involve the binding of a pathogenic microorganism to a target cell via the interaction between a specialized ligand and a host cell membrane molecule. Depending on the nature of the pathogen, this interaction can result either in an internalization of the microorganism (viruses and some bacteria) or in an attachment to the surface of the host cell. In some cases, bacterial pathogens produce toxins that bind to the surface of host cells. In this review, the term receptor will be assigned to the host cell molecule which participates in the binding of a pathogen ligand to a target cell membrane.

Despite the large diversity of receptors (carbohydrates, lipids or proteins) used by viral and bacterial ligands to adhere to host cells, proteins belonging to the immunoglobulin-like superfamily (White and Littman, 1989), to the integrin superfamily (Isberg and Tran Van Nhieu, 1994; White, 1993) or to membrane ectopeptidases are often chosen as pathogen targets. This chapter will focus on the interaction between pathogens and ectopeptidases. Members of this family of peptidases are expressed on the surface of a wide variety of cells and tissues and two of them have been described as being involved in the binding of unrelated pathogen ligands, pathogens which infect either vertebrates or invertebrates.

To date (see *Table 25.1*), aminopeptidase N (AP-N), a protein identical to CD13 in

Table 25.1. Ectopeptidases reported as acting as receptor, co-receptor or binding proteins for viruses and toxins

Host cell reporter/ binding protein[a]	Pathogen	Pathogen-binding protein	Host specificity
APN/CD13	Coronavirus TGEV	Spike protein S	Porcine
	Coronavirus HCV-229E	Spike protein S	Human
	Human cytomegalovirus		Human
	Bacillus thuringiensis	Toxin CryIA(c)	Lepidopteran
DPPIV/CD26	HIV	gp120	Human

[a] References for receptor characterization: Delmas *et al.* (1992), Yeager *et al.* (1992) for coronaviruses; Söderberg *et al.* (1993) for human cytomegalovirus; Knight *et al.* (1994) for *Bacillus thuringiensis* and Callebaut *et al.* (1993) for HIV.

Cell-Surface Peptidases in Health and Disease, edited by A.J. Kenny and C.M. Boustead.

humans, has been reported to act as a receptor for two coronaviruses, for the toxin CryIA(c) of *Bacillus thuringiensis,* and to enhance human cytomegalovirus binding to the cell surface. Dipeptidyl peptidase IV/CD26 has been suggested to play a role in human immunodeficiency virus (HIV) entry into $CD4^+$ cells. This review will describe how ectopeptidases have been identified as receptors in these different infectious systems, define the ligands of the pathogens that bind to these molecules and discuss the biological relevance of the interactions between these ligands and ectopeptidases.

25.2 CD13/AP-N as a receptor for coronaviruses TGEV and HCV-229E

Coronaviruses are a group of enveloped, plus-strand RNA viruses which cause diseases in humans and in animals. They generally exhibit a strong species specificity, natural coronavirus infection being mostly associated with a disease in one host species (Wege *et al.*, 1982). They also have a high restriction in tissue tropism and mainly infect respiratory and/or intestinal epithelia, even if some of them can spread in other cell types and organs. Porcine transmissible gastroenteritis virus (TGEV) replicates in the enterocytes which cover the intestinal villi, but a natural variant of TGEV was also reported to multiply in the epithelial layer of the respiratory tract. TGEV infection causes atrophy of the intestinal villi, leading to severe intestinal disorders, often fatal to animals younger than 2 weeks old (Enjuanes and Van der Zeijst, 1995). Human coronavirus 229E (HCV-229E) infects the epithelial cells of the respiratory tract and may cause 20% of colds (McIntosh, 1990).

25.2.1 Identification of the TGEV receptor

To identify the TGEV receptor, monoclonal antibodies were raised against a porcine cell line and selected for their ability to protect against infection *in vitro* (Delmas *et al.*, 1992). Six protective antibodies were shown to recognize a 150 kDa membrane protein in the cell line culture system, whereas three polypeptides of 150, 95 and 50 kDa were co-precipitated from intestinal brush border membrane preparations. N-terminal sequencing revealed that these polypeptides represented either the uncleaved form or the natural cleavage products of pig AP-N (pAP-N). To obtain direct evidence for the role of AP-N as a TGEV receptor, a specific binding between TGEV and AP-N was demonstrated by biochemical methods. Furthermore, recombinant expression of pAP-N in non-susceptible cells was able to confer virus susceptibility.

25.2.2 Identification of the HCV-229E receptor

A similar strategy was used in parallel studies by the group of K. Holmes to identify the receptor of HCV-229E (Yeager *et al.*, 1992). A monoclonal antibody produced against membrane proteins derived from human cell lines was selected for its capacity to protect cells from virus infection. As TGEV uses AP-N as a receptor and as susceptibility to HCV-229E depends on a gene located on human chromosome 15 where the AP-N gene was mapped, experiments were carried out which demonstrated that the protective antibody recognized human AP-N (hAP-N). Then, it was shown that cells naturally non-susceptible to HCV-229E could become infected after expression of recombinant hAP-N.

25.2.3 Binding domains on AP-N and on the coronavirus spike protein involved in AP-N/virus docking

On the AP-N side. TGEV uses pAP-N, but not hAP-N as a receptor, although the two proteins show 80% amino acid identity. Taking advantage of this observation, chimeric molecules constructed from pAP-N and hAP-N were analysed for their capacity to confer virus susceptibility (Delmas *et al.*, 1994). A C-terminal region located between amino acids 717 and 813 was found to be required for TGEV receptor function. This domain also contains epitopes for TGEV-blocking antibodies. Sequence comparison showed that this region has a higher degree of divergence (i.e. 29% compared with 20% for the whole molecule). This domain is distant from the AP-N catalytic site defined by the zinc-binding signature HEXXH (residues 383–387). Furthermore, the enzymic activity of AP-N appears not to be involved in the process of infection. AP-N inhibitors with K_i values in the nanomolar range did not block TGEV infection. A form of AP-N devoid of enzymic activity by replacement of the two histidines of the catalytic site was able to confer TGEV susceptibility. Earlier experiments using a mutant hAP-N which had a 39-amino-acid deletion that included the active site, showed that this mutated form of AP-N was not able to confer HCV-229E susceptibility (Yeager *et al.*, 1992). These different findings for the two viruses are probably due to the fact that the deletion induces large conformational changes which alter the virus binding site.

On the viral spike S protein side. Coronaviruses have spikes approximately 20 nm long and 7 nm wide which form a 'corona' around the virion characteristic of this family of viruses. The main functions of the spikes are to bind to specific host cell membrane receptors and to induce fusion between viral and host membranes. Spikes are made up of the S protein associated in trimers by non-covalent bonds. Primary sequence determination of the S proteins allowed identification of four structural domains. The amino-terminal domain is assumed to form the globular structure of the spike whereas the C-terminal domain contains a region with two α helices in a coiled-coil structure, a transmembrane stretch and a cytoplasmic part rich in cysteine residues (Cavanagh, 1995).

Mapping of the S domain recognizing AP-N was carried out by co-expression of AP-N with different TGEV S domains using the baculovirus expression system (Godet *et al.*, 1994). An S polypeptide (150 amino acid residues long) was found to form a complex with AP-N, thus showing that receptor-binding determinants are located mainly in the domain between residues 506 and 655. Interestingly, there is only 25% amino acid identity between this TGEV S domain and its HCV-229E S counterpart which potentially recognizes hAP-N.

25.3 CD13/AP-N and human cytomegalovirus infection

Early events that mediate the attachment of herpes viruses and particularly the human cytomegalovirus (HCMV) are not well characterized, even if receptors of some members of the herpes virus family have already been identified. As for other herpes viruses, initial HCMV attachment is mediated by the binding of virus components to cell-surface glycosaminoglycans, mainly heparan sulphate (Mettenleiter, 1994). Cellular proteins are probably required for virus binding and fusion with the host cell. CD13/hAP-N was assumed to be involved in HCMV infection on the basis of the following observations: anti-hAP-N antibodies inhibited infection and binding of virions to susceptible cells; a higher HCMV binding was observed in hAP-N-transfected cells compared to controls and cells expressing a mutant of hAP-N, deleted around the catalytic site, remained also susceptible to HCMV infection (Söderberg *et al.*, 1993).

25.4 AP-N as a receptor for toxin CryIA(c) of *Bacillus thuringiensis*

Bacillus thuringiensis is a Gram-positive bacterium producing parasporal, insecticidal crystal protein (Cry) inclusions. The Cry proteins are synthesised as inactive protoxins, which are activated upon ingestion by proteases in the gut of susceptible insects. Subsequently, the toxin binds to high affinity sites on the surface of the midgut epithelial cells. After this binding event, Cry toxins such as CryIA(c) and CryIIIA integrate into the lipid bilayers and form ion channels. As a consequence, target cells usually swell and burst, causing a loss of gut integrity and a rapid larval death (Gill *et al.*, 1992; Knowles, 1994). Since the 1960s, *B. thuringiensis* expressing Cry toxins have been used as commercial insecticides. The advantages of these toxins as 'green' insecticides are their specificity and the slow rate of development of resistance by insects in the field.

A large number of toxin genes of different *B. thuringiensis* strains were cloned and classified into four major groups according to their deduced amino acid sequences. Despite sequence divergence, Cry toxins usually share five segments of significant homology and their mode of action appears to be very similar (Feitelson *et al.*, 1992; Knowles, 1994). Moreover, a certain correlation was observed between Cry sequences and their host range: for instance, CryI toxins are Lepidoptera specific and CryIII Coleoptera specific.

With the aim of understanding both the biochemical basis for toxin–host specificity and channel formation in the membrane, recent reports have described the purification and the identification of the toxin binding targets for two CryIA toxins. The receptor of CryIA(b) shares sequence similarities with the cadherin protein family (Vadlamudi *et al.*, 1995) whereas CryIA(c) uses AP-N as a receptor.

25.4.1 Characterization of a CryIA(c) receptor

Purification and identification of a CryIA(c) insect receptor as AP-N. Most of the *in vitro* studies undertaken to identify the target of CryIA(c) toxin have used brush border membrane preparations purified from the gut of susceptible insects. Using different biochemical approaches to purify the Cry binding protein combined with partial N-terminal sequencing, three different groups have identified the CryIA(c) receptor from lepidopterans *Manduca sexta* (Knight *et al.*, 1994; Sangadala *et al.*, 1994) and *Heliothis virescens* (Gill *et al*., 1995) as AP-N. Validation of the binding was carried out by protein blotting of the receptor identified as AP-N and revealing it with the labelled toxin. Using an optical biosensor approach, these results were then extended to the three CryIA toxins CryIA(a), -(b) and -(c), which share at least 80% amino acid identity, and which were shown to bind *M. sexta* AP-N with K_d values ranging from 2×10^{-8} to 5×10^{-8} M (Masson *et al*., 1995). However, it must be noted that Vadlamudi and co-workers (1995) identified another CryIA(b) binding protein as a 210 kDa protein from the cadherin superfamily, a family of proteins with no known homology with AP-Ns. It has yet to be elucidated whether or not CryIA(b) and CryIA(a) (Lee *et al*., 1995) use two receptors *in vivo*, the 210 kDa protein and/or AP-N.

Structure of lepidopteran AP-N. Cloning of AP-N cDNA from both *M. sexta* and *H. virescens* revealed that the C-terminal sequence possessed a potential glycosylphosphatidylinositol anchor signal peptide (Gill *et al*., 1995; Knight *et al*., 1995). Confirmation was obtained by partial release of an AP-N activity from membranes by phosphatidylinositol-specific phospholipase C treatment (Knight *et al*., 1995). This observation shows a different mode of AP-N anchoring in the cell membrane compared to that of mammalian AP-Ns which are inserted via a hydrophobic amino acid stretch in their N-terminal domain (type II membrane protein).

25.4.2 Interaction between CryIA toxins and AP-N

On the AP-N side. Mapping of the toxin binding site on the AP-N primary structure is not yet available. By competition and ligand blotting assays performed on brush border membrane vesicles, toxins were shown to bind to a single site as well as to distinct classes of toxin-binding proteins (see Masson *et al*., 1995, and references therein). Using an optical biosensor and purified AP-N, CryIA(a) and CryIA(b) were shown to bind to a single site on AP-N whereas CryIA(c) bound to two distinct sites (Masson *et al*., 1995). An interesting feature is that CryIA(c), but not CryIA(a) or CryIA(b) binding to brush border membranes or to AP-N can be inhibited by the aminosugar *N*-acetyl galactosamine, but not by *N*-acetyl glucosamine or other sugars (Knight *et al*., 1994; Masson *et al*., 1995). Since *N*-acetyl galactosamine is a component of AP-N, this last property has been proposed to explain why insect resistance to CryIA(b) did not always confer CryIA(c) resistance (Masson *et al*., 1995).

On the CryIA(c) side. The three-dimensional structure of the activated CryIA(a) toxin by X-ray crystallography at 2.25 Å resolution was recently described (Grochulski *et al*., 1995). Since CryIA(a) and CryIA(c) have approximately 83% amino acid identity (Feitelson *et al*., 1992), the three domains defined on CryIA(a) are presumably well conserved in CryIA(c). Domain II was first assumed to be involved in recognition and binding to cell-surface receptors on the basis of the X-ray data and genetics studies. However, studies have also suggested that domain III played a role in determining binding specificity (Grochulski *et al*., 1995, and references therein). Additional data were obtained with hybrids obtained between CryIA(c) and CryIA(a) and used in slot blot assays and ligand blot experiments. They showed that residues 451–623, comprising essentially domain III, were essential and sufficient to target binding either to AP-N or to a 210 kDa protein (Lee *et al.*, 1995).

25.5 CD26/dipeptidyl peptidase IV and HIV infection

Infection by HIV is initiated by the binding of the virus envelope protein gp120 to human CD4. The different domains of gp120 and CD4 involved in this process have been well characterized by structural and/or mutational analyses. However, although human CD4 is necessary, its expression is not always sufficient to allow HIV entry and replication. Maddon *et al.* (1986) first pointed out that CD4 expression in human cells conferred HIV susceptibility, but not in non-human cells. Virus binding occurred in both types of cells but fusion and entry of HIV was only observed in human cells. During the last 10 years, additional studies have been performed suggesting that the inability of HIV to enter $CD4^{+}$HIV-resistant cells was probably due to the lack of an additional factor(s) in non-human cells. Recently, Feng and co-workers (1996) identified a cofactor for HIV entry and fusion as a putative G protein-coupled receptor with seven transmembrane segments, which they named fusin. Recombinant fusin enabled CD4-expressing non-human cell types to permit HIV entry, preferentially for T-cell line-tropic HIV isolates. Subsequently, an additional CD4 entry cofactor for macrophage-tropic strains of HIV was identified as CC-CKR-5, a receptor for several β-chemokines RANTES, MIP-1α and MIP-1β (Alkhatib *et al.*, 1996; Deng *et al.*, 1996; Dragic *et al.*, 1996). However, although the nature of essential CD4 cofactors is likely to be found, the cell-surface peptidase CD26 (dipeptidyl peptidase IV; DPP-IV) has retained particular attention because it was also proposed to play a role in HIV entry as a cofactor and to correlate with HIV replication.

25.5.1 Assays to identify CD26 (DPP-IV) as a CD4 co-receptor for HIV entry

Proteolytic cleavage of viral surface envelope glycoprotein is very often implicated in fusion between the viral and the host cell membranes, not only in retroviruses but also in other groups of viruses such as myxoviruses. A small domain of gp120, the third hypervariable region of gp120 (V3 loop) which is not directly involved in the interaction between gp120 and CD4, was shown to be critical for several biological HIV phenotypes. Although there are arguments for a cleavage of the V3 loop during early steps of HIV entry, there is no direct evidence that this domain is actually cleaved during HIV entry into the cell. Callebaut and co-workers (1993) proposed that the additional factor to CD4 could be CD26 (DPP-IV) because potential DPP-IV cleavage sites exist in the V3 loop and also on the basis of the following data: (i) a specific DPP-IV inhibitor and an anti-CD26 antibody blocked HIV entry, (ii) co-expression of human CD4 and CD26 in murine cells rendered them HIV susceptible. These results have been the subject of a major controversy, because of the assay used to quantify HIV entry and the inability of other groups to reproduce these results (see Lazaro *et al.*, 1994; Oravecz *et al.*, 1995, for references).

25.5.2 Correlation between HIV replication and CD26 expression

A selective decline of CD26 expression in T lymphocytes of HIV-infected subjects, in either the $CD4^+$ or the $CD8^+$ cell populations was observed by Blazquez *et al.* (1992). Polymerase chain reaction studies showed that HIV was present mainly in the $CD26^-$ subset of $CD4^+$ cells, suggesting a critical role of CD26 molecule in HIV infection. Moreover, Oravecz *et al.* (1995) have shown that CD26 expression in a $CD4^+$ T-cell line correlates with the strain phenotype of HIV-1. A specific decrease in the expression of CD26 was revealed after macrophage-tropic, but not T-cell-tropic virus infection. In addition, the level of CD26 expression in derived cell subclones correlates with the entry rate and cytopathic effect of macrophage-tropic variants, resulting in preferential survival of cells with low CD26 expression levels after infection. Using recombinant viruses, they showed that the V3 loop plays a critical role in this selection process. These observations will probably constitute the basis of new strategies to better define the role of CD26 in the infection process of HIV.

25.6 Biological relevance of the interaction between pathogens and ectopeptidases

The fact that unrelated pathogens such as bacteria or viruses use ectopeptidases as targets raises the question of whether or not the intrinsic enzymic activity of the receptor plays a role in the infectious process. As mentioned above, for TGEV and HCMV, the enzymic activity of AP-N has been clearly differentiated from the viral binding activity. In the case of CryIA(c), AP-N, as an exopeptidase, could participate in an N-terminal trimming of the toxin during the binding event or during the integration of the toxin into the membrane, but it cannot be involved by itself in the activation process of the 130 kDa protoxin to the 65 kDa active form of the toxin. Thus, the fact that AP-N is used as a receptor by both the enteropathogenic TGEV and CryIA(c) probably reflects its abundance at the apical side of the gut epithelial cells, and is not related to its enzymic activity.

Concerning HIV infection, the idea that an ectopeptidase acts as a putative entry factor is still a very attractive hypothesis and would explain differences of HIV phenotypes associated with the V3 loop. No doubt, investigations will be carried out to pursue the work of Oravecz and co-workers to increase the understanding of the relationship between the level of CD26 expression in $CD4^+$ cells, the progression of the disease and the dominant virus type in infected individuals. Identification of fusin and CC-CKR5 as CD4 receptor cofactors will probably facilitate the analysis of the contribution of CD26 to the infectious process *in vitro*, but also *in vivo* through the development of small transgenic animal models.

Acknowledgements

The author thanks M. Bremont, H. Laude and A. Bouroche (INRA, Jouy-en-josas) for critical reading of the manuscript.

References

Alkhatib G, Combadiere C, Broder CC, Feng Y, Kennedy PE, Murphy PM, Berger EA. (1996) CC CKR5: a RANTES, MIP-1α, MIP-1β receptor as a fusion cofactor for macrophage-tropic HIV-1. *Science* **272:** 1955–1958.

Blazquez MV, Madueno JA, Gonzalez R, Jurado R, Bachovchin WW, Pena J, Munoz E. (1992) Selective decrease of CD26 expression in T cells from HIV-1-infected individuals. *J. Immunol.* **149:** 3073–3077.

Callebaut C, Krust B, Jacotot E, Hovanessian AG. (1993) T cell activation antigen, CD26, as a cofactor for entry of HIV in CD4⁺ cells. *Science* **262:** 2045–2050.

Cavanagh D. (1995) The coronavirus surface glycoprotein. In: *The Coronaviridae* (ed. S.G. Siddell), pp. 73–113. Plenum Press, New York.

Delmas B, Gelfi J, L'Haridon R, Vogel L, Sjöström H, Norén O, Laude H. (1992) Aminopeptidase-N is a major receptor for the enteropathogenic coronavirus TGEV. *Nature* **357:** 417–419.

Delmas B, Gelfi J, Kut E, Sjöström H, Norén O, Laude H. (1994) Determinants essential for the transmissible gastroenteritis virus-receptor interaction reside within a domain of aminopeptidase-N that is distinct from the enzymatic site. *J. Virol.* **68:** 5216–5224.

Deng H, Liu R, Ellmeier W *et al.* (1996) Identification of a major co-receptor for primary isolates of HIV-1. *Nature* **381:** 661–666.

Dragic T, Litwin V, Allaway GP *et al.* (1996) HIV-1 entry into CD4⁺ cells is mediated by the chemokine receptor CC-CKR-5. *Nature* **381:** 667–673.

Enjuanes L, Van der Zeijst BAM. (1995) Molecular basis of transmissible gastroenteritis virus epidemiology. In: *The Coronaviridae* (ed. S.G. Siddell), pp. 337–376. Plenum Press, New York.

Feitelson JS, Payne J, Kim L. (1992) *Bacillus thuringiensis:* insects and beyond. *Bio/Technology* **10:** 271–275.

Feng Y, Broder CC, Kennedy PE, Berger EA. (1996) HIV-1 entry cofactor: functional cDNA cloning of a seven-transmembrane, G protein-coupled receptor. *Science* **272:** 872–877.

Gill SS, Cowles EA, Pietrantonio PV. (1992) The mode of action of *Bacillus thuringiensis* endotoxins. *Annu. Rev. Entomol.* **37:** 615–636.

Gill SS, Cowles EA, Francis V. (1995) Identification, isolation, and cloning of a *Bacillus thuringiensis* CryIAc toxin-binding protein from the midgut of the lepidopteran insect *Heliothis virescens. J. Biol. Chem.* **270:** 27277–27282.

Godet M, Grosclaude J, Delmas B, Laude H. (1994) Major receptor-binding and neutralization determinants are located within the same domain of the transmissible gastroenteritis virus (coronavirus) spike protein. *J. Virol.* **68:** 8008–8016.

Grochulski P, Masson L, Borisova S, Pusztai-Carey M, Schwartz J-L, Brousseau R, Cygler M. (1995) *Bacillus thuringiensis* CryIA(a) insecticidal toxin: crystal structure and channel formation. *J. Mol. Biol.* **254:** 447–464.

Isberg RR, Tran Van Nhieu G. (1994) Binding and internalization of microorganisms by integrin receptors. *Trends Microb.* **2:** 10–14.

Knight PJK, Crickmore N, Ellar DJ. (1994) The receptor for *Bacillus thuringiensis* CryIA(c) delta-endotoxin in the brush border membrane of the lepidopteran *Manduca sexta* is aminopeptidase-N. *Mol. Microbiol.* **11:** 429–436.

Knight PJK, Knowles BH, Ellar DJ. (1995) Molecular cloning of an insect aminopeptidase N that serves as a receptor for *Bacillus thuringiensis* CryIA(c) toxin. *J. Biol. Chem.* **270:** 17765–17770.

Knowles BH. (1994) Mechanism of action of *Bacillus thuringiensis* insecticidal δ-endotoxins. *Adv. Insect Physiol.* **24:** 275–308.

Lazaro I, Naniche D, Signoret N, Bernard A-M, Marguet D, Klatzmann D, Dragic T, Alizon M, Sattentau Q. (1994) Factors involved in entry of the human immunodeficiency virus type 1 into permissive cells: lack of evidence of a role for CD26. *J. Virol.* **68:** 6535–6546.

Lee MK, Young BA, Dean DH. (1995) Domain III exchanges of *Bacillus thuringiensis* CryIA toxins affect binding to different gypsy moth midgut receptors. *Biochem. Biophys. Res. Commun.* **216:** 306–312.

Maddon PJ, Dalgleish AG, McDougal JS, Clapham PR, Weiss RA, Axel R. (1986) The T4 gene encodes the AIDS virus receptor and is expressed in the immune system and brain. *Cell* **47:** 333–348.

Masson L, Lu Y, Mazza A, Brousseau R, Adang MJ. (1995) The CryIA(c) receptor purified from *Manduca sexta* displays multiple specificities. *J. Biol. Chem.* **270:** 20309–20315.

McIntosh K. (1990) Coronaviruses. In: *Virology*, Vol.1, 2nd Edn (eds B.N. Fields and D.M. Knipe), pp. 857–864. Raven Press, New York.

Mettenleiter TC. (1994) Initiation and spread of α-herpesvirus infections. *Trends Microb.* **2:** 2–4.

Oravecz T, Roderiquez G, Koffi J, Wang J, Ditto M, Bou-Habib DC, Lusso P, Norcross MA. (1995) CD26 expression correlates with entry, replication and cytopathicity of monocytotropic HIV-1 strains in a T-cell line. *Nature Medicine* **1:** 919–926.

Sangadala S, Walters FS, English LH, Adang MJ. (1994) A mixture of *Manduca sexta* aminopeptidase and phosphatase enhances *Bacillus thuringiensis* insecticidal CryIA(c) toxin binding and $^{86}Rb^{+}$-K^{+} efflux *in vitro*. *J. Biol. Chem.* **269:** 10088–10092.

Söderberg C, Giugni TD, Zaia JA, Larsson S, Wahlberg JM, Möller E. (1993) CD13 (human aminopeptidase-N) mediates human cytomegalovirus infection. *J. Virol.* **67:** 6576–6585.

Vadlamudi RK, Weber E, Ji I, Ji TA, Bulla LA Jr. (1995) Cloning and expression of a receptor for an insecticidal toxin of *Bacillus thuringiensis*. *J. Biol. Chem.* **270:** 5490–5494.

Wege H, Siddell SG, ter Meulen V. (1982) The biology and pathogenesis of coronaviruses. *Curr. Top. Microbiol. Immunol.* **99:** 165–200.

White JM. (1993) Integrins as virus receptors. *Curr. Biol.* **3:** 596–599.

White JM, Littman DR. (1989) Viral receptors of the immunoglobulin superfamily. *Cell* **56:** 725–728.

Yeager CL, Ashmun RA, Williams RK, Cardellichio CB, Shapiro LH, Look AT, Holmes KV. (1992) Human aminopeptidase-N is a receptor for human coronavirus 229E. *Nature* **357:** 420–422.

Index